FEYNMAN

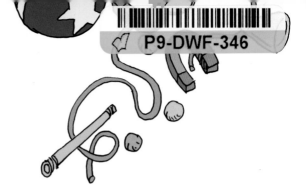

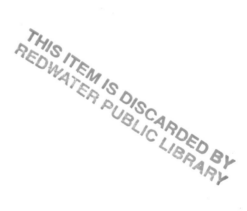

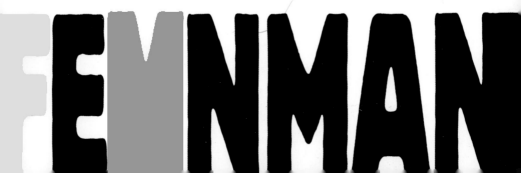

FEYNMAN

WRITTEN BY
JIM OTTAVIANI

ART BY
LELAND MYRICK

COLORING BY
HILARY SYCAMORE

:01

First Second
NEW YORK

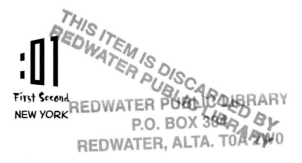

First Second

New York

www.firstsecondbooks.com

Text copyright © 2011 by Jim Ottaviani

Illustrations copyright © 2011 by Leland Myrick

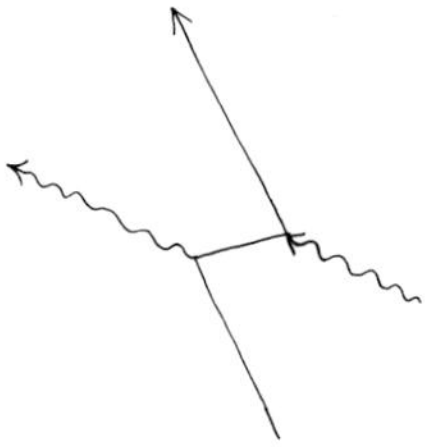

Published by First Second

First Second is an imprint of Roaring Brook Press,
a division of Holtzbrinck Publishing Holdings Limited Partnership
175 Fifth Avenue, New York, New York 10010

Library of Congress Cataloging-in-Publication Data

Ottaviani, Jim.
  Feynman / written by Jim Ottaviani ; art by Leland Myrick ; coloring by Hilary Sycamore. — 1st ed.
    p. cm.
  ISBN 978-1-59643-259-8
1. Feynman, Richard P. (Richard Phillips), 1918–1988. 2.
Physicists—United States—Biography. I. Title.
  QC16.F49O88 2011
  530.092—dc22
  [B]

2010036260

Hardcover ISBN: 978-1-59643-259-8
Paperback ISBN: 978-1-59643-827-9

First Second books are available for special promotions and premiums.
For details, contact: Director of Special Markets, Holtzbrinck Publishers.
Coloring by Hilary Sycamore and Sky Blue Ink with assistance from Marion Vitus

Book design by Marion Vitus
Printed in China

First Hardcover Edition 2011
First Paperback Edition 2013

Hardcover: 5 7 9 10 8 6
Paperback: 5 7 9 10 8 6 4

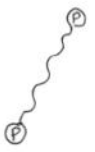

THANKS TO CARL FEYNMAN, MICHELLE FEYNMAN, AND RALPH LEIGHTON FOR THEIR GENEROSITY OVER THE YEARS. AND IF THE LAWS OF PHYSICS ALLOWED, I'D GO BACK TO THANK WHOEVER IT WAS THAT FIRST SHOWED ME "SURELY YOU'RE JOKING, MR. FEYNMAN!" THEN I WOULD READ IT AGAIN, FOR THE FIRST TIME.

—J.O.

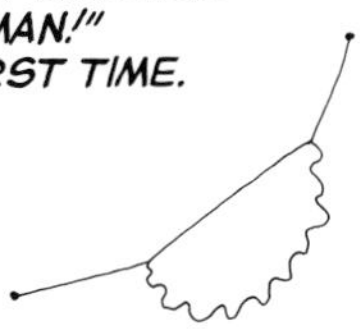

FOR MARIA.

—L.M.

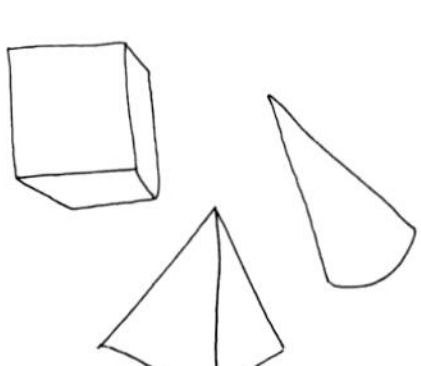

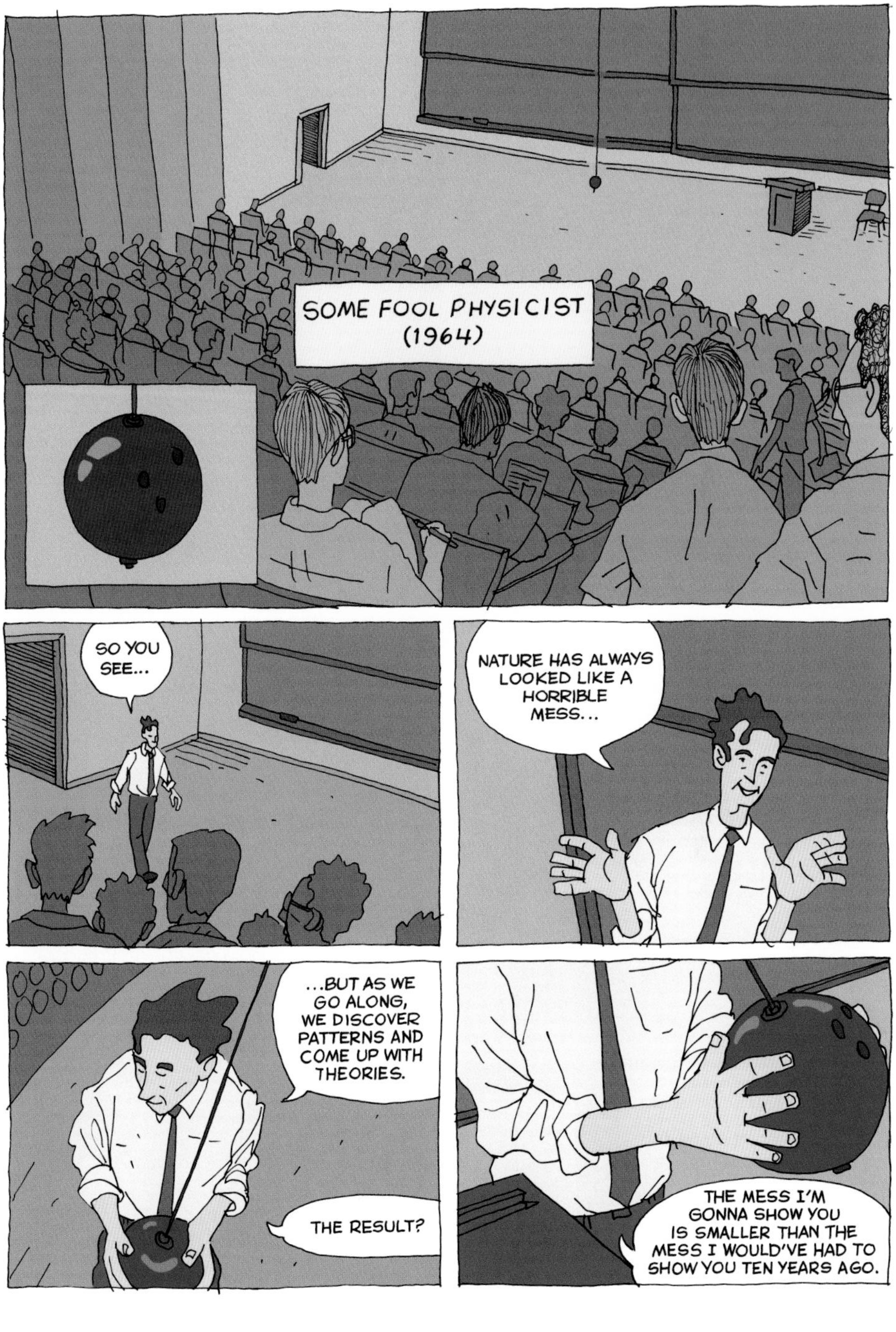
SOME FOOL PHYSICIST (1964)
SO YOU SEE...
NATURE HAS ALWAYS LOOKED LIKE A HORRIBLE MESS...
...BUT AS WE GO ALONG, WE DISCOVER PATTERNS AND COME UP WITH THEORIES.
THE RESULT?
THE MESS I'M GONNA SHOW YOU IS SMALLER THAN THE MESS I WOULD'VE HAD TO SHOW YOU TEN YEARS AGO.

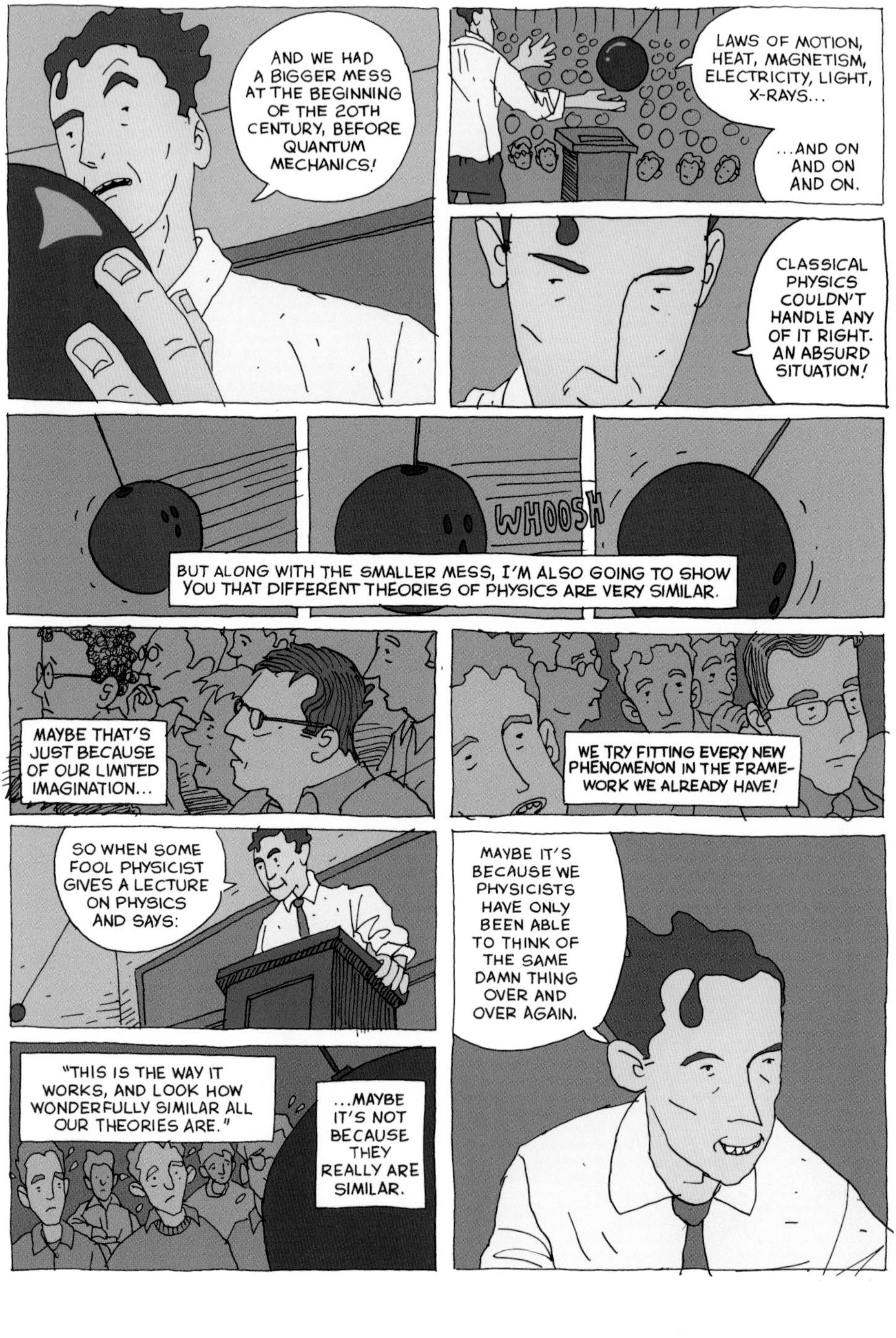
AND WE HAD A BIGGER MESS AT THE BEGINNING OF THE 20TH CENTURY, BEFORE QUANTUM MECHANICS!
LAWS OF MOTION, HEAT, MAGNETISM, ELECTRICITY, LIGHT, X-RAYS...
...AND ON AND ON AND ON.
CLASSICAL PHYSICS COULDN'T HANDLE ANY OF IT RIGHT. AN ABSURD SITUATION!
WHOOSH
BUT ALONG WITH THE SMALLER MESS, I'M ALSO GOING TO SHOW YOU THAT DIFFERENT THEORIES OF PHYSICS ARE VERY SIMILAR.
MAYBE THAT'S JUST BECAUSE OF OUR LIMITED IMAGINATION...
WE TRY FITTING EVERY NEW PHENOMENON IN THE FRAMEWORK WE ALREADY HAVE!
SO WHEN SOME FOOL PHYSICIST GIVES A LECTURE ON PHYSICS AND SAYS:
"THIS IS THE WAY IT WORKS, AND LOOK HOW WONDERFULLY SIMILAR ALL OUR THEORIES ARE."
...MAYBE IT'S NOT BECAUSE THEY REALLY ARE SIMILAR.
MAYBE IT'S BECAUSE WE PHYSICISTS HAVE ONLY BEEN ABLE TO THINK OF THE SAME DAMN THING OVER AND OVER AGAIN.

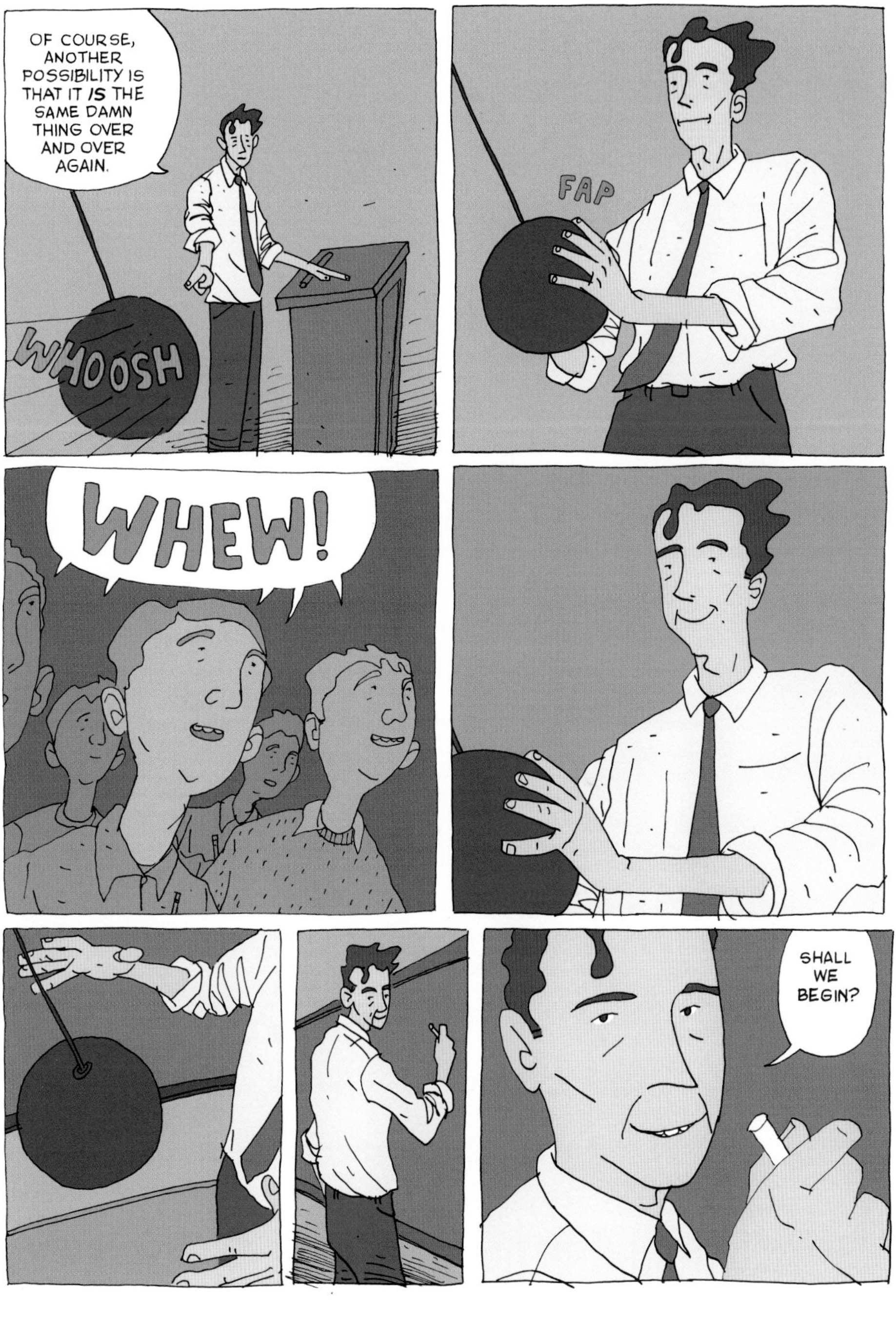
OF COURSE, ANOTHER POSSIBILITY IS THAT IT *IS* THE SAME DAMN THING OVER AND OVER AGAIN.
WHOOSH
FAP
WHEW!
SHALL WE BEGIN?

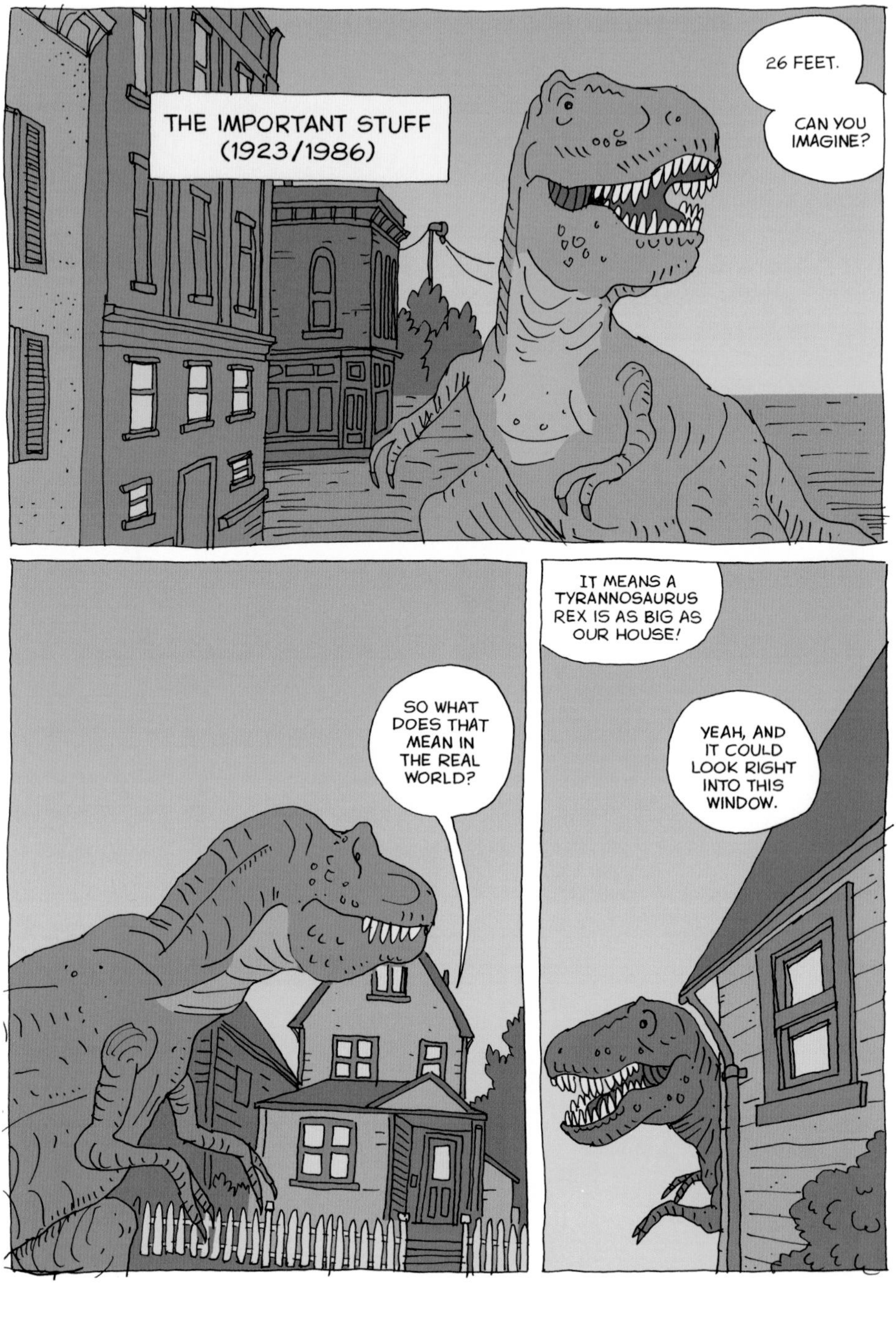
THE IMPORTANT STUFF (1923/1986)
26 FEET.
CAN YOU IMAGINE?
SO WHAT DOES THAT MEAN IN THE REAL WORLD?
IT MEANS A TYRANNOSAURUS REX IS AS BIG AS OUR HOUSE!
YEAH, AND IT COULD LOOK RIGHT INTO THIS WINDOW.

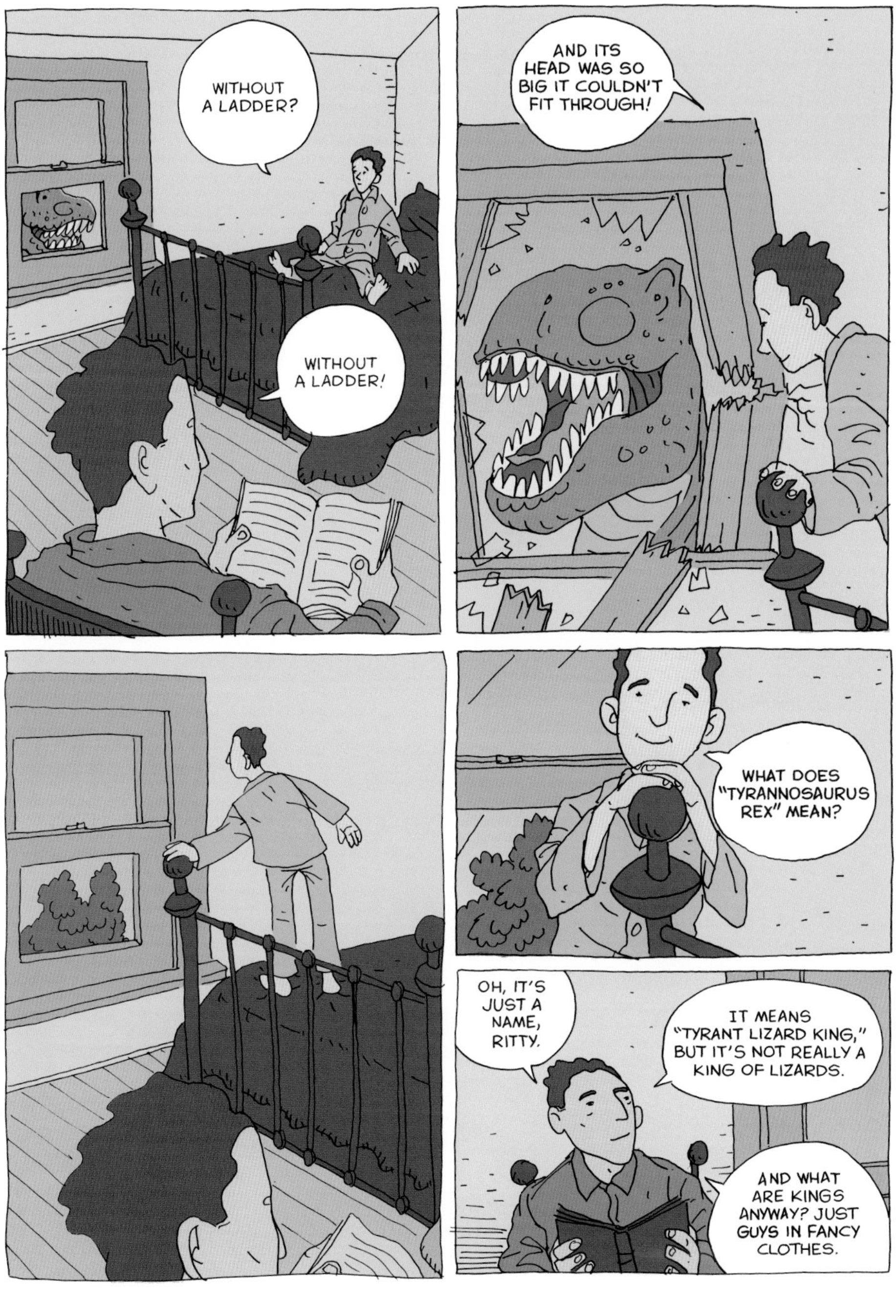
WITHOUT A LADDER?
WITHOUT A LADDER!
AND ITS HEAD WAS SO BIG IT COULDN'T FIT THROUGH!
WHAT DOES "TYRANNOSAURUS REX" MEAN?
OH, IT'S JUST A NAME, RITTY.
IT MEANS "TYRANT LIZARD KING," BUT IT'S NOT REALLY A KING OF LIZARDS.
AND WHAT ARE KINGS ANYWAY? JUST GUYS IN FANCY CLOTHES.

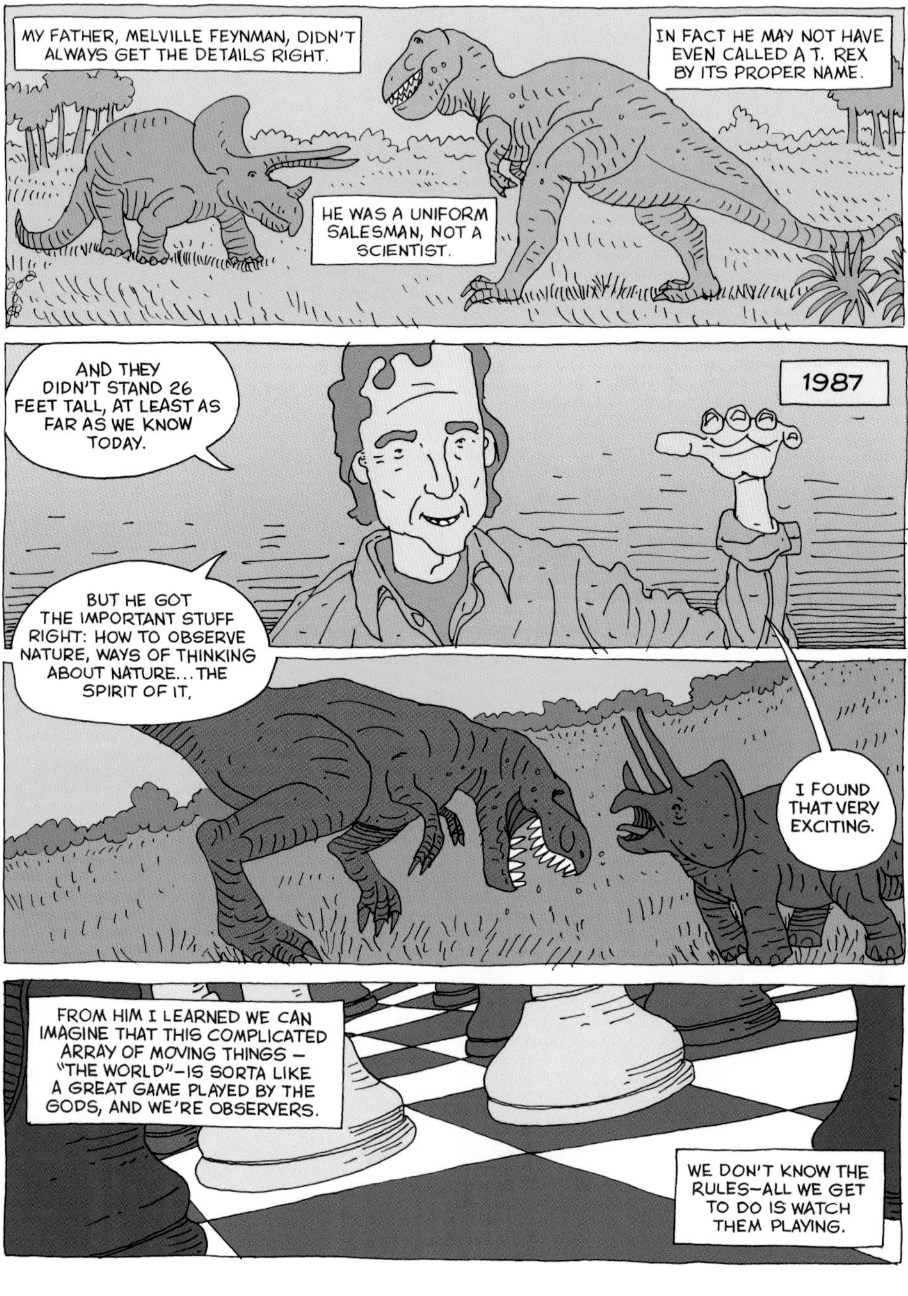
MY FATHER, MELVILLE FEYNMAN, DIDN'T ALWAYS GET THE DETAILS RIGHT.
IN FACT HE MAY NOT HAVE EVEN CALLED A T. REX BY ITS PROPER NAME.
HE WAS A UNIFORM SALESMAN, NOT A SCIENTIST.
AND THEY DIDN'T STAND 26 FEET TALL, AT LEAST AS FAR AS WE KNOW TODAY.
1987
BUT HE GOT THE IMPORTANT STUFF RIGHT: HOW TO OBSERVE NATURE, WAYS OF THINKING ABOUT NATURE... THE SPIRIT OF IT,
I FOUND THAT VERY EXCITING.
FROM HIM I LEARNED WE CAN IMAGINE THAT THIS COMPLICATED ARRAY OF MOVING THINGS — "THE WORLD"—IS SORTA LIKE A GREAT GAME PLAYED BY THE GODS, AND WE'RE OBSERVERS.
WE DON'T KNOW THE RULES—ALL WE GET TO DO IS WATCH THEM PLAYING.

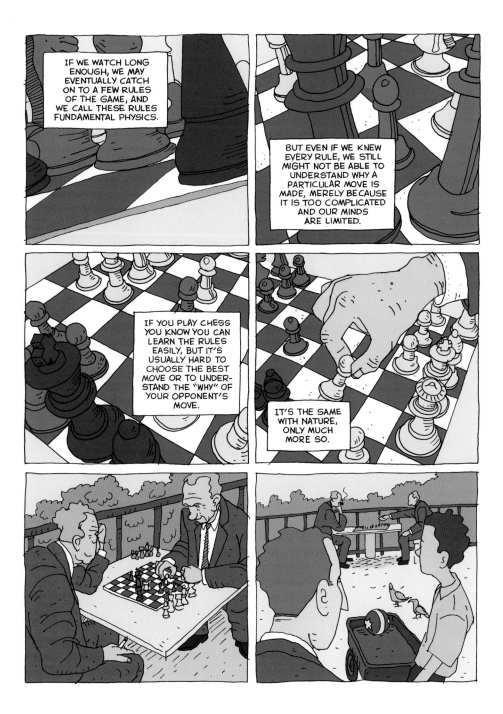

IF WE WATCH LONG ENOUGH, WE MAY EVENTUALLY CATCH ON TO A FEW RULES OF THE GAME, AND WE CALL THESE RULES FUNDAMENTAL PHYSICS.

BUT EVEN IF WE KNEW EVERY RULE, WE STILL MIGHT NOT BE ABLE TO UNDERSTAND WHY A PARTICULAR MOVE IS MADE, MERELY BECAUSE IT IS TOO COMPLICATED AND OUR MINDS ARE LIMITED.

IF YOU PLAY CHESS YOU KNOW YOU CAN LEARN THE RULES EASILY, BUT IT'S USUALLY HARD TO CHOOSE THE BEST MOVE OR TO UNDERSTAND THE "WHY" OF YOUR OPPONENT'S MOVE.

IT'S THE SAME WITH NATURE, ONLY MUCH MORE SO.

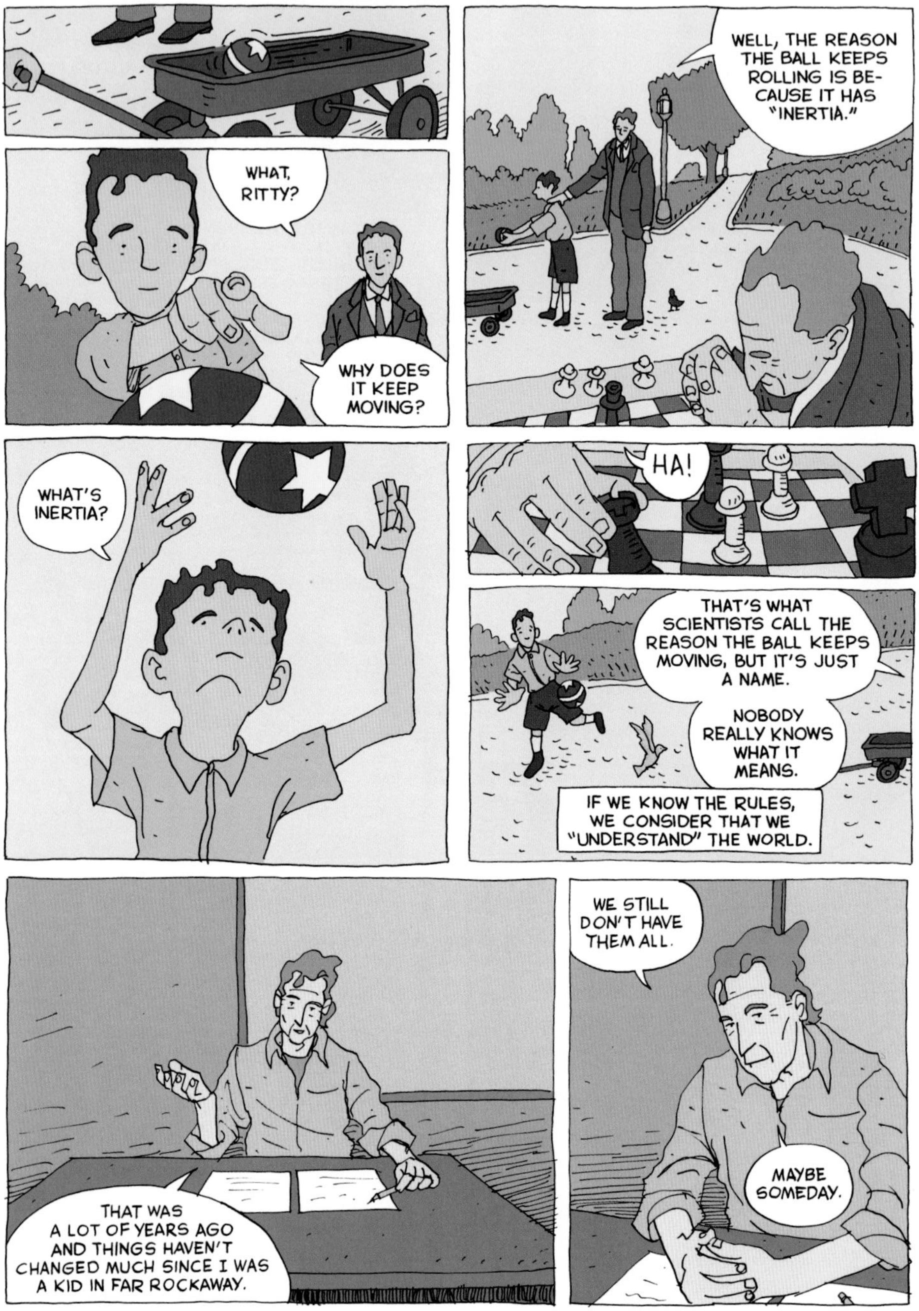
WHAT, RITTY?
WHY DOES IT KEEP MOVING?
WELL, THE REASON THE BALL KEEPS ROLLING IS BE-CAUSE IT HAS "INERTIA."
WHAT'S INERTIA?
HA!
THAT'S WHAT SCIENTISTS CALL THE REASON THE BALL KEEPS MOVING, BUT IT'S JUST A NAME.
NOBODY REALLY KNOWS WHAT IT MEANS.
IF WE KNOW THE RULES, WE CONSIDER THAT WE "UNDERSTAND" THE WORLD.
THAT WAS A LOT OF YEARS AGO AND THINGS HAVEN'T CHANGED MUCH SINCE I WAS A KID IN FAR ROCKAWAY.
WE STILL DON'T HAVE THEM ALL.
MAYBE SOMEDAY.

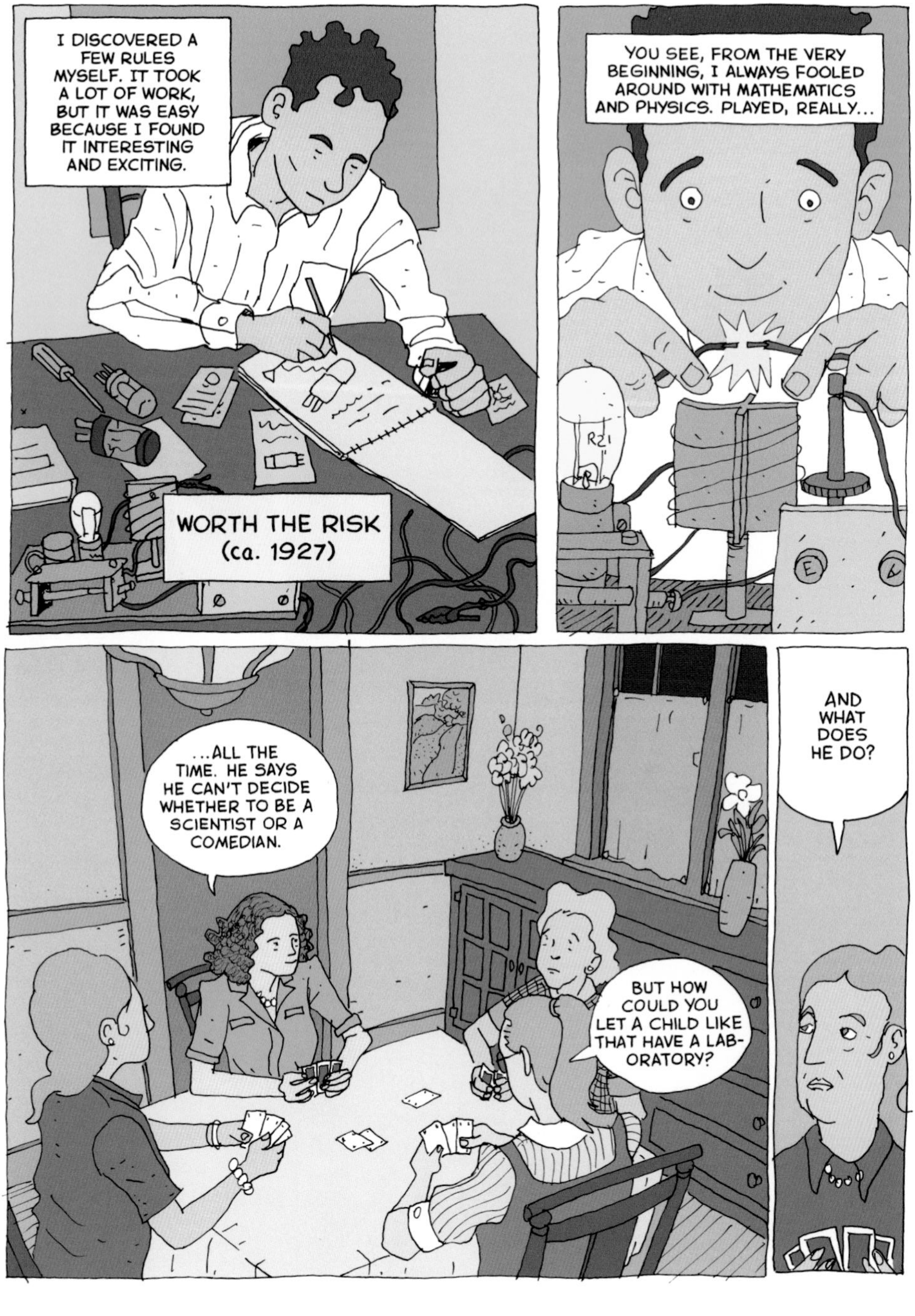
I DISCOVERED A FEW RULES MYSELF. IT TOOK A LOT OF WORK, BUT IT WAS EASY BECAUSE I FOUND IT INTERESTING AND EXCITING.
WORTH THE RISK (ca. 1927)
YOU SEE, FROM THE VERY BEGINNING, I ALWAYS FOOLED AROUND WITH MATHEMATICS AND PHYSICS. PLAYED, REALLY...
R21
E
...ALL THE TIME. HE SAYS HE CAN'T DECIDE WHETHER TO BE A SCIENTIST OR A COMEDIAN.
BUT HOW COULD YOU LET A CHILD LIKE THAT HAVE A LAB-ORATORY?
AND WHAT DOES HE DO?

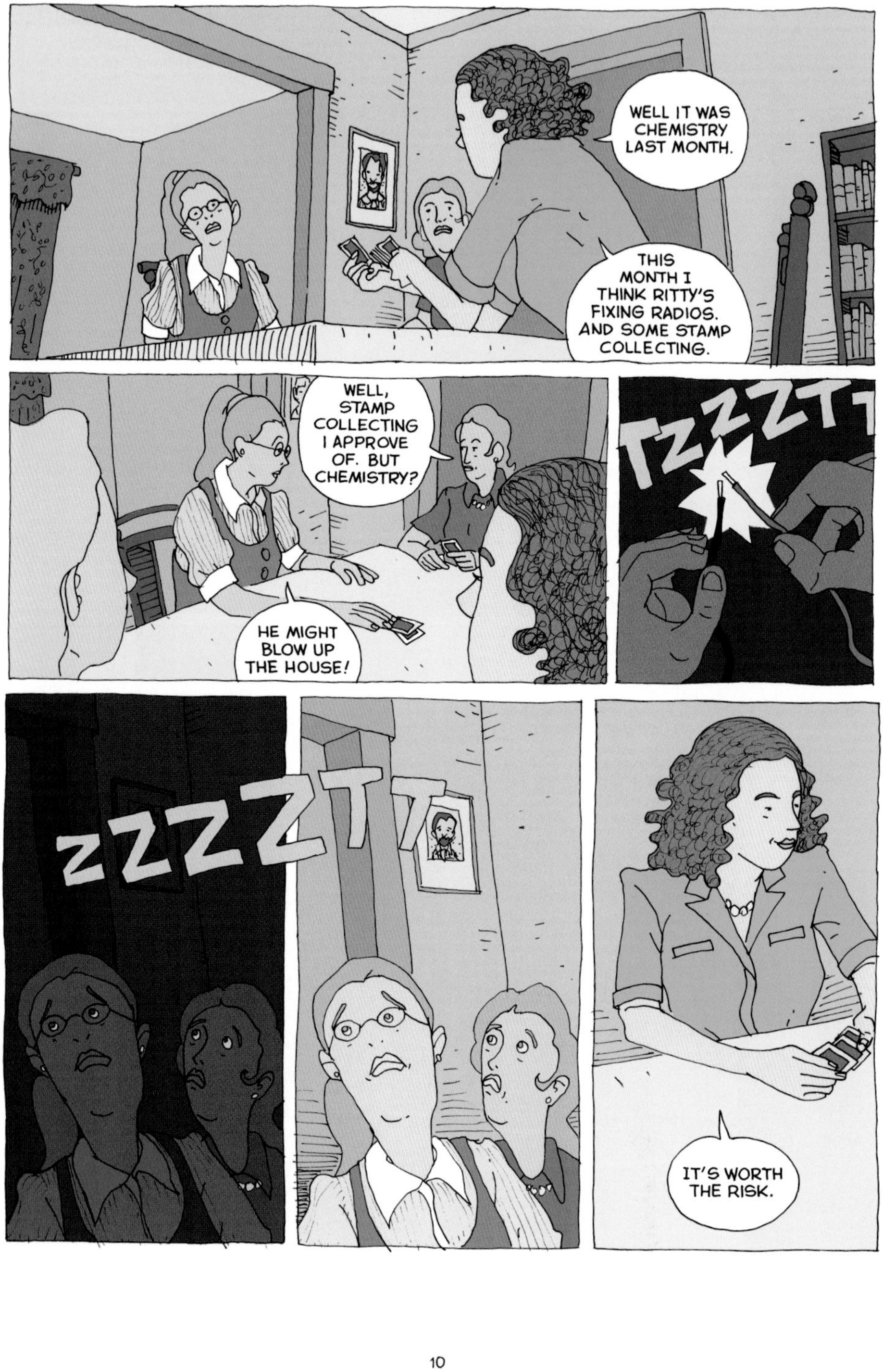
WELL IT WAS CHEMISTRY LAST MONTH.
THIS MONTH I THINK RITTY'S FIXING RADIOS. AND SOME STAMP COLLECTING.
WELL, STAMP COLLECTING I APPROVE OF. BUT CHEMISTRY?
HE MIGHT BLOW UP THE HOUSE!
TZZZTT
zZZZTTT
IT'S WORTH THE RISK.

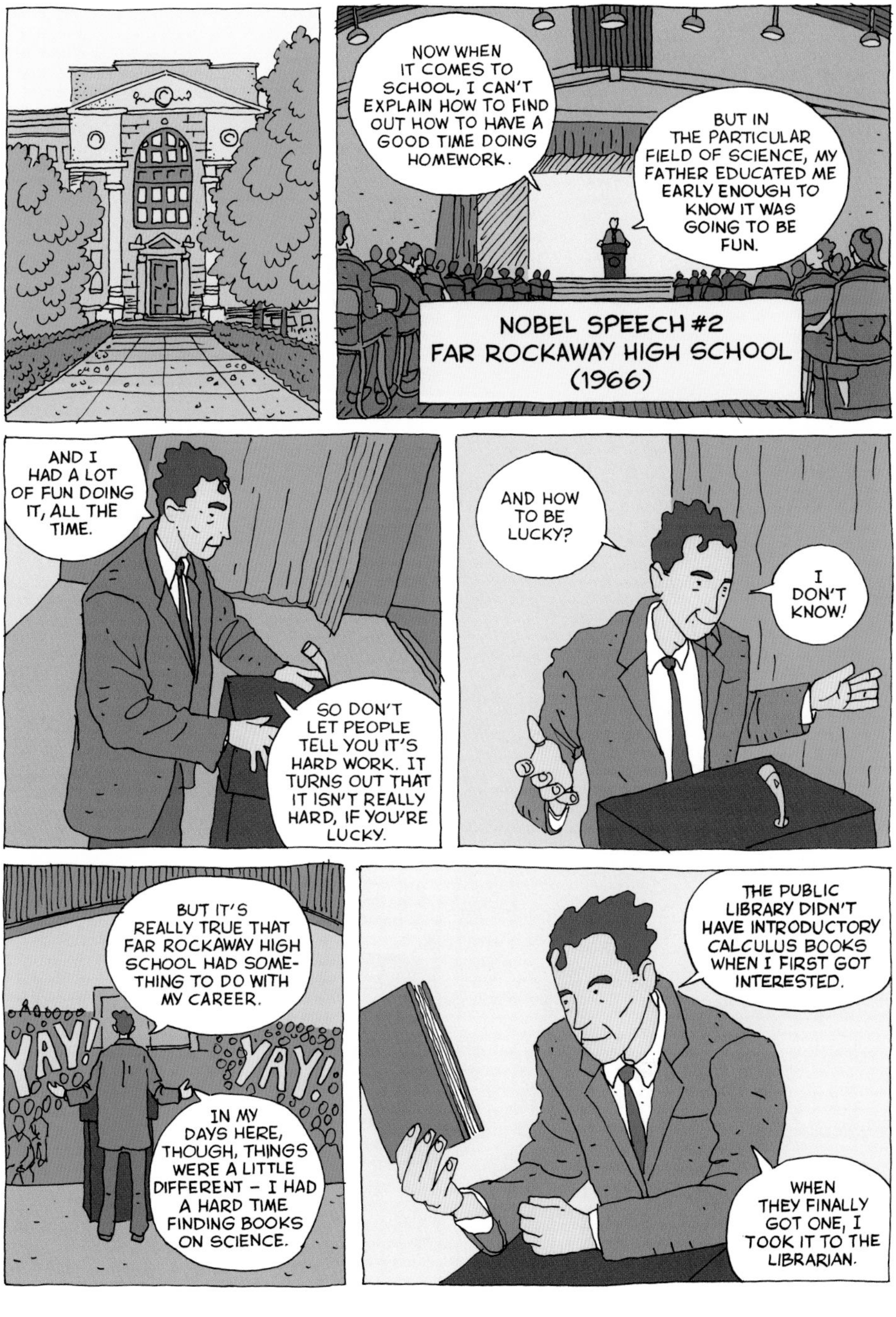
NOW WHEN IT COMES TO SCHOOL, I CAN'T EXPLAIN HOW TO FIND OUT HOW TO HAVE A GOOD TIME DOING HOMEWORK.
BUT IN THE PARTICULAR FIELD OF SCIENCE, MY FATHER EDUCATED ME EARLY ENOUGH TO KNOW IT WAS GOING TO BE FUN.
NOBEL SPEECH #2 FAR ROCKAWAY HIGH SCHOOL (1966)
AND I HAD A LOT OF FUN DOING IT, ALL THE TIME.
SO DON'T LET PEOPLE TELL YOU IT'S HARD WORK. IT TURNS OUT THAT IT ISN'T REALLY HARD, IF YOU'RE LUCKY.
AND HOW TO BE LUCKY?
I DON'T KNOW!
BUT IT'S REALLY TRUE THAT FAR ROCKAWAY HIGH SCHOOL HAD SOMETHING TO DO WITH MY CAREER.
YAY!
YAY!
IN MY DAYS HERE, THOUGH, THINGS WERE A LITTLE DIFFERENT – I HAD A HARD TIME FINDING BOOKS ON SCIENCE.
THE PUBLIC LIBRARY DIDN'T HAVE INTRODUCTORY CALCULUS BOOKS WHEN I FIRST GOT INTERESTED.
WHEN THEY FINALLY GOT ONE, I TOOK IT TO THE LIBRARIAN.

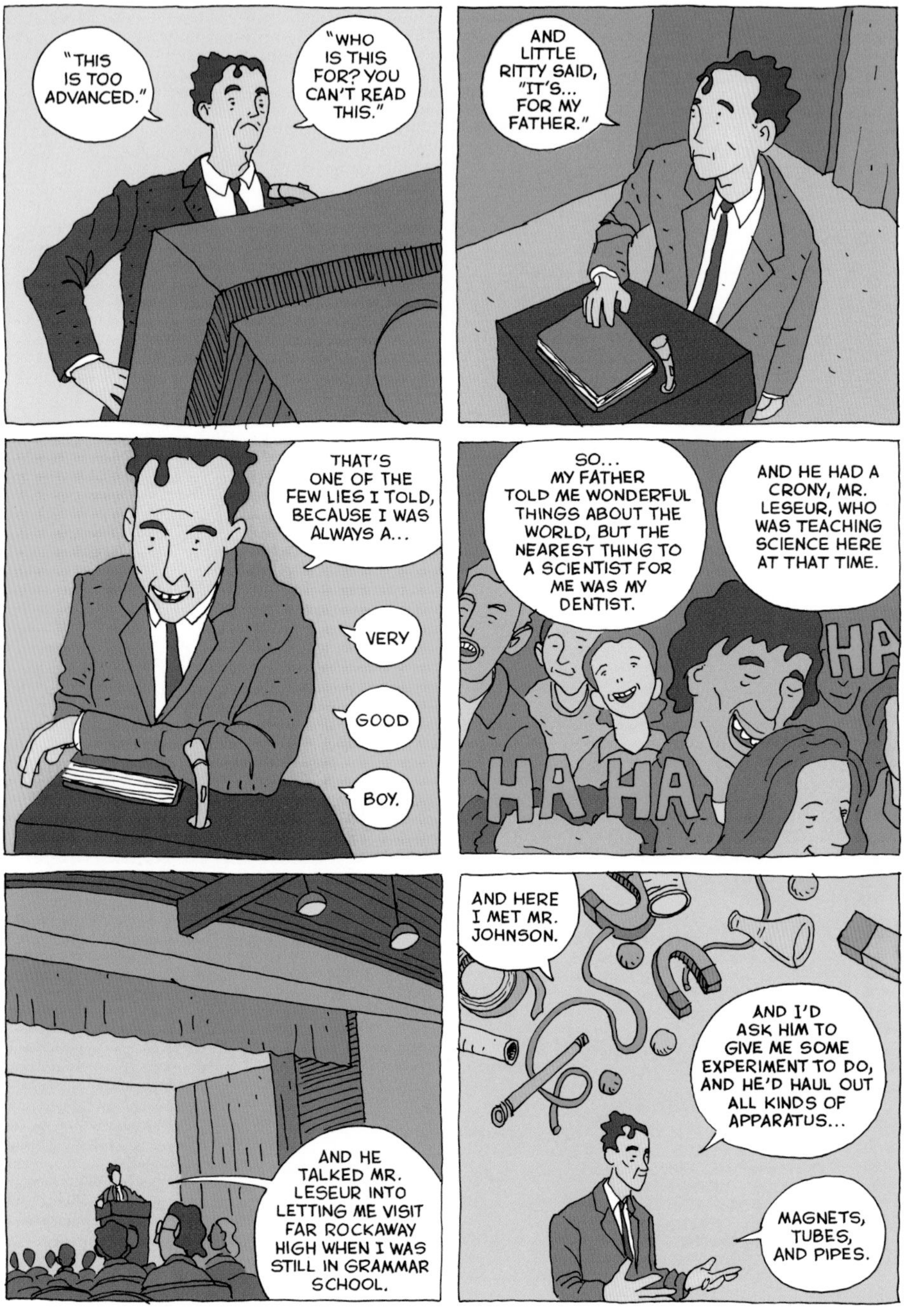
"THIS IS TOO ADVANCED."
"WHO IS THIS FOR? YOU CAN'T READ THIS."
AND LITTLE RITTY SAID, "IT'S... FOR MY FATHER."
THAT'S ONE OF THE FEW LIES I TOLD, BECAUSE I WAS ALWAYS A...
VERY
GOOD
BOY.
SO... MY FATHER TOLD ME WONDERFUL THINGS ABOUT THE WORLD, BUT THE NEAREST THING TO A SCIENTIST FOR ME WAS MY DENTIST.
AND HE HAD A CRONY, MR. LESEUR, WHO WAS TEACHING SCIENCE HERE AT THAT TIME.
HA
HA HA
AND HE TALKED MR. LESEUR INTO LETTING ME VISIT FAR ROCKAWAY HIGH WHEN I WAS STILL IN GRAMMAR SCHOOL.
AND HERE I MET MR. JOHNSON.
AND I'D ASK HIM TO GIVE ME SOME EXPERIMENT TO DO, AND HE'D HAUL OUT ALL KINDS OF APPARATUS...
MAGNETS, TUBES, AND PIPES.

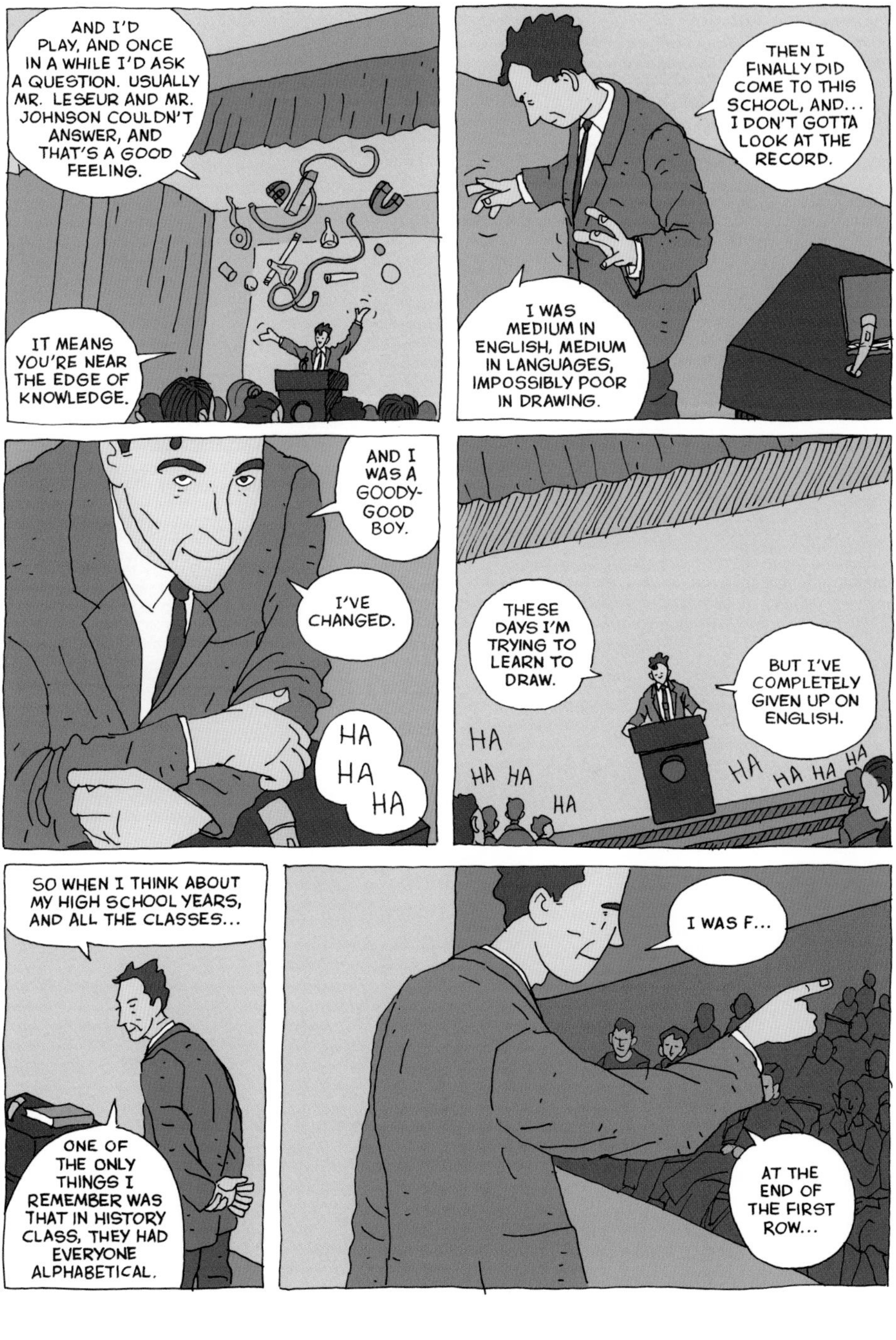
AND I'D PLAY, AND ONCE IN A WHILE I'D ASK A QUESTION. USUALLY MR. LESEUR AND MR. JOHNSON COULDN'T ANSWER, AND THAT'S A GOOD FEELING.
IT MEANS YOU'RE NEAR THE EDGE OF KNOWLEDGE.
THEN I FINALLY DID COME TO THIS SCHOOL, AND... I DON'T GOTTA LOOK AT THE RECORD.
I WAS MEDIUM IN ENGLISH, MEDIUM IN LANGUAGES, IMPOSSIBLY POOR IN DRAWING.
AND I WAS A GOODY-GOOD BOY.
I'VE CHANGED.
HA HA HA
THESE DAYS I'M TRYING TO LEARN TO DRAW.
BUT I'VE COMPLETELY GIVEN UP ON ENGLISH.
HA HA HA HA
HA HA HA HA
SO WHEN I THINK ABOUT MY HIGH SCHOOL YEARS, AND ALL THE CLASSES...
ONE OF THE ONLY THINGS I REMEMBER WAS THAT IN HISTORY CLASS, THEY HAD EVERYONE ALPHABETICAL.
I WAS F...
AT THE END OF THE FIRST ROW...

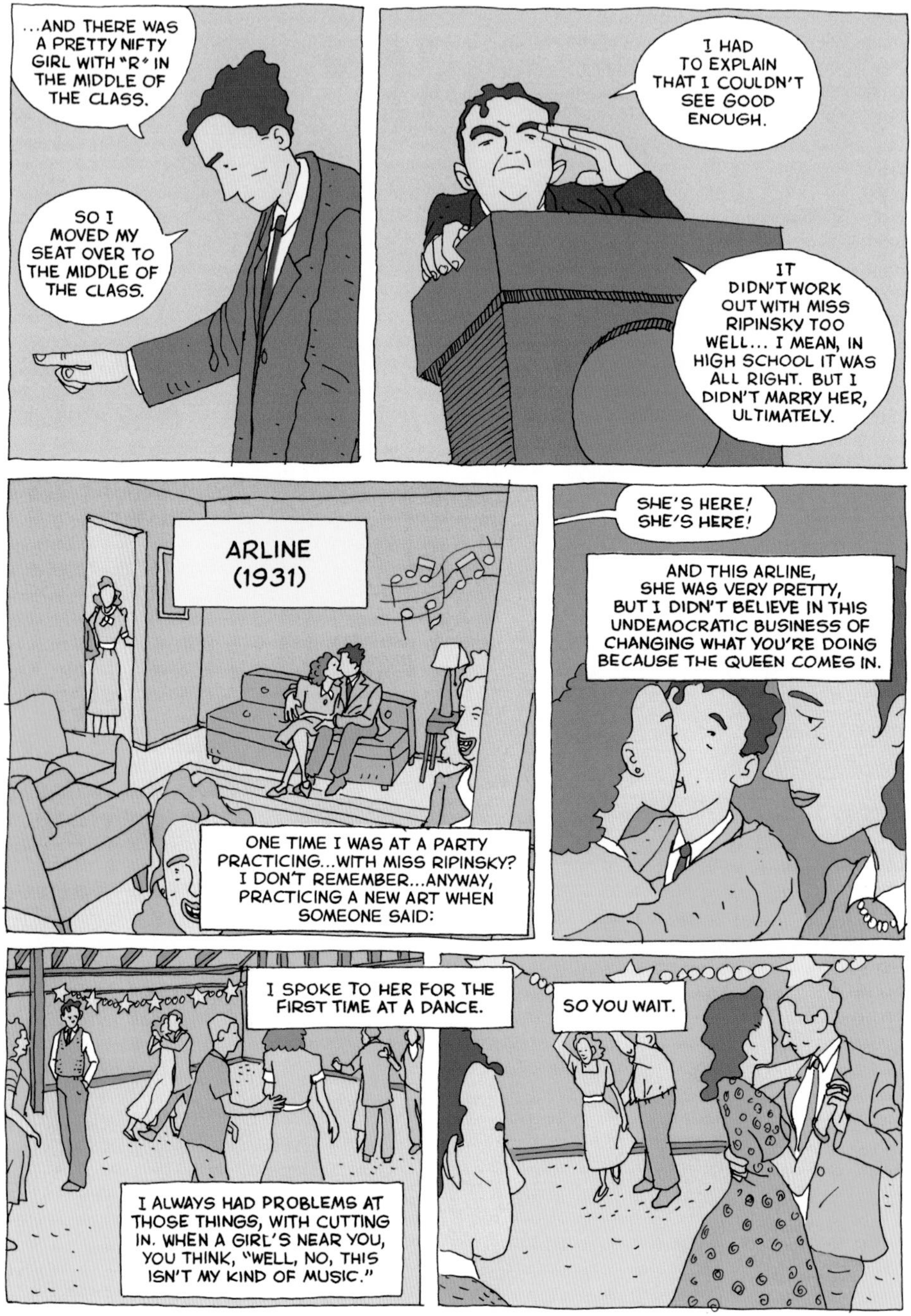
...AND THERE WAS A PRETTY NIFTY GIRL WITH "R" IN THE MIDDLE OF THE CLASS.
SO I MOVED MY SEAT OVER TO THE MIDDLE OF THE CLASS.
I HAD TO EXPLAIN THAT I COULDN'T SEE GOOD ENOUGH.
IT DIDN'T WORK OUT WITH MISS RIPINSKY TOO WELL... I MEAN, IN HIGH SCHOOL IT WAS ALL RIGHT. BUT I DIDN'T MARRY HER, ULTIMATELY.
ARLINE (1931)
ONE TIME I WAS AT A PARTY PRACTICING...WITH MISS RIPINSKY? I DON'T REMEMBER...ANYWAY, PRACTICING A NEW ART WHEN SOMEONE SAID:
SHE'S HERE! SHE'S HERE!
AND THIS ARLINE, SHE WAS VERY PRETTY, BUT I DIDN'T BELIEVE IN THIS UNDEMOCRATIC BUSINESS OF CHANGING WHAT YOU'RE DOING BECAUSE THE QUEEN COMES IN.
I SPOKE TO HER FOR THE FIRST TIME AT A DANCE.
I ALWAYS HAD PROBLEMS AT THOSE THINGS, WITH CUTTING IN. WHEN A GIRL'S NEAR YOU, YOU THINK, "WELL, NO, THIS ISN'T MY KIND OF MUSIC."
SO YOU WAIT.

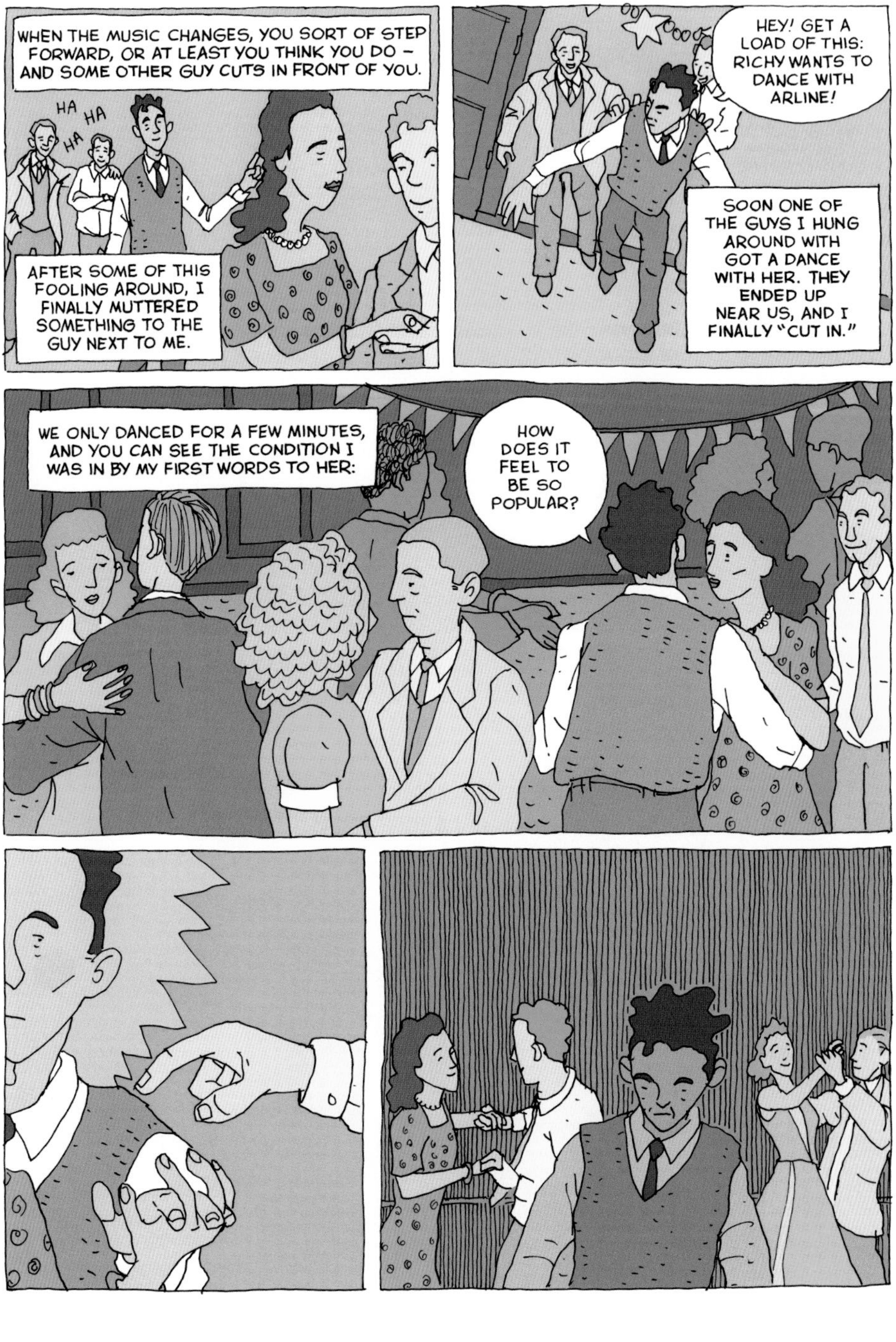
WHEN THE MUSIC CHANGES, YOU SORT OF STEP FORWARD, OR AT LEAST YOU THINK YOU DO – AND SOME OTHER GUY CUTS IN FRONT OF YOU.
HA HA HA
AFTER SOME OF THIS FOOLING AROUND, I FINALLY MUTTERED SOMETHING TO THE GUY NEXT TO ME.
HEY! GET A LOAD OF THIS: RICHY WANTS TO DANCE WITH ARLINE!
SOON ONE OF THE GUYS I HUNG AROUND WITH GOT A DANCE WITH HER. THEY ENDED UP NEAR US, AND I FINALLY "CUT IN."
WE ONLY DANCED FOR A FEW MINUTES, AND YOU CAN SEE THE CONDITION I WAS IN BY MY FIRST WORDS TO HER:
HOW DOES IT FEEL TO BE SO POPULAR?

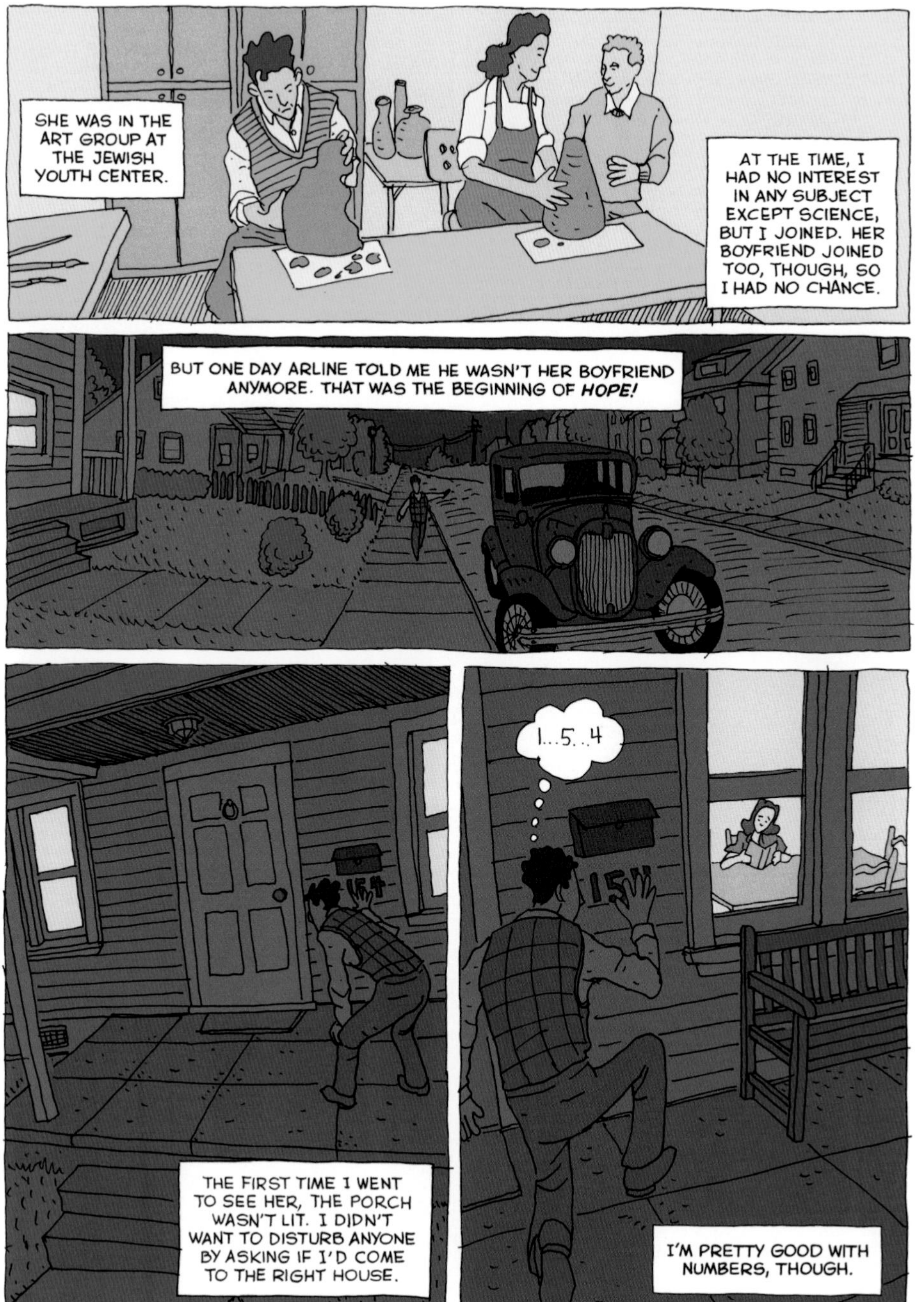
SHE WAS IN THE ART GROUP AT THE JEWISH YOUTH CENTER.
AT THE TIME, I HAD NO INTEREST IN ANY SUBJECT EXCEPT SCIENCE, BUT I JOINED. HER BOYFRIEND JOINED TOO, THOUGH, SO I HAD NO CHANCE.
BUT ONE DAY ARLINE TOLD ME HE WASN'T HER BOYFRIEND ANYMORE. THAT WAS THE BEGINNING OF ***HOPE!***
THE FIRST TIME I WENT TO SEE HER, THE PORCH WASN'T LIT. I DIDN'T WANT TO DISTURB ANYONE BY ASKING IF I'D COME TO THE RIGHT HOUSE.
1...5...4
I'M PRETTY GOOD WITH NUMBERS, THOUGH.

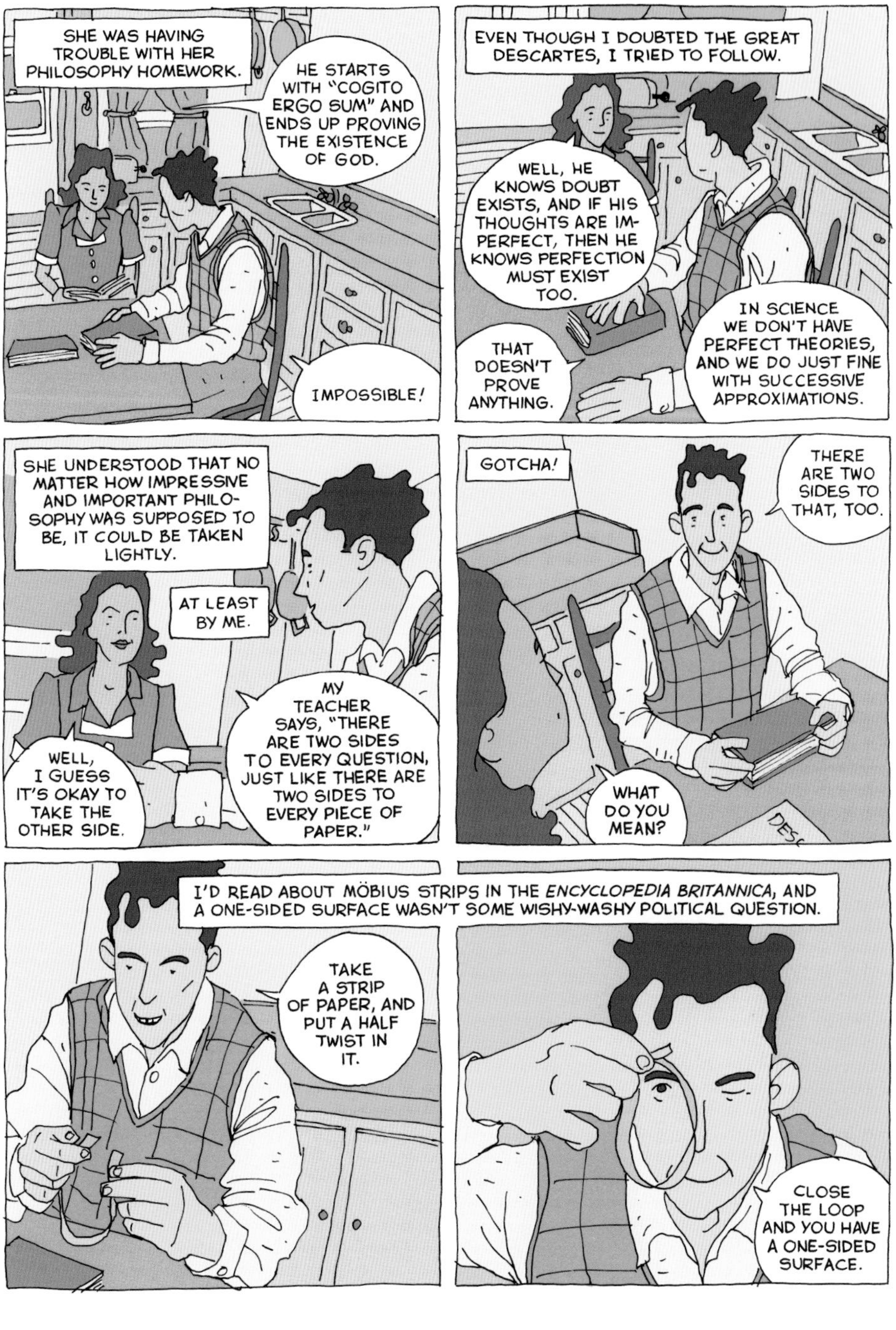
SHE WAS HAVING TROUBLE WITH HER PHILOSOPHY HOMEWORK.
HE STARTS WITH "COGITO ERGO SUM" AND ENDS UP PROVING THE EXISTENCE OF GOD.
IMPOSSIBLE!
EVEN THOUGH I DOUBTED THE GREAT DESCARTES, I TRIED TO FOLLOW.
WELL, HE KNOWS DOUBT EXISTS, AND IF HIS THOUGHTS ARE IMPERFECT, THEN HE KNOWS PERFECTION MUST EXIST TOO.
THAT DOESN'T PROVE ANYTHING.
IN SCIENCE WE DON'T HAVE PERFECT THEORIES, AND WE DO JUST FINE WITH SUCCESSIVE APPROXIMATIONS.
SHE UNDERSTOOD THAT NO MATTER HOW IMPRESSIVE AND IMPORTANT PHILOSOPHY WAS SUPPOSED TO BE, IT COULD BE TAKEN LIGHTLY.
AT LEAST BY ME.
WELL, I GUESS IT'S OKAY TO TAKE THE OTHER SIDE.
MY TEACHER SAYS, "THERE ARE TWO SIDES TO EVERY QUESTION, JUST LIKE THERE ARE TWO SIDES TO EVERY PIECE OF PAPER."
GOTCHA!
THERE ARE TWO SIDES TO THAT, TOO.
WHAT DO YOU MEAN?
I'D READ ABOUT MÖBIUS STRIPS IN THE *ENCYCLOPEDIA BRITANNICA*, AND A ONE-SIDED SURFACE WASN'T SOME WISHY-WASHY POLITICAL QUESTION.
TAKE A STRIP OF PAPER, AND PUT A HALF TWIST IN IT.
CLOSE THE LOOP AND YOU HAVE A ONE-SIDED SURFACE.

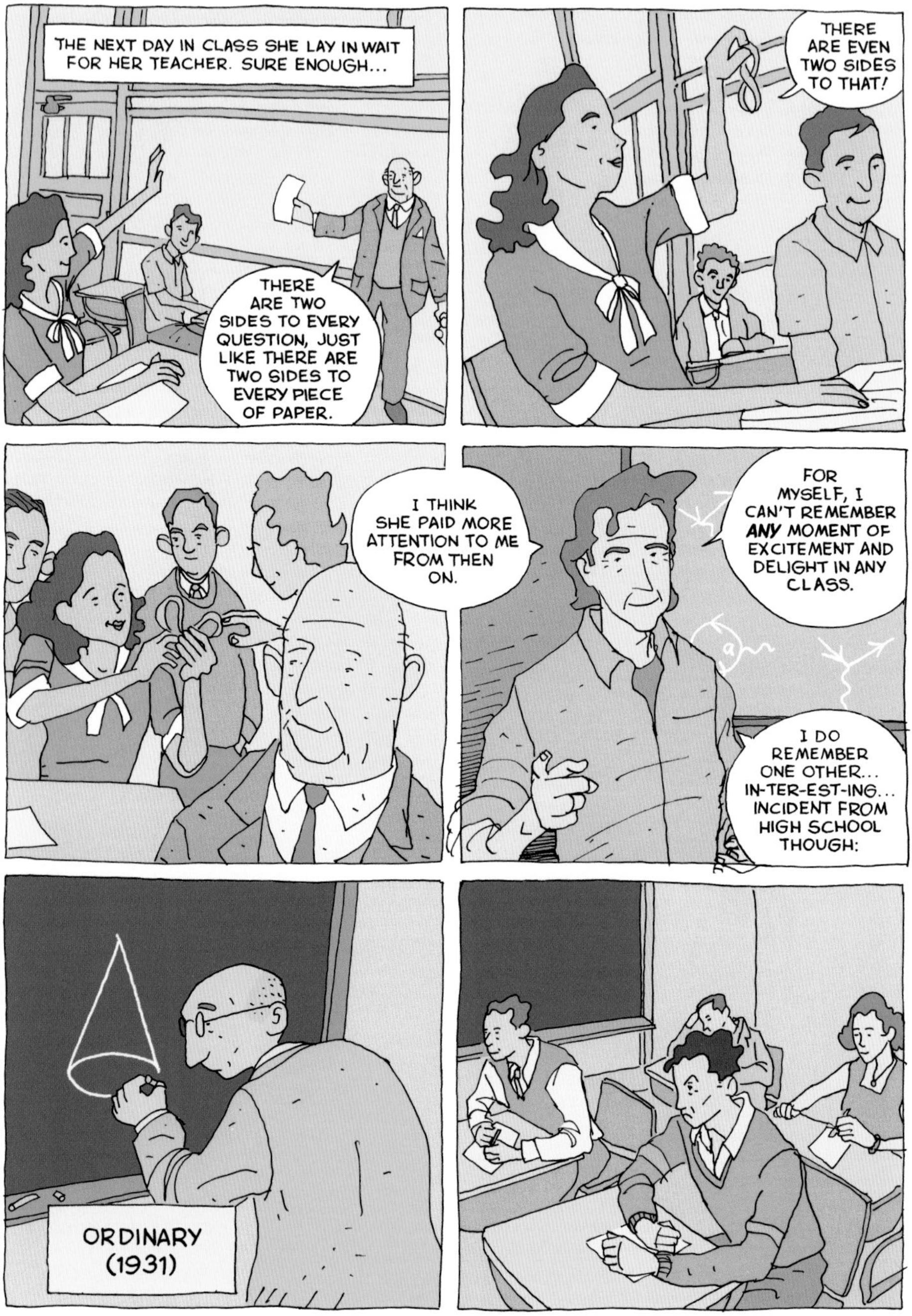
THE NEXT DAY IN CLASS SHE LAY IN WAIT FOR HER TEACHER. SURE ENOUGH...
THERE ARE TWO SIDES TO EVERY QUESTION, JUST LIKE THERE ARE TWO SIDES TO EVERY PIECE OF PAPER.
THERE ARE EVEN TWO SIDES TO THAT!
I THINK SHE PAID MORE ATTENTION TO ME FROM THEN ON.
FOR MYSELF, I CAN'T REMEMBER ANY MOMENT OF EXCITEMENT AND DELIGHT IN ANY CLASS.
I DO REMEMBER ONE OTHER... IN-TER-EST-ING... INCIDENT FROM HIGH SCHOOL THOUGH:
ORDINARY (1931)

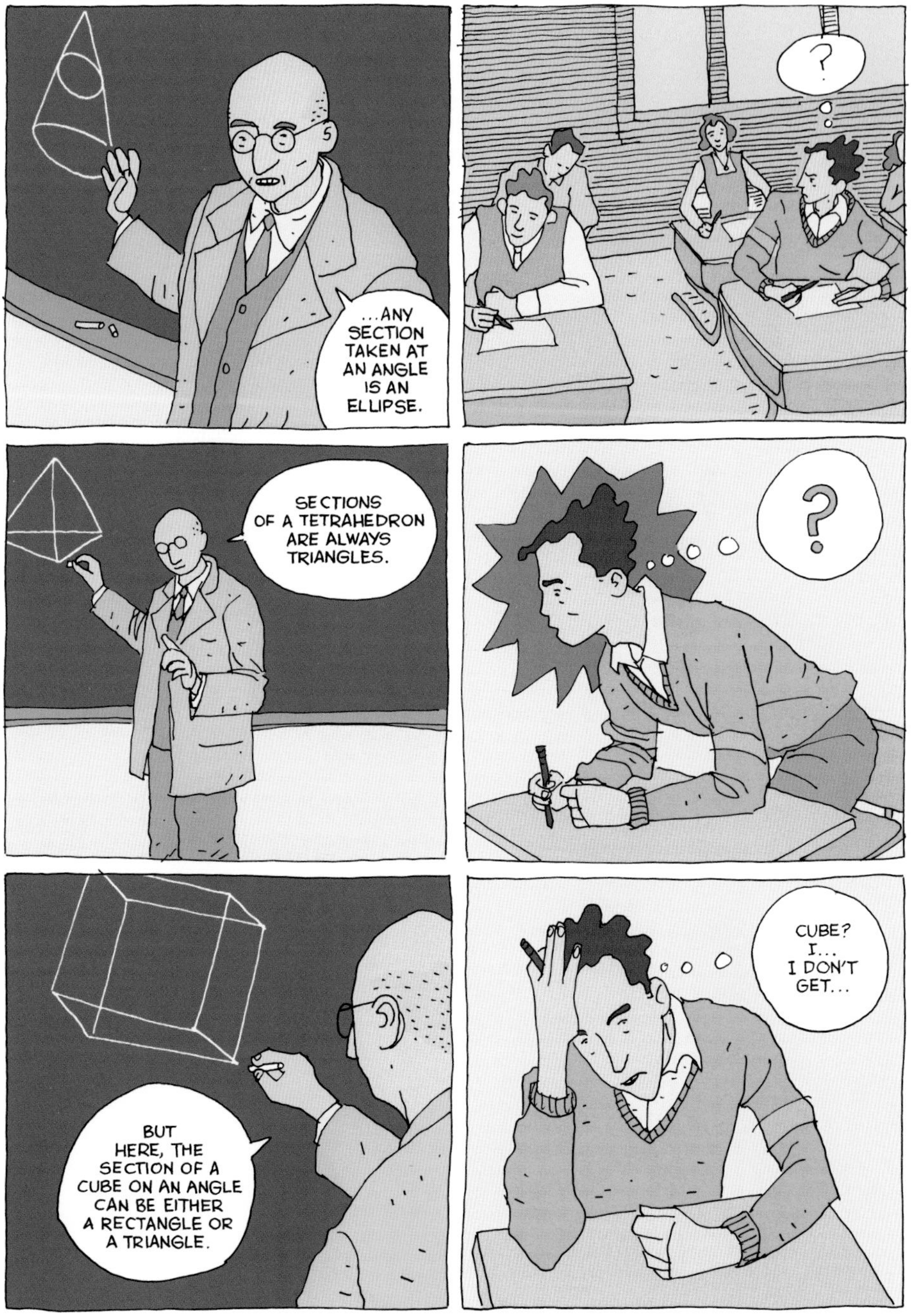
...ANY SECTION TAKEN AT AN ANGLE IS AN ELLIPSE.
?
SECTIONS OF A TETRAHEDRON ARE ALWAYS TRIANGLES.
?
BUT HERE, THE SECTION OF A CUBE ON AN ANGLE CAN BE EITHER A RECTANGLE OR A TRIANGLE.
CUBE? I... I DON'T GET...

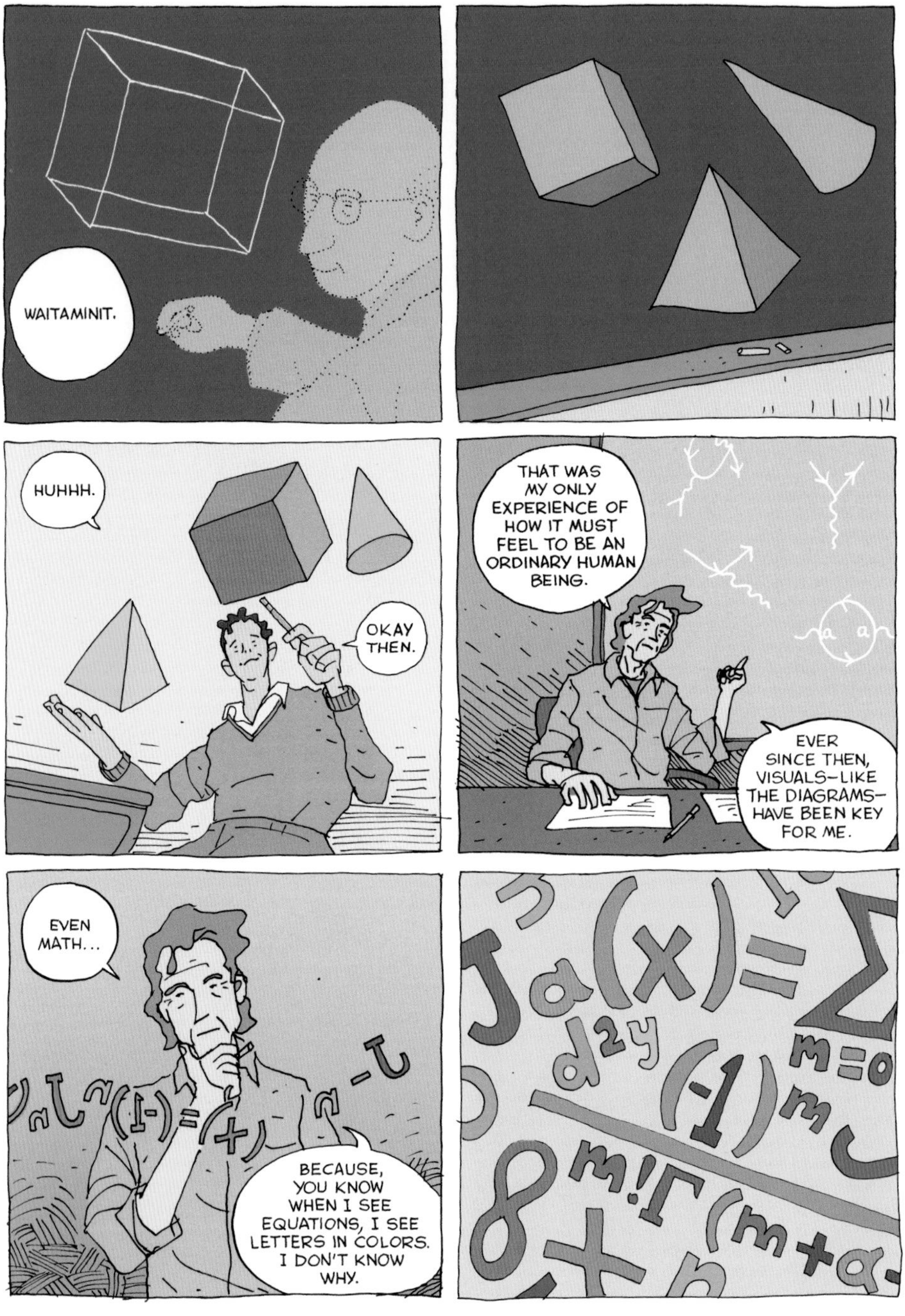
WAITAMINIT.
HUHHH.
OKAY THEN.
THAT WAS MY ONLY EXPERIENCE OF HOW IT MUST FEEL TO BE AN ORDINARY HUMAN BEING.
EVER SINCE THEN, VISUALS—LIKE THE DIAGRAMS—HAVE BEEN KEY FOR ME.
EVEN MATH...
BECAUSE, YOU KNOW WHEN I SEE EQUATIONS, I SEE LETTERS IN COLORS. I DON'T KNOW WHY.

$4n^2)/4x^2]v = 0 \ x^2 C_n'' + xC_n' + (x^2 - n^2)C_n = 0 \ C_{n-1} - C_{n+1} = 2(dC_n/dx) \ C_{n-1} + C_{n+1} = (2n/x) \ C_n J_v(r) = 2 \ (r/2)^v / [\Gamma(v + J_v(r) = 2 (/ [\Gamma(v + \frac{1}{2}) \ \Gamma(\frac{1}{2})] \int_0^{\pi/2}(\sin \theta)^{2v} \cos (r \cos \theta) \ d\theta \ \frac{1}{2}) \ \Gamma(\frac{1}{2})] \int_0^{\pi/2}(\sin \theta)^{2v} \cos (r \cos \theta) \ d\theta \ J_y(r) = 2 \ (r/2)^v / [\Gamma(v + \frac{1}{2}) \ \Gamma(\frac{1}{2} \int_0^{\pi/2}(\sin \theta)^{2v} \cos (r \cos \theta) \ d\theta \ d\Theta / d\theta^2 = -\lambda^2\Theta d^2R / dr^2 + 1 / rdR / dr + (k^2 - \lambda^2 / r^2)R = 0 \ r^2 / R(d^2R / dr^2 + 1 / rdR / dr) + k^2r^2 = -1 / \Theta d \ d\theta^2 \ J_n = (x/2)^n/2^n\Gamma(n+1)\{1 - (x/2)^2/2(2n+2) + (x/2)^4/2 \ 4(2n+2)(2n+4) \ v'' + [1 + (1 - 4n^2)/4x^2]v = 0 \ x^2C_n'' + xC_n' + (x^2 - n^2)C_n \ C_{n-1} - C_{n+1} = 2(dC_n/dx) \ C_{n-1} + C_{n+1} = (2n/x) \ C_n J_v(r) = 2 \ (r/2)^v / [\Gamma(v +J_v(r) = 2 (r/2)^v / [\Gamma(v + \frac{1}{2}) \ \Gamma(\frac{1}{2})] \int_0^{\pi/2}(\sin \theta)^{2v} \cos (r \ \theta) \ d\theta \ \frac{1}{2}) \ \Gamma(\frac{1}{2})] \int_0^{\pi/2} (\sin \theta)^{2v}\cos (r \cos \theta) \ d\theta \ J_v(r) = 2 \ (r/2)^v / [\Gamma(v + \frac{1}{2}) \ \Gamma(\frac{1}{2})] \int_0^{\pi/2}(\sin \theta)^{2v} \cos (r \cos \theta) \ d\theta d\Theta / d\theta^2 = \lambda^2\Theta d^2R / dr^2 + 1 / rdR / dr + (k^2 - \lambda^2 / r^2)R = 0 \ r^2 / R(d^2R / dr^2 + 1 / rdR / dr) + k^2r^2 = -1 / \Theta d^2\Theta / d\theta^2 \ J_n = (x/2)^n/2^n\Gamma(n+1)\{1 - (x/2)^2/2(2n + 2) + (x/2)^4/2 \ 4(2n+2)(2n+4) \ v'' + [1 + (1 - 4n^2)/4x^2]v = 0 \ x^2C_n'' + xC_n' + (x^2 - n^2)C_n = 0 \ C_{n-1} - C_{n+1} = 2(dC_n/dx) \ C_{n-1} + C_{n+1} = (2n/x) \ C_n J_v(r) = 2 \ (r/2)^v / [\Gamma(v +J_v(r) = 2n \ (r/2)^v / [\Gamma(v + \frac{1}{2}) \ \Gamma(\frac{1}{2})] \int_0^{\pi/2}(\sin \theta)^{2v} \cos (r \cos \theta) \ d\theta \ \frac{1}{2}) \ \Gamma(\frac{1}{2})] \int_0^{\pi/2}(\sin \theta)^{2v} \cos (r \cos \theta) \ d\theta \ J_v(r) = 2 \ (r/2)^v / [\Gamma(v + \frac{1}{2} \ \Gamma(\frac{1}{2})] \int_0^{\pi/2}(\sin \theta)^{2v} \cos (r \cos \theta) \ d\theta \ d\Theta / d\theta^2 = -\lambda^2\Theta d^2R / dr^2 + 1 / rdR / dr + (k^2 - \lambda^2 / r^2)R = 0 \ r^2 / R(d^2R / dr^2 + 1 / rdR / dr) + k^2r^2 = -\Theta d^2\Theta / d\theta^2 \ J_n = (x/2)^n/2^n\Gamma(n+1) \{1 - (x/2)^2/2(2n+2) + (x/2)^4/2 \ 4(2n+2)(2n+4) \ v'' + [1 + (1 - 4n^2)/4x^2]v = 0 \ x^2C_n'' + xC_n' + (x^2 -$

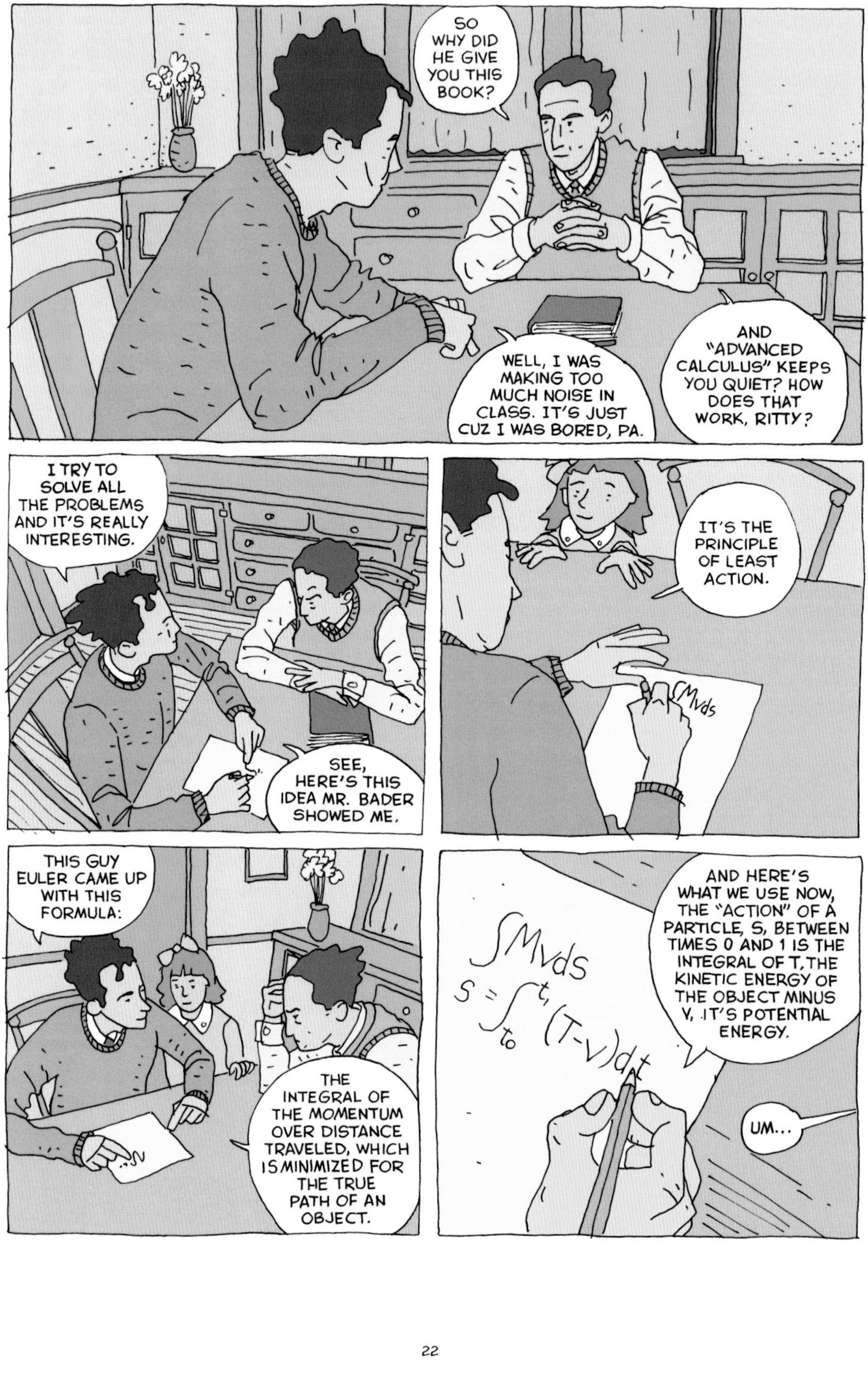
SO WHY DID HE GIVE YOU THIS BOOK?
WELL, I WAS MAKING TOO MUCH NOISE IN CLASS. IT'S JUST CUZ I WAS BORED, PA.
AND "ADVANCED CALCULUS" KEEPS YOU QUIET? HOW DOES THAT WORK, RITTY?
I TRY TO SOLVE ALL THE PROBLEMS AND IT'S REALLY INTERESTING.
SEE, HERE'S THIS IDEA MR. BADER SHOWED ME.
IT'S THE PRINCIPLE OF LEAST ACTION.
∫Mvds
THIS GUY EULER CAME UP WITH THIS FORMULA:
THE INTEGRAL OF THE MOMENTUM OVER DISTANCE TRAVELED, WHICH IS MINIMIZED FOR THE TRUE PATH OF AN OBJECT.
∫Mvds
S = ∫_to^t1 (T-v)dt
AND HERE'S WHAT WE USE NOW, THE "ACTION" OF A PARTICLE, S, BETWEEN TIMES 0 AND 1 IS THE INTEGRAL OF T, THE KINETIC ENERGY OF THE OBJECT MINUS V, IT'S POTENTIAL ENERGY.
UM...

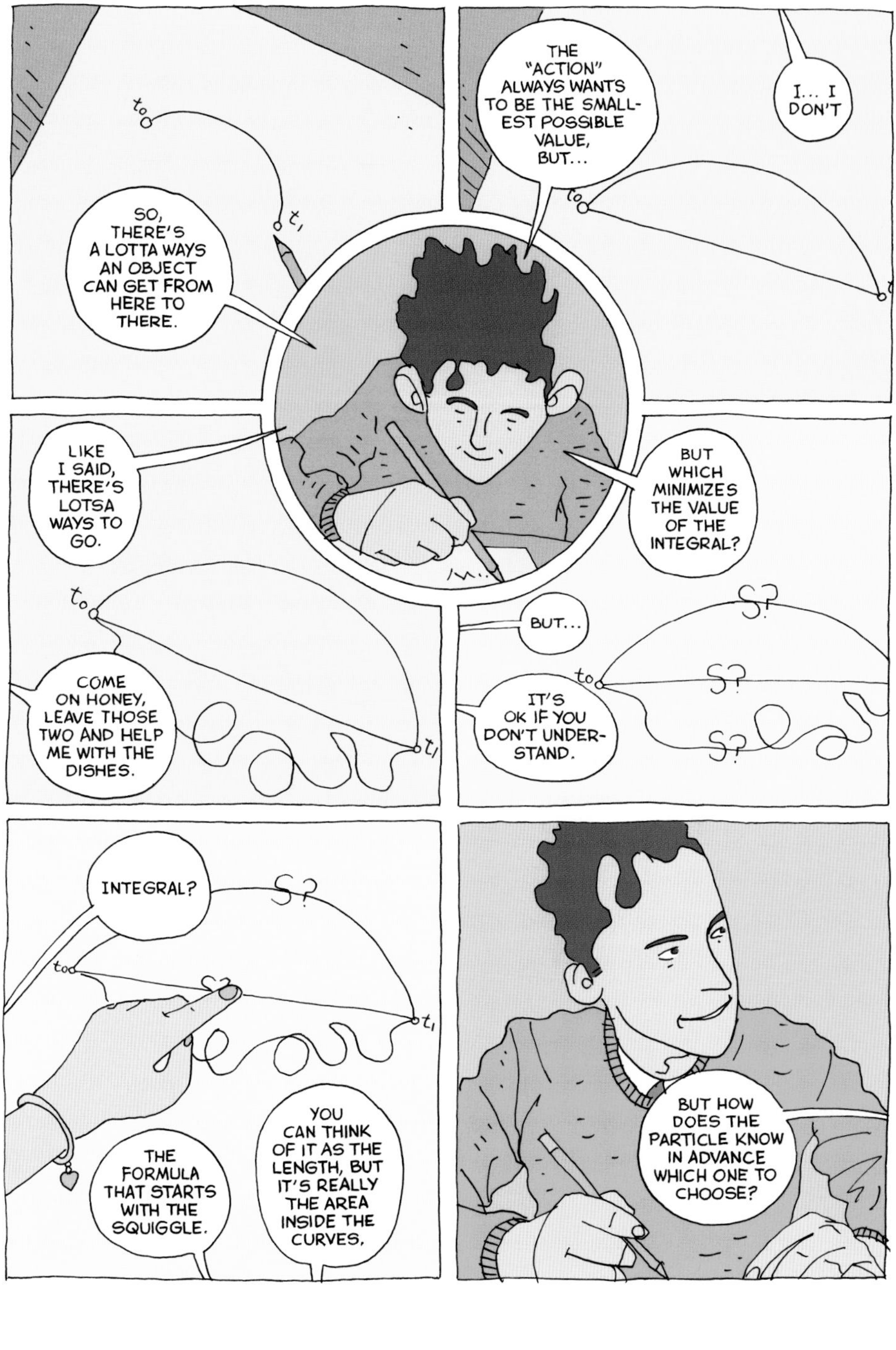
to
t1
SO, THERE'S A LOTTA WAYS AN OBJECT CAN GET FROM HERE TO THERE.
THE "ACTION" ALWAYS WANTS TO BE THE SMALLEST POSSIBLE VALUE, BUT...
I... I DON'T
to
LIKE I SAID, THERE'S LOTSA WAYS TO GO.
BUT WHICH MINIMIZES THE VALUE OF THE INTEGRAL?
to
t1
COME ON HONEY, LEAVE THOSE TWO AND HELP ME WITH THE DISHES.
BUT...
IT'S OK IF YOU DON'T UNDERSTAND.
S?
S?
S?
to
INTEGRAL?
S?
to
t1
THE FORMULA THAT STARTS WITH THE SQUIGGLE.
YOU CAN THINK OF IT AS THE LENGTH, BUT IT'S REALLY THE AREA INSIDE THE CURVES.
BUT HOW DOES THE PARTICLE KNOW IN ADVANCE WHICH ONE TO CHOOSE?

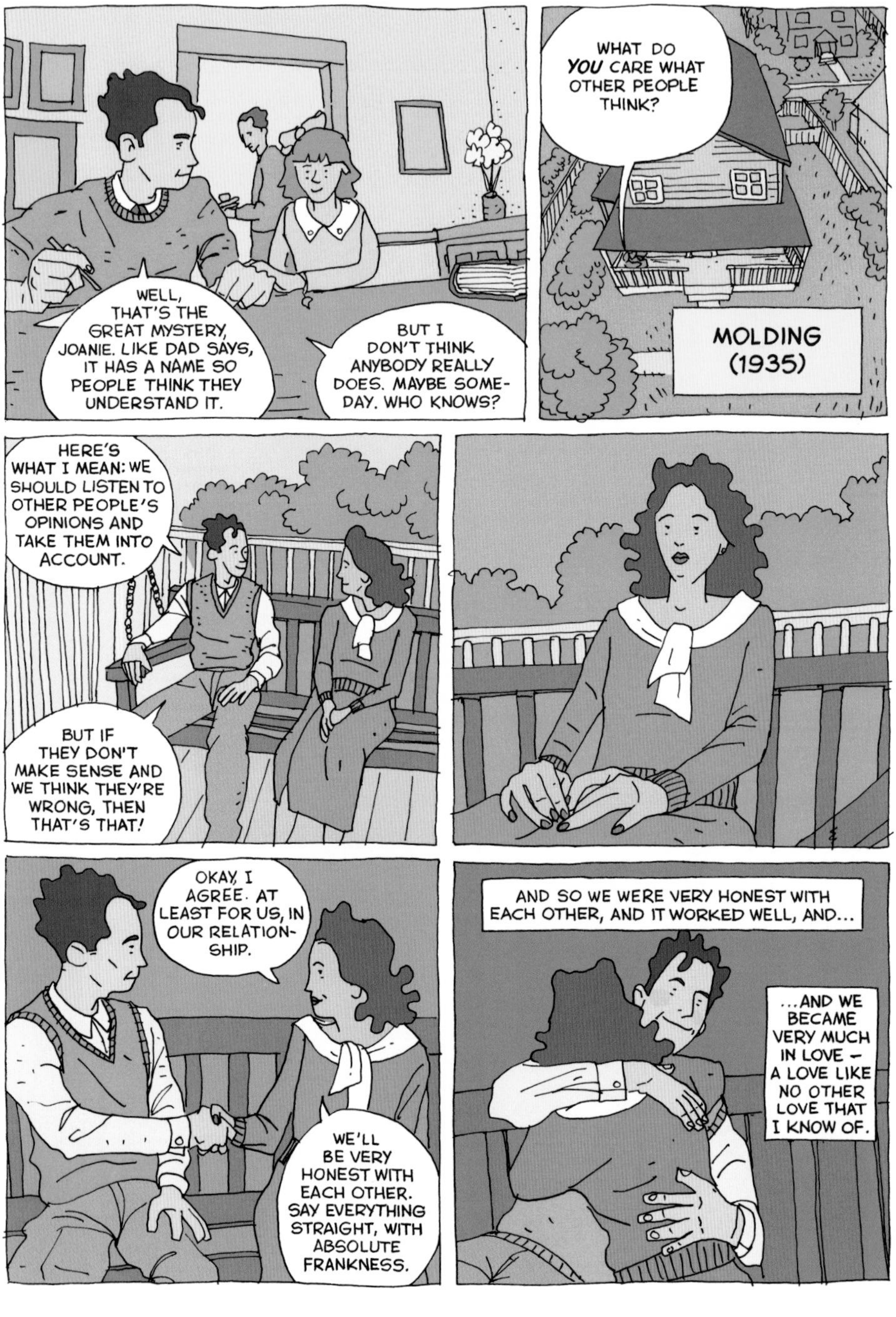
WELL, THAT'S THE GREAT MYSTERY, JOANIE. LIKE DAD SAYS, IT HAS A NAME SO PEOPLE THINK THEY UNDERSTAND IT.
BUT I DON'T THINK ANYBODY REALLY DOES. MAYBE SOMEDAY. WHO KNOWS?
MOLDING (1935)
WHAT DO YOU CARE WHAT OTHER PEOPLE THINK?
HERE'S WHAT I MEAN: WE SHOULD LISTEN TO OTHER PEOPLE'S OPINIONS AND TAKE THEM INTO ACCOUNT.
BUT IF THEY DON'T MAKE SENSE AND WE THINK THEY'RE WRONG, THEN THAT'S THAT!
OKAY, I AGREE. AT LEAST FOR US, IN OUR RELATIONSHIP.
WE'LL BE VERY HONEST WITH EACH OTHER. SAY EVERYTHING STRAIGHT, WITH ABSOLUTE FRANKNESS.
AND SO WE WERE VERY HONEST WITH EACH OTHER, AND IT WORKED WELL, AND...
...AND WE BECAME VERY MUCH IN LOVE – A LOVE LIKE NO OTHER LOVE THAT I KNOW OF.

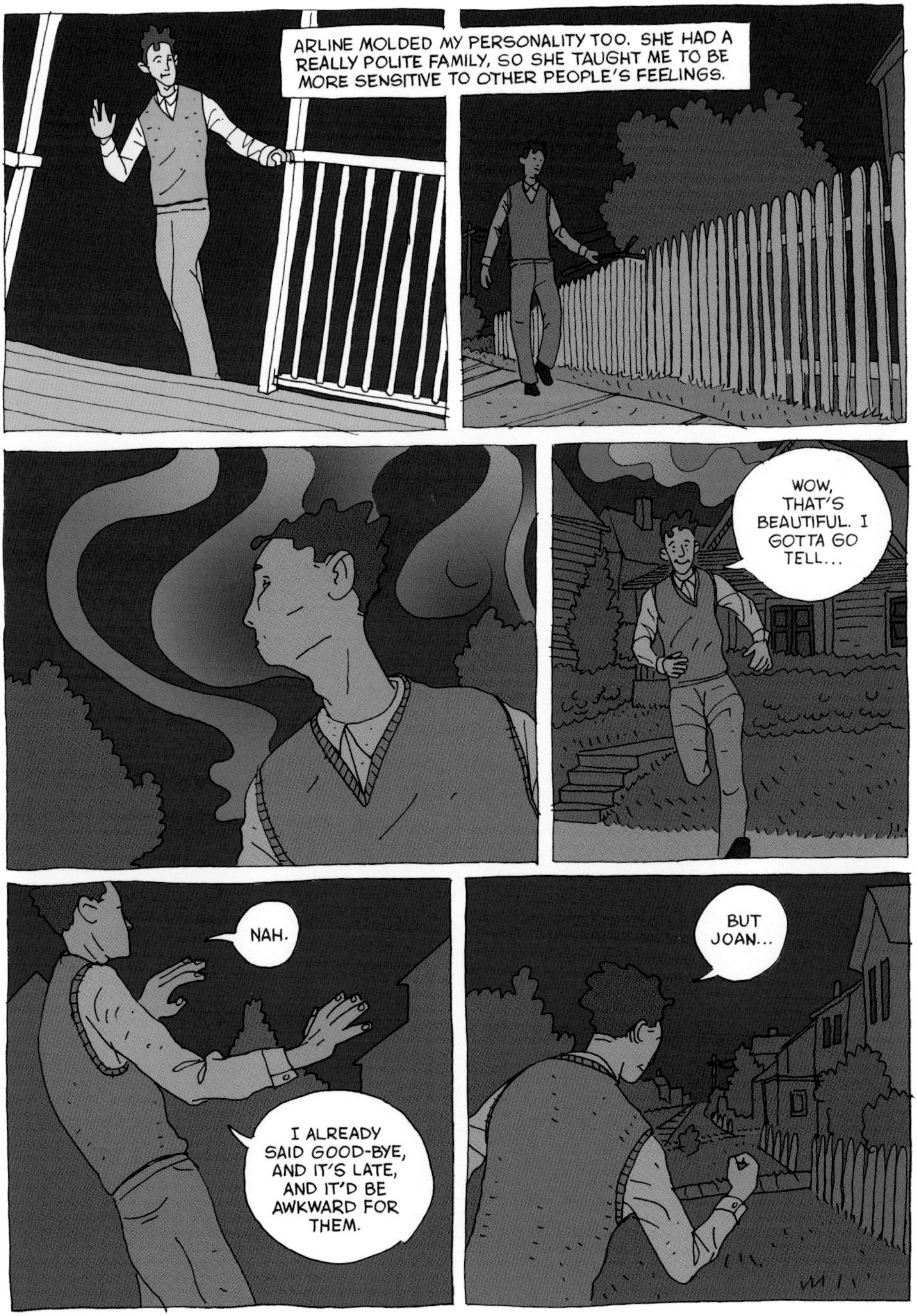
ARLINE MOLDED MY PERSONALITY TOO. SHE HAD A REALLY POLITE FAMILY, SO SHE TAUGHT ME TO BE MORE SENSITIVE TO OTHER PEOPLE'S FEELINGS.
WOW, THAT'S BEAUTIFUL. I GOTTA GO TELL...
NAH.
I ALREADY SAID GOOD-BYE, AND IT'S LATE, AND IT'D BE AWKWARD FOR THEM.
BUT JOAN...

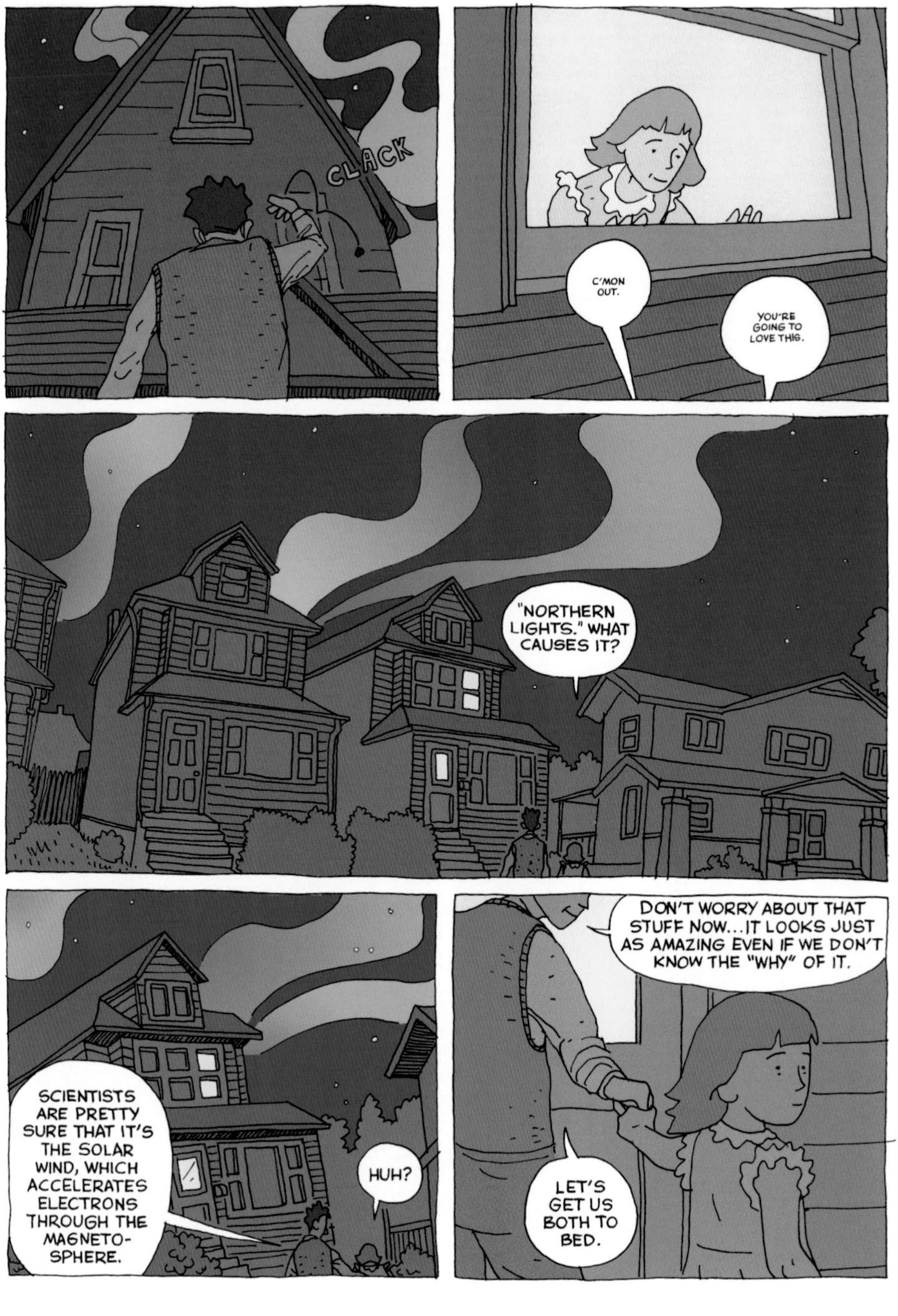
CLACK
C'MON OUT.
YOU'RE GOING TO LOVE THIS.
"NORTHERN LIGHTS." WHAT CAUSES IT?
SCIENTISTS ARE PRETTY SURE THAT IT'S THE SOLAR WIND, WHICH ACCELERATES ELECTRONS THROUGH THE MAGNETO-SPHERE.
HUH?
DON'T WORRY ABOUT THAT STUFF NOW... IT LOOKS JUST AS AMAZING EVEN IF WE DON'T KNOW THE "WHY" OF IT.
LET'S GET US BOTH TO BED.

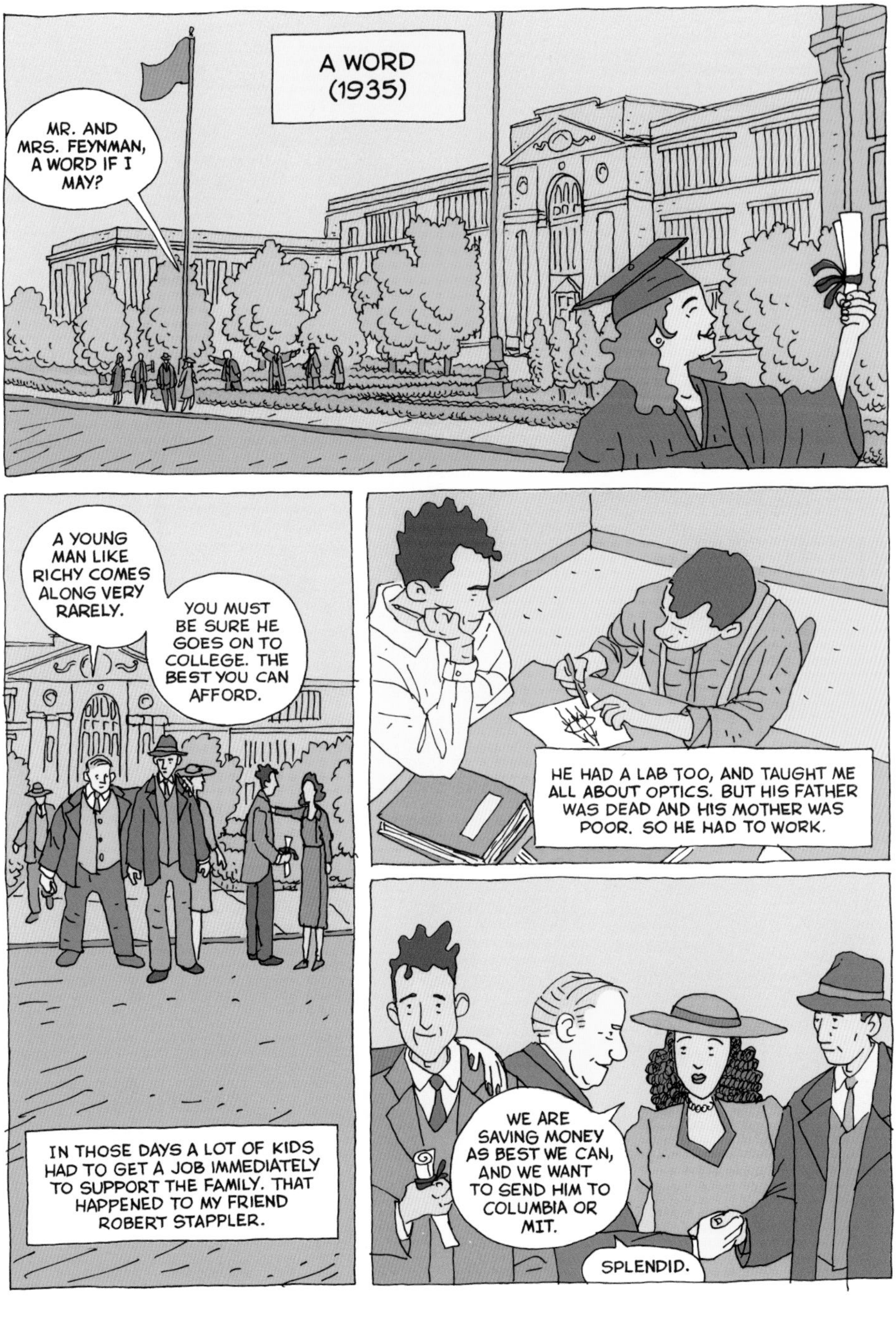
A WORD (1935)
MR. AND MRS. FEYNMAN, A WORD IF I MAY?
A YOUNG MAN LIKE RICHY COMES ALONG VERY RARELY.
YOU MUST BE SURE HE GOES ON TO COLLEGE. THE BEST YOU CAN AFFORD.
IN THOSE DAYS A LOT OF KIDS HAD TO GET A JOB IMMEDIATELY TO SUPPORT THE FAMILY. THAT HAPPENED TO MY FRIEND ROBERT STAPPLER.
HE HAD A LAB TOO, AND TAUGHT ME ALL ABOUT OPTICS. BUT HIS FATHER WAS DEAD AND HIS MOTHER WAS POOR. SO HE HAD TO WORK.
WE ARE SAVING MONEY AS BEST WE CAN, AND WE WANT TO SEND HIM TO COLUMBIA OR MIT.
SPLENDID.

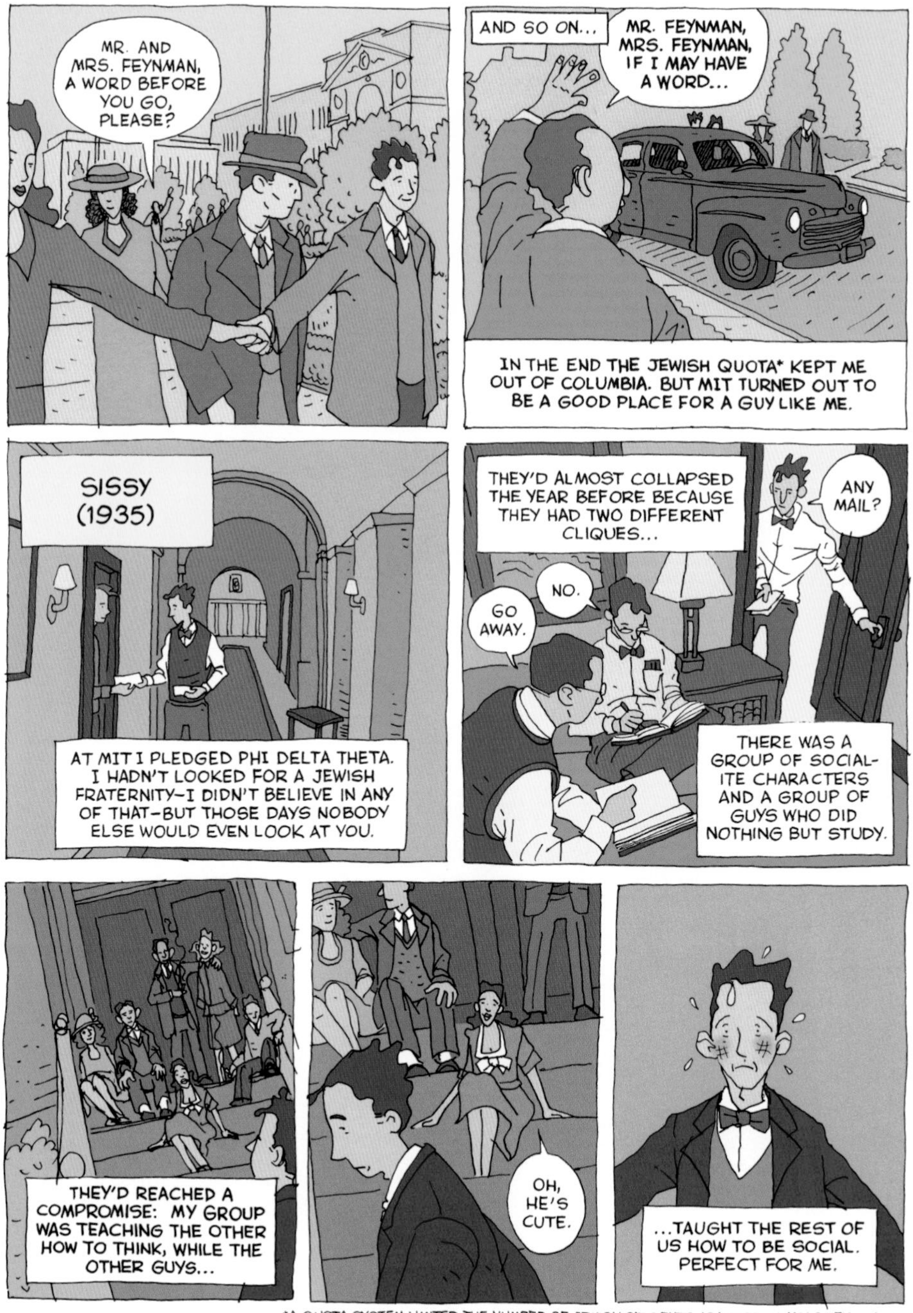
MR. AND MRS. FEYNMAN, A WORD BEFORE YOU GO, PLEASE?
AND SO ON...
MR. FEYNMAN, MRS. FEYNMAN, IF I MAY HAVE A WORD...
IN THE END THE JEWISH QUOTA* KEPT ME OUT OF COLUMBIA. BUT MIT TURNED OUT TO BE A GOOD PLACE FOR A GUY LIKE ME.
SISSY (1935)
AT MIT I PLEDGED PHI DELTA THETA. I HADN'T LOOKED FOR A JEWISH FRATERNITY–I DIDN'T BELIEVE IN ANY OF THAT–BUT THOSE DAYS NOBODY ELSE WOULD EVEN LOOK AT YOU.
THEY'D ALMOST COLLAPSED THE YEAR BEFORE BECAUSE THEY HAD TWO DIFFERENT CLIQUES...
ANY MAIL?
NO.
GO AWAY.
THERE WAS A GROUP OF SOCIAL-ITE CHARACTERS AND A GROUP OF GUYS WHO DID NOTHING BUT STUDY.
THEY'D REACHED A COMPROMISE: MY GROUP WAS TEACHING THE OTHER HOW TO THINK, WHILE THE OTHER GUYS...
OH, HE'S CUTE.
...TAUGHT THE REST OF US HOW TO BE SOCIAL. PERFECT FOR ME.

*A QUOTA SYSTEM LIMITED THE NUMBER OF JEWISH STUDENTS ADMITTED IN ANY GIVEN CLASS.

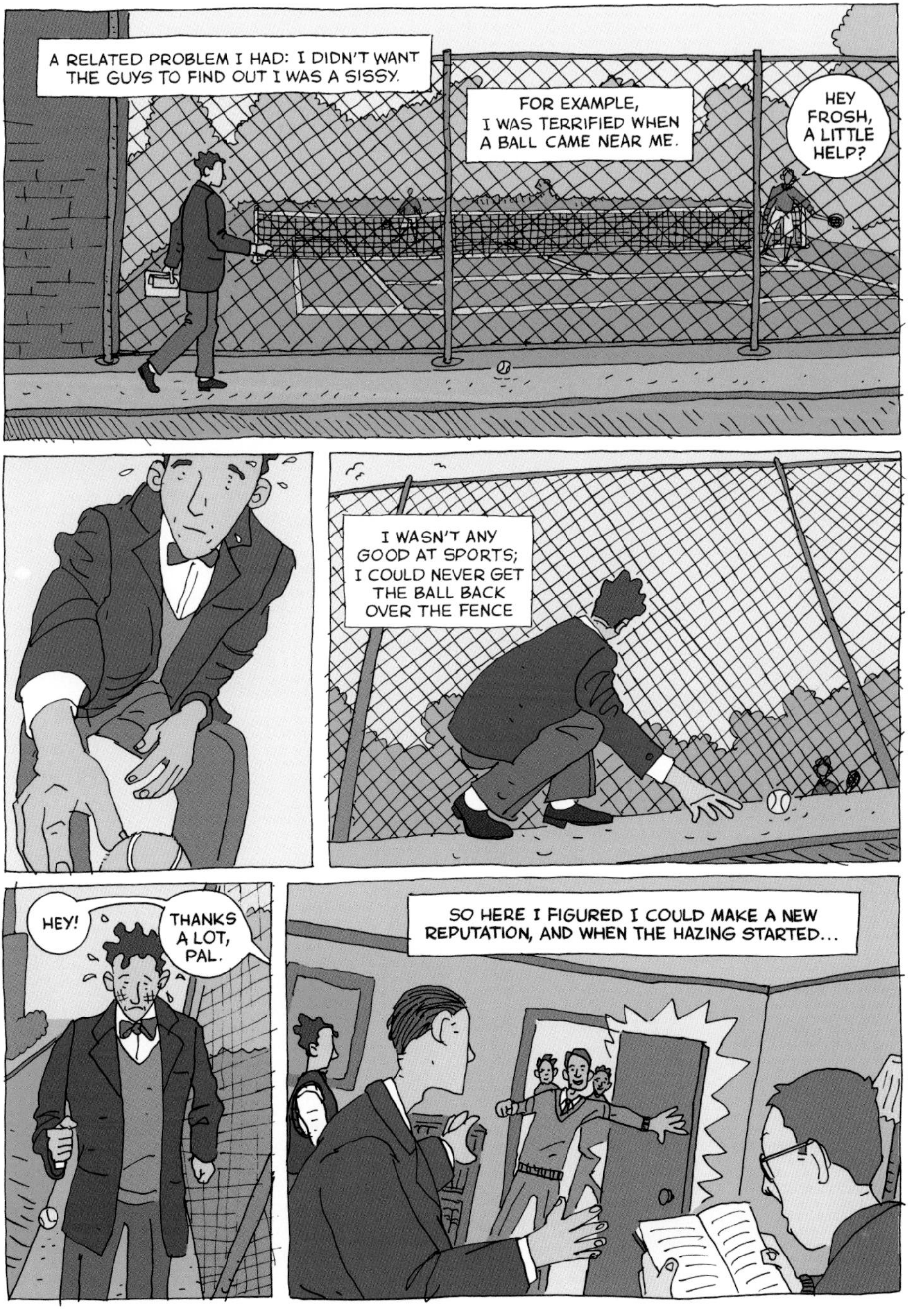
A RELATED PROBLEM I HAD: I DIDN'T WANT THE GUYS TO FIND OUT I WAS A SISSY.
FOR EXAMPLE, I WAS TERRIFIED WHEN A BALL CAME NEAR ME.
HEY FROSH, A LITTLE HELP?
I WASN'T ANY GOOD AT SPORTS; I COULD NEVER GET THE BALL BACK OVER THE FENCE
HEY!
THANKS A LOT, PAL.
SO HERE I FIGURED I COULD MAKE A NEW REPUTATION, AND WHEN THE HAZING STARTED...

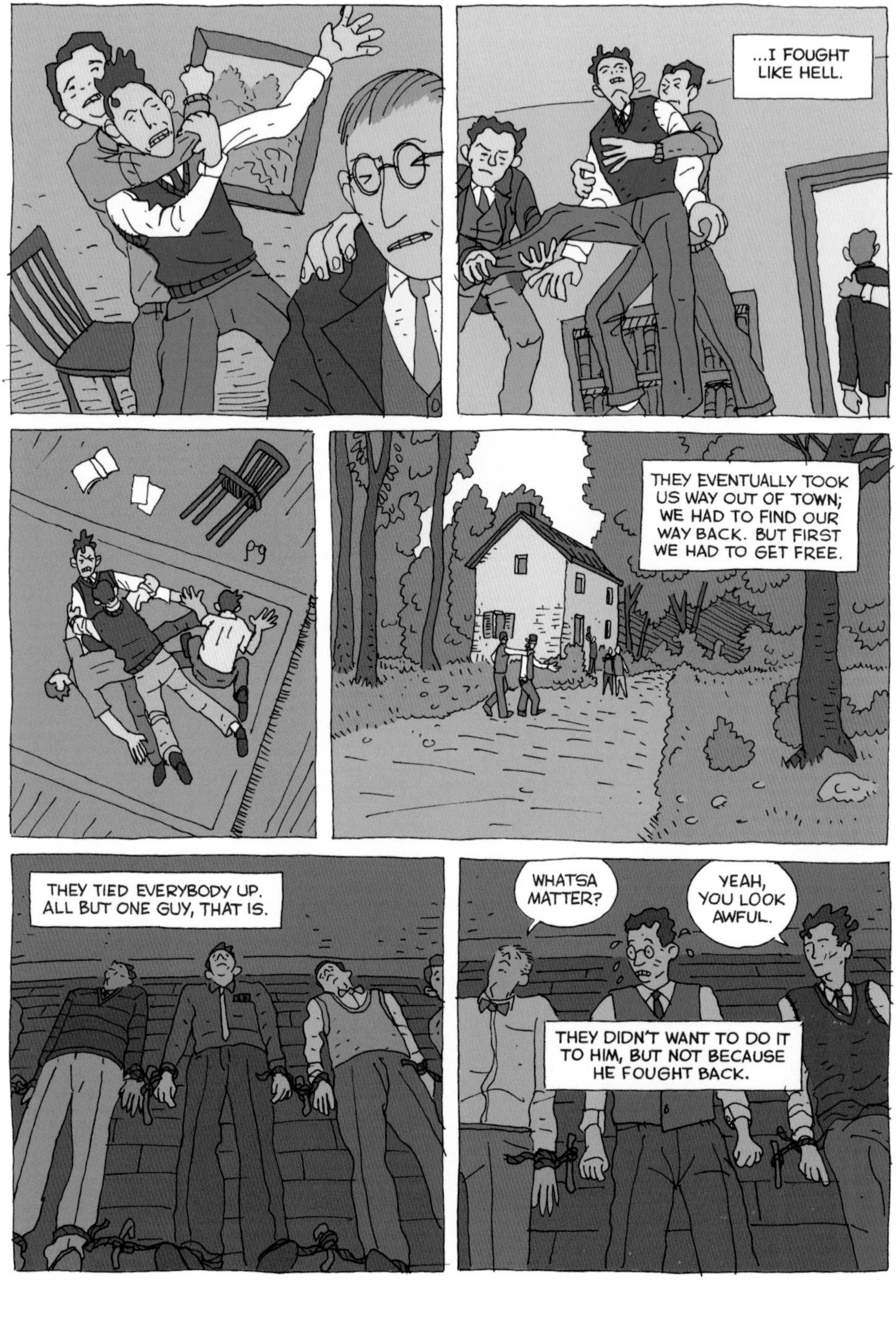
...I FOUGHT LIKE HELL.
THEY EVENTUALLY TOOK US WAY OUT OF TOWN; WE HAD TO FIND OUR WAY BACK. BUT FIRST WE HAD TO GET FREE.
THEY TIED EVERYBODY UP. ALL BUT ONE GUY, THAT IS.
WHATSA MATTER?
YEAH, YOU LOOK AWFUL.
THEY DIDN'T WANT TO DO IT TO HIM, BUT NOT BECAUSE HE FOUGHT BACK.

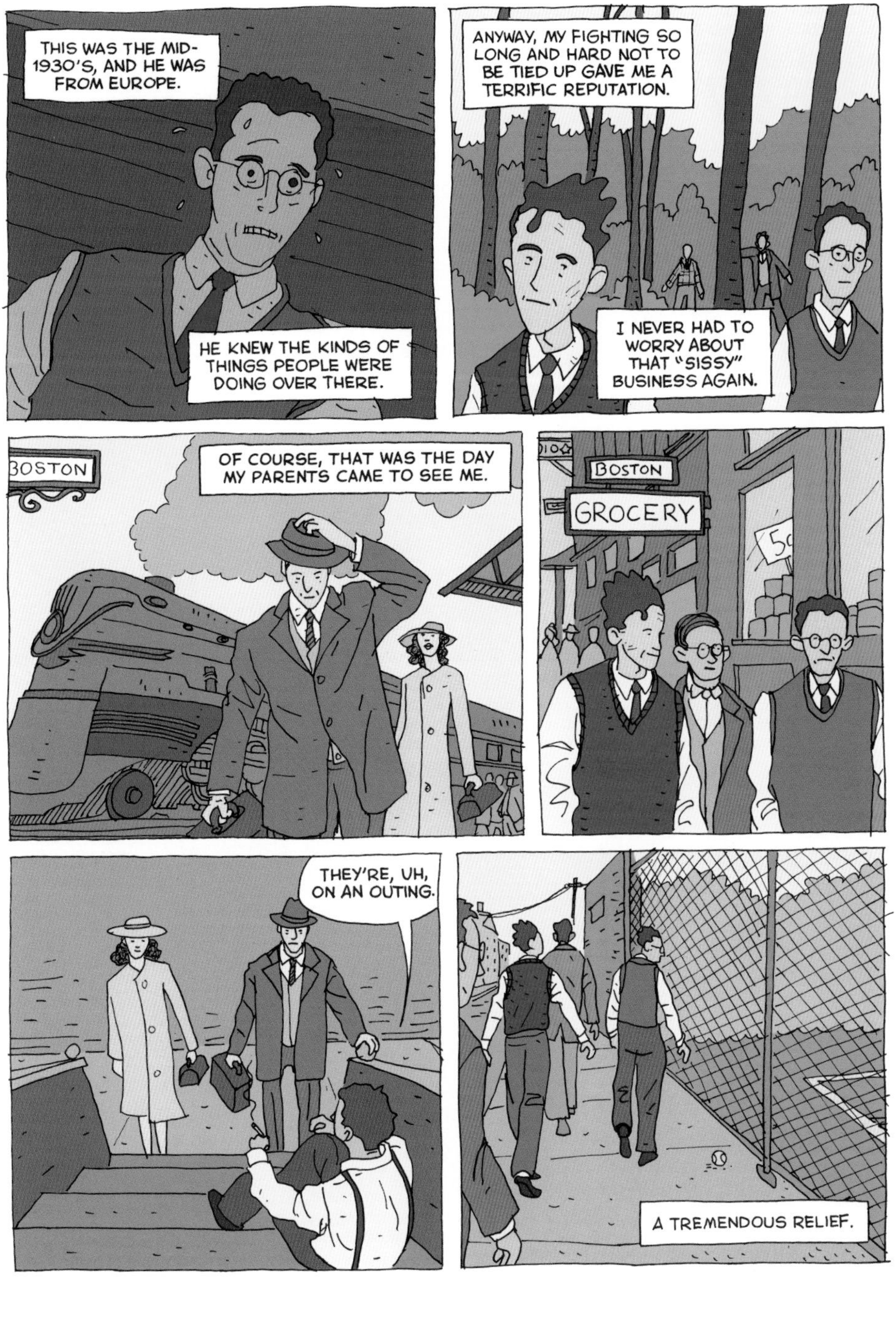
THIS WAS THE MID-1930'S, AND HE WAS FROM EUROPE.
HE KNEW THE KINDS OF THINGS PEOPLE WERE DOING OVER THERE.
ANYWAY, MY FIGHTING SO LONG AND HARD NOT TO BE TIED UP GAVE ME A TERRIFIC REPUTATION.
I NEVER HAD TO WORRY ABOUT THAT "SISSY" BUSINESS AGAIN.
BOSTON
OF COURSE, THAT WAS THE DAY MY PARENTS CAME TO SEE ME.
BOSTON
GROCERY
THEY'RE, UH, ON AN OUTING.
A TREMENDOUS RELIEF.

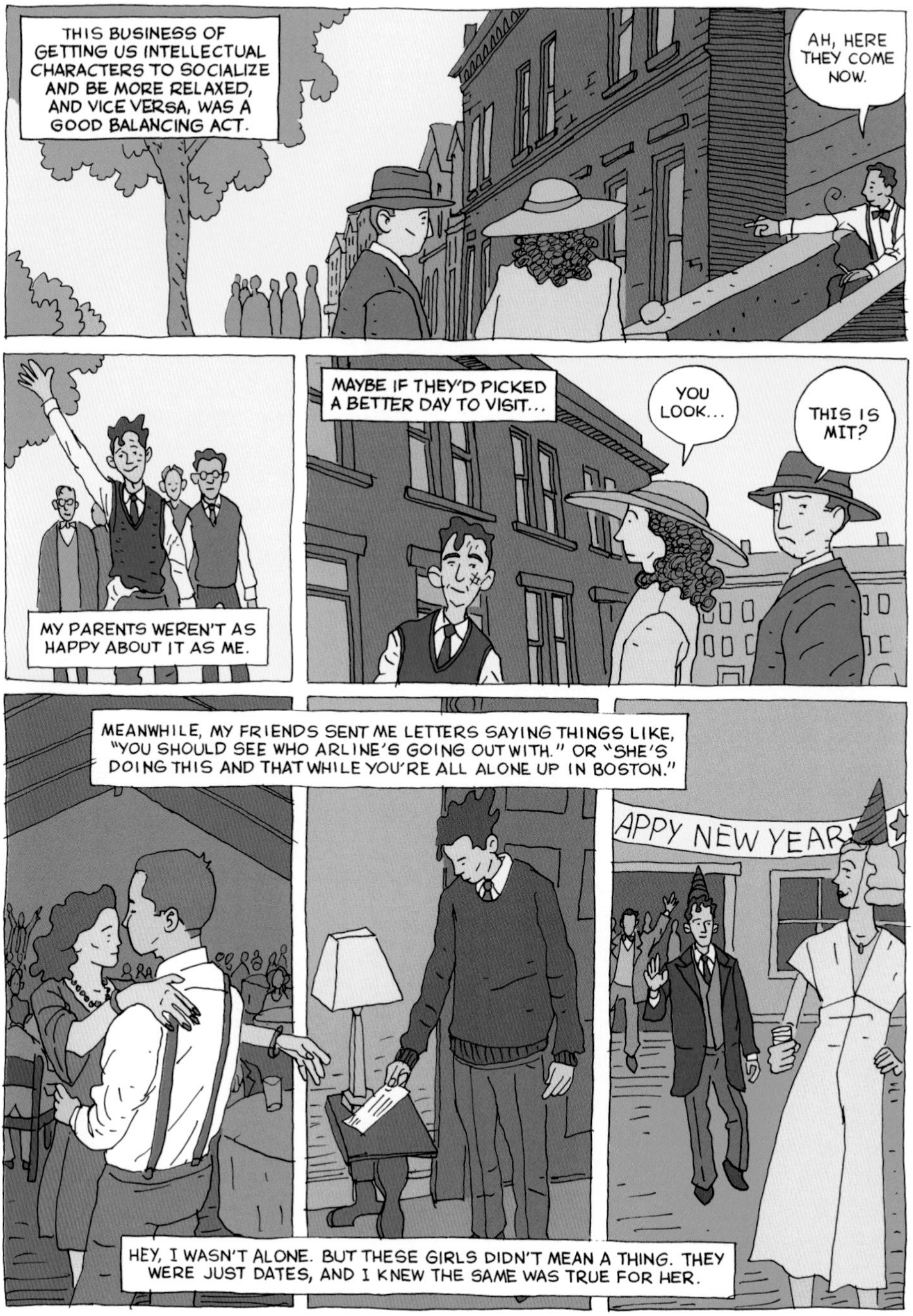
THIS BUSINESS OF GETTING US INTELLECTUAL CHARACTERS TO SOCIALIZE AND BE MORE RELAXED, AND VICE VERSA, WAS A GOOD BALANCING ACT.
AH, HERE THEY COME NOW.
MY PARENTS WEREN'T AS HAPPY ABOUT IT AS ME.
MAYBE IF THEY'D PICKED A BETTER DAY TO VISIT...
YOU LOOK...
THIS IS MIT?
MEANWHILE, MY FRIENDS SENT ME LETTERS SAYING THINGS LIKE, "YOU SHOULD SEE WHO ARLINE'S GOING OUT WITH." OR "SHE'S DOING THIS AND THAT WHILE YOU'RE ALL ALONE UP IN BOSTON."
APPY NEW YEAR
HEY, I WASN'T ALONE. BUT THESE GIRLS DIDN'T MEAN A THING. THEY WERE JUST DATES, AND I KNEW THE SAME WAS TRUE FOR HER.

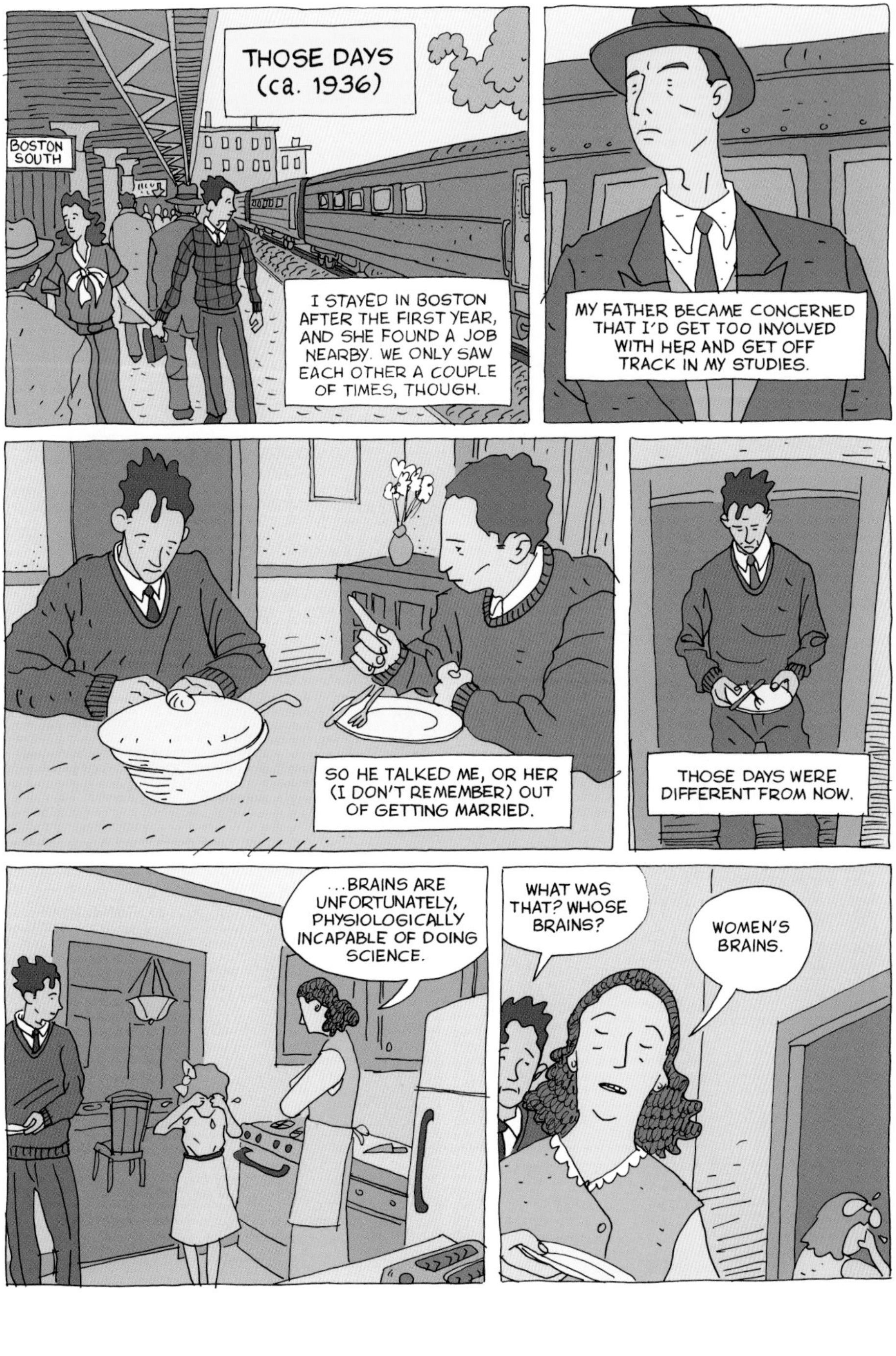
THOSE DAYS (ca. 1936)
BOSTON SOUTH
I STAYED IN BOSTON AFTER THE FIRST YEAR, AND SHE FOUND A JOB NEARBY. WE ONLY SAW EACH OTHER A COUPLE OF TIMES, THOUGH.
MY FATHER BECAME CONCERNED THAT I'D GET TOO INVOLVED WITH HER AND GET OFF TRACK IN MY STUDIES.
SO HE TALKED ME, OR HER (I DON'T REMEMBER) OUT OF GETTING MARRIED.
THOSE DAYS WERE DIFFERENT FROM NOW.
...BRAINS ARE UNFORTUNATELY, PHYSIOLOGICALLY INCAPABLE OF DOING SCIENCE.
WHAT WAS THAT? WHOSE BRAINS?
WOMEN'S BRAINS.

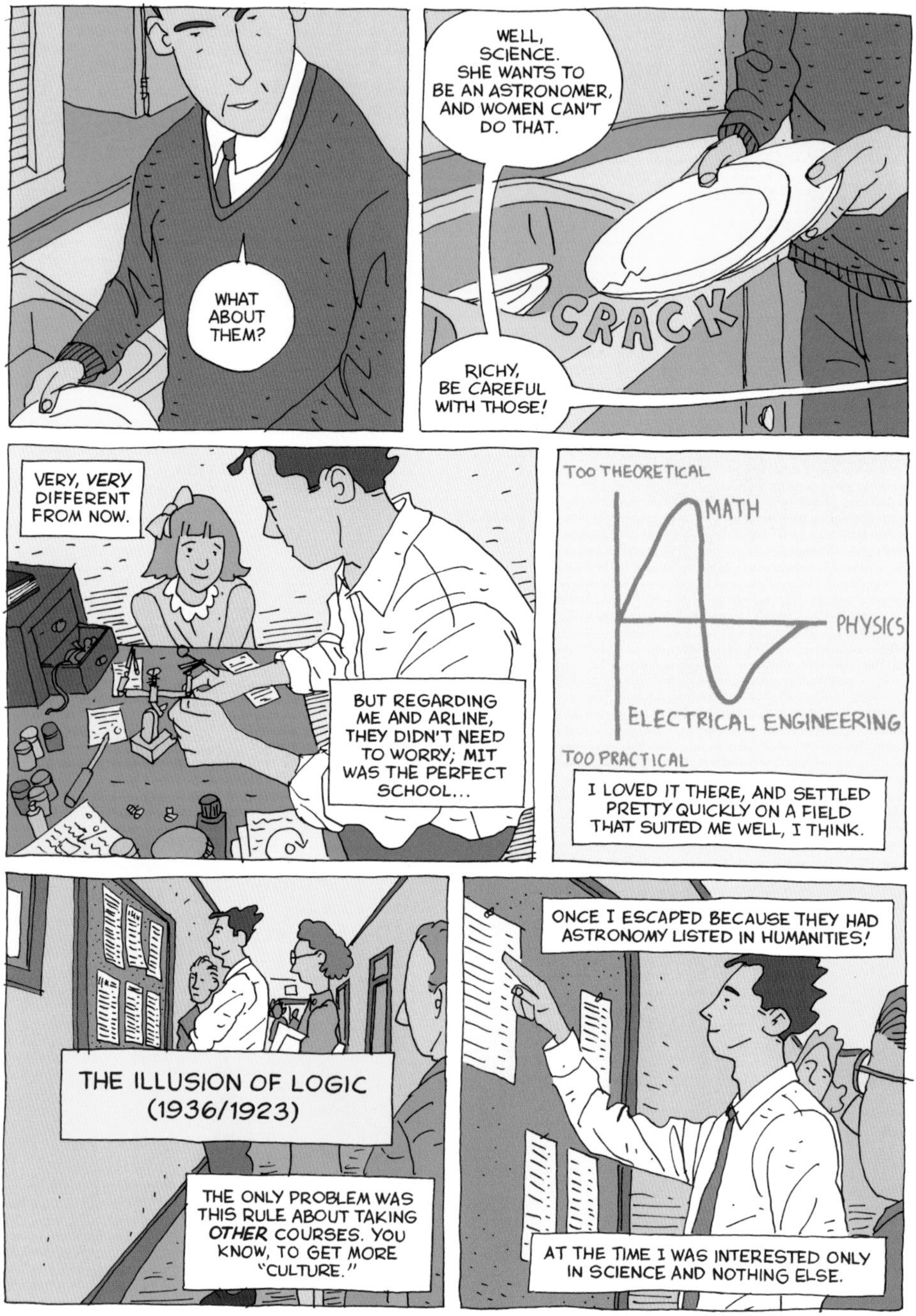
WHAT ABOUT THEM?
WELL, SCIENCE. SHE WANTS TO BE AN ASTRONOMER, AND WOMEN CAN'T DO THAT.
CRACK
RICHY, BE CAREFUL WITH THOSE!
VERY, VERY DIFFERENT FROM NOW.
BUT REGARDING ME AND ARLINE, THEY DIDN'T NEED TO WORRY; MIT WAS THE PERFECT SCHOOL...
TOO THEORETICAL
MATH
PHYSICS
ELECTRICAL ENGINEERING
TOO PRACTICAL
I LOVED IT THERE, AND SETTLED PRETTY QUICKLY ON A FIELD THAT SUITED ME WELL, I THINK.
THE ILLUSION OF LOGIC (1936/1923)
THE ONLY PROBLEM WAS THIS RULE ABOUT TAKING OTHER COURSES. YOU KNOW, TO GET MORE "CULTURE."
ONCE I ESCAPED BECAUSE THEY HAD ASTRONOMY LISTED IN HUMANITIES!
AT THE TIME I WAS INTERESTED ONLY IN SCIENCE AND NOTHING ELSE.

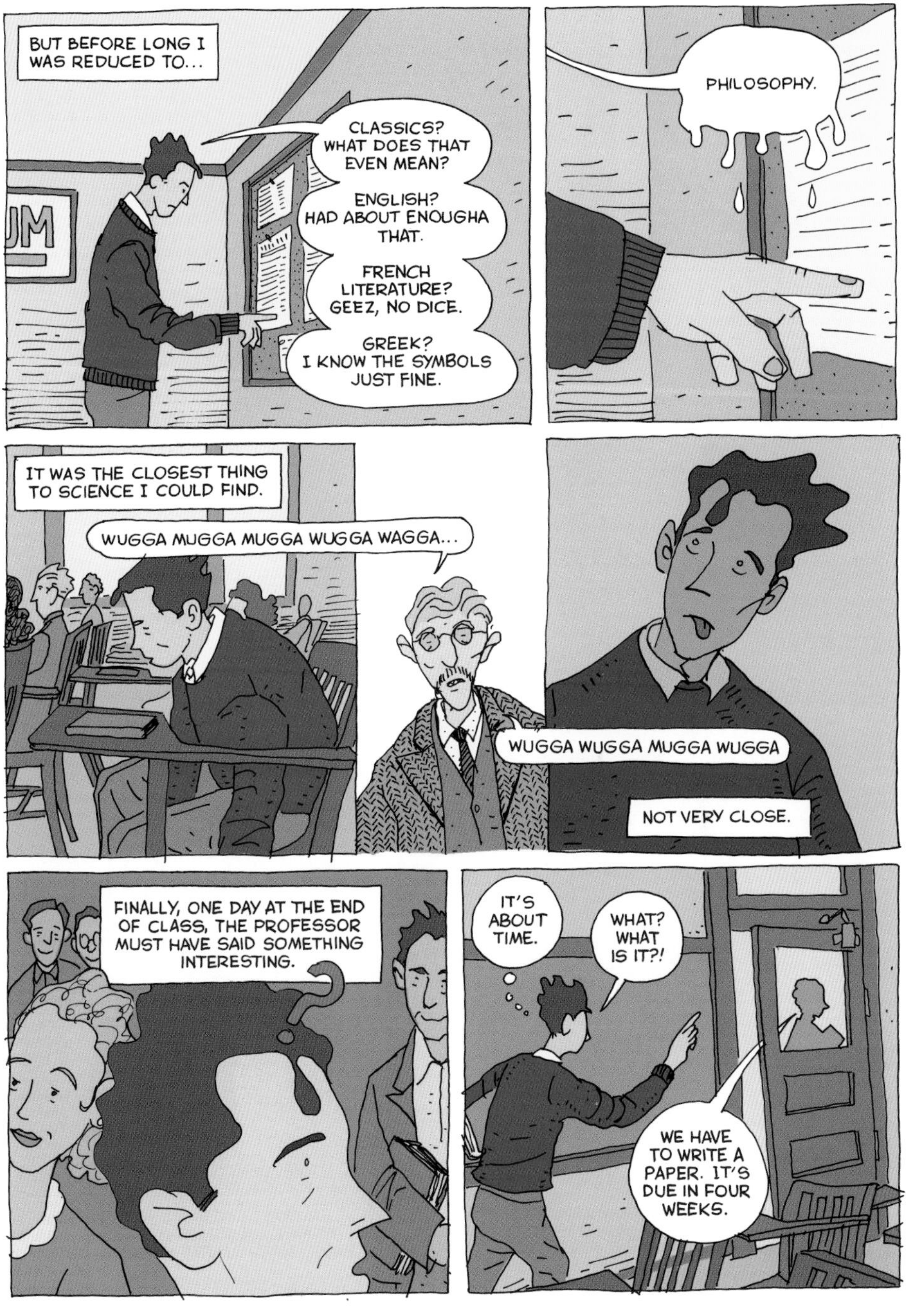
BUT BEFORE LONG I WAS REDUCED TO...
CLASSICS? WHAT DOES THAT EVEN MEAN?
ENGLISH? HAD ABOUT ENOUGHA THAT.
FRENCH LITERATURE? GEEZ, NO DICE.
GREEK? I KNOW THE SYMBOLS JUST FINE.
PHILOSOPHY.
IT WAS THE CLOSEST THING TO SCIENCE I COULD FIND.
WUGGA MUGGA MUGGA WUGGA WAGGA...
WUGGA WUGGA MUGGA WUGGA
NOT VERY CLOSE.
FINALLY, ONE DAY AT THE END OF CLASS, THE PROFESSOR MUST HAVE SAID SOMETHING INTERESTING.
IT'S ABOUT TIME.
WHAT? WHAT IS IT?!
WE HAVE TO WRITE A PAPER. IT'S DUE IN FOUR WEEKS.

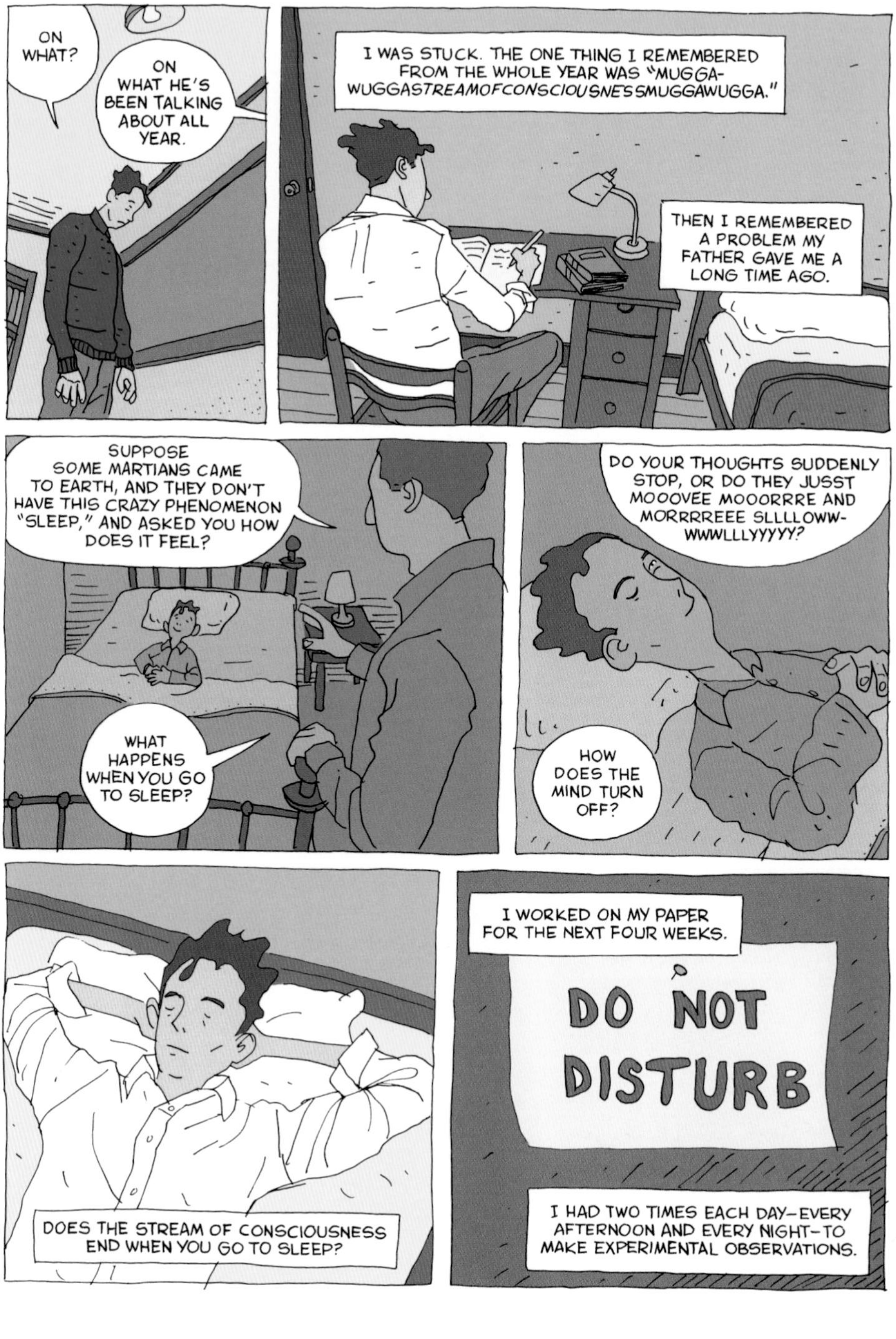
ON WHAT?
ON WHAT HE'S BEEN TALKING ABOUT ALL YEAR.
I WAS STUCK. THE ONE THING I REMEMBERED FROM THE WHOLE YEAR WAS "MUGGA-WUGGASTREAMOFCONSCIOUSNESSMUGGAWUGGA."
THEN I REMEMBERED A PROBLEM MY FATHER GAVE ME A LONG TIME AGO.
SUPPOSE SOME MARTIANS CAME TO EARTH, AND THEY DON'T HAVE THIS CRAZY PHENOMENON "SLEEP," AND ASKED YOU HOW DOES IT FEEL?
WHAT HAPPENS WHEN YOU GO TO SLEEP?
DO YOUR THOUGHTS SUDDENLY STOP, OR DO THEY JUSST MOOOVEE MOOORRRE AND MORRRREEE SLLLLOWW-WWWLLLYYYYY?
HOW DOES THE MIND TURN OFF?
DOES THE STREAM OF CONSCIOUSNESS END WHEN YOU GO TO SLEEP?
I WORKED ON MY PAPER FOR THE NEXT FOUR WEEKS.
DO NOT DISTURB
I HAD TWO TIMES EACH DAY—EVERY AFTERNOON AND EVERY NIGHT—TO MAKE EXPERIMENTAL OBSERVATIONS.

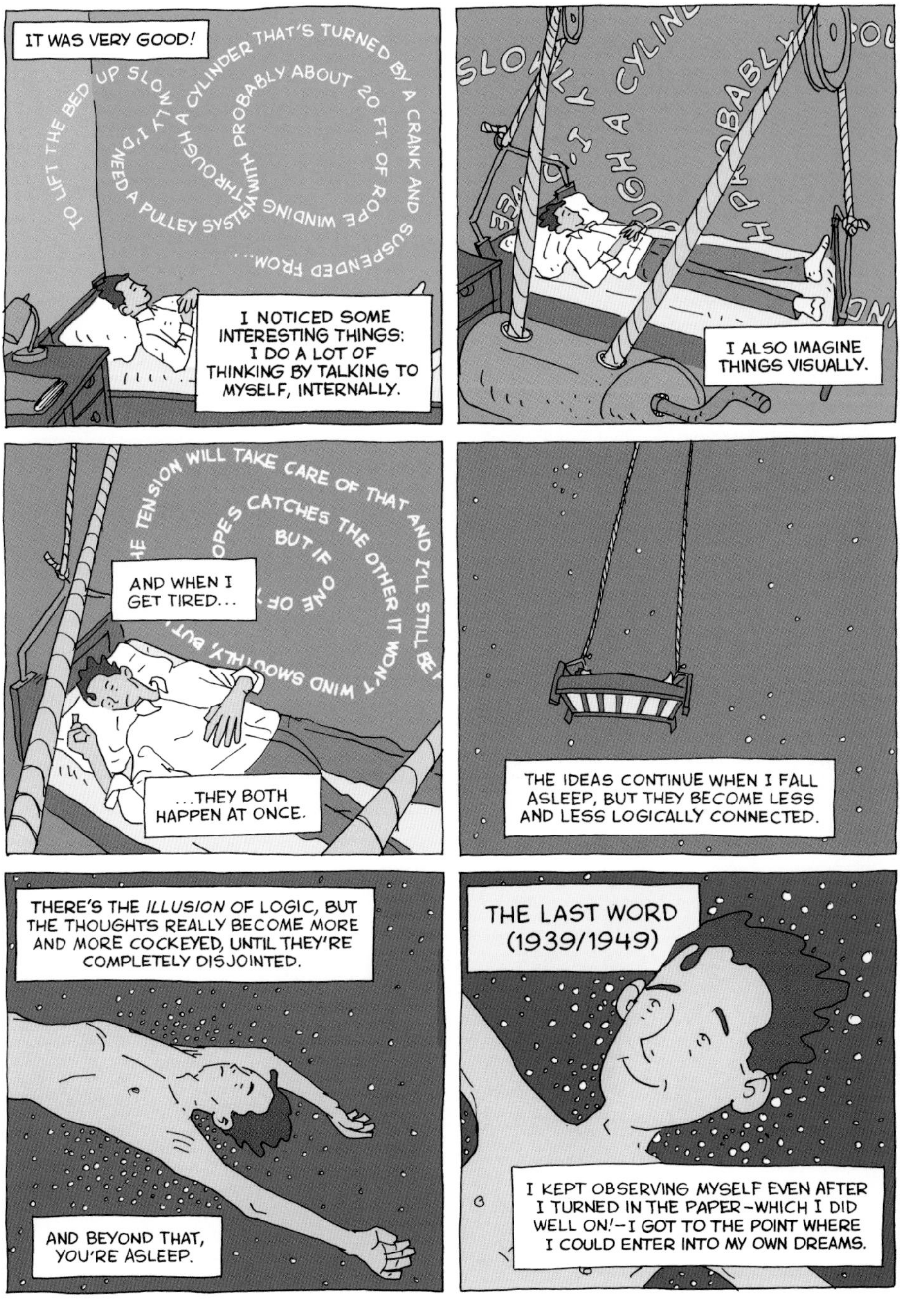
IT WAS VERY GOOD!
TO LIFT THE BED UP SLOWLY, I'D NEED A PULLEY SYSTEM
WITH PROBABLY ABOUT 20 FT. OF ROPE WINDING
THROUGH A CYLINDER THAT'S TURNED BY A CRANK AND SUSPENDED FROM...
I NOTICED SOME INTERESTING THINGS: I DO A LOT OF THINKING BY TALKING TO MYSELF, INTERNALLY.
I ALSO IMAGINE THINGS VISUALLY.
AND WHEN I GET TIRED...
TENSION WILL TAKE CARE OF THAT AND I'LL STILL BE
OPES CATCHES THE OTHER IT WON'T WIND SMOOTHLY, BUT
BUT IF ONE OF
...THEY BOTH HAPPEN AT ONCE.
THE IDEAS CONTINUE WHEN I FALL ASLEEP, BUT THEY BECOME LESS AND LESS LOGICALLY CONNECTED.
THERE'S THE *ILLUSION* OF LOGIC, BUT THE THOUGHTS REALLY BECOME MORE AND MORE COCKEYED, UNTIL THEY'RE COMPLETELY DISJOINTED.
AND BEYOND THAT, YOU'RE ASLEEP.
THE LAST WORD (1939/1949)
I KEPT OBSERVING MYSELF EVEN AFTER I TURNED IN THE PAPER–WHICH I DID WELL ON!–I GOT TO THE POINT WHERE I COULD ENTER INTO MY OWN DREAMS.

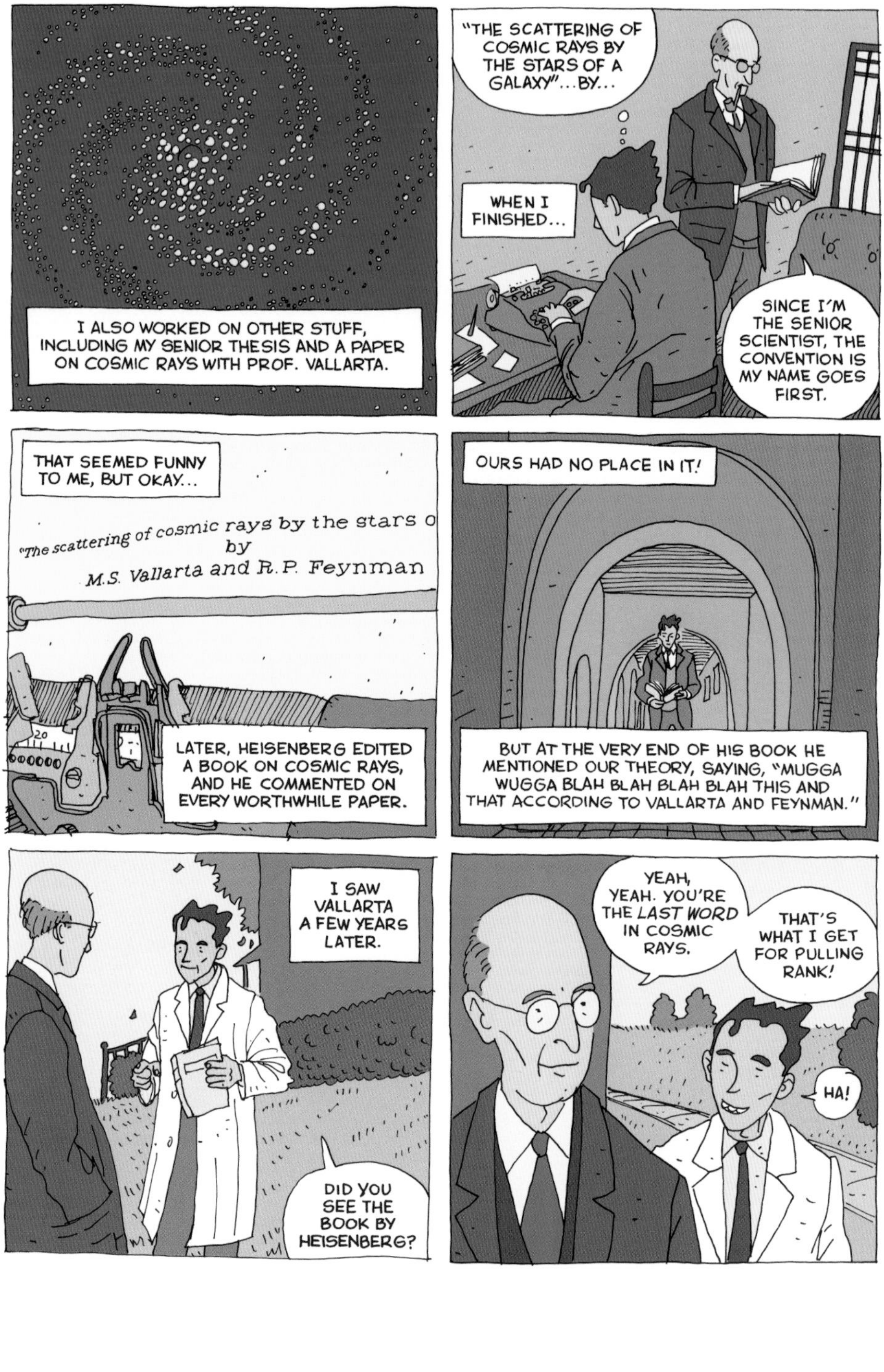
I ALSO WORKED ON OTHER STUFF, INCLUDING MY SENIOR THESIS AND A PAPER ON COSMIC RAYS WITH PROF. VALLARTA.
"THE SCATTERING OF COSMIC RAYS BY THE STARS OF A GALAXY"... BY...
WHEN I FINISHED...
SINCE I'M THE SENIOR SCIENTIST, THE CONVENTION IS MY NAME GOES FIRST.
THAT SEEMED FUNNY TO ME, BUT OKAY...
"The scattering of cosmic rays by the stars o
by
M.S. Vallarta and R.P. Feynman
LATER, HEISENBERG EDITED A BOOK ON COSMIC RAYS, AND HE COMMENTED ON EVERY WORTHWHILE PAPER.
OURS HAD NO PLACE IN IT!
BUT AT THE VERY END OF HIS BOOK HE MENTIONED OUR THEORY, SAYING, "MUGGA WUGGA BLAH BLAH BLAH BLAH THIS AND THAT ACCORDING TO VALLARTA AND FEYNMAN."
I SAW VALLARTA A FEW YEARS LATER.
DID YOU SEE THE BOOK BY HEISENBERG?
YEAH, YEAH. YOU'RE THE *LAST WORD* IN COSMIC RAYS.
THAT'S WHAT I GET FOR PULLING RANK!
HA!

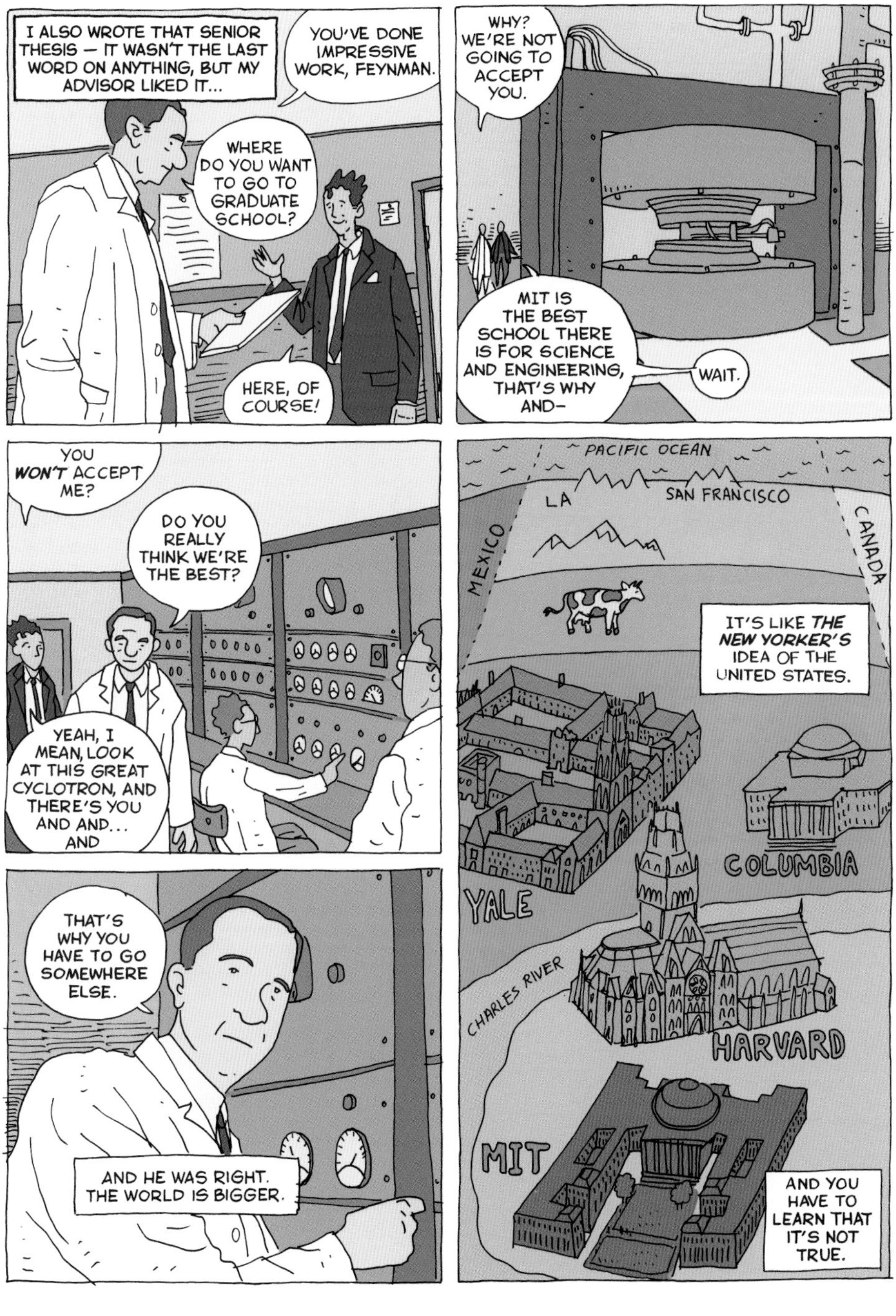
I ALSO WROTE THAT SENIOR THESIS — IT WASN'T THE LAST WORD ON ANYTHING, BUT MY ADVISOR LIKED IT...
YOU'VE DONE IMPRESSIVE WORK, FEYNMAN.
WHERE DO YOU WANT TO GO TO GRADUATE SCHOOL?
HERE, OF COURSE!
WHY? WE'RE NOT GOING TO ACCEPT YOU.
MIT IS THE BEST SCHOOL THERE IS FOR SCIENCE AND ENGINEERING, THAT'S WHY AND–
WAIT.
YOU *WON'T* ACCEPT ME?
DO YOU REALLY THINK WE'RE THE BEST?
YEAH, I MEAN, LOOK AT THIS GREAT CYCLOTRON, AND THERE'S YOU AND AND... AND
THAT'S WHY YOU HAVE TO GO SOMEWHERE ELSE.
AND HE WAS RIGHT. THE WORLD IS BIGGER.
PACIFIC OCEAN
LA
SAN FRANCISCO
MEXICO
CANADA
IT'S LIKE *THE NEW YORKER'S* IDEA OF THE UNITED STATES.
COLUMBIA
YALE
CHARLES RIVER
HARVARD
MIT
AND YOU HAVE TO LEARN THAT IT'S NOT TRUE.

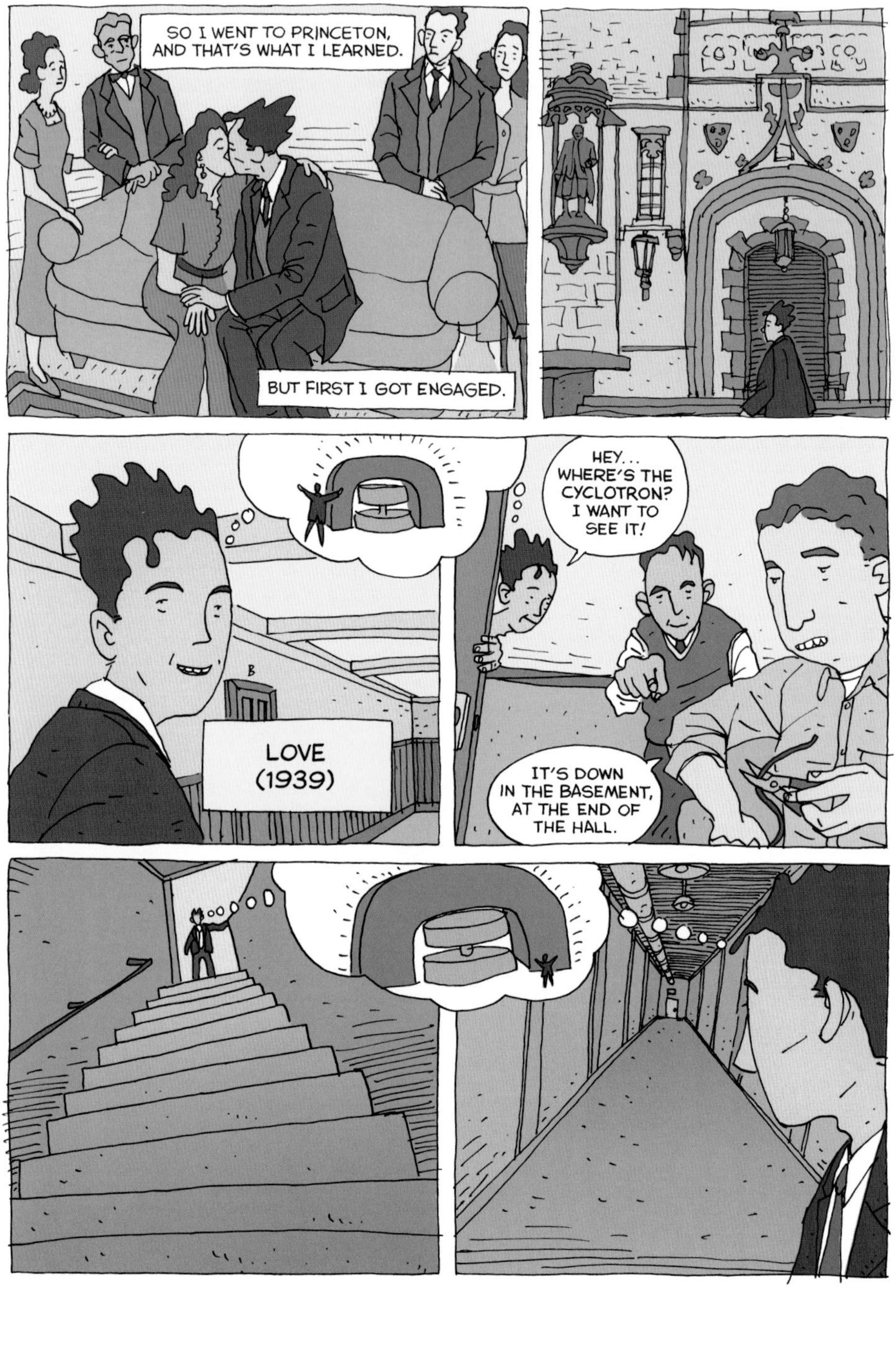
SO I WENT TO PRINCETON, AND THAT'S WHAT I LEARNED.
BUT FIRST I GOT ENGAGED.
LOVE (1939)
HEY... WHERE'S THE CYCLOTRON? I WANT TO SEE IT!
IT'S DOWN IN THE BASEMENT, AT THE END OF THE HALL.

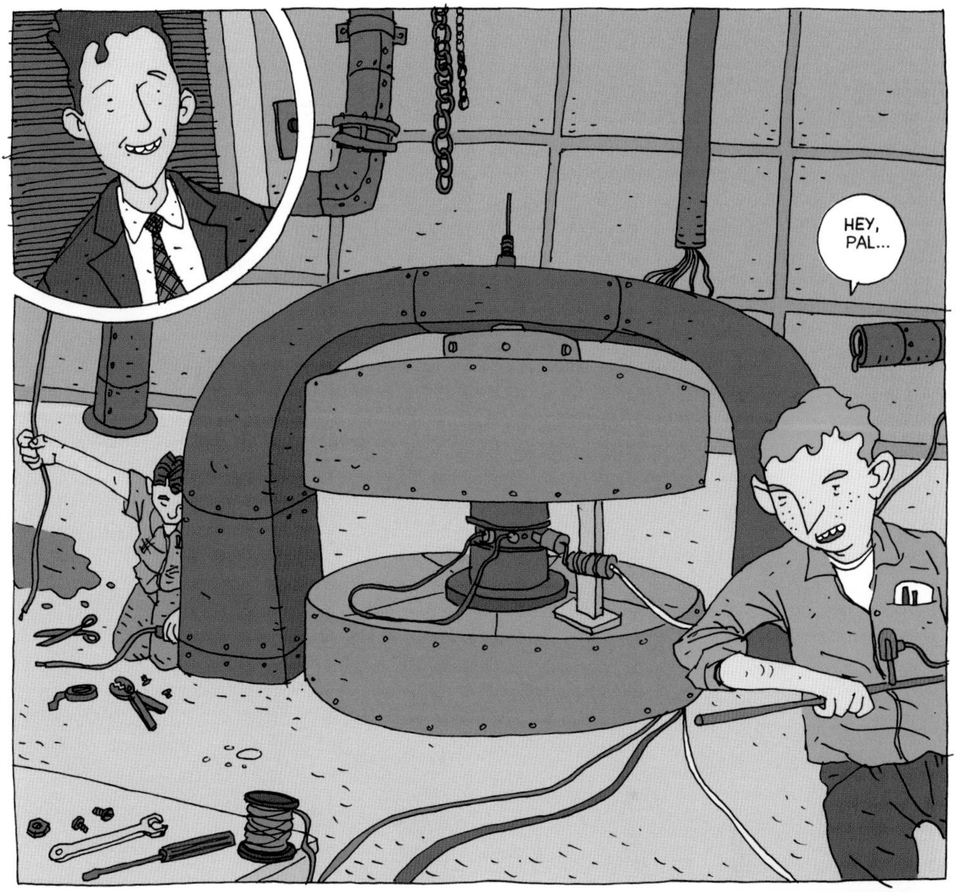
HEY, PAL...

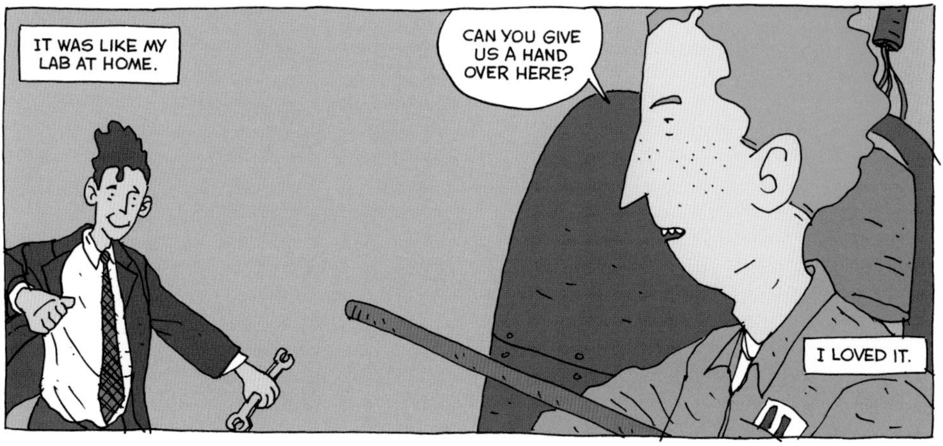
IT WAS LIKE MY LAB AT HOME.
CAN YOU GIVE US A HAND OVER HERE?
I LOVED IT.

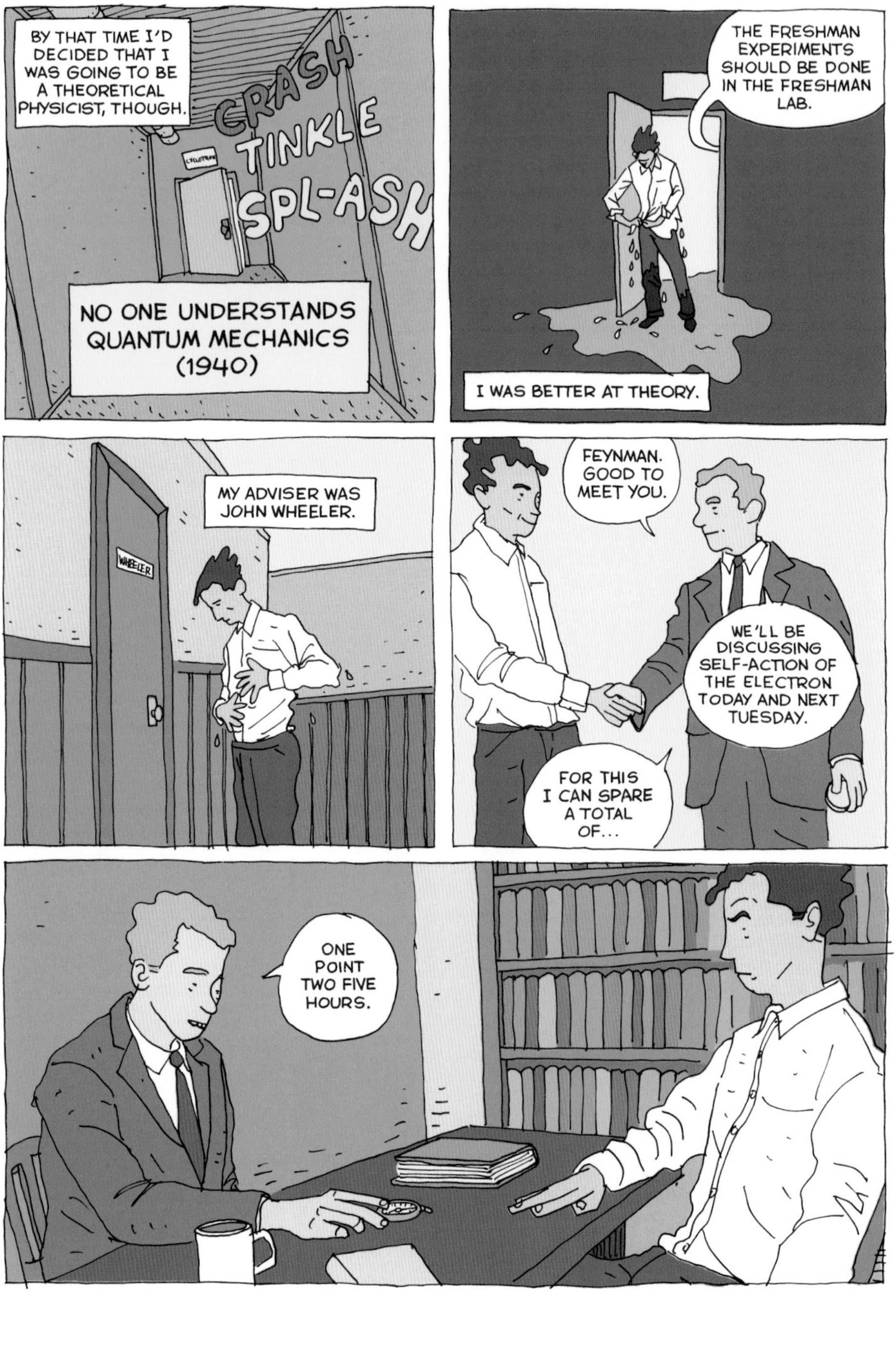
BY THAT TIME I'D DECIDED THAT I WAS GOING TO BE A THEORETICAL PHYSICIST, THOUGH.
CRASH
TINKLE
SPL-ASH
NO ONE UNDERSTANDS QUANTUM MECHANICS (1940)
THE FRESHMAN EXPERIMENTS SHOULD BE DONE IN THE FRESHMAN LAB.
I WAS BETTER AT THEORY.
MY ADVISER WAS JOHN WHEELER.
WHEELER
FEYNMAN. GOOD TO MEET YOU.
WE'LL BE DISCUSSING SELF-ACTION OF THE ELECTRON TODAY AND NEXT TUESDAY.
FOR THIS I CAN SPARE A TOTAL OF...
ONE POINT TWO FIVE HOURS.

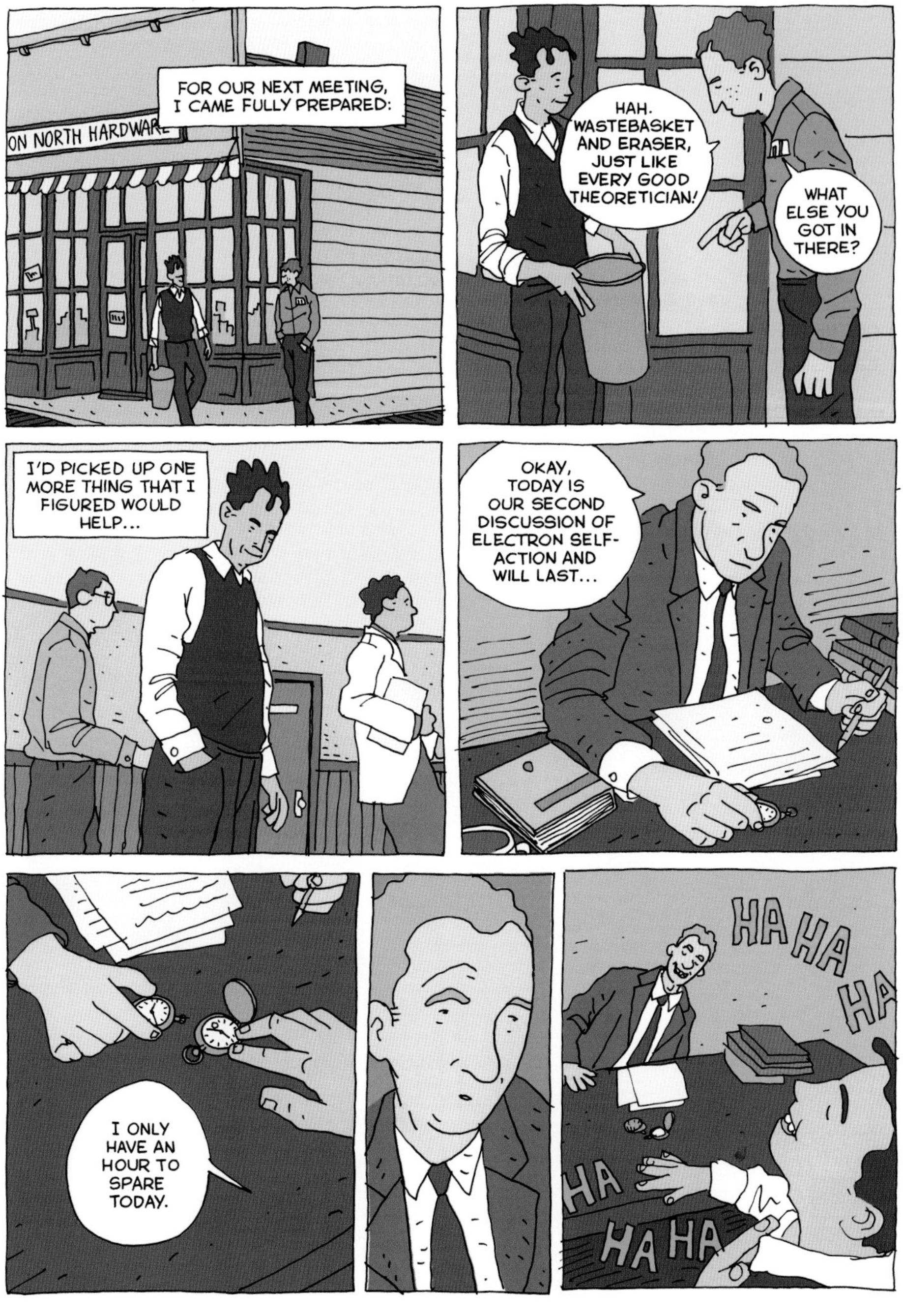

FOR OUR NEXT MEETING, I CAME FULLY PREPARED:
ON NORTH HARDWARE
HAH. WASTEBASKET AND ERASER, JUST LIKE EVERY GOOD THEORETICIAN!
WHAT ELSE YOU GOT IN THERE?
I'D PICKED UP ONE MORE THING THAT I FIGURED WOULD HELP...
OKAY, TODAY IS OUR SECOND DISCUSSION OF ELECTRON SELF-ACTION AND WILL LAST...
I ONLY HAVE AN HOUR TO SPARE TODAY.
HA HA HA HA
HA
HA HA

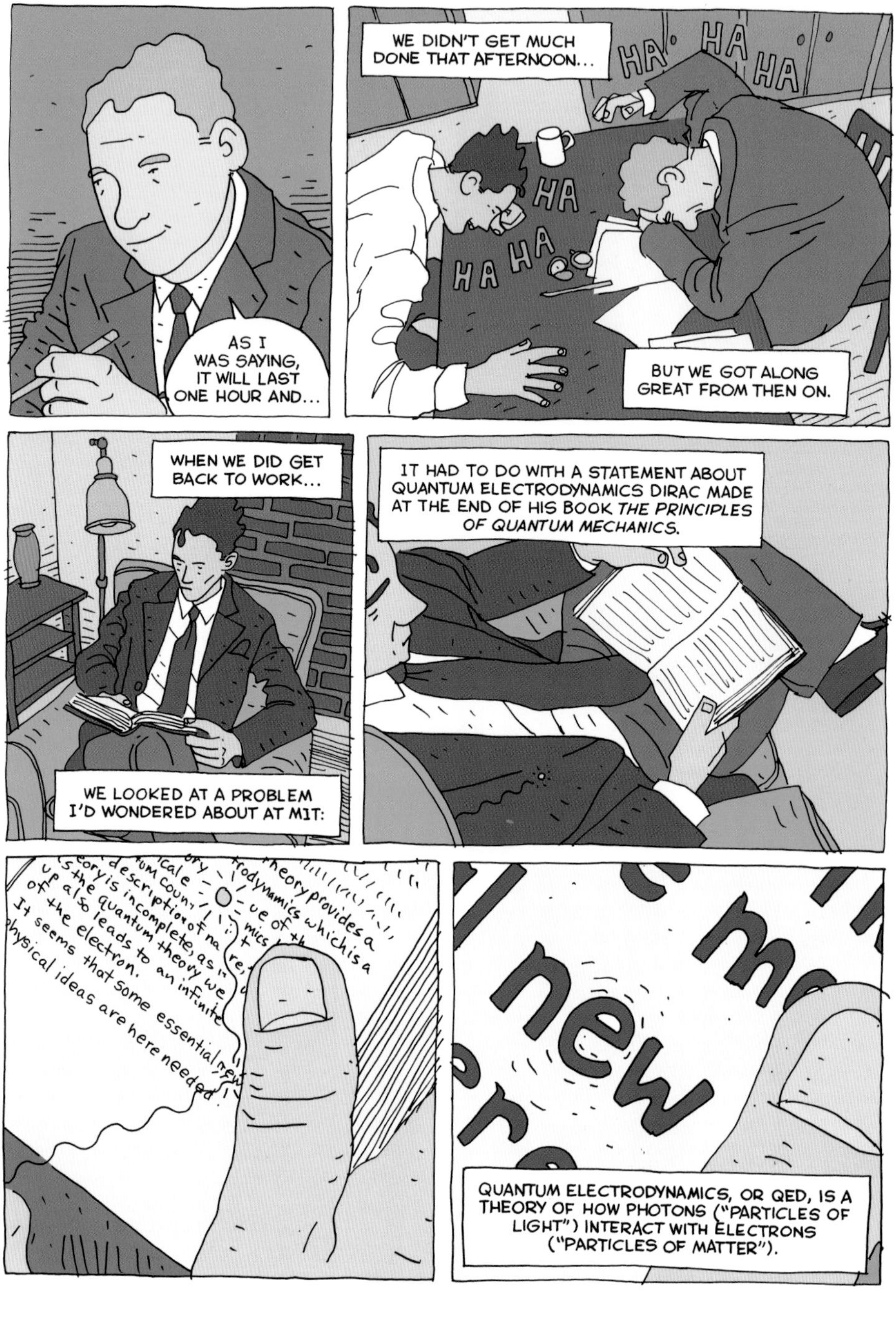
AS I WAS SAYING, IT WILL LAST ONE HOUR AND...
WE DIDN'T GET MUCH DONE THAT AFTERNOON...
HA HA HA
HA
HA HA
BUT WE GOT ALONG GREAT FROM THEN ON.
WHEN WE DID GET BACK TO WORK...
WE LOOKED AT A PROBLEM I'D WONDERED ABOUT AT MIT:
IT HAD TO DO WITH A STATEMENT ABOUT QUANTUM ELECTRODYNAMICS DIRAC MADE AT THE END OF HIS BOOK *THE PRINCIPLES OF QUANTUM MECHANICS.*
the electron.
It seems that some essential new
physical ideas are here needed.
new
QUANTUM ELECTRODYNAMICS, OR QED, IS A THEORY OF HOW PHOTONS ("PARTICLES OF LIGHT") INTERACT WITH ELECTRONS ("PARTICLES OF MATTER").

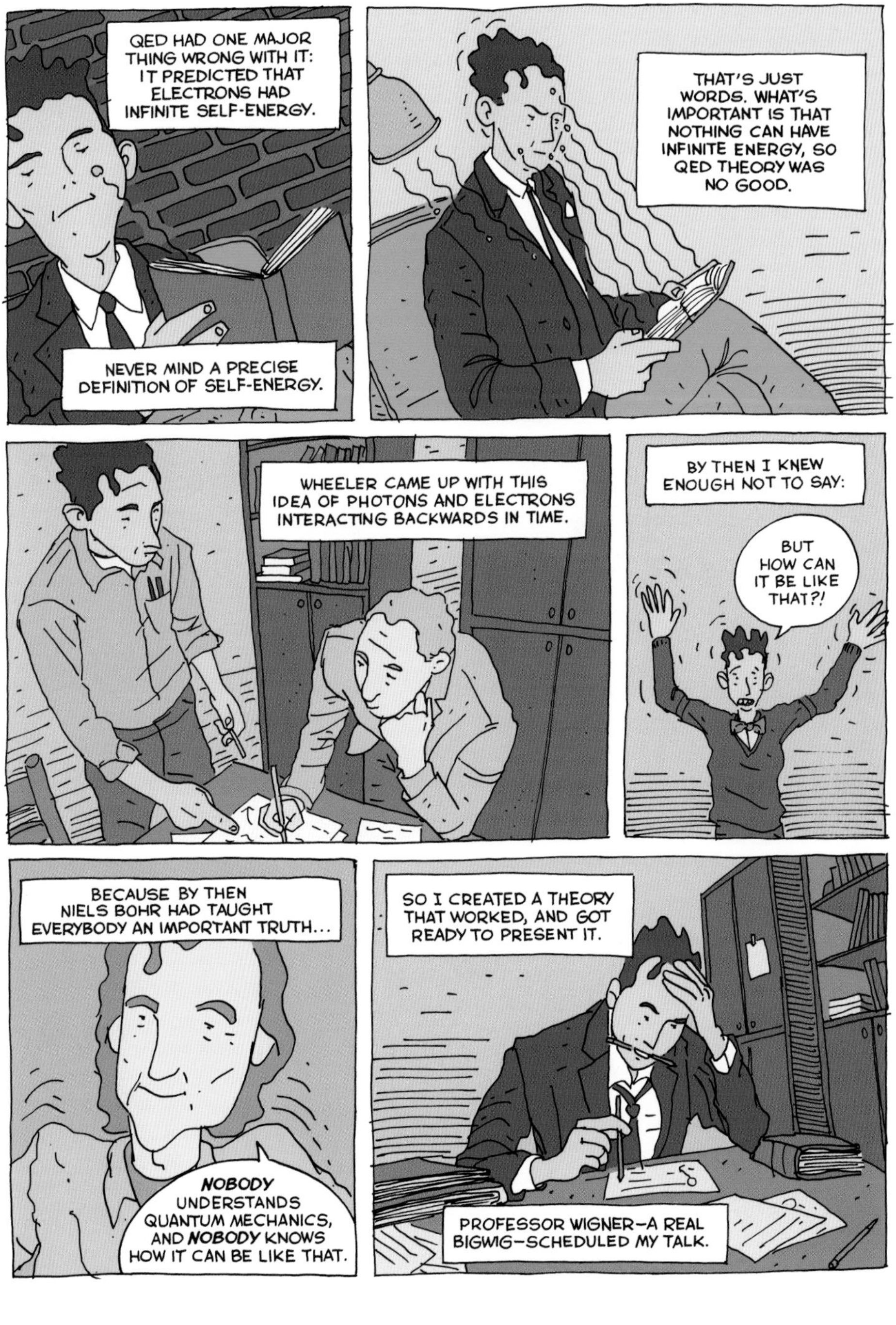
QED HAD ONE MAJOR THING WRONG WITH IT: IT PREDICTED THAT ELECTRONS HAD INFINITE SELF-ENERGY.
NEVER MIND A PRECISE DEFINITION OF SELF-ENERGY.
THAT'S JUST WORDS. WHAT'S IMPORTANT IS THAT NOTHING CAN HAVE INFINITE ENERGY, SO QED THEORY WAS NO GOOD.
WHEELER CAME UP WITH THIS IDEA OF PHOTONS AND ELECTRONS INTERACTING BACKWARDS IN TIME.
BY THEN I KNEW ENOUGH NOT TO SAY:
BUT HOW CAN IT BE LIKE THAT?!
BECAUSE BY THEN NIELS BOHR HAD TAUGHT EVERYBODY AN IMPORTANT TRUTH...
NOBODY UNDERSTANDS QUANTUM MECHANICS, AND NOBODY KNOWS HOW IT CAN BE LIKE THAT.
SO I CREATED A THEORY THAT WORKED, AND GOT READY TO PRESENT IT.
PROFESSOR WIGNER—A REAL BIGWIG—SCHEDULED MY TALK.

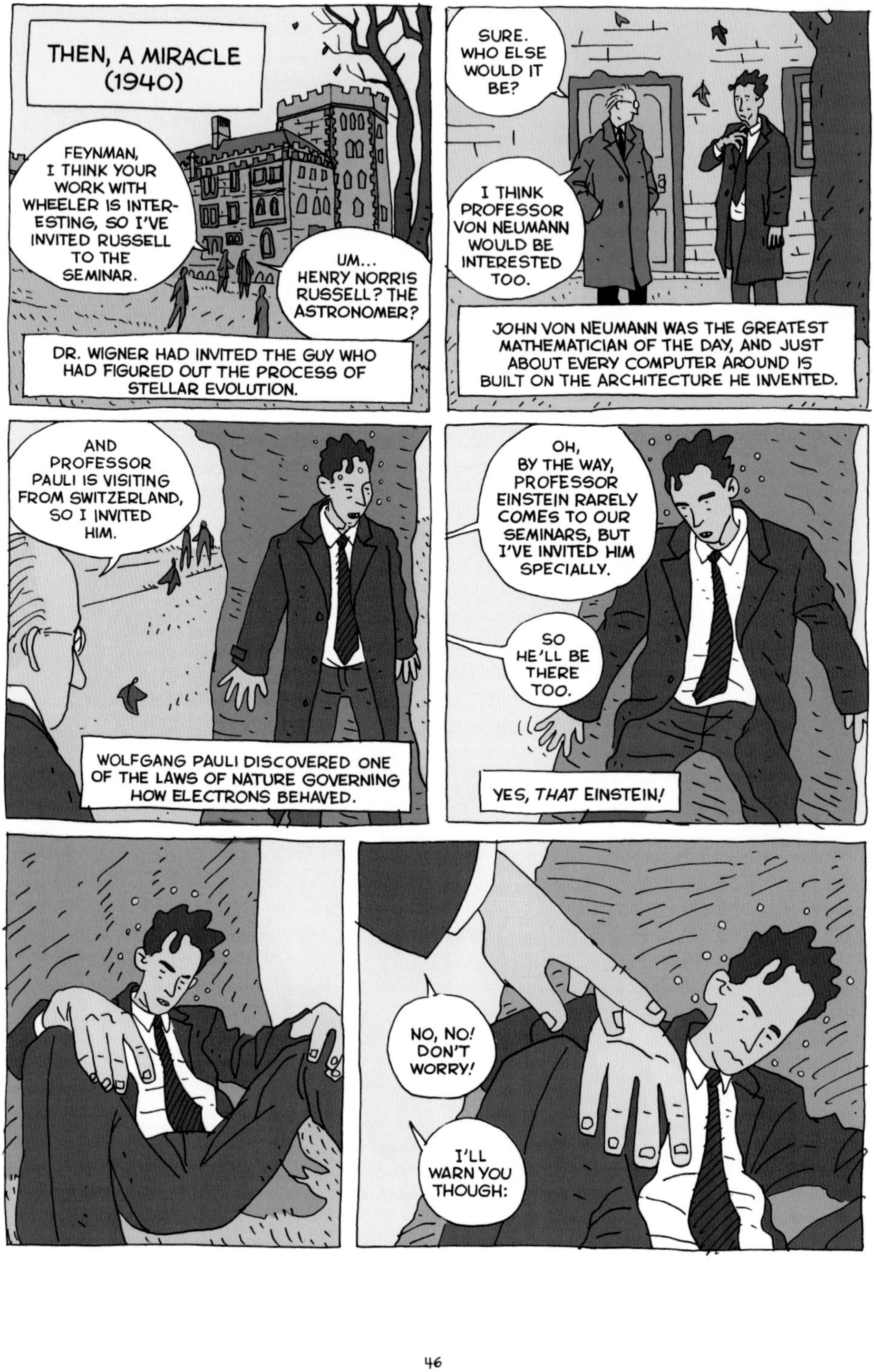
THEN, A MIRACLE (1940)
FEYNMAN, I THINK YOUR WORK WITH WHEELER IS INTER-ESTING, SO I'VE INVITED RUSSELL TO THE SEMINAR.
UM... HENRY NORRIS RUSSELL? THE ASTRONOMER?
DR. WIGNER HAD INVITED THE GUY WHO HAD FIGURED OUT THE PROCESS OF STELLAR EVOLUTION.
SURE. WHO ELSE WOULD IT BE?
I THINK PROFESSOR VON NEUMANN WOULD BE INTERESTED TOO.
JOHN VON NEUMANN WAS THE GREATEST MATHEMATICIAN OF THE DAY, AND JUST ABOUT EVERY COMPUTER AROUND IS BUILT ON THE ARCHITECTURE HE INVENTED.
AND PROFESSOR PAULI IS VISITING FROM SWITZERLAND, SO I INVITED HIM.
WOLFGANG PAULI DISCOVERED ONE OF THE LAWS OF NATURE GOVERNING HOW ELECTRONS BEHAVED.
OH, BY THE WAY, PROFESSOR EINSTEIN RARELY COMES TO OUR SEMINARS, BUT I'VE INVITED HIM SPECIALLY.
SO HE'LL BE THERE TOO.
YES, *THAT* EINSTEIN!
NO, NO! DON'T WORRY!
I'LL WARN YOU THOUGH:

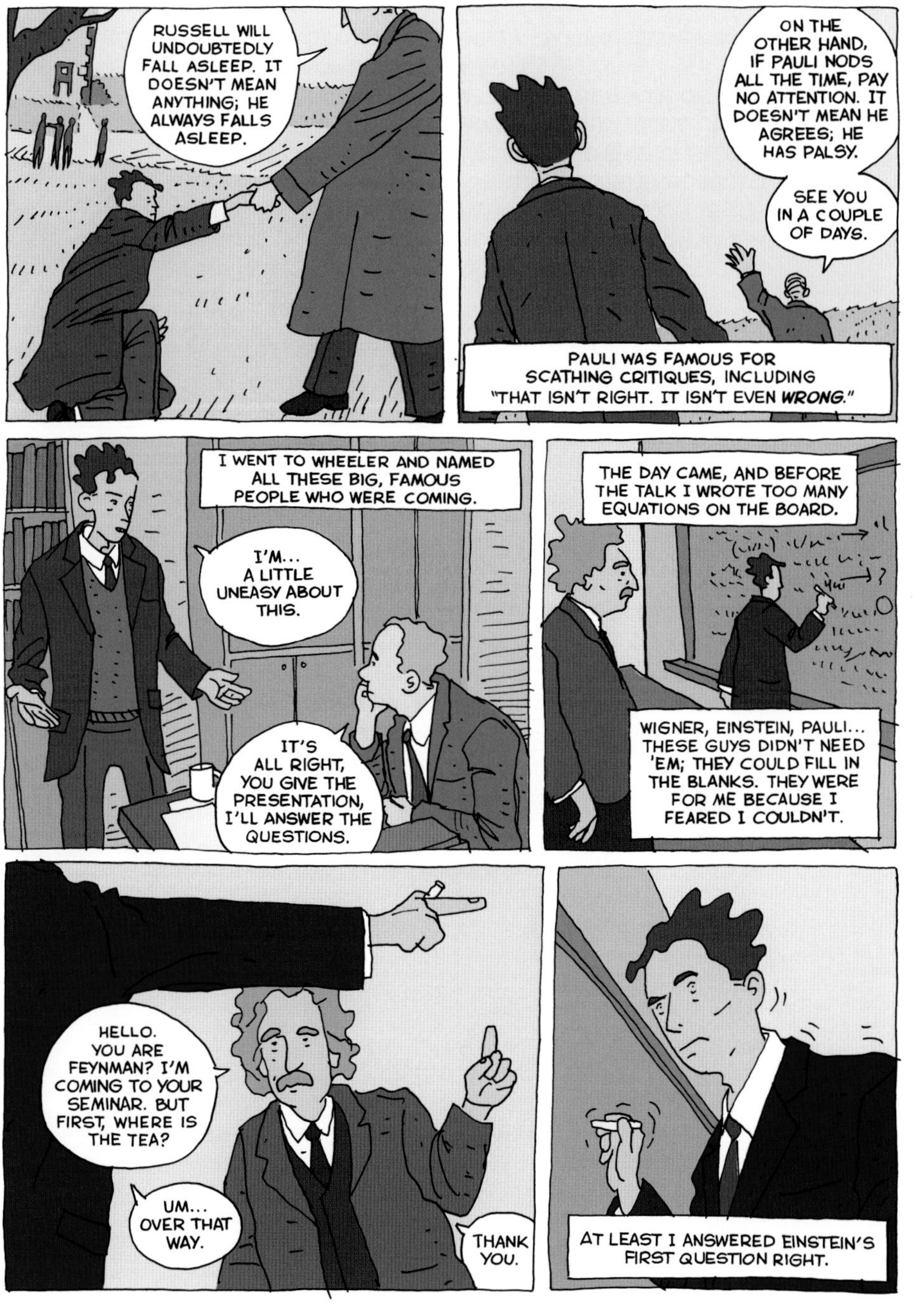
RUSSELL WILL UNDOUBTEDLY FALL ASLEEP. IT DOESN'T MEAN ANYTHING; HE ALWAYS FALLS ASLEEP.
ON THE OTHER HAND, IF PAULI NODS ALL THE TIME, PAY NO ATTENTION. IT DOESN'T MEAN HE AGREES; HE HAS PALSY.
SEE YOU IN A COUPLE OF DAYS.
PAULI WAS FAMOUS FOR SCATHING CRITIQUES, INCLUDING "THAT ISN'T RIGHT. IT ISN'T EVEN *WRONG*."
I WENT TO WHEELER AND NAMED ALL THESE BIG, FAMOUS PEOPLE WHO WERE COMING.
I'M... A LITTLE UNEASY ABOUT THIS.
IT'S ALL RIGHT, YOU GIVE THE PRESENTATION, I'LL ANSWER THE QUESTIONS.
THE DAY CAME, AND BEFORE THE TALK I WROTE TOO MANY EQUATIONS ON THE BOARD.
WIGNER, EINSTEIN, PAULI... THESE GUYS DIDN'T NEED 'EM; THEY COULD FILL IN THE BLANKS. THEY WERE FOR ME BECAUSE I FEARED I COULDN'T.
HELLO. YOU ARE FEYNMAN? I'M COMING TO YOUR SEMINAR. BUT FIRST, WHERE IS THE TEA?
UM... OVER THAT WAY.
THANK YOU.
AT LEAST I ANSWERED EINSTEIN'S FIRST QUESTION RIGHT.

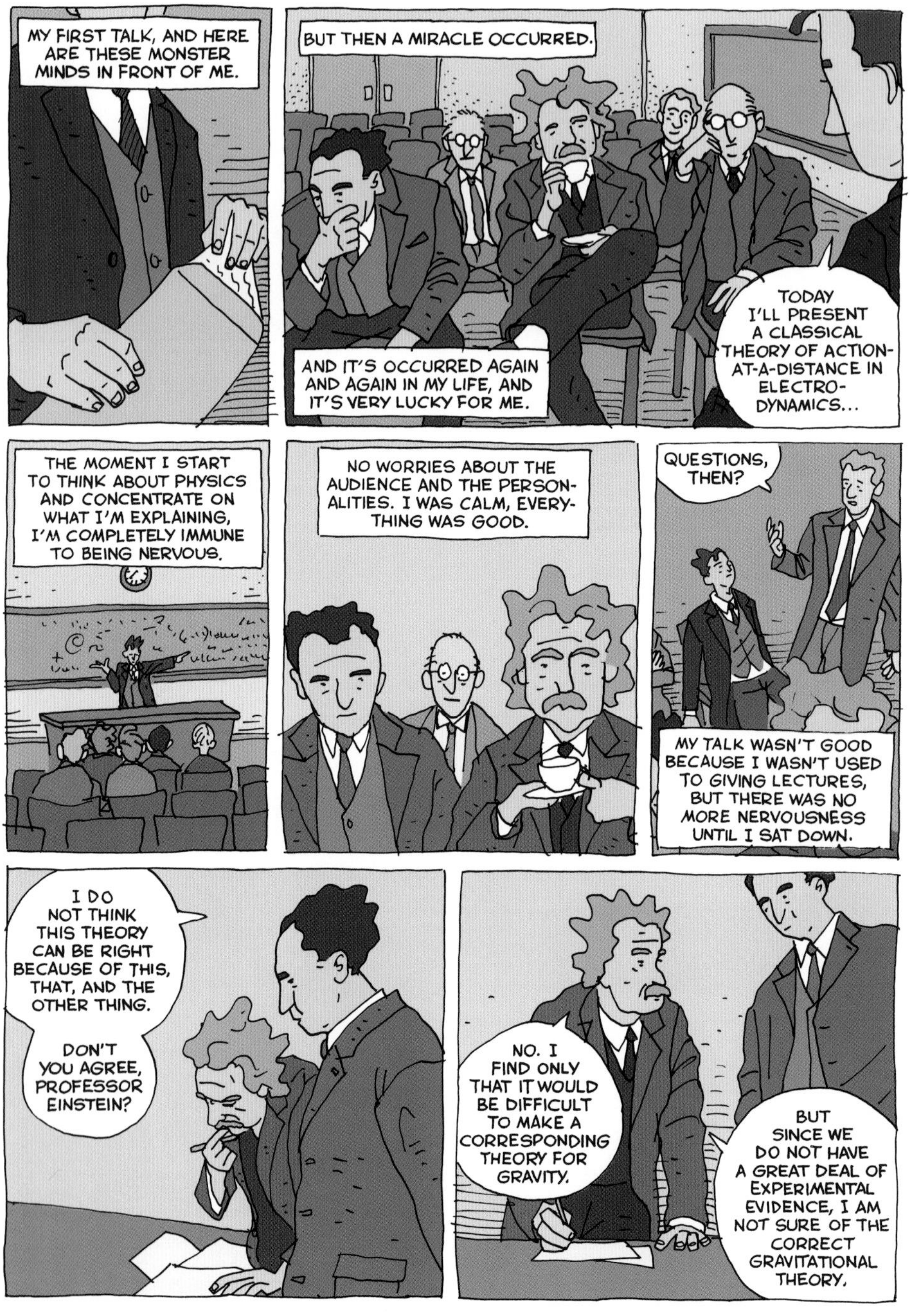

MY FIRST TALK, AND HERE ARE THESE MONSTER MINDS IN FRONT OF ME.
BUT THEN A MIRACLE OCCURRED.
AND IT'S OCCURRED AGAIN AND AGAIN IN MY LIFE, AND IT'S VERY LUCKY FOR ME.
TODAY I'LL PRESENT A CLASSICAL THEORY OF ACTION-AT-A-DISTANCE IN ELECTRO-DYNAMICS...
THE MOMENT I START TO THINK ABOUT PHYSICS AND CONCENTRATE ON WHAT I'M EXPLAINING, I'M COMPLETELY IMMUNE TO BEING NERVOUS.
NO WORRIES ABOUT THE AUDIENCE AND THE PERSON-ALITIES. I WAS CALM, EVERY-THING WAS GOOD.
QUESTIONS, THEN?
MY TALK WASN'T GOOD BECAUSE I WASN'T USED TO GIVING LECTURES, BUT THERE WAS NO MORE NERVOUSNESS UNTIL I SAT DOWN.
I DO NOT THINK THIS THEORY CAN BE RIGHT BECAUSE OF THIS, THAT, AND THE OTHER THING.
DON'T YOU AGREE, PROFESSOR EINSTEIN?
NO. I FIND ONLY THAT IT WOULD BE DIFFICULT TO MAKE A CORRESPONDING THEORY FOR GRAVITY.
BUT SINCE WE DO NOT HAVE A GREAT DEAL OF EXPERIMENTAL EVIDENCE, I AM NOT SURE OF THE CORRECT GRAVITATIONAL THEORY.

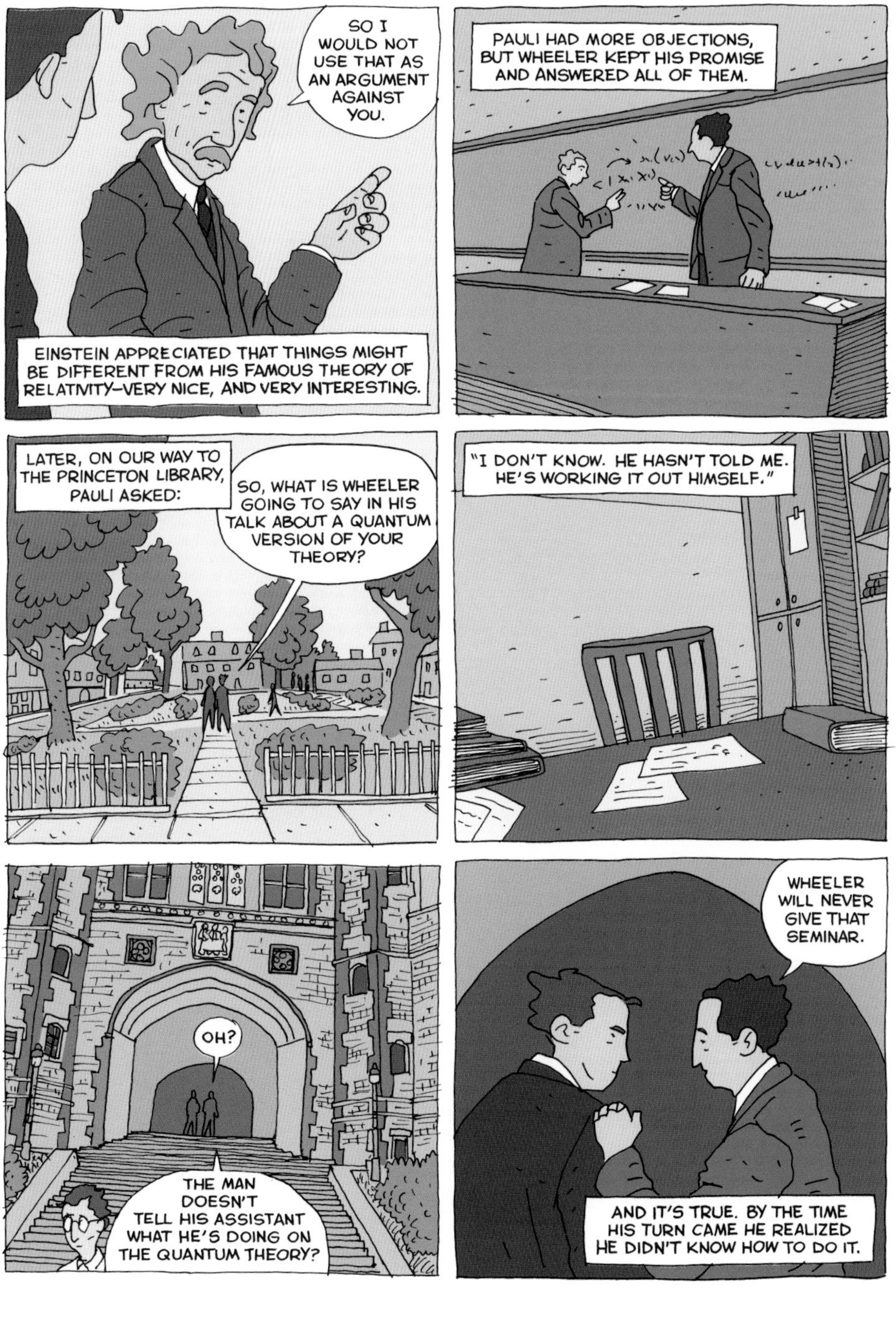
SO I WOULD NOT USE THAT AS AN ARGUMENT AGAINST YOU.
EINSTEIN APPRECIATED THAT THINGS MIGHT BE DIFFERENT FROM HIS FAMOUS THEORY OF RELATIVITY—VERY NICE, AND VERY INTERESTING.
PAULI HAD MORE OBJECTIONS, BUT WHEELER KEPT HIS PROMISE AND ANSWERED ALL OF THEM.
LATER, ON OUR WAY TO THE PRINCETON LIBRARY, PAULI ASKED:
SO, WHAT IS WHEELER GOING TO SAY IN HIS TALK ABOUT A QUANTUM VERSION OF YOUR THEORY?
"I DON'T KNOW. HE HASN'T TOLD ME. HE'S WORKING IT OUT HIMSELF."
OH?
THE MAN DOESN'T TELL HIS ASSISTANT WHAT HE'S DOING ON THE QUANTUM THEORY?
WHEELER WILL NEVER GIVE THAT SEMINAR.
AND IT'S TRUE. BY THE TIME HIS TURN CAME HE REALIZED HE DIDN'T KNOW HOW TO DO IT.

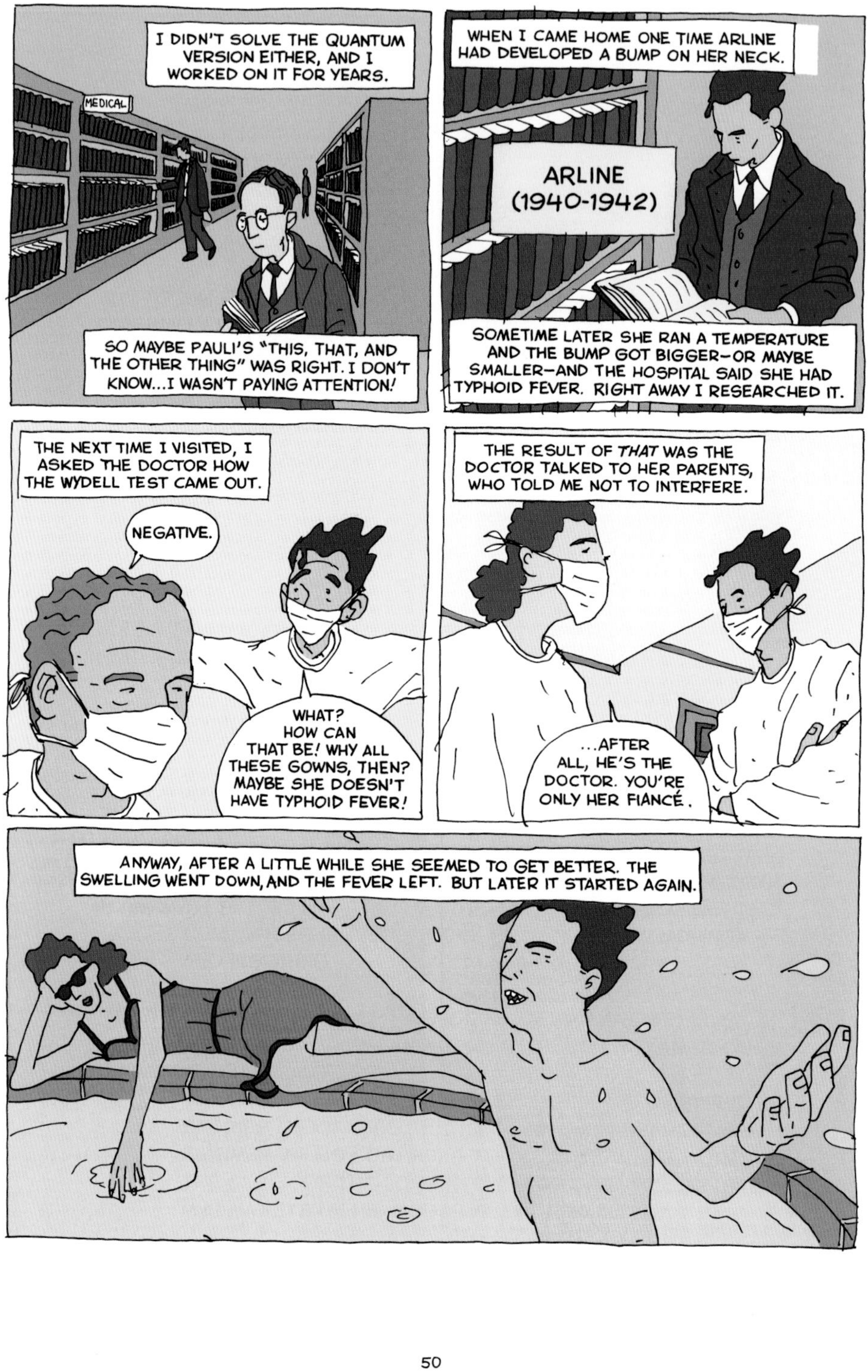
I DIDN'T SOLVE THE QUANTUM VERSION EITHER, AND I WORKED ON IT FOR YEARS.
MEDICAL
SO MAYBE PAULI'S "THIS, THAT, AND THE OTHER THING" WAS RIGHT. I DON'T KNOW...I WASN'T PAYING ATTENTION!
WHEN I CAME HOME ONE TIME ARLINE HAD DEVELOPED A BUMP ON HER NECK.
ARLINE (1940-1942)
SOMETIME LATER SHE RAN A TEMPERATURE AND THE BUMP GOT BIGGER—OR MAYBE SMALLER—AND THE HOSPITAL SAID SHE HAD TYPHOID FEVER. RIGHT AWAY I RESEARCHED IT.
THE NEXT TIME I VISITED, I ASKED THE DOCTOR HOW THE WYDELL TEST CAME OUT.
NEGATIVE.
WHAT? HOW CAN THAT BE! WHY ALL THESE GOWNS, THEN? MAYBE SHE DOESN'T HAVE TYPHOID FEVER!
THE RESULT OF *THAT* WAS THE DOCTOR TALKED TO HER PARENTS, WHO TOLD ME NOT TO INTERFERE.
...AFTER ALL, HE'S THE DOCTOR. YOU'RE ONLY HER FIANCÉ.
ANYWAY, AFTER A LITTLE WHILE SHE SEEMED TO GET BETTER. THE SWELLING WENT DOWN, AND THE FEVER LEFT. BUT LATER IT STARTED AGAIN.

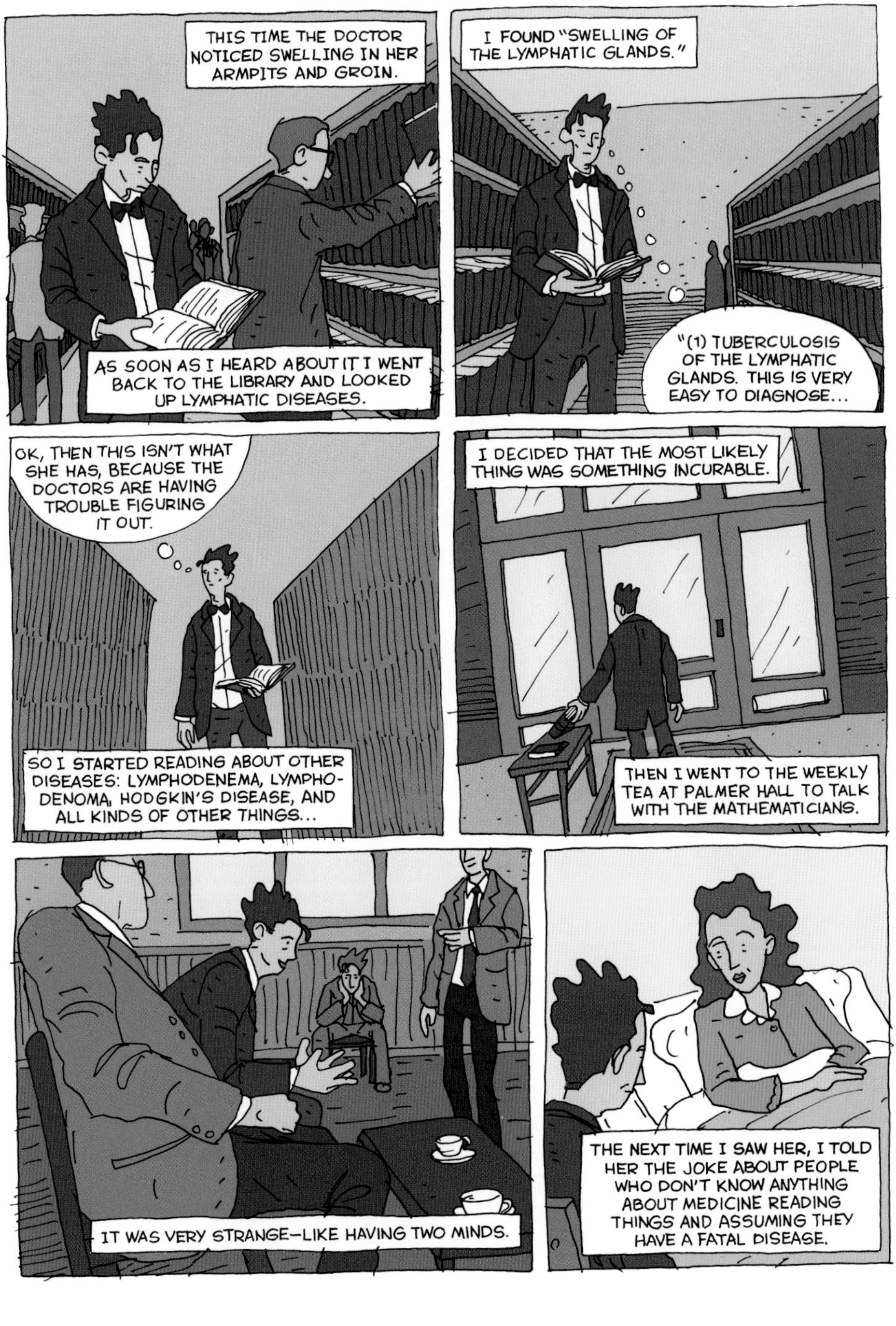
THIS TIME THE DOCTOR NOTICED SWELLING IN HER ARMPITS AND GROIN.
AS SOON AS I HEARD ABOUT IT I WENT BACK TO THE LIBRARY AND LOOKED UP LYMPHATIC DISEASES.
I FOUND "SWELLING OF THE LYMPHATIC GLANDS."
"(1) TUBERCULOSIS OF THE LYMPHATIC GLANDS. THIS IS VERY EASY TO DIAGNOSE...
OK, THEN THIS ISN'T WHAT SHE HAS, BECAUSE THE DOCTORS ARE HAVING TROUBLE FIGURING IT OUT.
SO I STARTED READING ABOUT OTHER DISEASES: LYMPHODENEMA, LYMPHO-DENOMA, HODGKIN'S DISEASE, AND ALL KINDS OF OTHER THINGS...
I DECIDED THAT THE MOST LIKELY THING WAS SOMETHING INCURABLE.
THEN I WENT TO THE WEEKLY TEA AT PALMER HALL TO TALK WITH THE MATHEMATICIANS.
IT WAS VERY STRANGE—LIKE HAVING TWO MINDS.
THE NEXT TIME I SAW HER, I TOLD HER THE JOKE ABOUT PEOPLE WHO DON'T KNOW ANYTHING ABOUT MEDICINE READING THINGS AND ASSUMING THEY HAVE A FATAL DISEASE.

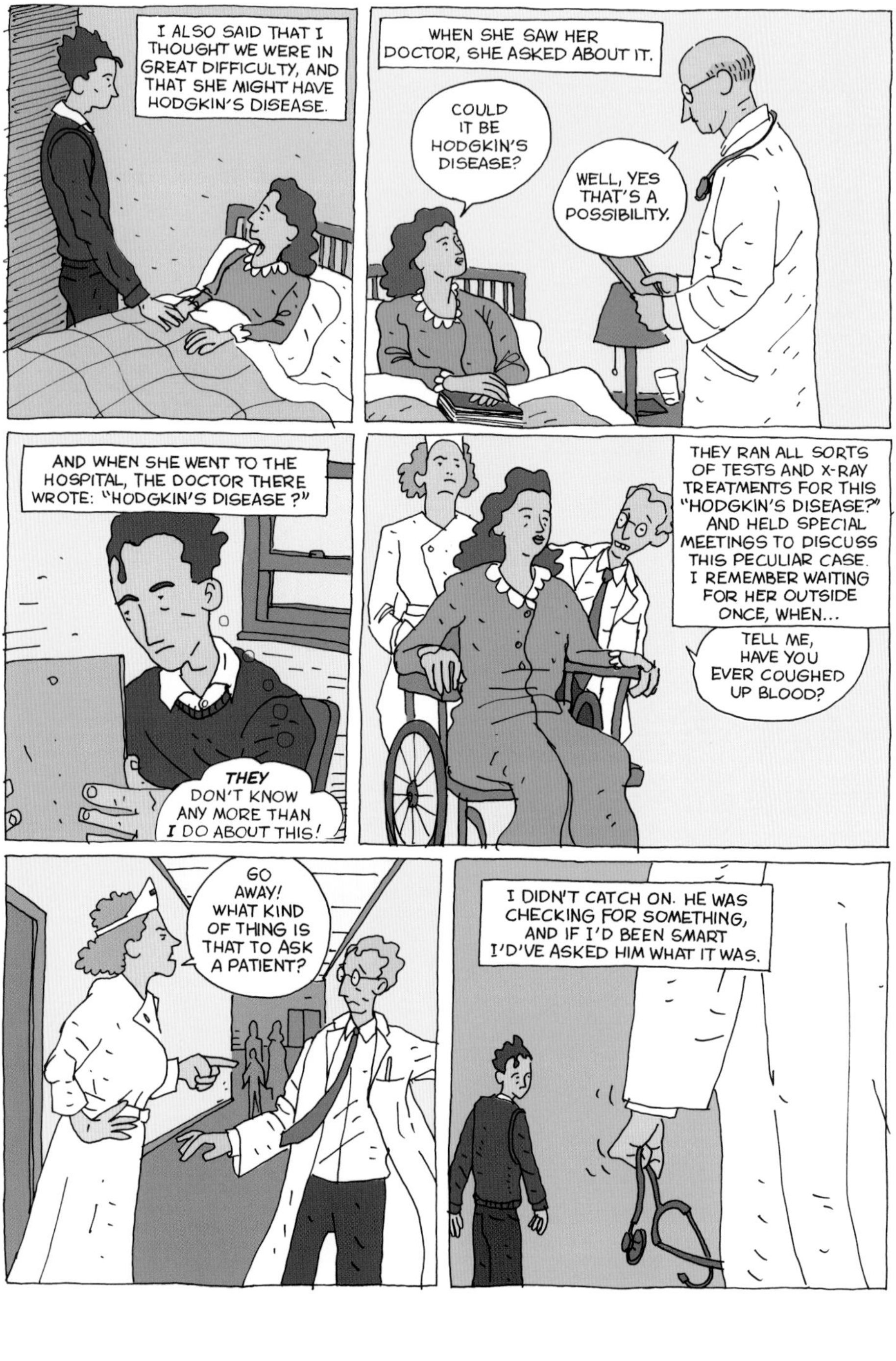
I ALSO SAID THAT I THOUGHT WE WERE IN GREAT DIFFICULTY, AND THAT SHE MIGHT HAVE HODGKIN'S DISEASE.
WHEN SHE SAW HER DOCTOR, SHE ASKED ABOUT IT.
COULD IT BE HODGKIN'S DISEASE?
WELL, YES THAT'S A POSSIBILITY.
AND WHEN SHE WENT TO THE HOSPITAL, THE DOCTOR THERE WROTE: "HODGKIN'S DISEASE ?"
THEY DON'T KNOW ANY MORE THAN ***I*** DO ABOUT THIS!
THEY RAN ALL SORTS OF TESTS AND X-RAY TREATMENTS FOR THIS "HODGKIN'S DISEASE?" AND HELD SPECIAL MEETINGS TO DISCUSS THIS PECULIAR CASE. I REMEMBER WAITING FOR HER OUTSIDE ONCE, WHEN...
TELL ME, HAVE YOU EVER COUGHED UP BLOOD?
GO AWAY! WHAT KIND OF THING IS THAT TO ASK A PATIENT?
I DIDN'T CATCH ON. HE WAS CHECKING FOR SOMETHING, AND IF I'D BEEN SMART I'D'VE ASKED HIM WHAT IT WAS.

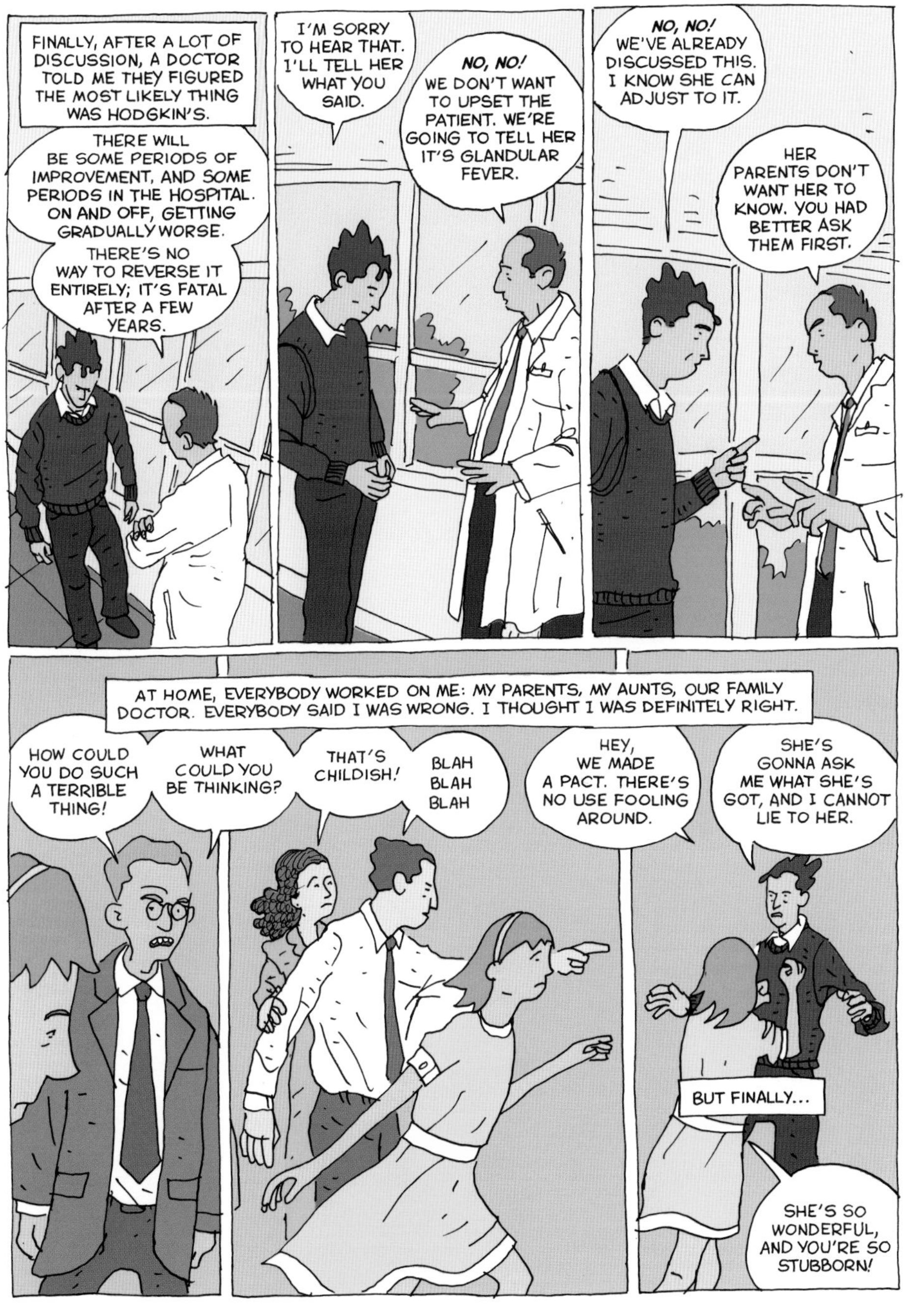

FINALLY, AFTER A LOT OF DISCUSSION, A DOCTOR TOLD ME THEY FIGURED THE MOST LIKELY THING WAS HODGKIN'S.
THERE WILL BE SOME PERIODS OF IMPROVEMENT, AND SOME PERIODS IN THE HOSPITAL. ON AND OFF, GETTING GRADUALLY WORSE.
THERE'S NO WAY TO REVERSE IT ENTIRELY; IT'S FATAL AFTER A FEW YEARS.
I'M SORRY TO HEAR THAT. I'LL TELL HER WHAT YOU SAID.
NO, NO! WE DON'T WANT TO UPSET THE PATIENT. WE'RE GOING TO TELL HER IT'S GLANDULAR FEVER.
NO, NO! WE'VE ALREADY DISCUSSED THIS. I KNOW SHE CAN ADJUST TO IT.
HER PARENTS DON'T WANT HER TO KNOW. YOU HAD BETTER ASK THEM FIRST.
AT HOME, EVERYBODY WORKED ON ME: MY PARENTS, MY AUNTS, OUR FAMILY DOCTOR. EVERYBODY SAID I WAS WRONG. I THOUGHT I WAS DEFINITELY RIGHT.
HOW COULD YOU DO SUCH A TERRIBLE THING!
WHAT COULD YOU BE THINKING?
THAT'S CHILDISH!
BLAH BLAH BLAH
HEY, WE MADE A PACT. THERE'S NO USE FOOLING AROUND.
SHE'S GONNA ASK ME WHAT SHE'S GOT, AND I CANNOT LIE TO HER.
BUT FINALLY...
SHE'S SO WONDERFUL, AND YOU'RE SO STUBBORN!

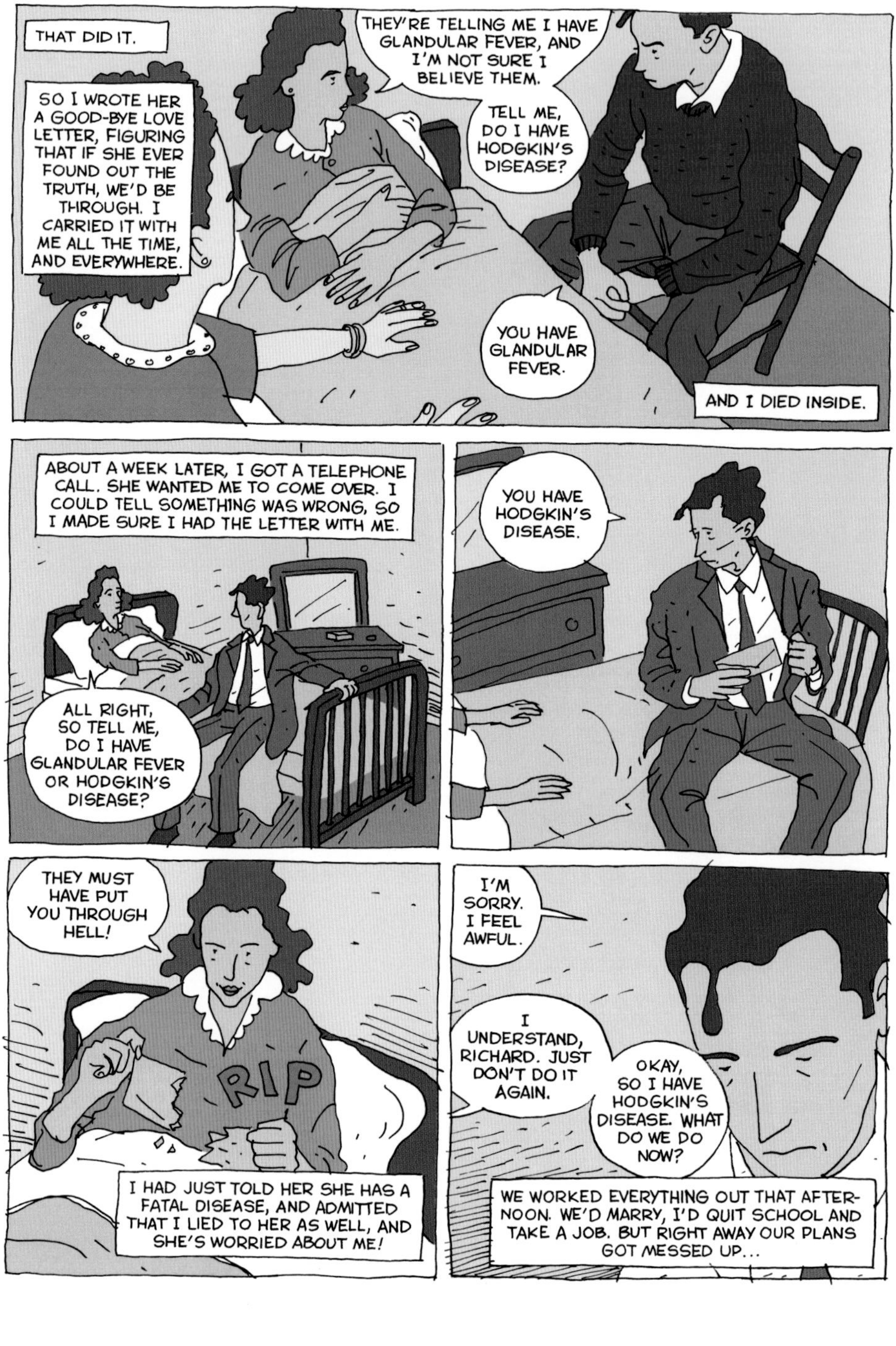
THAT DID IT.
SO I WROTE HER A GOOD-BYE LOVE LETTER, FIGURING THAT IF SHE EVER FOUND OUT THE TRUTH, WE'D BE THROUGH. I CARRIED IT WITH ME ALL THE TIME, AND EVERYWHERE.
THEY'RE TELLING ME I HAVE GLANDULAR FEVER, AND I'M NOT SURE I BELIEVE THEM.
TELL ME, DO I HAVE HODGKIN'S DISEASE?
YOU HAVE GLANDULAR FEVER.
AND I DIED INSIDE.
ABOUT A WEEK LATER, I GOT A TELEPHONE CALL. SHE WANTED ME TO COME OVER. I COULD TELL SOMETHING WAS WRONG, SO I MADE SURE I HAD THE LETTER WITH ME.
ALL RIGHT, SO TELL ME, DO I HAVE GLANDULAR FEVER OR HODGKIN'S DISEASE?
YOU HAVE HODGKIN'S DISEASE.
THEY MUST HAVE PUT YOU THROUGH HELL!
RIP
I HAD JUST TOLD HER SHE HAS A FATAL DISEASE, AND ADMITTED THAT I LIED TO HER AS WELL, AND SHE'S WORRIED ABOUT ME!
I'M SORRY. I FEEL AWFUL.
I UNDERSTAND, RICHARD. JUST DON'T DO IT AGAIN.
OKAY, SO I HAVE HODGKIN'S DISEASE. WHAT DO WE DO NOW?
WE WORKED EVERYTHING OUT THAT AFTERNOON. WE'D MARRY, I'D QUIT SCHOOL AND TAKE A JOB. BUT RIGHT AWAY OUR PLANS GOT MESSED UP...

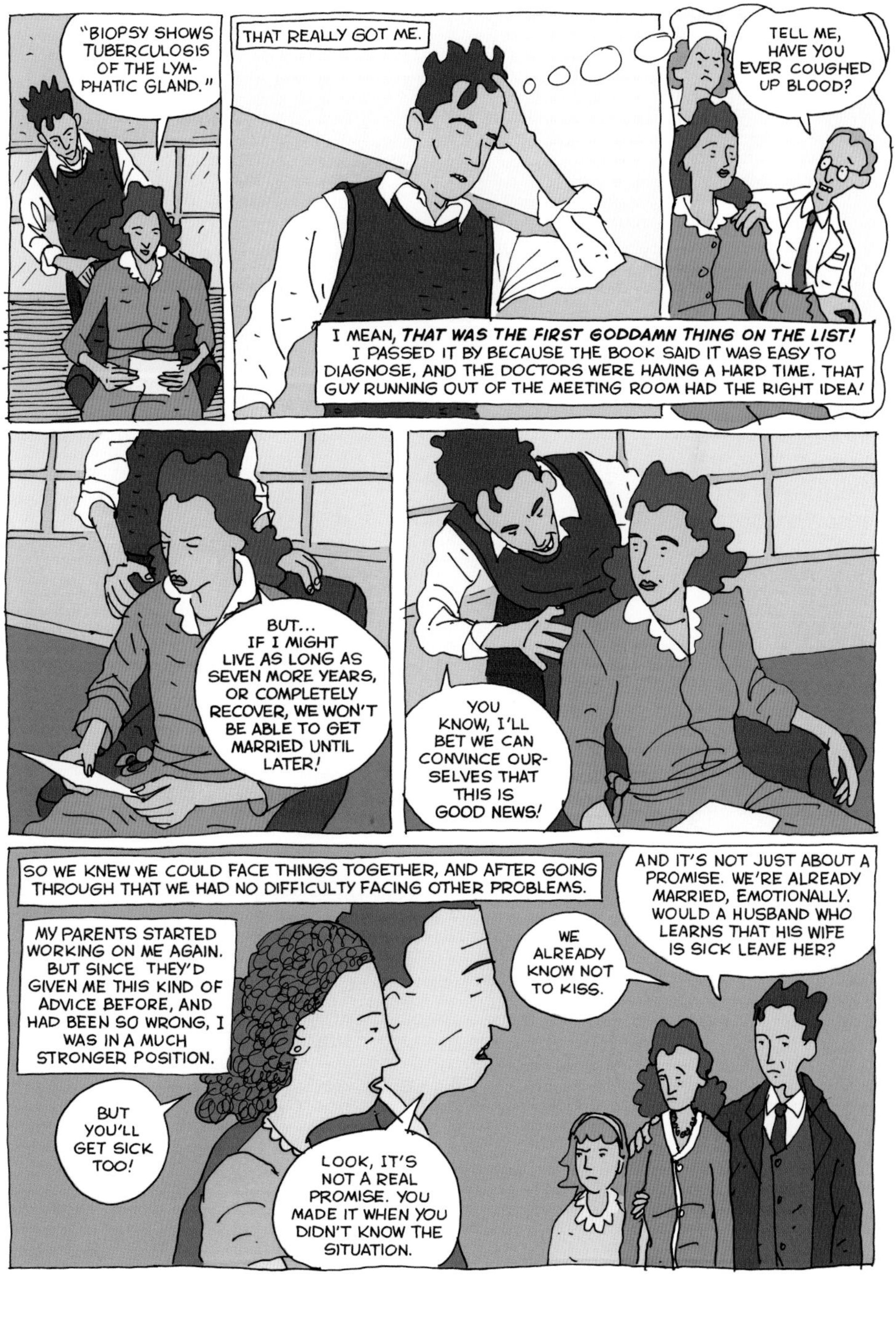
"BIOPSY SHOWS TUBERCULOSIS OF THE LYMPHATIC GLAND."
THAT REALLY GOT ME.
TELL ME, HAVE YOU EVER COUGHED UP BLOOD?
I MEAN, ***THAT WAS THE FIRST GODDAMN THING ON THE LIST!*** I PASSED IT BY BECAUSE THE BOOK SAID IT WAS EASY TO DIAGNOSE, AND THE DOCTORS WERE HAVING A HARD TIME. THAT GUY RUNNING OUT OF THE MEETING ROOM HAD THE RIGHT IDEA!
BUT... IF I MIGHT LIVE AS LONG AS SEVEN MORE YEARS, OR COMPLETELY RECOVER, WE WON'T BE ABLE TO GET MARRIED UNTIL LATER!
YOU KNOW, I'LL BET WE CAN CONVINCE OURSELVES THAT THIS IS GOOD NEWS!
SO WE KNEW WE COULD FACE THINGS TOGETHER, AND AFTER GOING THROUGH THAT WE HAD NO DIFFICULTY FACING OTHER PROBLEMS.
MY PARENTS STARTED WORKING ON ME AGAIN. BUT SINCE THEY'D GIVEN ME THIS KIND OF ADVICE BEFORE, AND HAD BEEN SO WRONG, I WAS IN A MUCH STRONGER POSITION.
BUT YOU'LL GET SICK TOO!
LOOK, IT'S NOT A REAL PROMISE. YOU MADE IT WHEN YOU DIDN'T KNOW THE SITUATION.
WE ALREADY KNOW NOT TO KISS.
AND IT'S NOT JUST ABOUT A PROMISE. WE'RE ALREADY MARRIED, EMOTIONALLY. WOULD A HUSBAND WHO LEARNS THAT HIS WIFE IS SICK LEAVE HER?

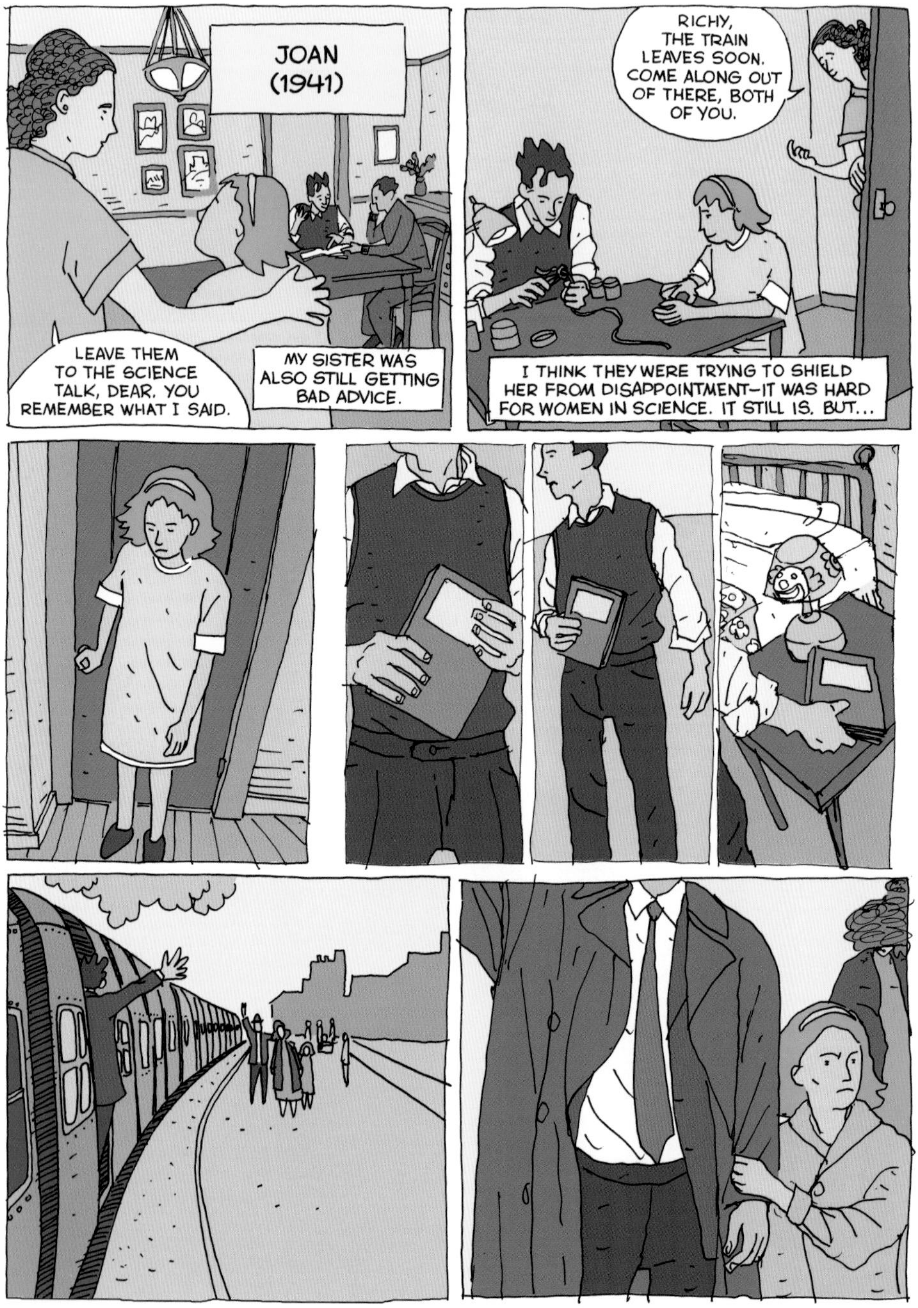
JOAN
(1941)
LEAVE THEM TO THE SCIENCE TALK, DEAR. YOU REMEMBER WHAT I SAID.
MY SISTER WAS ALSO STILL GETTING BAD ADVICE.
RICHY, THE TRAIN LEAVES SOON. COME ALONG OUT OF THERE, BOTH OF YOU.
I THINK THEY WERE TRYING TO SHIELD HER FROM DISAPPOINTMENT—IT WAS HARD FOR WOMEN IN SCIENCE. IT STILL IS. BUT...

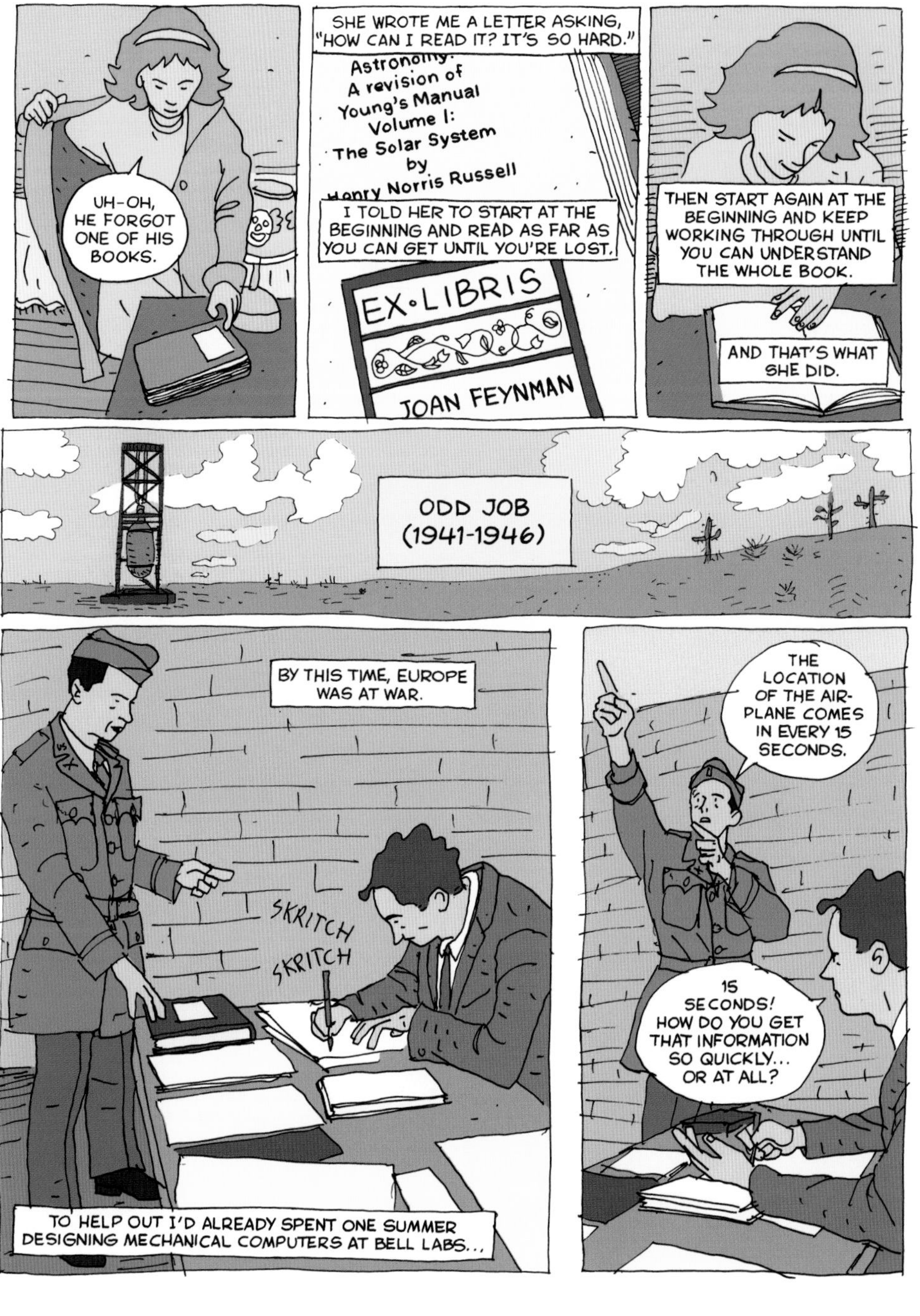
UH-OH, HE FORGOT ONE OF HIS BOOKS.
SHE WROTE ME A LETTER ASKING, "HOW CAN I READ IT? IT'S SO HARD."
A revision of Young's Manual Volume I: The Solar System by
Henry Norris Russell
I TOLD HER TO START AT THE BEGINNING AND READ AS FAR AS YOU CAN GET UNTIL YOU'RE LOST.
EX • LIBRIS
JOAN FEYNMAN
THEN START AGAIN AT THE BEGINNING AND KEEP WORKING THROUGH UNTIL YOU CAN UNDERSTAND THE WHOLE BOOK.
AND THAT'S WHAT SHE DID.
ODD JOB (1941-1946)
BY THIS TIME, EUROPE WAS AT WAR.
SKRITCH SKRITCH
TO HELP OUT I'D ALREADY SPENT ONE SUMMER DESIGNING MECHANICAL COMPUTERS AT BELL LABS. . ,
THE LOCATION OF THE AIR-PLANE COMES IN EVERY 15 SECONDS.
15 SECONDS! HOW DO YOU GET THAT INFORMATION SO QUICKLY... OR AT ALL?

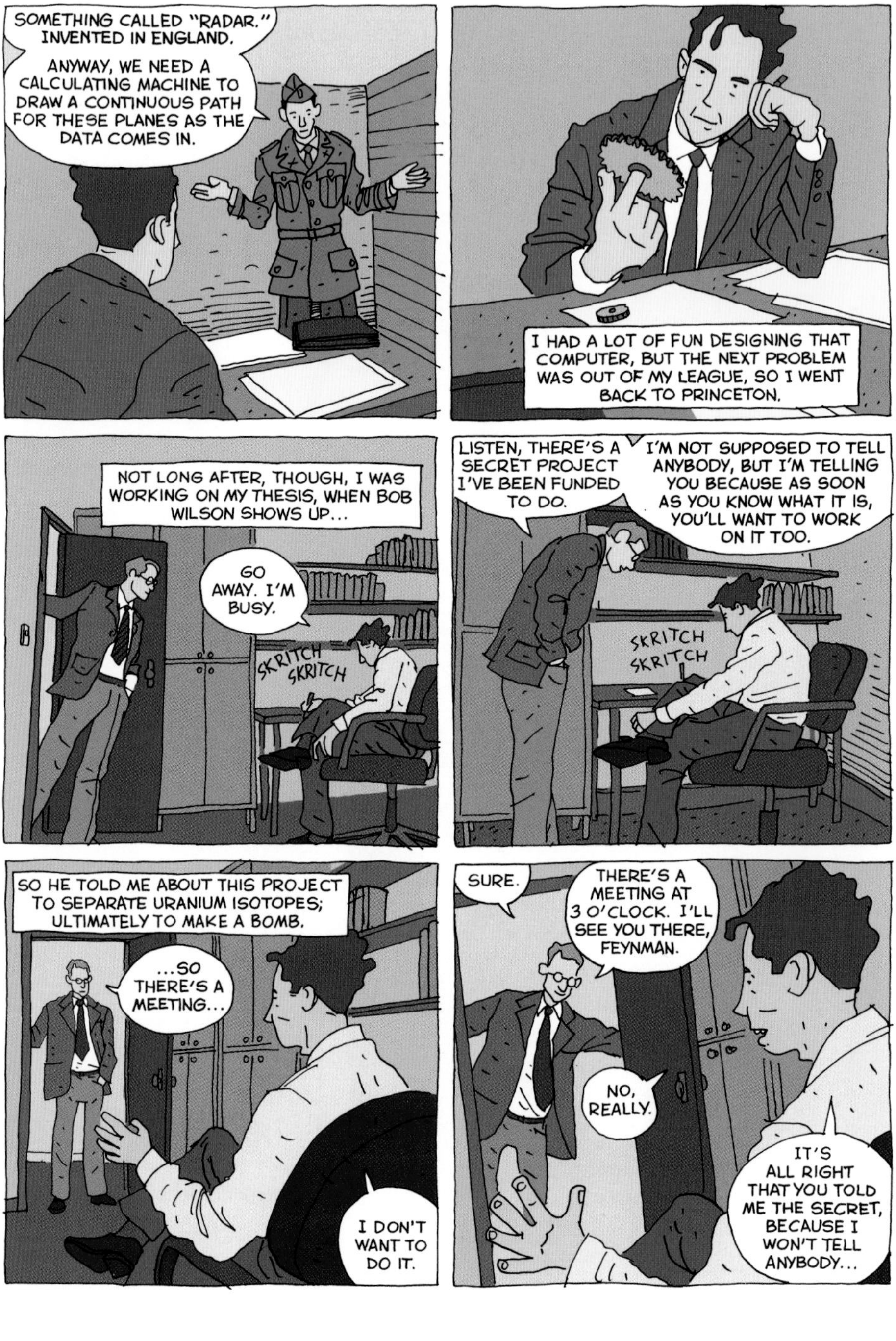
SOMETHING CALLED "RADAR." INVENTED IN ENGLAND.
ANYWAY, WE NEED A CALCULATING MACHINE TO DRAW A CONTINUOUS PATH FOR THESE PLANES AS THE DATA COMES IN.
I HAD A LOT OF FUN DESIGNING THAT COMPUTER, BUT THE NEXT PROBLEM WAS OUT OF MY LEAGUE, SO I WENT BACK TO PRINCETON.
NOT LONG AFTER, THOUGH, I WAS WORKING ON MY THESIS, WHEN BOB WILSON SHOWS UP...
GO AWAY. I'M BUSY.
SKRITCH SKRITCH
LISTEN, THERE'S A SECRET PROJECT I'VE BEEN FUNDED TO DO.
I'M NOT SUPPOSED TO TELL ANYBODY, BUT I'M TELLING YOU BECAUSE AS SOON AS YOU KNOW WHAT IT IS, YOU'LL WANT TO WORK ON IT TOO.
SKRITCH SKRITCH
SO HE TOLD ME ABOUT THIS PROJECT TO SEPARATE URANIUM ISOTOPES; ULTIMATELY TO MAKE A BOMB.
...SO THERE'S A MEETING...
I DON'T WANT TO DO IT.
SURE.
THERE'S A MEETING AT 3 O'CLOCK. I'LL SEE YOU THERE, FEYNMAN.
NO, REALLY.
IT'S ALL RIGHT THAT YOU TOLD ME THE SECRET, BECAUSE I WON'T TELL ANYBODY...

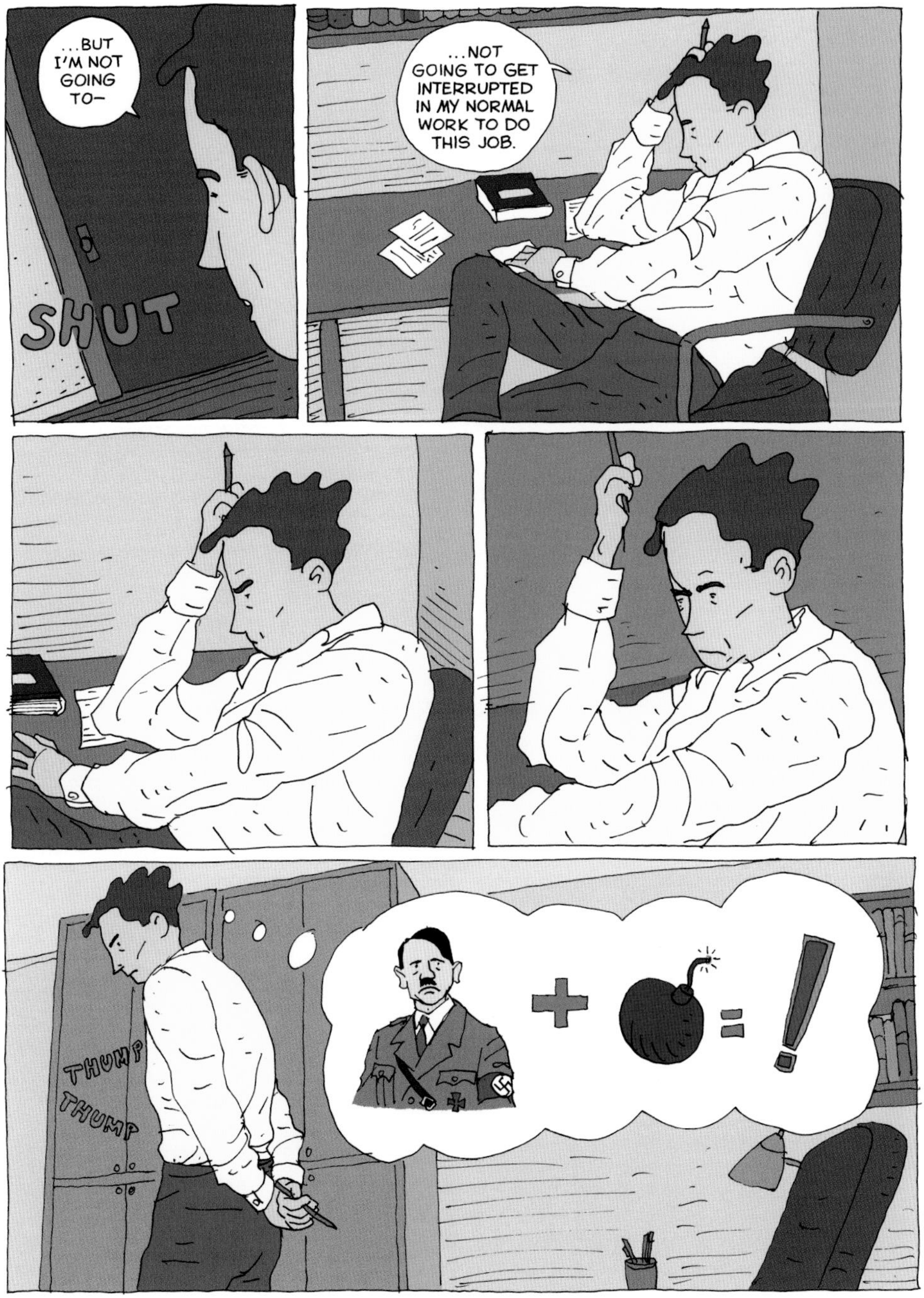

...BUT I'M NOT GOING TO—
SHUT
...NOT GOING TO GET INTERRUPTED IN MY NORMAL WORK TO DO THIS JOB.
THUMP
THUMP

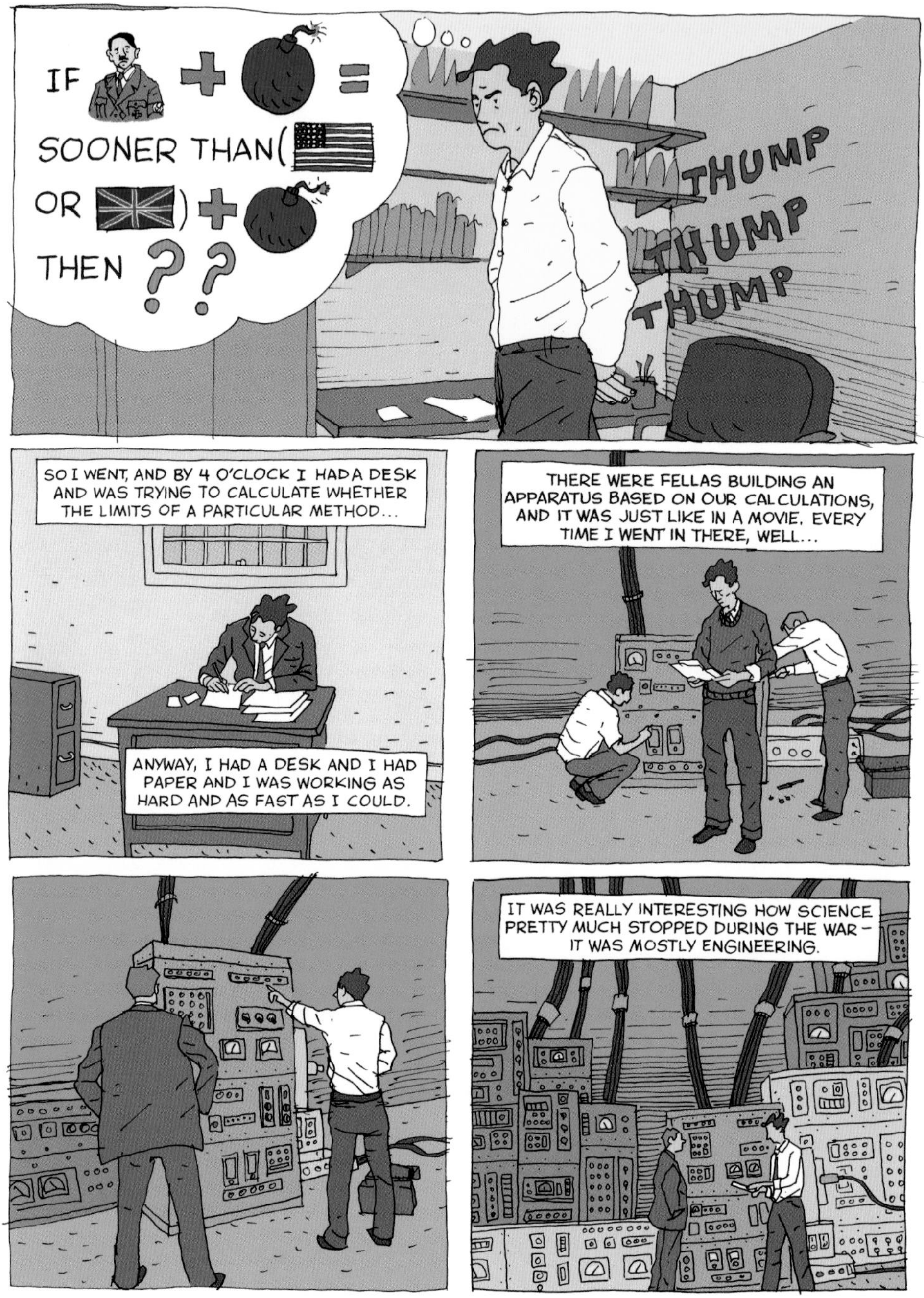
IF + = SOONER THAN(OR) + THEN ??
THUMP THUMP THUMP
SO I WENT, AND BY 4 O'CLOCK I HAD A DESK AND WAS TRYING TO CALCULATE WHETHER THE LIMITS OF A PARTICULAR METHOD...
ANYWAY, I HAD A DESK AND I HAD PAPER AND I WAS WORKING AS HARD AND AS FAST AS I COULD.
THERE WERE FELLAS BUILDING AN APPARATUS BASED ON OUR CALCULATIONS, AND IT WAS JUST LIKE IN A MOVIE. EVERY TIME I WENT IN THERE, WELL...
IT WAS REALLY INTERESTING HOW SCIENCE PRETTY MUCH STOPPED DURING THE WAR – IT WAS MOSTLY ENGINEERING.

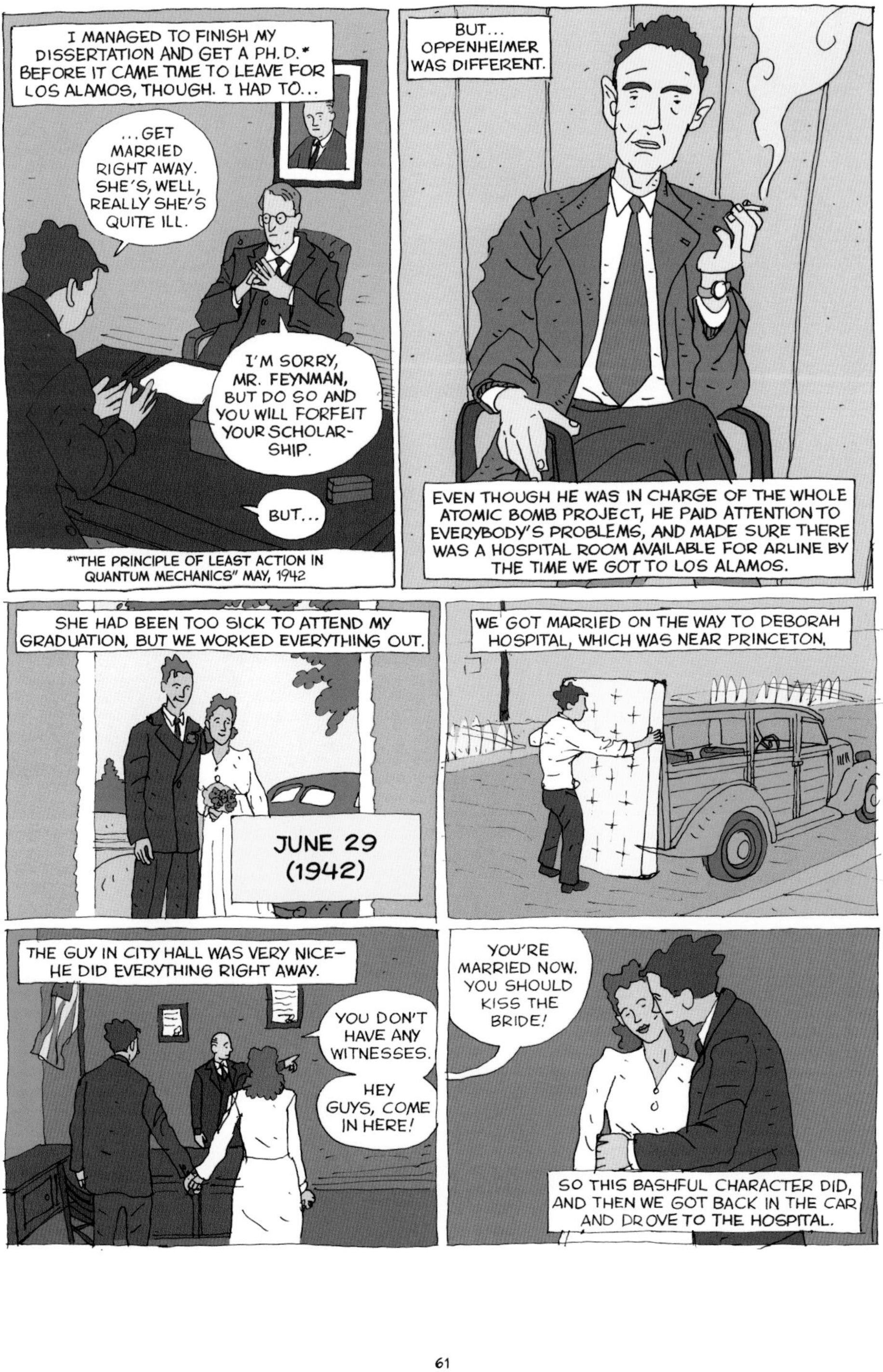
I MANAGED TO FINISH MY DISSERTATION AND GET A PH. D.* BEFORE IT CAME TIME TO LEAVE FOR LOS ALAMOS, THOUGH. I HAD TO...
...GET MARRIED RIGHT AWAY. SHE'S, WELL, REALLY SHE'S QUITE ILL.
I'M SORRY, MR. FEYNMAN, BUT DO SO AND YOU WILL FORFEIT YOUR SCHOLARSHIP.
BUT...
*"THE PRINCIPLE OF LEAST ACTION IN QUANTUM MECHANICS" MAY, 1942
BUT... OPPENHEIMER WAS DIFFERENT.
EVEN THOUGH HE WAS IN CHARGE OF THE WHOLE ATOMIC BOMB PROJECT, HE PAID ATTENTION TO EVERYBODY'S PROBLEMS, AND MADE SURE THERE WAS A HOSPITAL ROOM AVAILABLE FOR ARLINE BY THE TIME WE GOT TO LOS ALAMOS.
SHE HAD BEEN TOO SICK TO ATTEND MY GRADUATION, BUT WE WORKED EVERYTHING OUT.
JUNE 29 (1942)
WE GOT MARRIED ON THE WAY TO DEBORAH HOSPITAL, WHICH WAS NEAR PRINCETON.
THE GUY IN CITY HALL WAS VERY NICE— HE DID EVERYTHING RIGHT AWAY.
YOU DON'T HAVE ANY WITNESSES.
HEY GUYS, COME IN HERE!
YOU'RE MARRIED NOW. YOU SHOULD KISS THE BRIDE!
SO THIS BASHFUL CHARACTER DID, AND THEN WE GOT BACK IN THE CAR AND DROVE TO THE HOSPITAL.

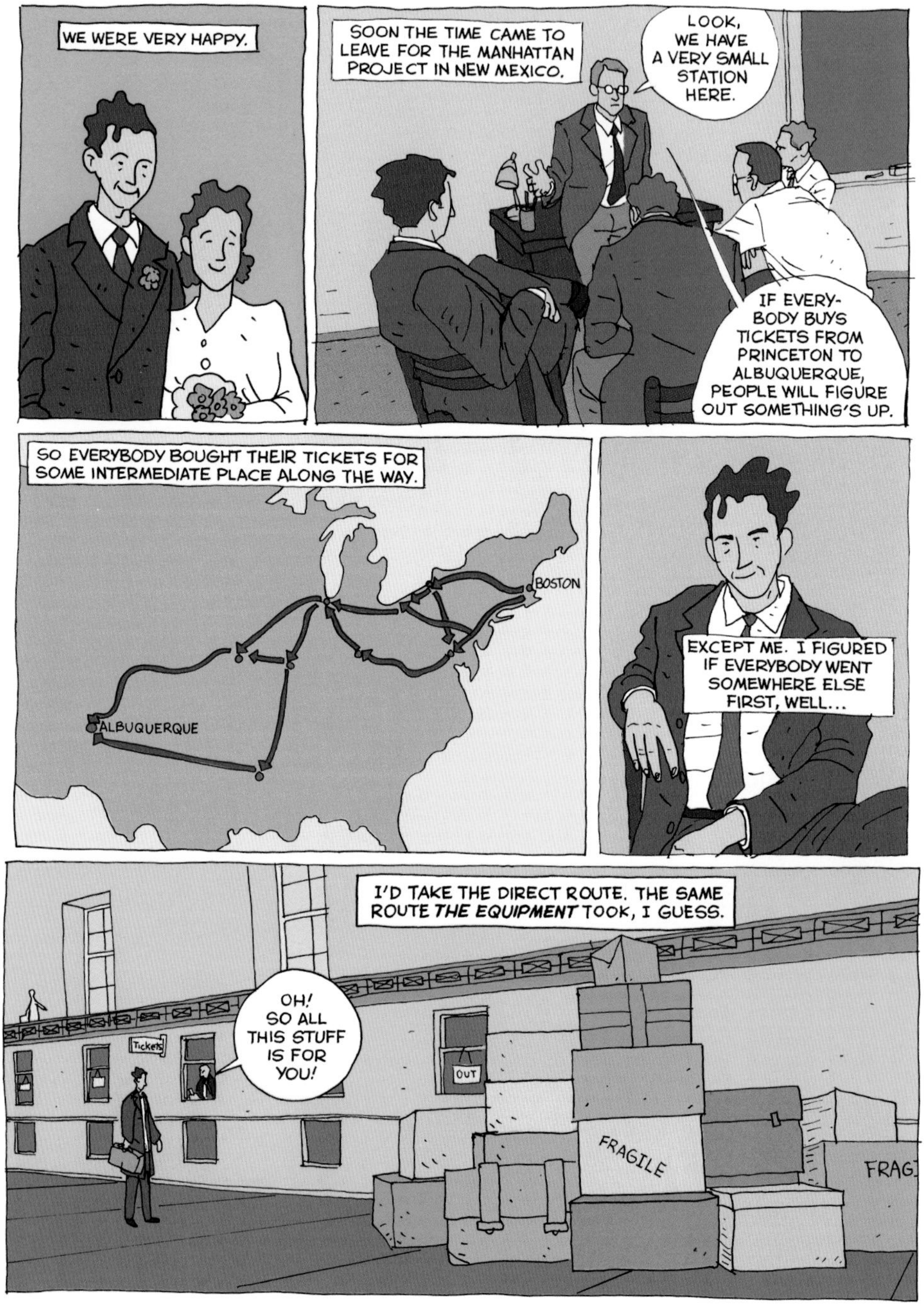

WE WERE VERY HAPPY.
SOON THE TIME CAME TO LEAVE FOR THE MANHATTAN PROJECT IN NEW MEXICO.
LOOK, WE HAVE A VERY SMALL STATION HERE.
IF EVERYBODY BUYS TICKETS FROM PRINCETON TO ALBUQUERQUE, PEOPLE WILL FIGURE OUT SOMETHING'S UP.
SO EVERYBODY BOUGHT THEIR TICKETS FOR SOME INTERMEDIATE PLACE ALONG THE WAY.
BOSTON
ALBUQUERQUE
EXCEPT ME. I FIGURED IF EVERYBODY WENT SOMEWHERE ELSE FIRST, WELL...
I'D TAKE THE DIRECT ROUTE. THE SAME ROUTE *THE EQUIPMENT* TOOK, I GUESS.
Tickets
OH! SO ALL THIS STUFF IS FOR YOU!
OUT
FRAGILE
FRAG

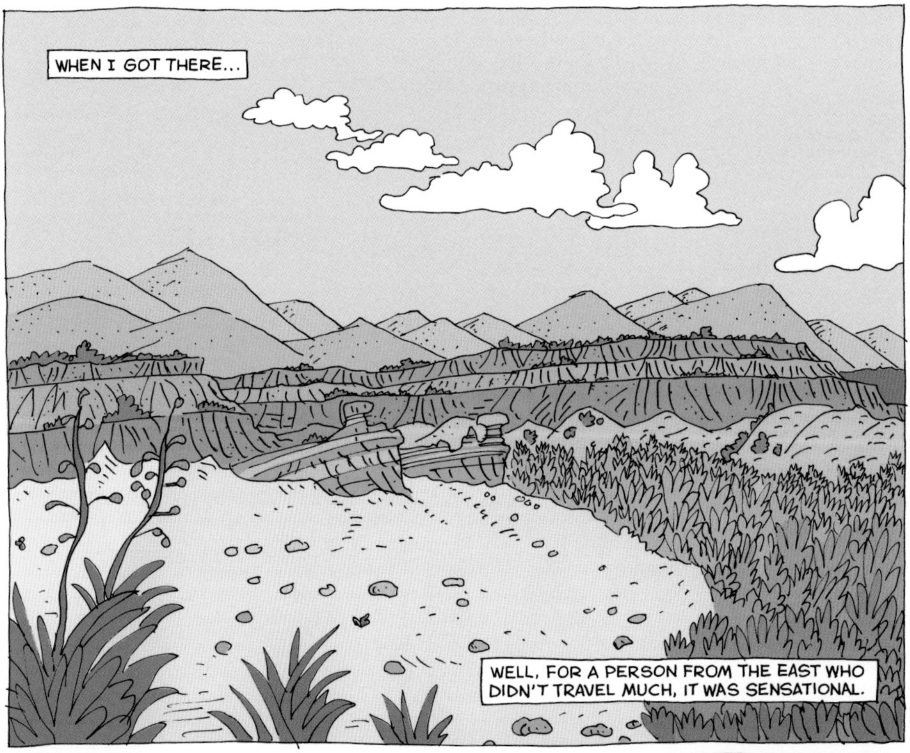
WHEN I GOT THERE...
WELL, FOR A PERSON FROM THE EAST WHO DIDN'T TRAVEL MUCH, IT WAS SENSATIONAL.

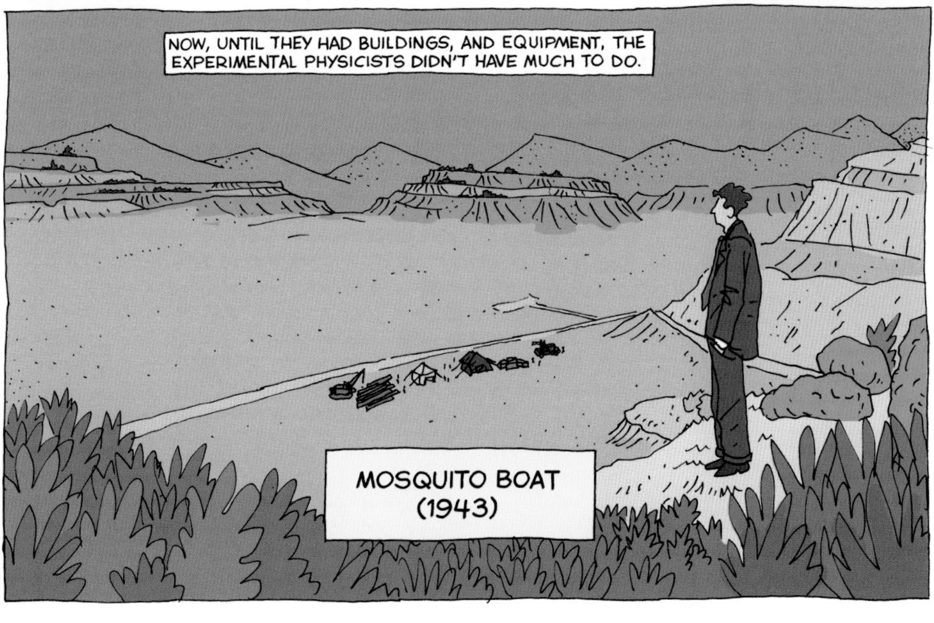
NOW, UNTIL THEY HAD BUILDINGS, AND EQUIPMENT, THE EXPERIMENTAL PHYSICISTS DIDN'T HAVE MUCH TO DO.
MOSQUITO BOAT (1943)

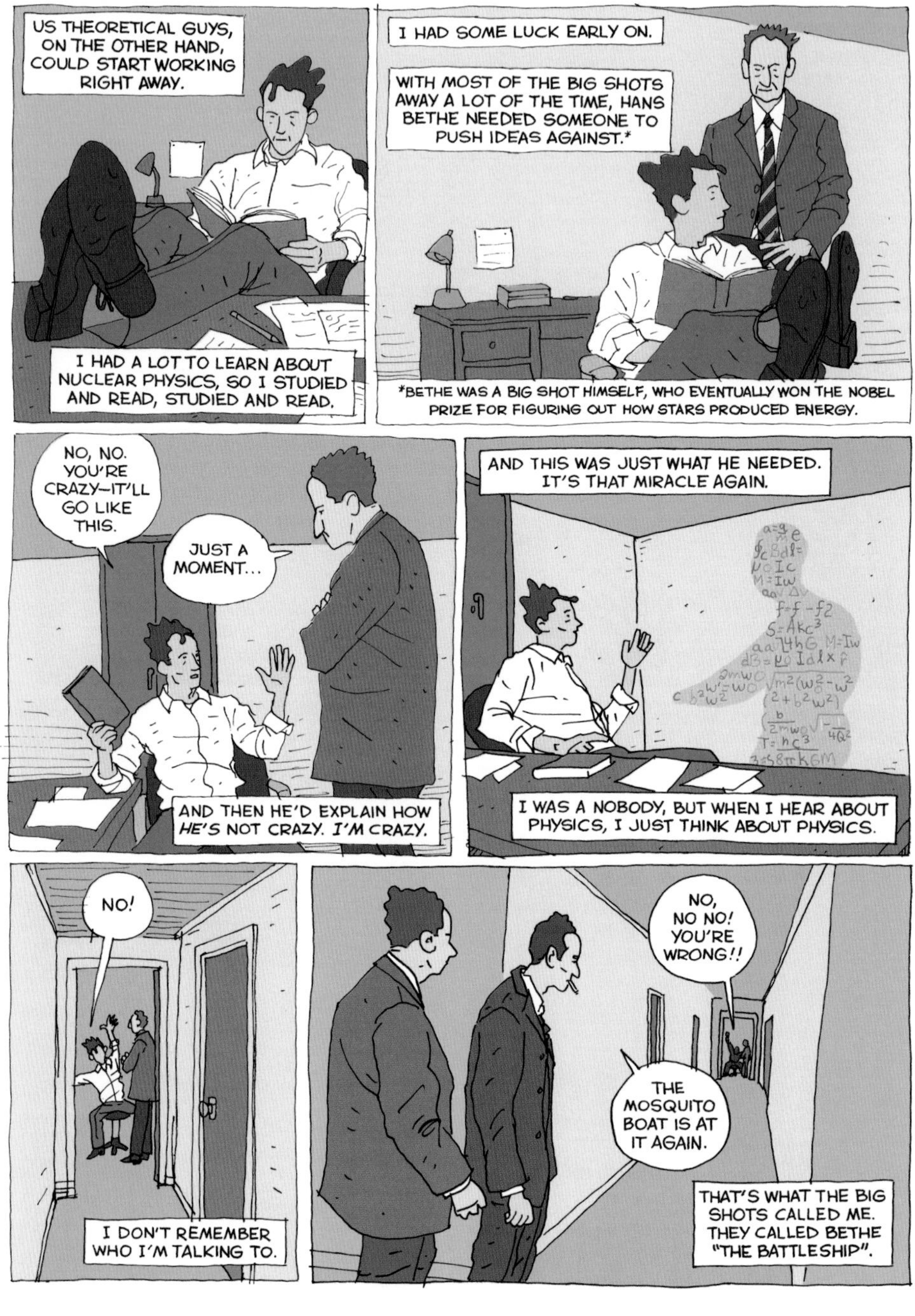
US THEORETICAL GUYS, ON THE OTHER HAND, COULD START WORKING RIGHT AWAY.
I HAD A LOT TO LEARN ABOUT NUCLEAR PHYSICS, SO I STUDIED AND READ, STUDIED AND READ.
I HAD SOME LUCK EARLY ON.
WITH MOST OF THE BIG SHOTS AWAY A LOT OF THE TIME, HANS BETHE NEEDED SOMEONE TO PUSH IDEAS AGAINST.*
*BETHE WAS A BIG SHOT HIMSELF, WHO EVENTUALLY WON THE NOBEL PRIZE FOR FIGURING OUT HOW STARS PRODUCED ENERGY.
NO, NO. YOU'RE CRAZY—IT'LL GO LIKE THIS.
JUST A MOMENT...
AND THEN HE'D EXPLAIN HOW HE'S NOT CRAZY. I'M CRAZY.
AND THIS WAS JUST WHAT HE NEEDED. IT'S THAT MIRACLE AGAIN.
I WAS A NOBODY, BUT WHEN I HEAR ABOUT PHYSICS, I JUST THINK ABOUT PHYSICS.
NO!
I DON'T REMEMBER WHO I'M TALKING TO.
NO, NO NO! YOU'RE WRONG!!
THE MOSQUITO BOAT IS AT IT AGAIN.
THAT'S WHAT THE BIG SHOTS CALLED ME. THEY CALLED BETHE "THE BATTLESHIP".

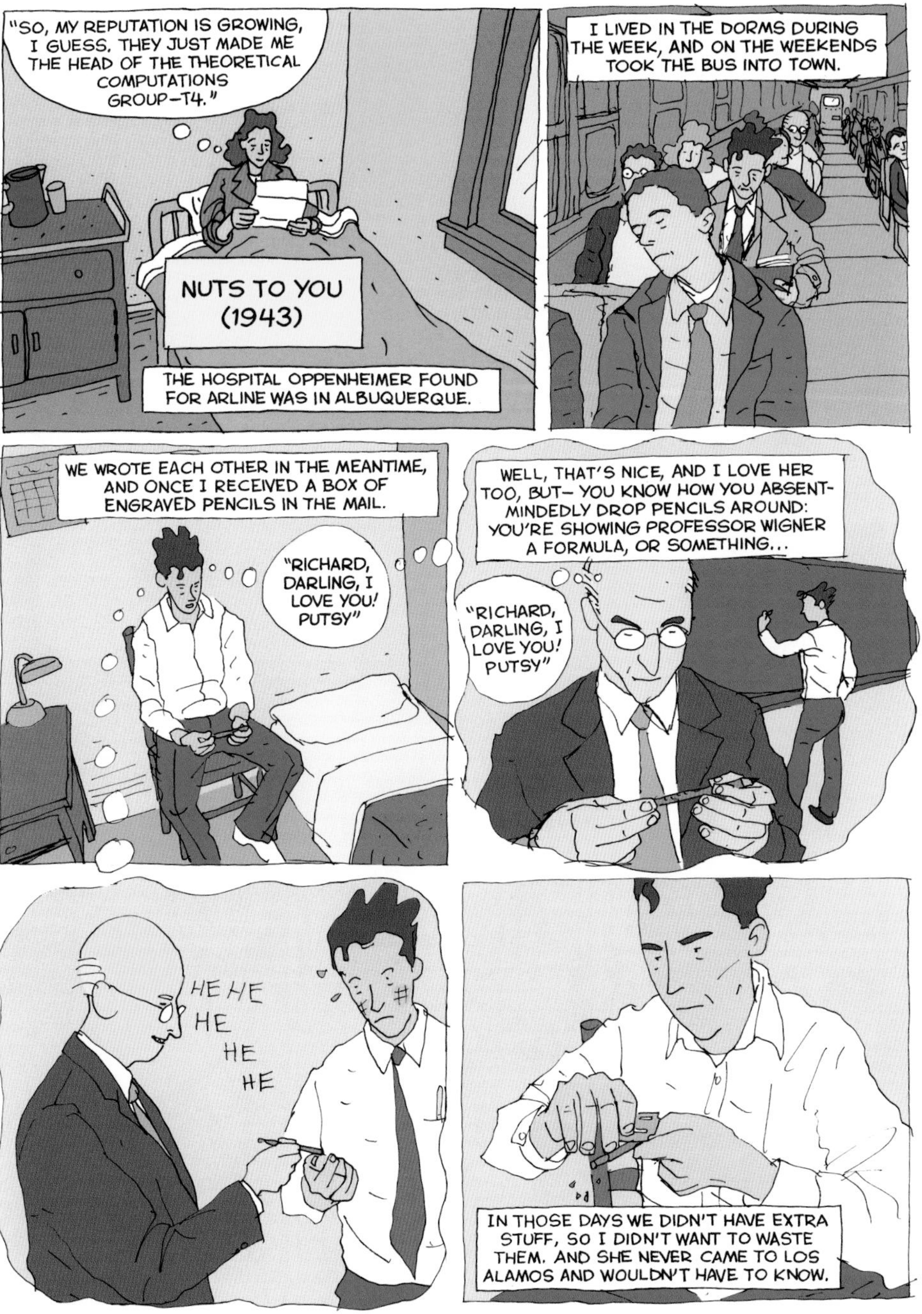
"SO, MY REPUTATION IS GROWING, I GUESS. THEY JUST MADE ME THE HEAD OF THE THEORETICAL COMPUTATIONS GROUP—T4."
NUTS TO YOU (1943)
THE HOSPITAL OPPENHEIMER FOUND FOR ARLINE WAS IN ALBUQUERQUE.
I LIVED IN THE DORMS DURING THE WEEK, AND ON THE WEEKENDS TOOK THE BUS INTO TOWN.
WE WROTE EACH OTHER IN THE MEANTIME, AND ONCE I RECEIVED A BOX OF ENGRAVED PENCILS IN THE MAIL.
"RICHARD, DARLING, I LOVE YOU! PUTSY"
WELL, THAT'S NICE, AND I LOVE HER TOO, BUT— YOU KNOW HOW YOU ABSENT-MINDEDLY DROP PENCILS AROUND: YOU'RE SHOWING PROFESSOR WIGNER A FORMULA, OR SOMETHING...
"RICHARD, DARLING, I LOVE YOU! PUTSY"
HE HE HE HE HE
IN THOSE DAYS WE DIDN'T HAVE EXTRA STUFF, SO I DIDN'T WANT TO WASTE THEM. AND SHE NEVER CAME TO LOS ALAMOS AND WOULDN'T HAVE TO KNOW.

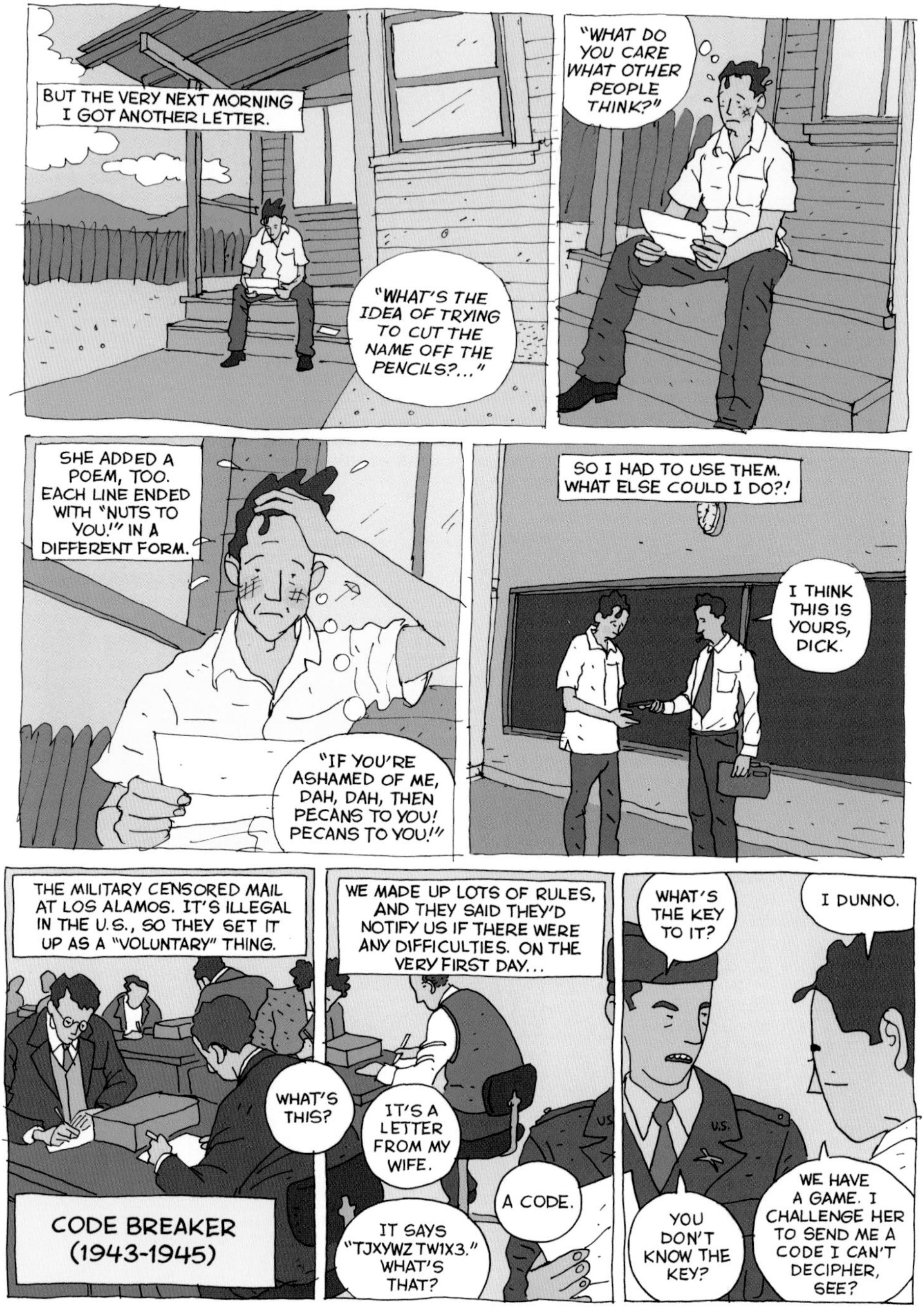

BUT THE VERY NEXT MORNING I GOT ANOTHER LETTER.
"WHAT'S THE IDEA OF TRYING TO CUT THE NAME OFF THE PENCILS?..."
"WHAT DO YOU CARE WHAT OTHER PEOPLE THINK?"
SHE ADDED A POEM, TOO. EACH LINE ENDED WITH "NUTS TO YOU!" IN A DIFFERENT FORM.
"IF YOU'RE ASHAMED OF ME, DAH, DAH, THEN PECANS TO YOU! PECANS TO YOU!"
SO I HAD TO USE THEM. WHAT ELSE COULD I DO?!
I THINK THIS IS YOURS, DICK.
THE MILITARY CENSORED MAIL AT LOS ALAMOS. IT'S ILLEGAL IN THE U.S., SO THEY SET IT UP AS A "VOLUNTARY" THING.
CODE BREAKER (1943-1945)
WE MADE UP LOTS OF RULES, AND THEY SAID THEY'D NOTIFY US IF THERE WERE ANY DIFFICULTIES. ON THE VERY FIRST DAY...
WHAT'S THIS?
IT'S A LETTER FROM MY WIFE.
IT SAYS "TJXYWZ TW1X3." WHAT'S THAT?
A CODE.
WHAT'S THE KEY TO IT?
I DUNNO.
YOU DON'T KNOW THE KEY?
WE HAVE A GAME. I CHALLENGE HER TO SEND ME A CODE I CAN'T DECIPHER, SEE?

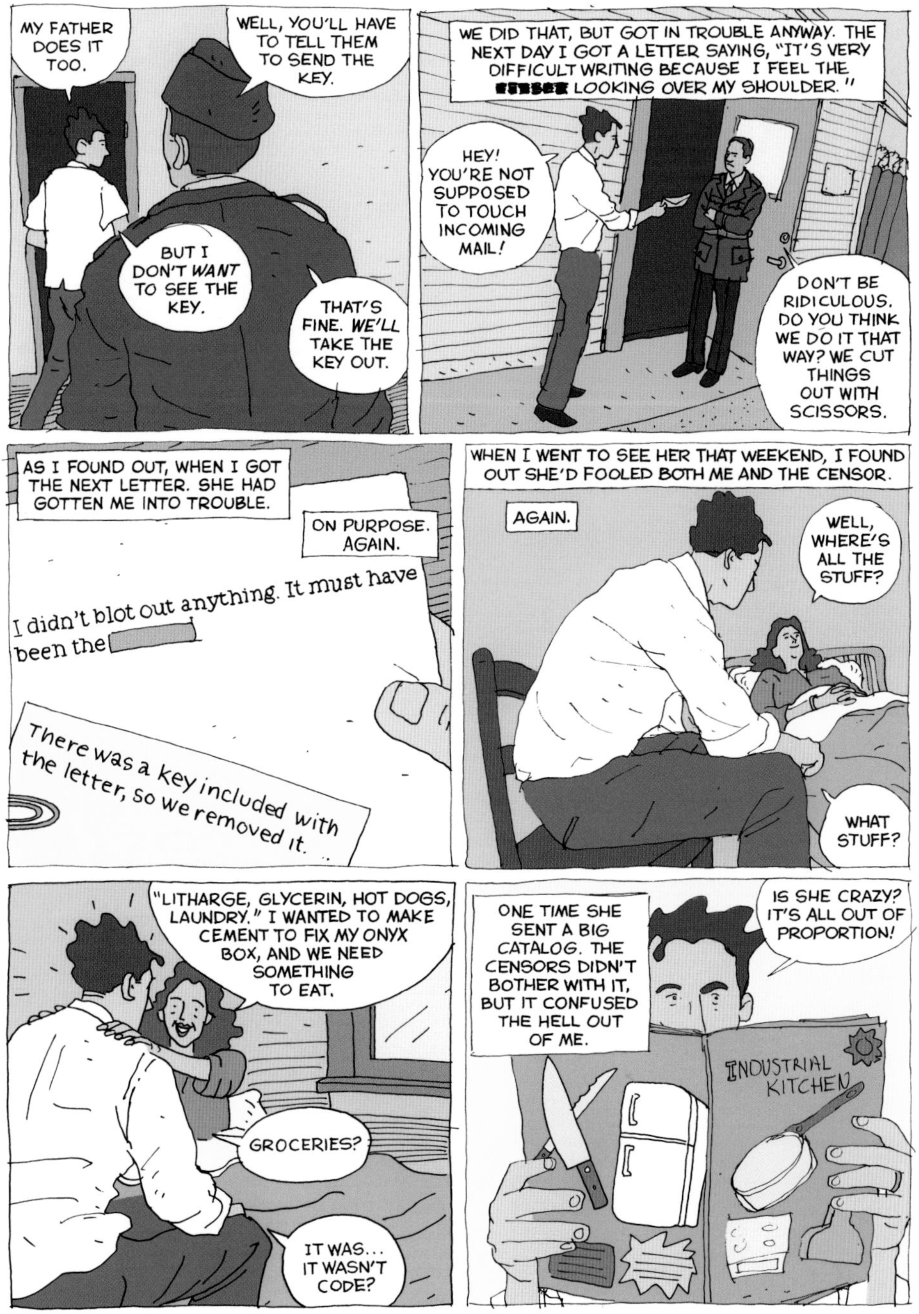
MY FATHER DOES IT TOO.
WELL, YOU'LL HAVE TO TELL THEM TO SEND THE KEY.
BUT I DON'T *WANT* TO SEE THE KEY.
THAT'S FINE. *WE'LL* TAKE THE KEY OUT.
WE DID THAT, BUT GOT IN TROUBLE ANYWAY. THE NEXT DAY I GOT A LETTER SAYING, "IT'S VERY DIFFICULT WRITING BECAUSE I FEEL THE [redacted] LOOKING OVER MY SHOULDER."
HEY! YOU'RE NOT SUPPOSED TO TOUCH INCOMING MAIL!
DON'T BE RIDICULOUS. DO YOU THINK WE DO IT THAT WAY? WE CUT THINGS OUT WITH SCISSORS.
AS I FOUND OUT, WHEN I GOT THE NEXT LETTER. SHE HAD GOTTEN ME INTO TROUBLE.
ON PURPOSE. AGAIN.
I didn't blot out anything. It must have been the
There was a key included with the letter, so we removed it.
WHEN I WENT TO SEE HER THAT WEEKEND, I FOUND OUT SHE'D FOOLED BOTH ME AND THE CENSOR.
AGAIN.
WELL, WHERE'S ALL THE STUFF?
WHAT STUFF?
"LITHARGE, GLYCERIN, HOT DOGS, LAUNDRY." I WANTED TO MAKE CEMENT TO FIX MY ONYX BOX, AND WE NEED SOMETHING TO EAT.
GROCERIES?
IT WAS... IT WASN'T CODE?
ONE TIME SHE SENT A BIG CATALOG. THE CENSORS DIDN'T BOTHER WITH IT, BUT IT CONFUSED THE HELL OUT OF ME.
IS SHE CRAZY? IT'S ALL OUT OF PROPORTION!
INDUSTRIAL KITCHEN

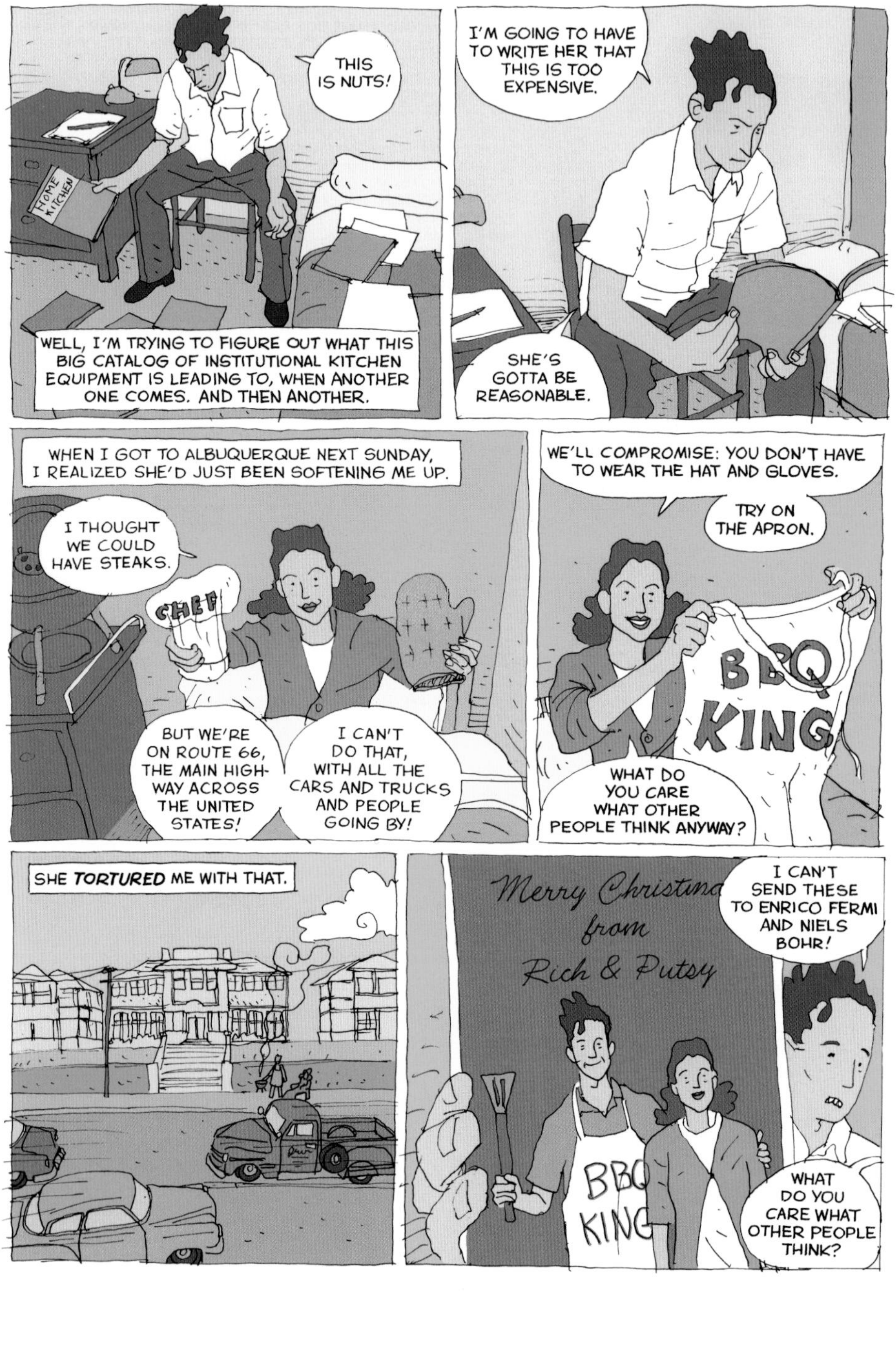
THIS IS NUTS!
WELL, I'M TRYING TO FIGURE OUT WHAT THIS BIG CATALOG OF INSTITUTIONAL KITCHEN EQUIPMENT IS LEADING TO, WHEN ANOTHER ONE COMES. AND THEN ANOTHER.
HOME KITCHEN
I'M GOING TO HAVE TO WRITE HER THAT THIS IS TOO EXPENSIVE.
SHE'S GOTTA BE REASONABLE.
WHEN I GOT TO ALBUQUERQUE NEXT SUNDAY, I REALIZED SHE'D JUST BEEN SOFTENING ME UP.
I THOUGHT WE COULD HAVE STEAKS.
CHEF
BUT WE'RE ON ROUTE 66, THE MAIN HIGHWAY ACROSS THE UNITED STATES!
I CAN'T DO THAT, WITH ALL THE CARS AND TRUCKS AND PEOPLE GOING BY!
WE'LL COMPROMISE: YOU DON'T HAVE TO WEAR THE HAT AND GLOVES.
TRY ON THE APRON.
BBQ KING
WHAT DO YOU CARE WHAT OTHER PEOPLE THINK ANYWAY?
SHE TORTURED ME WITH THAT.
Merry Christmas from Rich & Putsy
I CAN'T SEND THESE TO ENRICO FERMI AND NIELS BOHR!
BBQ KING
WHAT DO YOU CARE WHAT OTHER PEOPLE THINK?

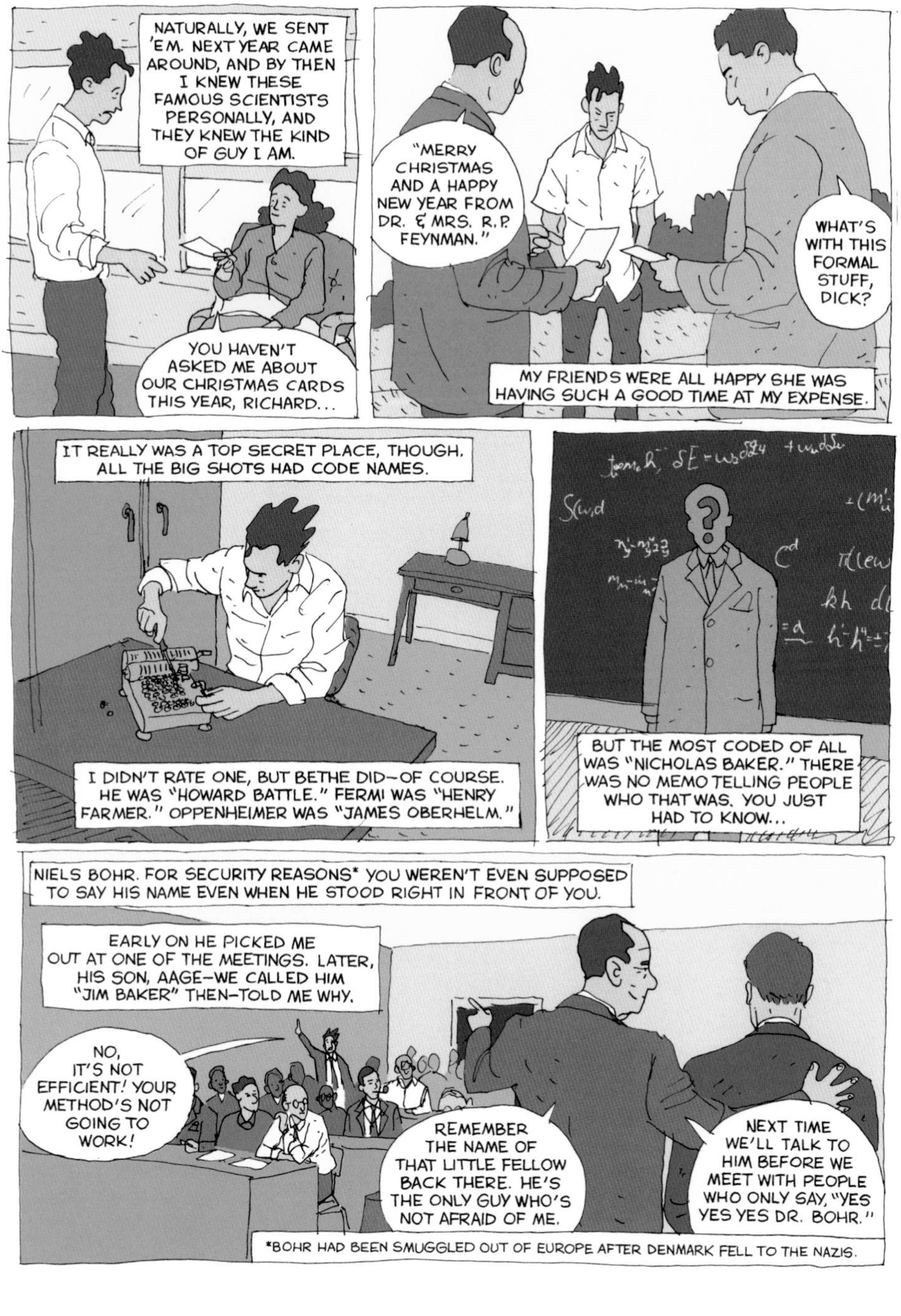
NATURALLY, WE SENT 'EM. NEXT YEAR CAME AROUND, AND BY THEN I KNEW THESE FAMOUS SCIENTISTS PERSONALLY, AND THEY KNEW THE KIND OF GUY I AM.
YOU HAVEN'T ASKED ME ABOUT OUR CHRISTMAS CARDS THIS YEAR, RICHARD...
"MERRY CHRISTMAS AND A HAPPY NEW YEAR FROM DR. & MRS. R.P. FEYNMAN."
WHAT'S WITH THIS FORMAL STUFF, DICK?
MY FRIENDS WERE ALL HAPPY SHE WAS HAVING SUCH A GOOD TIME AT MY EXPENSE.
IT REALLY WAS A TOP SECRET PLACE, THOUGH. ALL THE BIG SHOTS HAD CODE NAMES.
I DIDN'T RATE ONE, BUT BETHE DID—OF COURSE. HE WAS "HOWARD BATTLE." FERMI WAS "HENRY FARMER." OPPENHEIMER WAS "JAMES OBERHELM."
BUT THE MOST CODED OF ALL WAS "NICHOLAS BAKER." THERE WAS NO MEMO TELLING PEOPLE WHO THAT WAS. YOU JUST HAD TO KNOW...
NIELS BOHR. FOR SECURITY REASONS* YOU WEREN'T EVEN SUPPOSED TO SAY HIS NAME EVEN WHEN HE STOOD RIGHT IN FRONT OF YOU.
EARLY ON HE PICKED ME OUT AT ONE OF THE MEETINGS. LATER, HIS SON, AAGE—WE CALLED HIM "JIM BAKER" THEN—TOLD ME WHY.
NO, IT'S NOT EFFICIENT! YOUR METHOD'S NOT GOING TO WORK!
REMEMBER THE NAME OF THAT LITTLE FELLOW BACK THERE. HE'S THE ONLY GUY WHO'S NOT AFRAID OF ME.
NEXT TIME WE'LL TALK TO HIM BEFORE WE MEET WITH PEOPLE WHO ONLY SAY, "YES YES YES DR. BOHR."
*BOHR HAD BEEN SMUGGLED OUT OF EUROPE AFTER DENMARK FELL TO THE NAZIS.

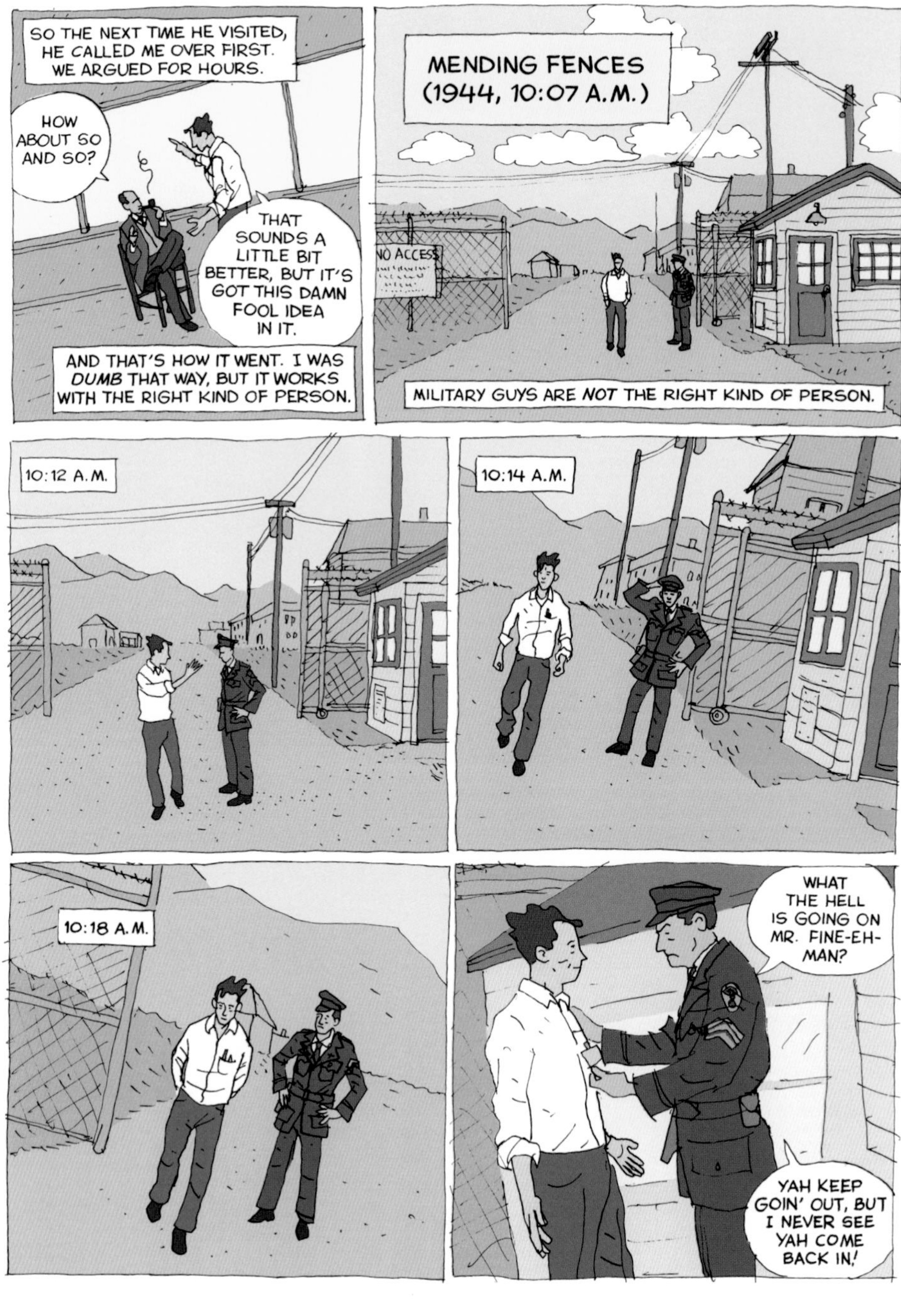
SO THE NEXT TIME HE VISITED, HE CALLED ME OVER FIRST. WE ARGUED FOR HOURS.
HOW ABOUT SO AND SO?
THAT SOUNDS A LITTLE BIT BETTER, BUT IT'S GOT THIS DAMN FOOL IDEA IN IT.
AND THAT'S HOW IT WENT. I WAS *DUMB* THAT WAY, BUT IT WORKS WITH THE RIGHT KIND OF PERSON.
MENDING FENCES (1944, 10:07 A.M.)
NO ACCESS
MILITARY GUYS ARE *NOT* THE RIGHT KIND OF PERSON.
10:12 A.M.
10:14 A.M.
10:18 A.M.
WHAT THE HELL IS GOING ON MR. FINE-EH-MAN?
YAH KEEP GOIN' OUT, BUT I NEVER SEE YAH COME BACK IN!

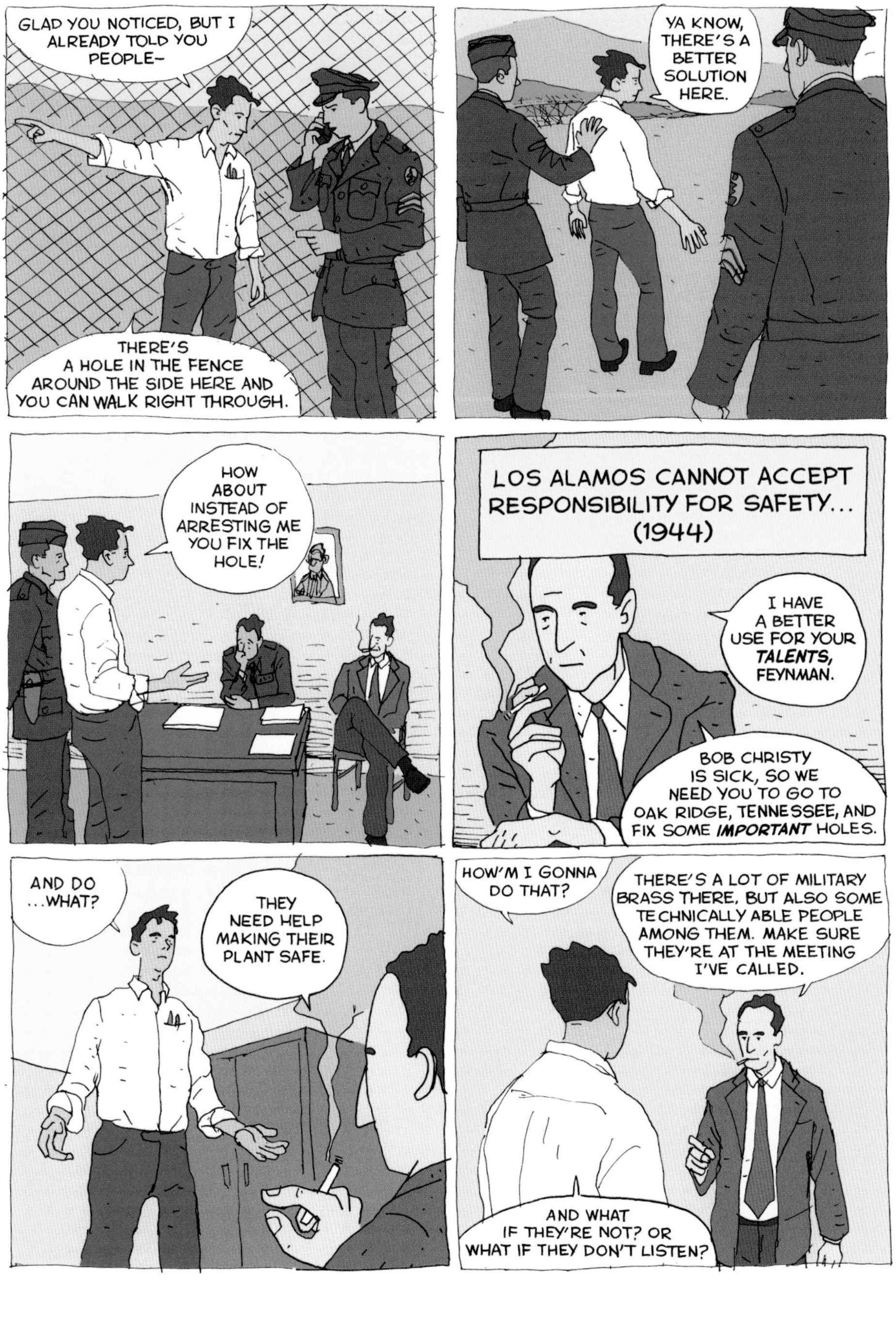
GLAD YOU NOTICED, BUT I ALREADY TOLD YOU PEOPLE–
THERE'S A HOLE IN THE FENCE AROUND THE SIDE HERE AND YOU CAN WALK RIGHT THROUGH.
YA KNOW, THERE'S A BETTER SOLUTION HERE.
HOW ABOUT INSTEAD OF ARRESTING ME YOU FIX THE HOLE!
LOS ALAMOS CANNOT ACCEPT RESPONSIBILITY FOR SAFETY... (1944)
I HAVE A BETTER USE FOR YOUR *TALENTS,* FEYNMAN.
BOB CHRISTY IS SICK, SO WE NEED YOU TO GO TO OAK RIDGE, TENNESSEE, AND FIX SOME *IMPORTANT* HOLES.
AND DO ...WHAT?
THEY NEED HELP MAKING THEIR PLANT SAFE.
HOW'M I GONNA DO THAT?
THERE'S A LOT OF MILITARY BRASS THERE, BUT ALSO SOME TECHNICALLY ABLE PEOPLE AMONG THEM. MAKE SURE THEY'RE AT THE MEETING I'VE CALLED.
AND WHAT IF THEY'RE NOT? OR WHAT IF THEY DON'T LISTEN?

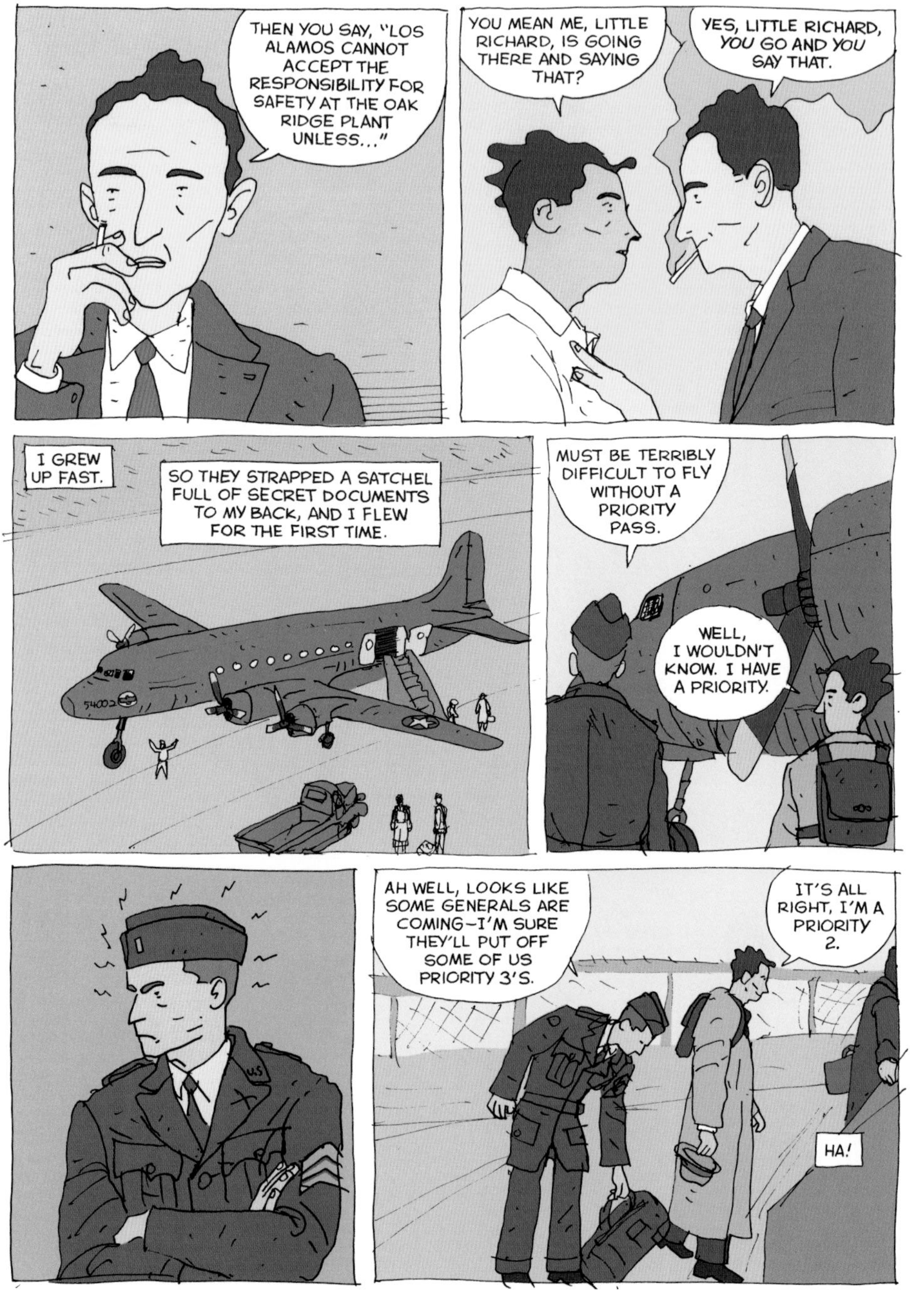
THEN YOU SAY, "LOS ALAMOS CANNOT ACCEPT THE RESPONSIBILITY FOR SAFETY AT THE OAK RIDGE PLANT UNLESS..."
YOU MEAN ME, LITTLE RICHARD, IS GOING THERE AND SAYING THAT?
YES, LITTLE RICHARD, *YOU* GO AND *YOU* SAY THAT.
I GREW UP FAST.
SO THEY STRAPPED A SATCHEL FULL OF SECRET DOCUMENTS TO MY BACK, AND I FLEW FOR THE FIRST TIME.
54002
MUST BE TERRIBLY DIFFICULT TO FLY WITHOUT A PRIORITY PASS.
WELL, I WOULDN'T KNOW. I HAVE A PRIORITY.
US
AH WELL, LOOKS LIKE SOME GENERALS ARE COMING—I'M SURE THEY'LL PUT OFF SOME OF US PRIORITY 3'S.
IT'S ALL RIGHT, I'M A PRIORITY 2.
HA!

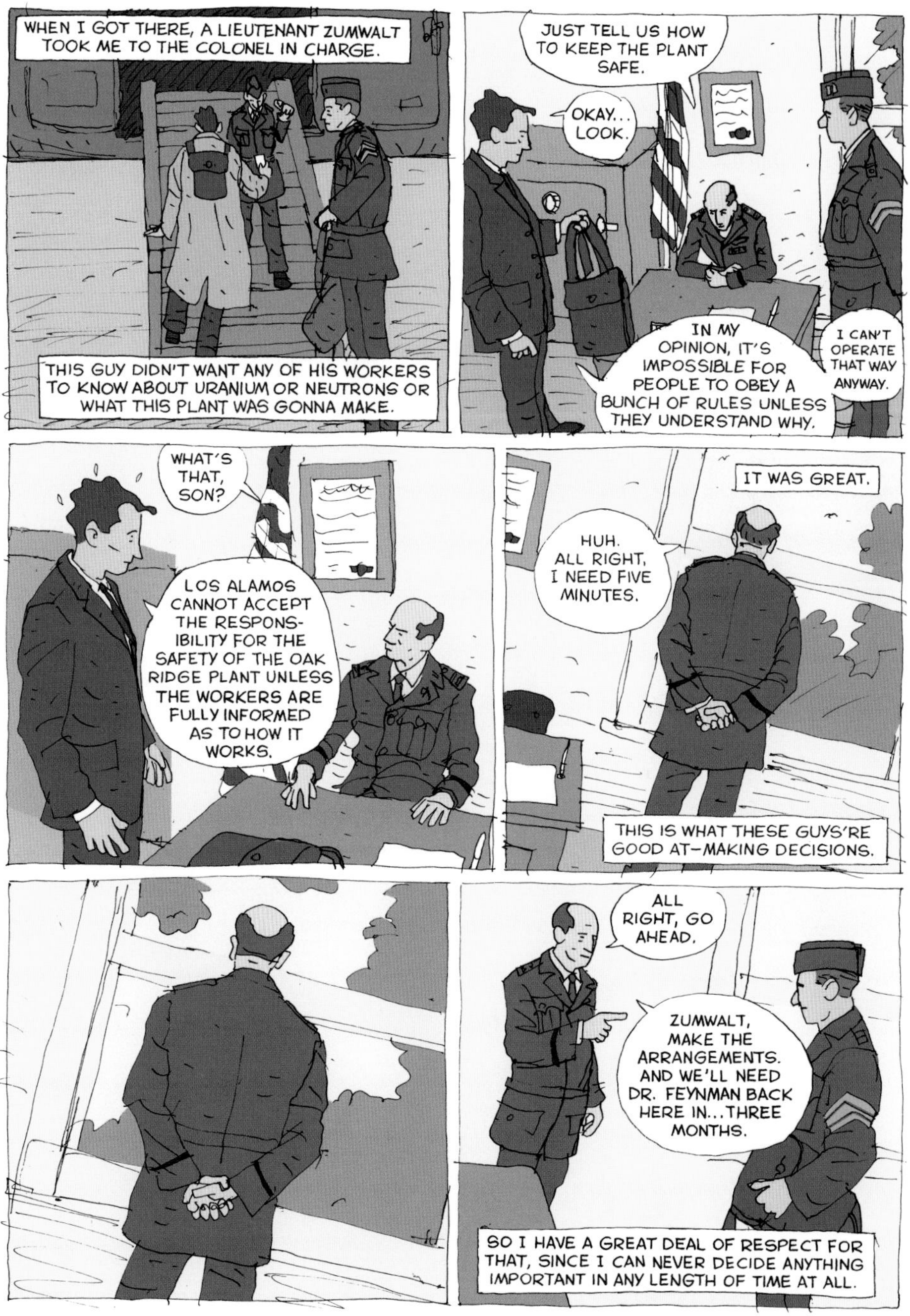
WHEN I GOT THERE, A LIEUTENANT ZUMWALT TOOK ME TO THE COLONEL IN CHARGE.
THIS GUY DIDN'T WANT ANY OF HIS WORKERS TO KNOW ABOUT URANIUM OR NEUTRONS OR WHAT THIS PLANT WAS GONNA MAKE.
JUST TELL US HOW TO KEEP THE PLANT SAFE.
OKAY... LOOK.
IN MY OPINION, IT'S IMPOSSIBLE FOR PEOPLE TO OBEY A BUNCH OF RULES UNLESS THEY UNDERSTAND WHY.
I CAN'T OPERATE THAT WAY ANYWAY.
WHAT'S THAT, SON?
LOS ALAMOS CANNOT ACCEPT THE RESPONS-IBILITY FOR THE SAFETY OF THE OAK RIDGE PLANT UNLESS THE WORKERS ARE FULLY INFORMED AS TO HOW IT WORKS.
IT WAS GREAT.
HUH. ALL RIGHT, I NEED FIVE MINUTES.
THIS IS WHAT THESE GUYS'RE GOOD AT–MAKING DECISIONS.
ALL RIGHT, GO AHEAD.
ZUMWALT, MAKE THE ARRANGEMENTS. AND WE'LL NEED DR. FEYNMAN BACK HERE IN... THREE MONTHS.
SO I HAVE A GREAT DEAL OF RESPECT FOR THAT, SINCE I CAN NEVER DECIDE ANYTHING IMPORTANT IN ANY LENGTH OF TIME AT ALL.

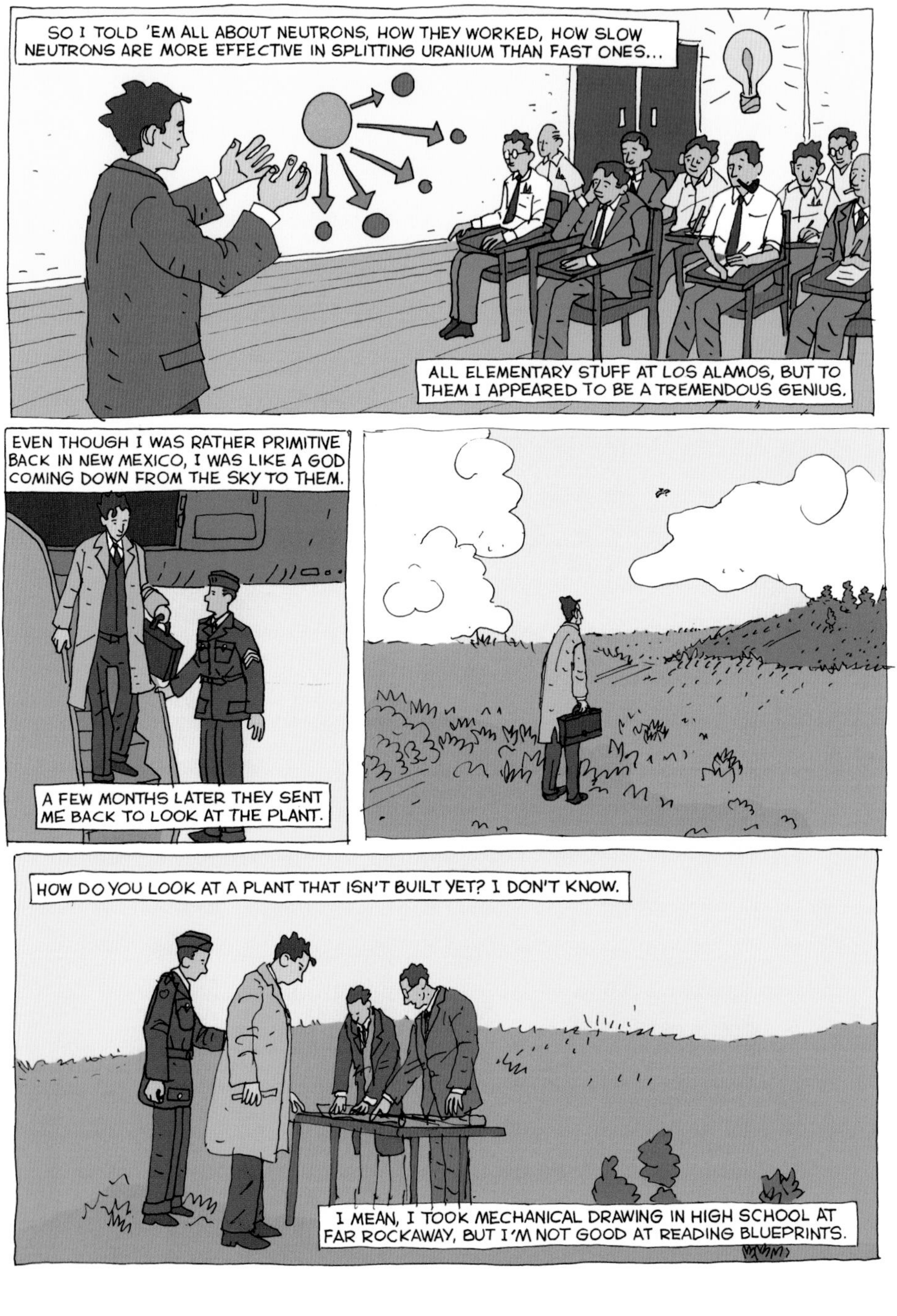
SO I TOLD 'EM ALL ABOUT NEUTRONS, HOW THEY WORKED, HOW SLOW NEUTRONS ARE MORE EFFECTIVE IN SPLITTING URANIUM THAN FAST ONES...
ALL ELEMENTARY STUFF AT LOS ALAMOS, BUT TO THEM I APPEARED TO BE A TREMENDOUS GENIUS.
EVEN THOUGH I WAS RATHER PRIMITIVE BACK IN NEW MEXICO, I WAS LIKE A GOD COMING DOWN FROM THE SKY TO THEM.
A FEW MONTHS LATER THEY SENT ME BACK TO LOOK AT THE PLANT.
HOW DO YOU LOOK AT A PLANT THAT ISN'T BUILT YET? I DON'T KNOW.
I MEAN, I TOOK MECHANICAL DRAWING IN HIGH SCHOOL AT FAR ROCKAWAY, BUT I'M NOT GOOD AT READING BLUEPRINTS.

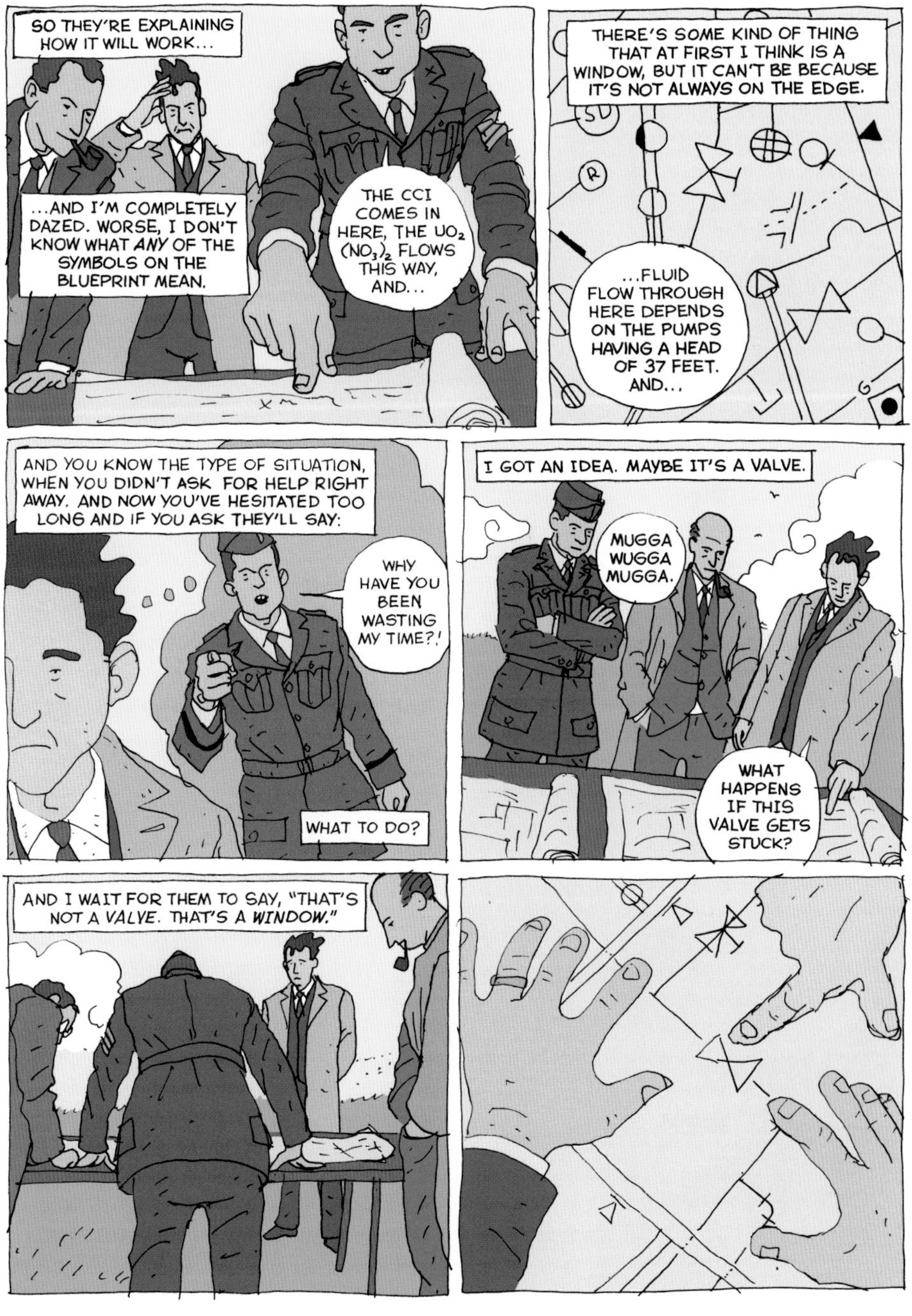
SO THEY'RE EXPLAINING HOW IT WILL WORK...
...AND I'M COMPLETELY DAZED. WORSE, I DON'T KNOW WHAT ANY OF THE SYMBOLS ON THE BLUEPRINT MEAN.
THE CCI COMES IN HERE, THE $UO_2(NO_3)_2$ FLOWS THIS WAY, AND...
THERE'S SOME KIND OF THING THAT AT FIRST I THINK IS A WINDOW, BUT IT CAN'T BE BECAUSE IT'S NOT ALWAYS ON THE EDGE.
...FLUID FLOW THROUGH HERE DEPENDS ON THE PUMPS HAVING A HEAD OF 37 FEET. AND...
AND YOU KNOW THE TYPE OF SITUATION, WHEN YOU DIDN'T ASK FOR HELP RIGHT AWAY. AND NOW YOU'VE HESITATED TOO LONG AND IF YOU ASK THEY'LL SAY:
WHY HAVE YOU BEEN WASTING MY TIME?!
WHAT TO DO?
I GOT AN IDEA. MAYBE IT'S A VALVE.
MUGGA WUGGA MUGGA.
WHAT HAPPENS IF THIS VALVE GETS STUCK?
AND I WAIT FOR THEM TO SAY, "THAT'S NOT A VALVE. THAT'S A WINDOW."

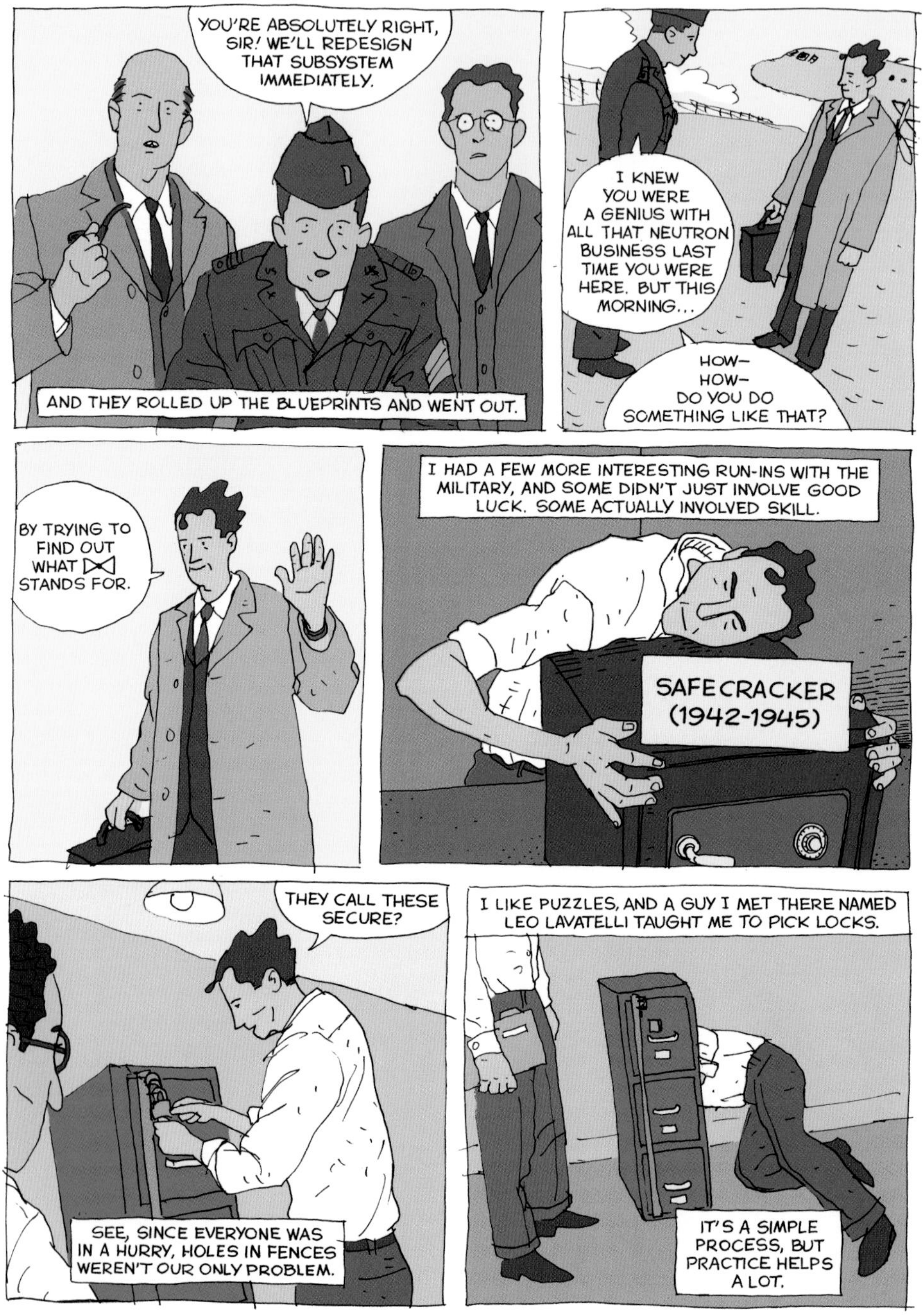
YOU'RE ABSOLUTELY RIGHT, SIR! WE'LL REDESIGN THAT SUBSYSTEM IMMEDIATELY.
AND THEY ROLLED UP THE BLUEPRINTS AND WENT OUT.
I KNEW YOU WERE A GENIUS WITH ALL THAT NEUTRON BUSINESS LAST TIME YOU WERE HERE. BUT THIS MORNING...
HOW— HOW— DO YOU DO SOMETHING LIKE THAT?
BY TRYING TO FIND OUT WHAT ⋈ STANDS FOR.
I HAD A FEW MORE INTERESTING RUN-INS WITH THE MILITARY, AND SOME DIDN'T JUST INVOLVE GOOD LUCK. SOME ACTUALLY INVOLVED SKILL.
SAFECRACKER (1942-1945)
THEY CALL THESE SECURE?
SEE, SINCE EVERYONE WAS IN A HURRY, HOLES IN FENCES WEREN'T OUR ONLY PROBLEM.
I LIKE PUZZLES, AND A GUY I MET THERE NAMED LEO LAVATELLI TAUGHT ME TO PICK LOCKS.
IT'S A SIMPLE PROCESS, BUT PRACTICE HELPS A LOT.

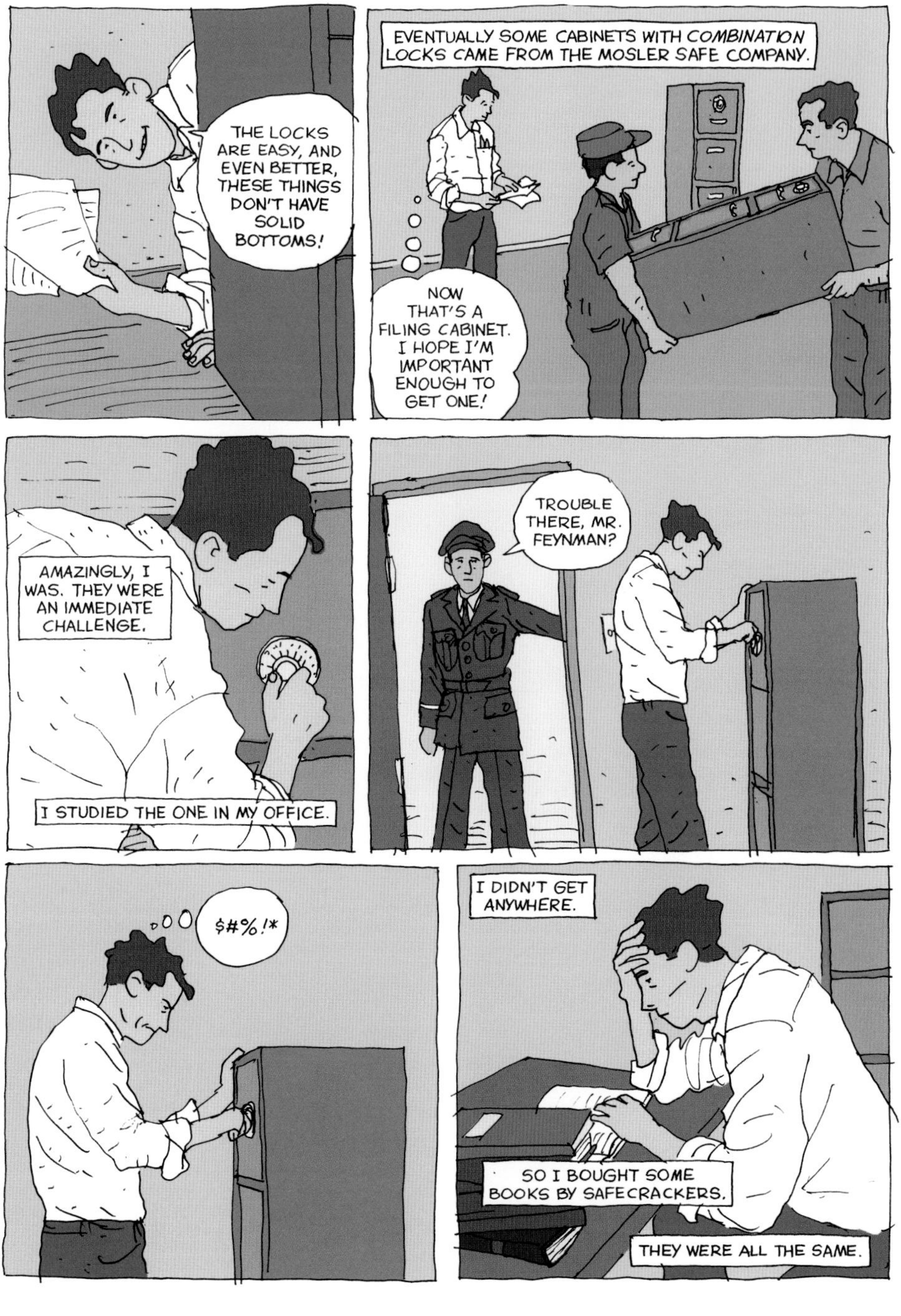
THE LOCKS ARE EASY, AND EVEN BETTER, THESE THINGS DON'T HAVE SOLID BOTTOMS!
EVENTUALLY SOME CABINETS WITH COMBINATION LOCKS CAME FROM THE MOSLER SAFE COMPANY.
NOW THAT'S A FILING CABINET. I HOPE I'M IMPORTANT ENOUGH TO GET ONE!
AMAZINGLY, I WAS. THEY WERE AN IMMEDIATE CHALLENGE.
I STUDIED THE ONE IN MY OFFICE.
TROUBLE THERE, MR. FEYNMAN?
$#%!*
I DIDN'T GET ANYWHERE.
SO I BOUGHT SOME BOOKS BY SAFECRACKERS.
THEY WERE ALL THE SAME.

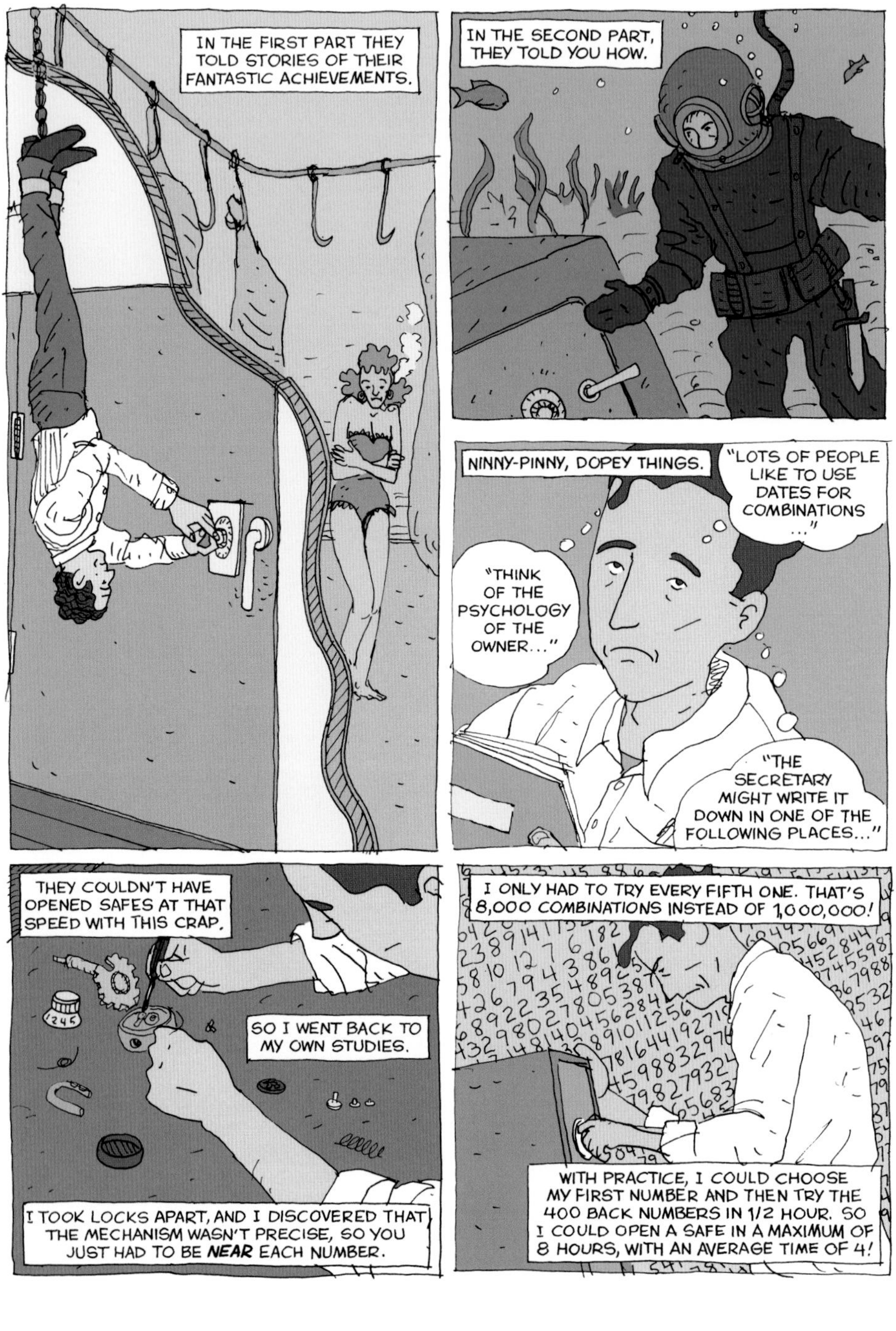
IN THE FIRST PART THEY TOLD STORIES OF THEIR FANTASTIC ACHIEVEMENTS.
IN THE SECOND PART, THEY TOLD YOU HOW.
NINNY-PINNY, DOPEY THINGS.
"LOTS OF PEOPLE LIKE TO USE DATES FOR COMBINATIONS ..."
"THINK OF THE PSYCHOLOGY OF THE OWNER..."
"THE SECRETARY MIGHT WRITE IT DOWN IN ONE OF THE FOLLOWING PLACES..."
THEY COULDN'T HAVE OPENED SAFES AT THAT SPEED WITH THIS CRAP.
SO I WENT BACK TO MY OWN STUDIES.
I TOOK LOCKS APART, AND I DISCOVERED THAT THE MECHANISM WASN'T PRECISE, SO YOU JUST HAD TO BE ***NEAR*** EACH NUMBER.
I ONLY HAD TO TRY EVERY FIFTH ONE. THAT'S 8,000 COMBINATIONS INSTEAD OF 1,000,000!
WITH PRACTICE, I COULD CHOOSE MY FIRST NUMBER AND THEN TRY THE 400 BACK NUMBERS IN 1/2 HOUR. SO I COULD OPEN A SAFE IN A MAXIMUM OF 8 HOURS, WITH AN AVERAGE TIME OF 4!

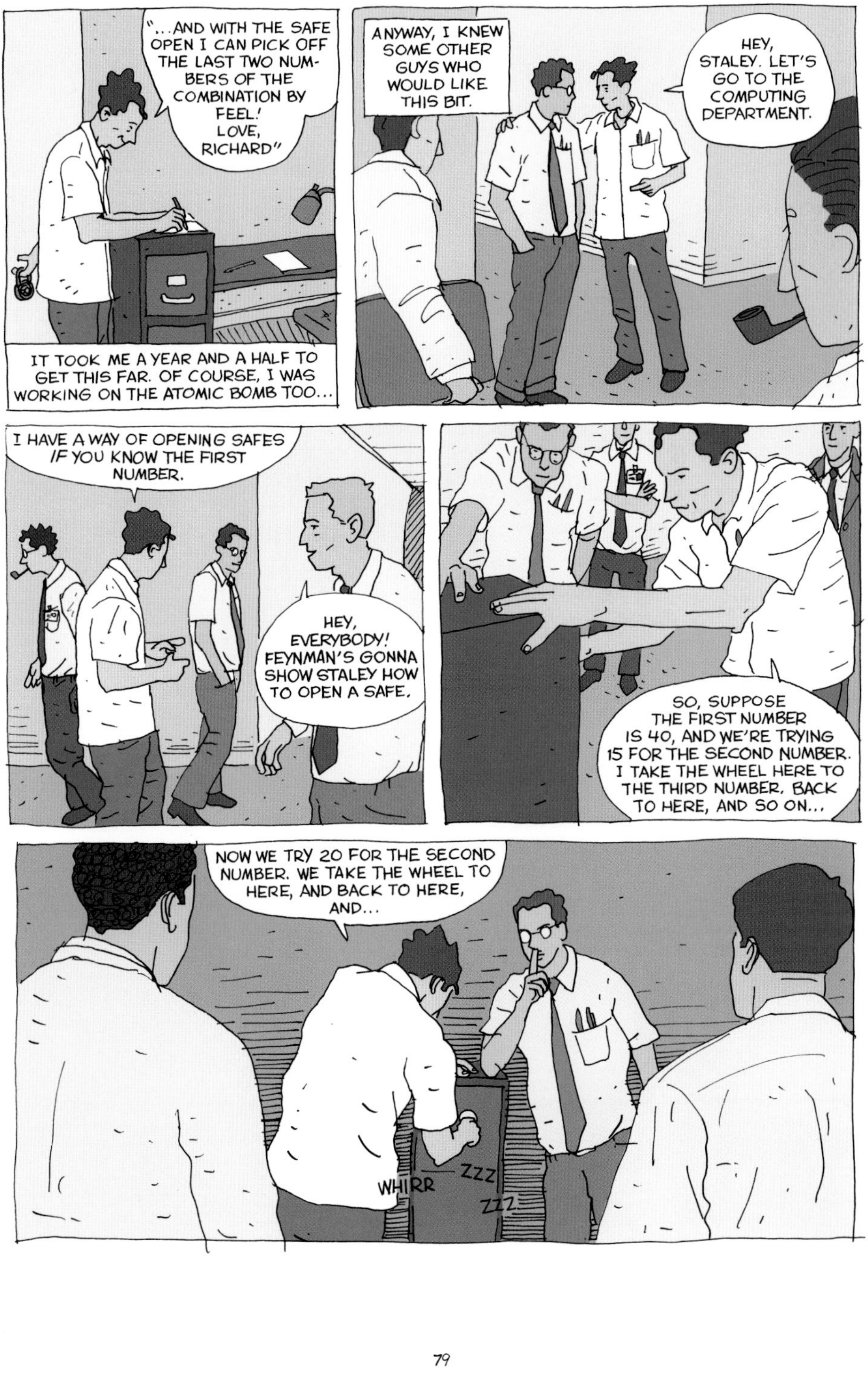
"...AND WITH THE SAFE OPEN I CAN PICK OFF THE LAST TWO NUMBERS OF THE COMBINATION BY FEEL! LOVE, RICHARD"
IT TOOK ME A YEAR AND A HALF TO GET THIS FAR. OF COURSE, I WAS WORKING ON THE ATOMIC BOMB TOO...
ANYWAY, I KNEW SOME OTHER GUYS WHO WOULD LIKE THIS BIT.
HEY, STALEY. LET'S GO TO THE COMPUTING DEPARTMENT.
I HAVE A WAY OF OPENING SAFES *IF* YOU KNOW THE FIRST NUMBER.
HEY, EVERYBODY! FEYNMAN'S GONNA SHOW STALEY HOW TO OPEN A SAFE.
SO, SUPPOSE THE FIRST NUMBER IS 40, AND WE'RE TRYING 15 FOR THE SECOND NUMBER. I TAKE THE WHEEL HERE TO THE THIRD NUMBER, BACK TO HERE, AND SO ON...
NOW WE TRY 20 FOR THE SECOND NUMBER. WE TAKE THE WHEEL TO HERE, AND BACK TO HERE, AND...
WHIRR
ZZZ
ZZZ

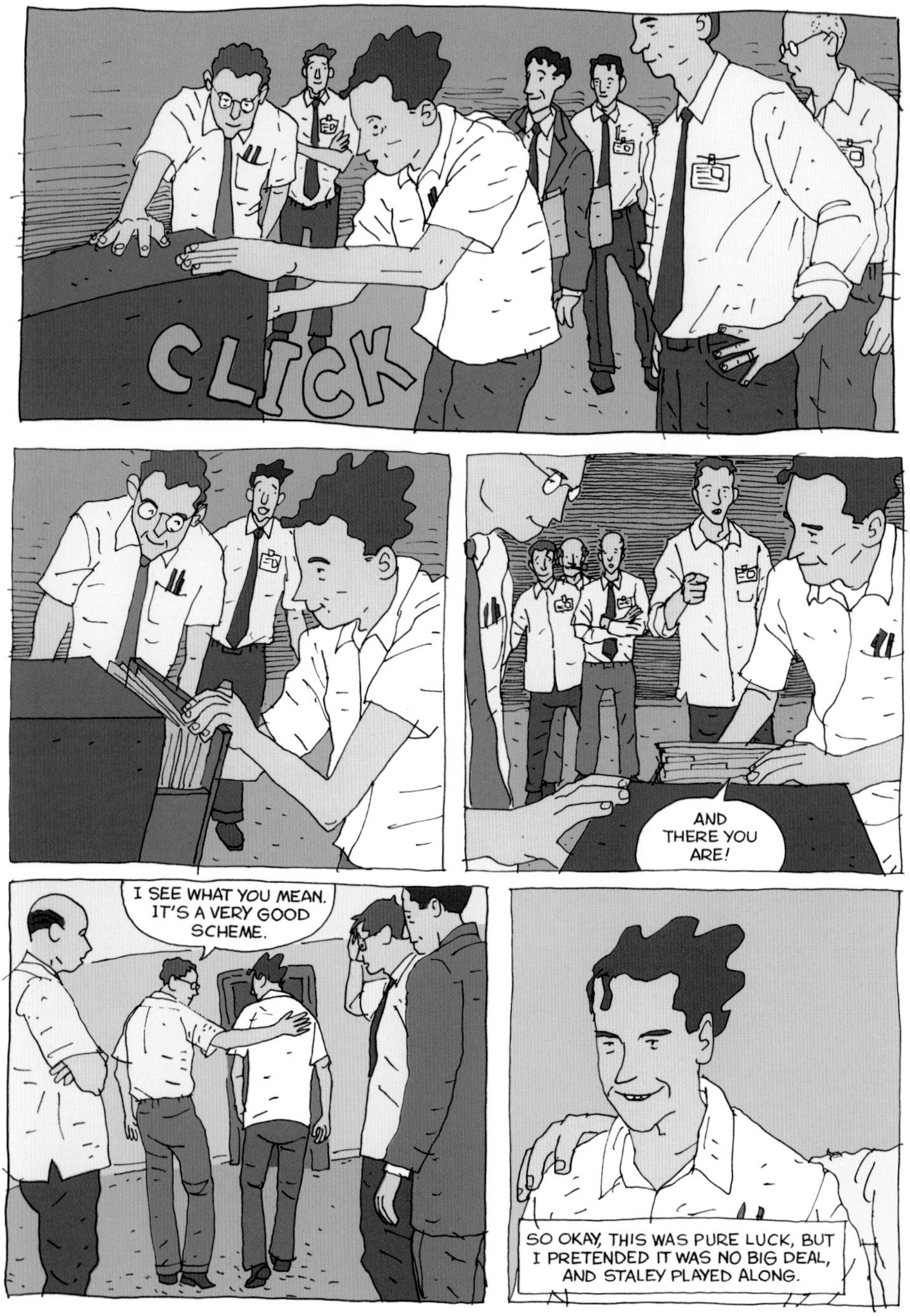
CLICK
AND THERE YOU ARE!
I SEE WHAT YOU MEAN. IT'S A VERY GOOD SCHEME.
SO OKAY, THIS WAS PURE LUCK, BUT I PRETENDED IT WAS NO BIG DEAL, AND STALEY PLAYED ALONG.

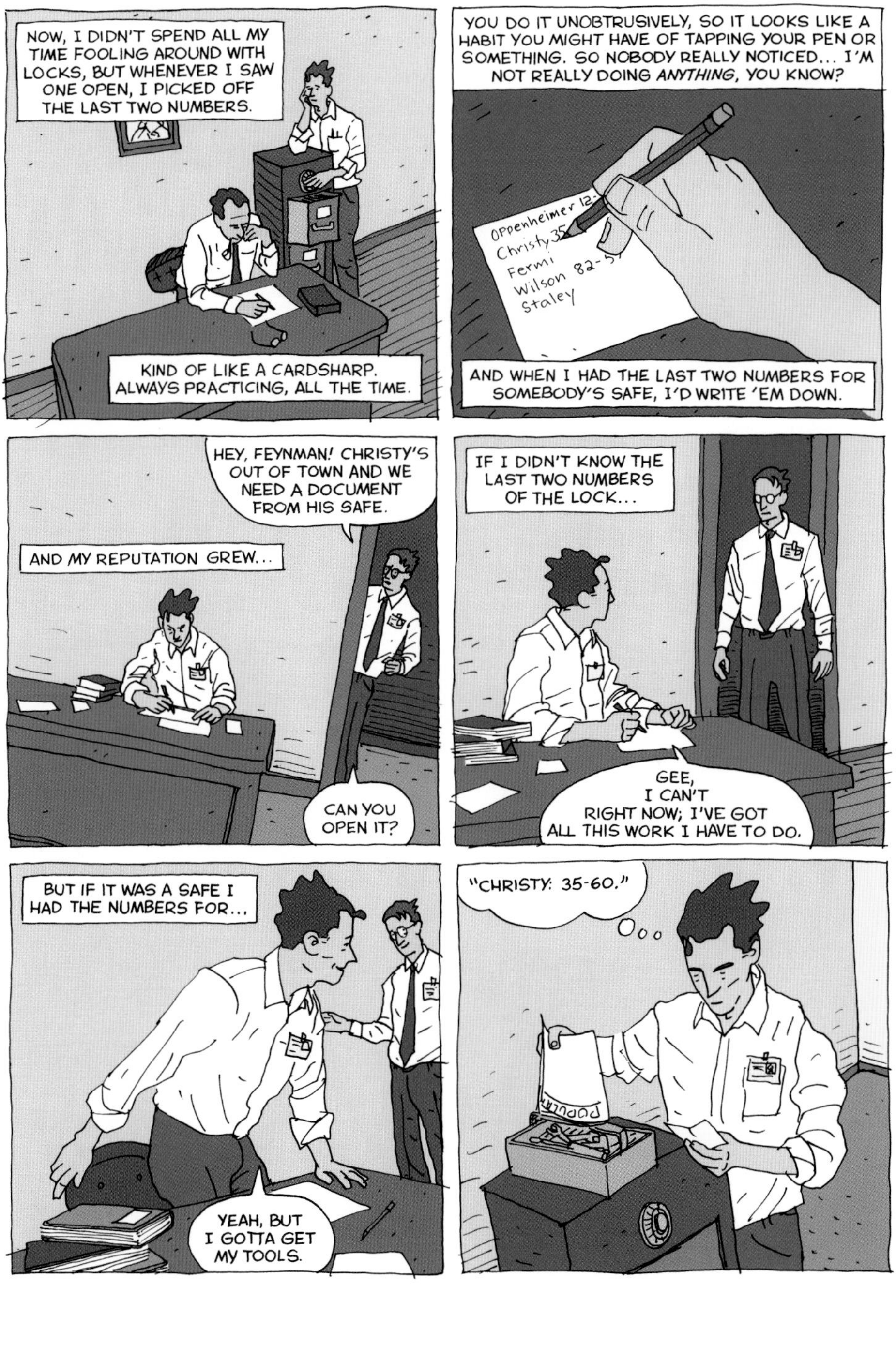
NOW, I DIDN'T SPEND ALL MY TIME FOOLING AROUND WITH LOCKS, BUT WHENEVER I SAW ONE OPEN, I PICKED OFF THE LAST TWO NUMBERS.
KIND OF LIKE A CARDSHARP. ALWAYS PRACTICING, ALL THE TIME.
YOU DO IT UNOBTRUSIVELY, SO IT LOOKS LIKE A HABIT YOU MIGHT HAVE OF TAPPING YOUR PEN OR SOMETHING. SO NOBODY REALLY NOTICED... I'M NOT REALLY DOING *ANYTHING*, YOU KNOW?
Oppenheimer 12-
Christy 35
Fermi
Wilson 82-5
Staley
AND WHEN I HAD THE LAST TWO NUMBERS FOR SOMEBODY'S SAFE, I'D WRITE 'EM DOWN.
AND MY REPUTATION GREW...
HEY, FEYNMAN! CHRISTY'S OUT OF TOWN AND WE NEED A DOCUMENT FROM HIS SAFE.
CAN YOU OPEN IT?
IF I DIDN'T KNOW THE LAST TWO NUMBERS OF THE LOCK...
GEE, I CAN'T RIGHT NOW; I'VE GOT ALL THIS WORK I HAVE TO DO.
BUT IF IT WAS A SAFE I HAD THE NUMBERS FOR...
YEAH, BUT I GOTTA GET MY TOOLS.
"CHRISTY: 35-60."

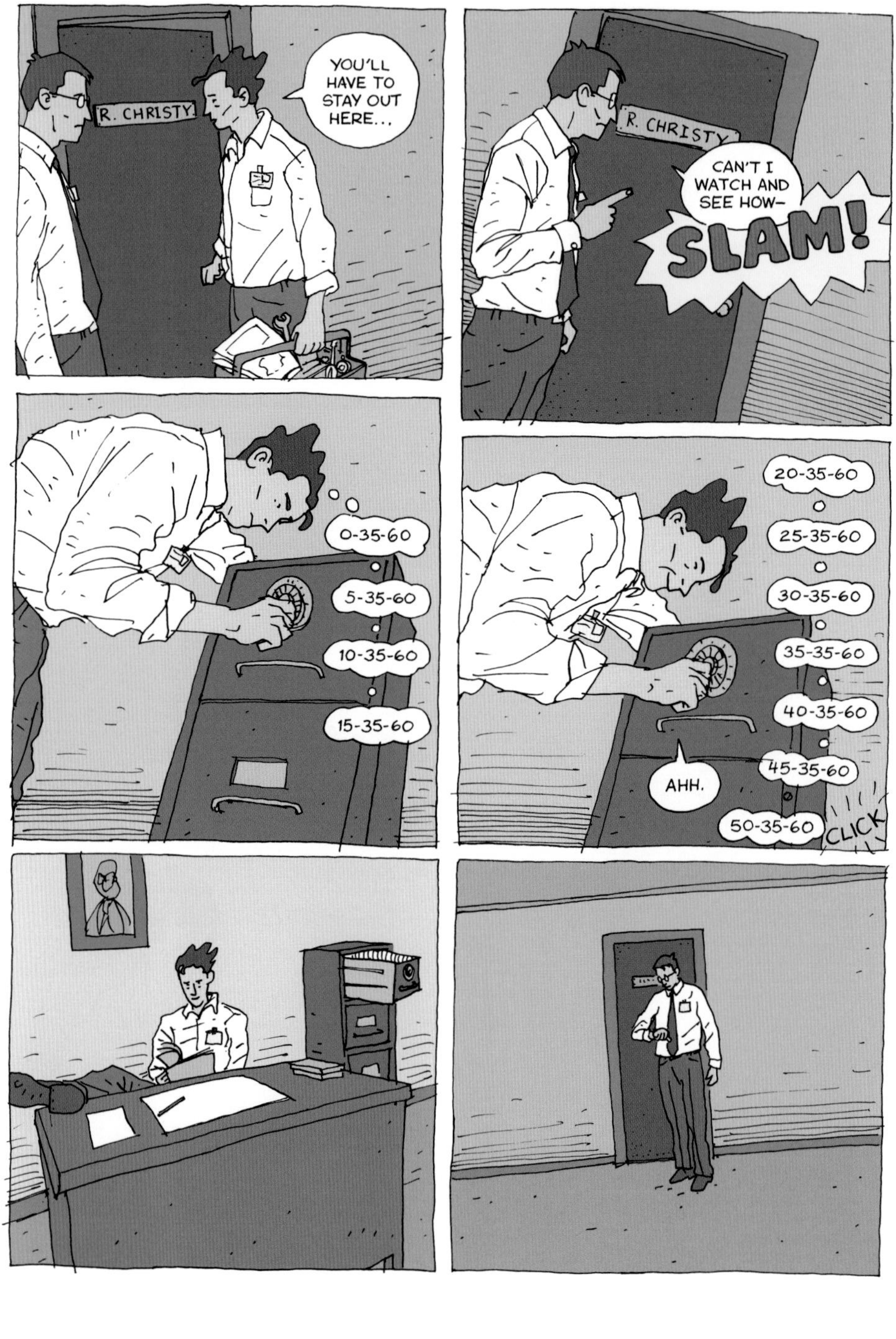
R. CHRISTY
YOU'LL HAVE TO STAY OUT HERE...
R. CHRISTY
CAN'T I WATCH AND SEE HOW–
SLAM!
0-35-60
5-35-60
10-35-60
15-35-60
20-35-60
25-35-60
30-35-60
35-35-60
40-35-60
45-35-60
AHH.
50-35-60
CLICK

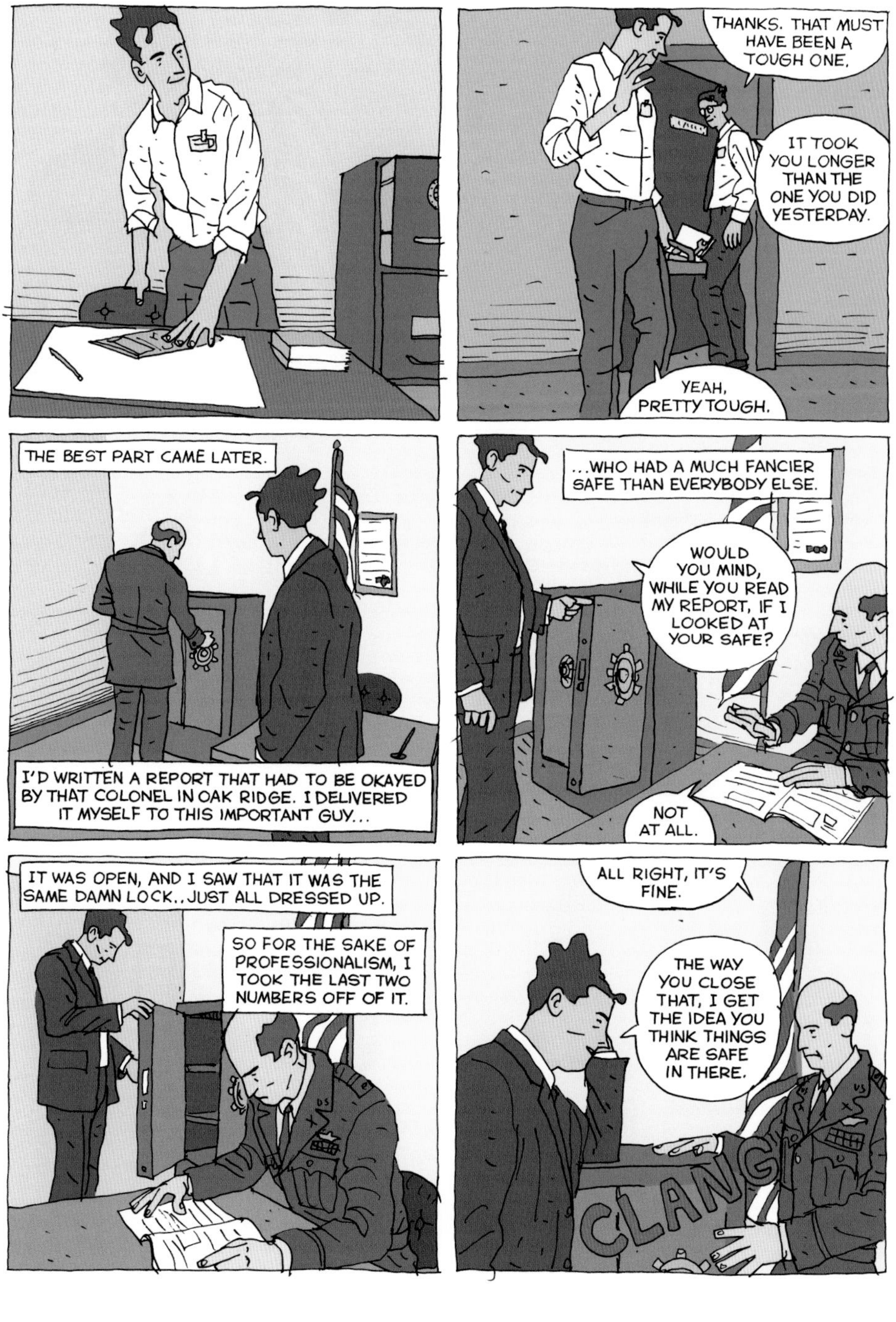
THANKS. THAT MUST HAVE BEEN A TOUGH ONE.
IT TOOK YOU LONGER THAN THE ONE YOU DID YESTERDAY.
YEAH, PRETTY TOUGH.
THE BEST PART CAME LATER.
I'D WRITTEN A REPORT THAT HAD TO BE OKAYED BY THAT COLONEL IN OAK RIDGE. I DELIVERED IT MYSELF TO THIS IMPORTANT GUY...
...WHO HAD A MUCH FANCIER SAFE THAN EVERYBODY ELSE.
WOULD YOU MIND, WHILE YOU READ MY REPORT, IF I LOOKED AT YOUR SAFE?
NOT AT ALL.
IT WAS OPEN, AND I SAW THAT IT WAS THE SAME DAMN LOCK.. JUST ALL DRESSED UP.
SO FOR THE SAKE OF PROFESSIONALISM, I TOOK THE LAST TWO NUMBERS OFF OF IT.
ALL RIGHT, IT'S FINE.
THE WAY YOU CLOSE THAT, I GET THE IDEA YOU THINK THINGS ARE SAFE IN THERE.
CLANG

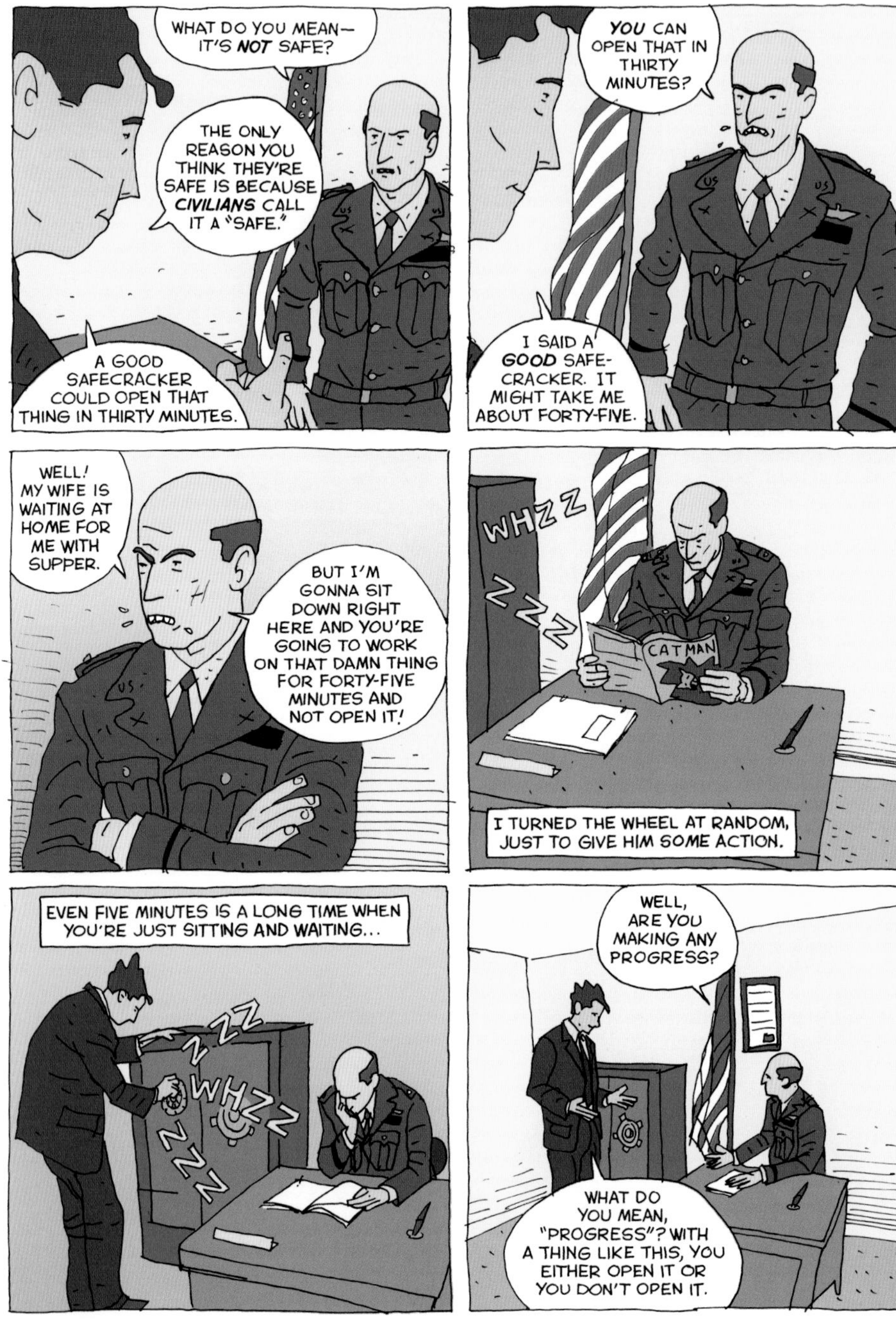
WHAT DO YOU MEAN— IT'S **NOT** SAFE?
THE ONLY REASON YOU THINK THEY'RE SAFE IS BECAUSE **CIVILIANS** CALL IT A "SAFE."
A GOOD SAFECRACKER COULD OPEN THAT THING IN THIRTY MINUTES.
YOU CAN OPEN THAT IN THIRTY MINUTES?
I SAID A **GOOD** SAFE-CRACKER. IT MIGHT TAKE ME ABOUT FORTY-FIVE.
WELL! MY WIFE IS WAITING AT HOME FOR ME WITH SUPPER.
BUT I'M GONNA SIT DOWN RIGHT HERE AND YOU'RE GOING TO WORK ON THAT DAMN THING FOR FORTY-FIVE MINUTES AND NOT OPEN IT!
WHZZ
ZZZ
CATMAN
I TURNED THE WHEEL AT RANDOM, JUST TO GIVE HIM SOME ACTION.
EVEN FIVE MINUTES IS A LONG TIME WHEN YOU'RE JUST SITTING AND WAITING...
ZZZ
WHZZ
ZZZ
WELL, ARE YOU MAKING ANY PROGRESS?
WHAT DO YOU MEAN, "PROGRESS"? WITH A THING LIKE THIS, YOU EITHER OPEN IT OR YOU DON'T OPEN IT.

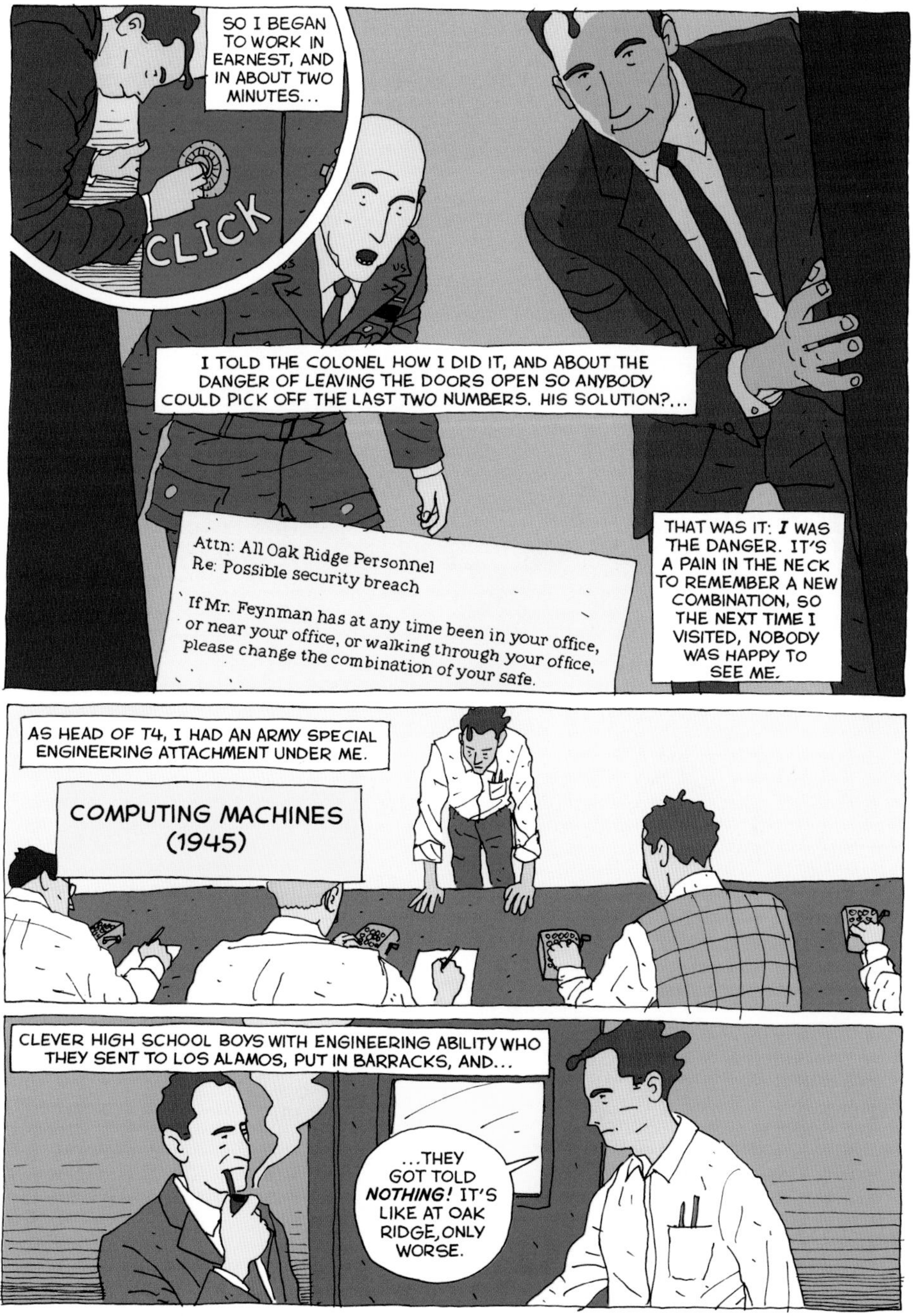

SO I BEGAN TO WORK IN EARNEST, AND IN ABOUT TWO MINUTES...
CLICK
I TOLD THE COLONEL HOW I DID IT, AND ABOUT THE DANGER OF LEAVING THE DOORS OPEN SO ANYBODY COULD PICK OFF THE LAST TWO NUMBERS. HIS SOLUTION?...
Attn: All Oak Ridge Personnel
Re: Possible security breach
If Mr. Feynman has at any time been in your office, or near your office, or walking through your office, please change the combination of your safe.
THAT WAS IT: *I* WAS THE DANGER. IT'S A PAIN IN THE NECK TO REMEMBER A NEW COMBINATION, SO THE NEXT TIME I VISITED, NOBODY WAS HAPPY TO SEE ME.
AS HEAD OF T4, I HAD AN ARMY SPECIAL ENGINEERING ATTACHMENT UNDER ME.
COMPUTING MACHINES (1945)
CLEVER HIGH SCHOOL BOYS WITH ENGINEERING ABILITY WHO THEY SENT TO LOS ALAMOS, PUT IN BARRACKS, AND...
...THEY GOT TOLD ***NOTHING!*** IT'S LIKE AT OAK RIDGE, ONLY WORSE.

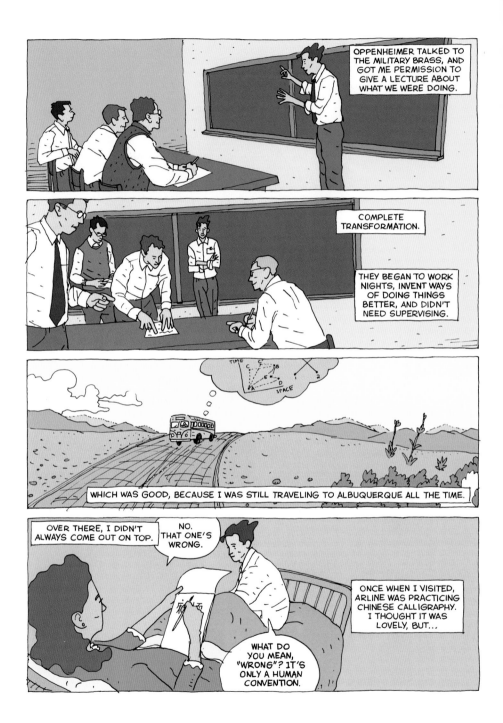

OPPENHEIMER TALKED TO THE MILITARY BRASS, AND GOT ME PERMISSION TO GIVE A LECTURE ABOUT WHAT WE WERE DOING.

COMPLETE TRANSFORMATION.

THEY BEGAN TO WORK NIGHTS, INVENT WAYS OF DOING THINGS BETTER, AND DIDN'T NEED SUPERVISING.

WHICH WAS GOOD, BECAUSE I WAS STILL TRAVELING TO ALBUQUERQUE ALL THE TIME.

OVER THERE, I DIDN'T ALWAYS COME OUT ON TOP.

NO. THAT ONE'S WRONG.

ONCE WHEN I VISITED, ARLINE WAS PRACTICING CHINESE CALLIGRAPHY. I THOUGHT IT WAS LOVELY, BUT...

WHAT DO YOU MEAN, "WRONG"? IT'S ONLY A HUMAN CONVENTION.

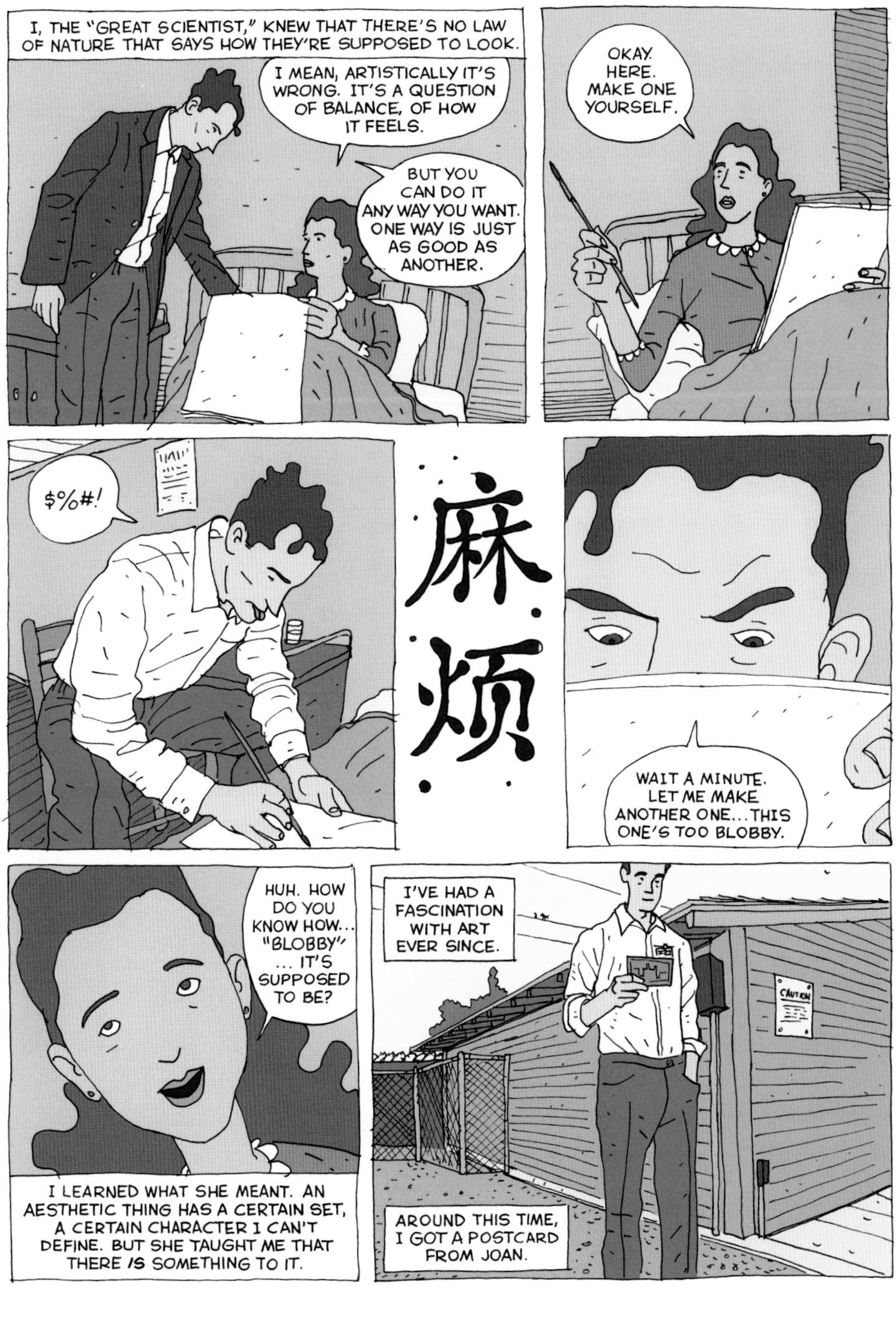
I, THE "GREAT SCIENTIST," KNEW THAT THERE'S NO LAW OF NATURE THAT SAYS HOW THEY'RE SUPPOSED TO LOOK.
I MEAN, ARTISTICALLY IT'S WRONG. IT'S A QUESTION OF BALANCE, OF HOW IT FEELS.
BUT YOU CAN DO IT ANY WAY YOU WANT. ONE WAY IS JUST AS GOOD AS ANOTHER.
OKAY. HERE. MAKE ONE YOURSELF.
$%#!
麻烦
WAIT A MINUTE. LET ME MAKE ANOTHER ONE... THIS ONE'S TOO BLOBBY.
HUH. HOW DO YOU KNOW HOW... "BLOBBY" ... IT'S SUPPOSED TO BE?
I LEARNED WHAT SHE MEANT. AN AESTHETIC THING HAS A CERTAIN SET, A CERTAIN CHARACTER I CAN'T DEFINE. BUT SHE TAUGHT ME THAT THERE *IS* SOMETHING TO IT.
I'VE HAD A FASCINATION WITH ART EVER SINCE.
CAUTION
AROUND THIS TIME, I GOT A POSTCARD FROM JOAN.

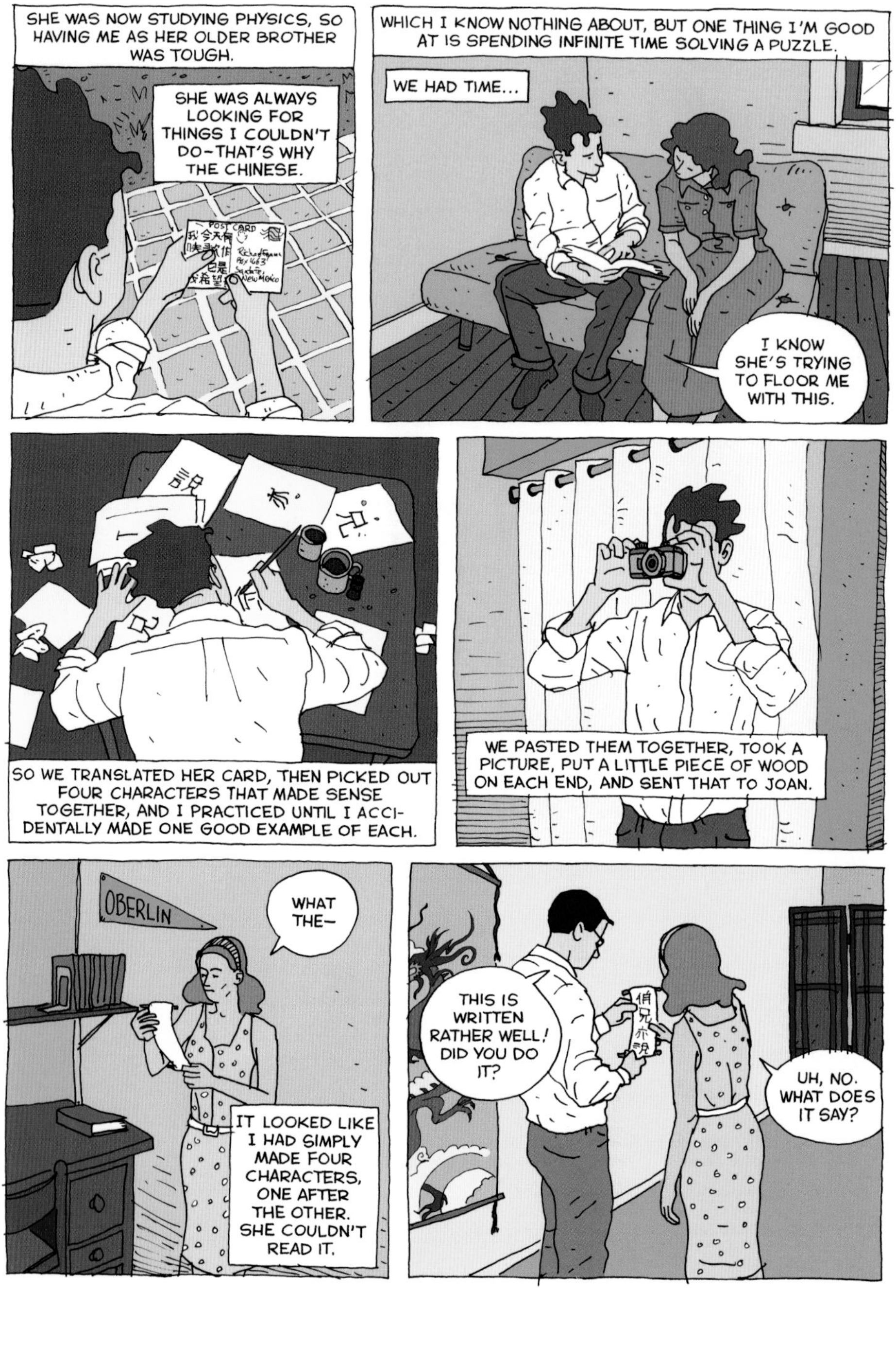
SHE WAS NOW STUDYING PHYSICS, SO HAVING ME AS HER OLDER BROTHER WAS TOUGH.
SHE WAS ALWAYS LOOKING FOR THINGS I COULDN'T DO–THAT'S WHY THE CHINESE.
POST CARD
WHICH I KNOW NOTHING ABOUT, BUT ONE THING I'M GOOD AT IS SPENDING INFINITE TIME SOLVING A PUZZLE.
WE HAD TIME...
I KNOW SHE'S TRYING TO FLOOR ME WITH THIS.
SO WE TRANSLATED HER CARD, THEN PICKED OUT FOUR CHARACTERS THAT MADE SENSE TOGETHER, AND I PRACTICED UNTIL I ACCIDENTALLY MADE ONE GOOD EXAMPLE OF EACH.
WE PASTED THEM TOGETHER, TOOK A PICTURE, PUT A LITTLE PIECE OF WOOD ON EACH END, AND SENT THAT TO JOAN.
OBERLIN
WHAT THE–
IT LOOKED LIKE I HAD SIMPLY MADE FOUR CHARACTERS, ONE AFTER THE OTHER. SHE COULDN'T READ IT.
THIS IS WRITTEN RATHER WELL! DID YOU DO IT?
UH, NO. WHAT DOES IT SAY?

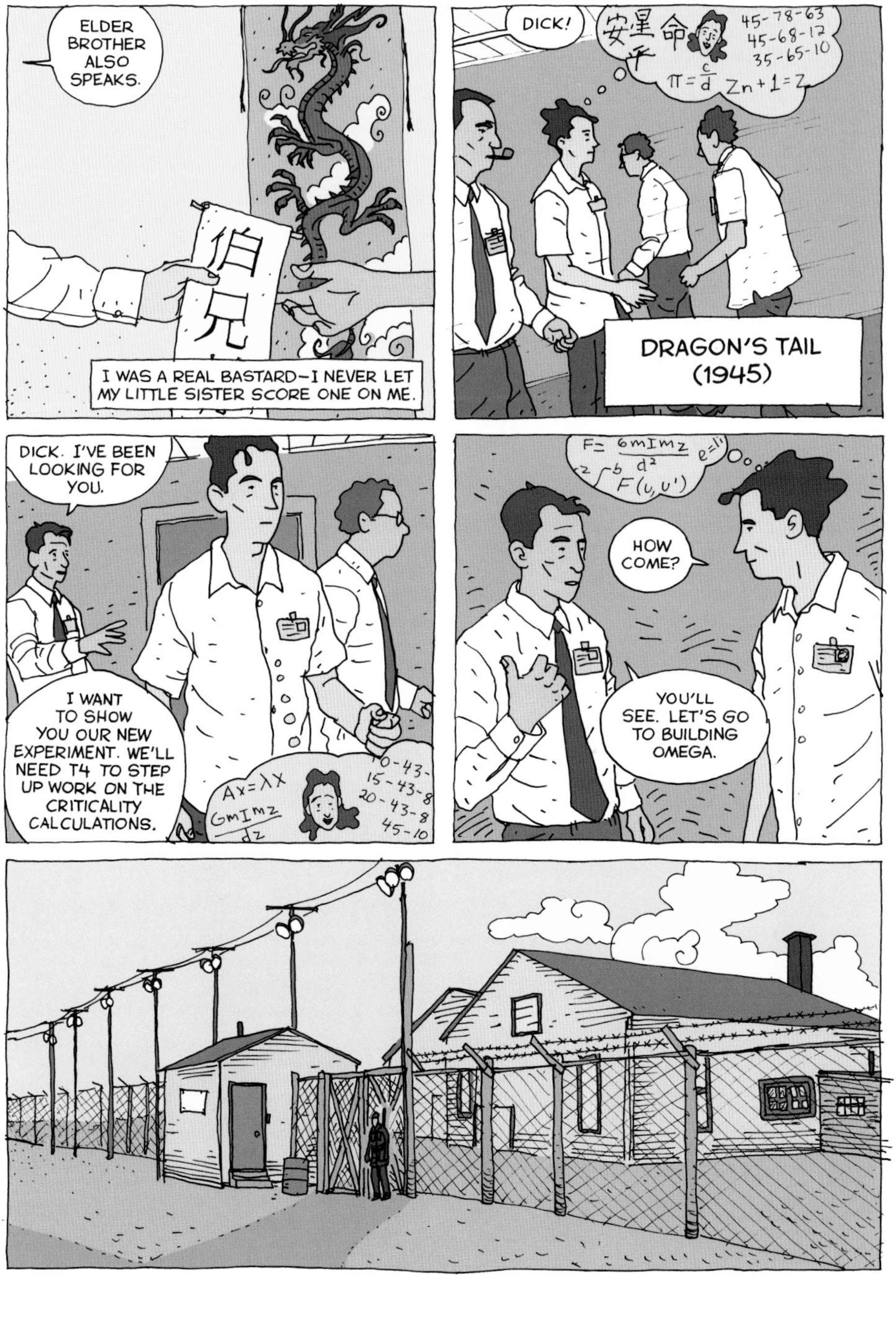
ELDER BROTHER ALSO SPEAKS.
伯兄
I WAS A REAL BASTARD—I NEVER LET MY LITTLE SISTER SCORE ONE ON ME.
DICK!
DRAGON'S TAIL (1945)
DICK. I'VE BEEN LOOKING FOR YOU.
I WANT TO SHOW YOU OUR NEW EXPERIMENT. WE'LL NEED T4 TO STEP UP WORK ON THE CRITICALITY CALCULATIONS.
HOW COME?
YOU'LL SEE. LET'S GO TO BUILDING OMEGA.

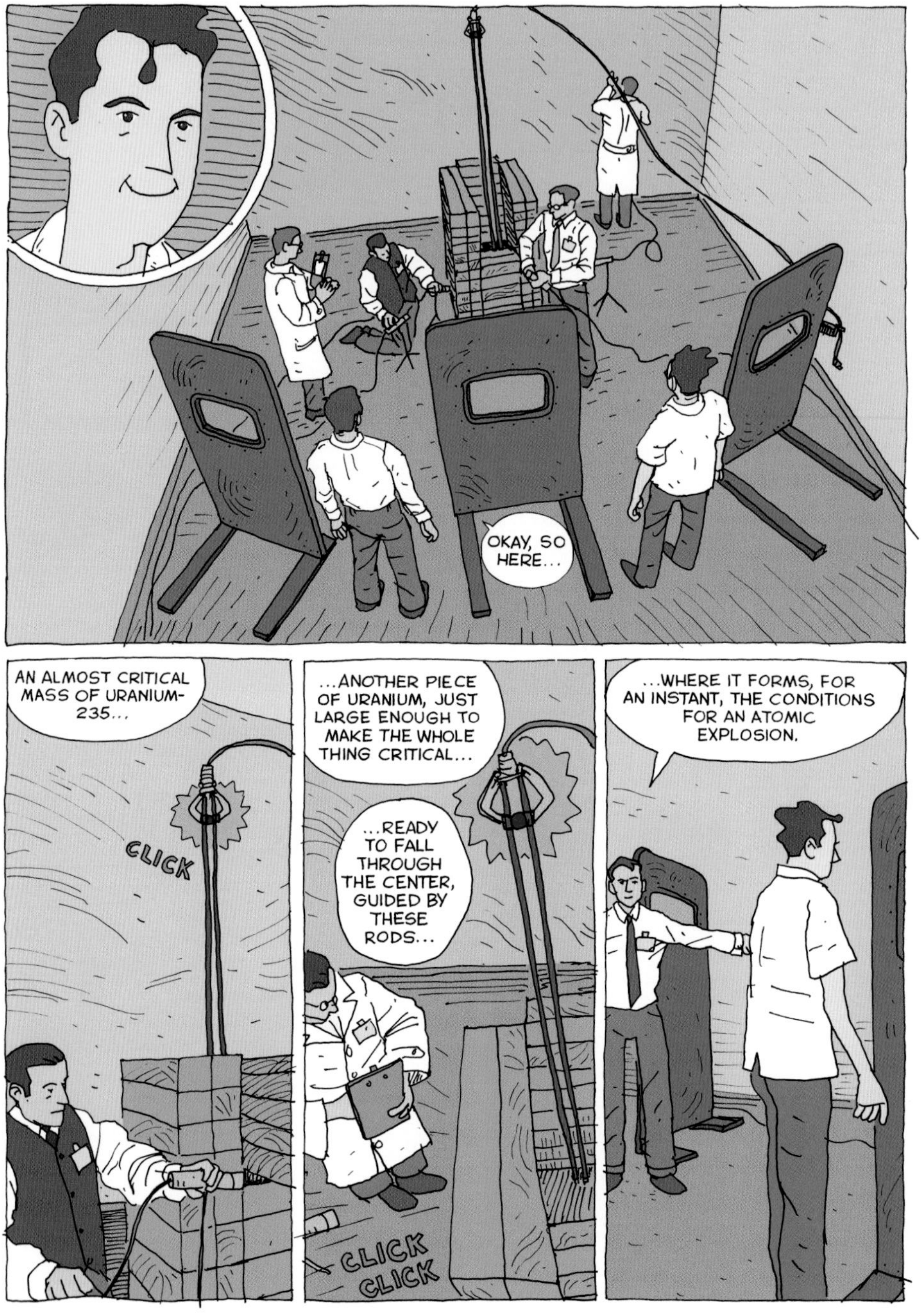
OKAY, SO HERE...
AN ALMOST CRITICAL MASS OF URANIUM-235...
CLICK
...ANOTHER PIECE OF URANIUM, JUST LARGE ENOUGH TO MAKE THE WHOLE THING CRITICAL...
...READY TO FALL THROUGH THE CENTER, GUIDED BY THESE RODS...
CLICK CLICK
...WHERE IT FORMS, FOR AN INSTANT, THE CONDITIONS FOR AN ATOMIC EXPLOSION.

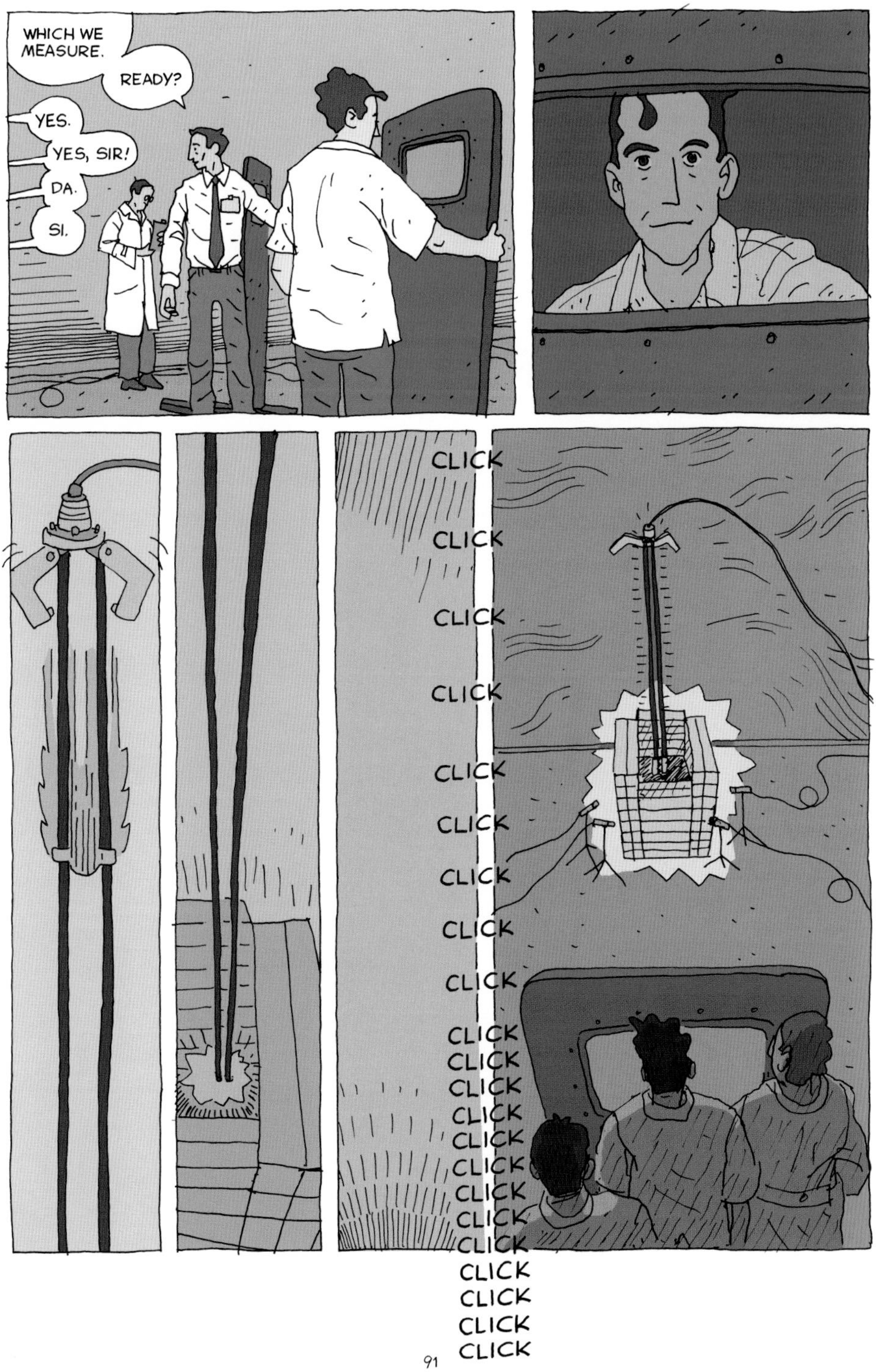
WHICH WE MEASURE.
READY?
YES.
YES, SIR!
DA.
SI.
CLICK
CLICK
CLICK
CLICK
CLICK
CLICK
CLICK
CLICK
CLICK
CLICK
CLICK
CLICK
CLICK
CLICK
CLICK
CLICK
CLICK
CLICK
CLICK
CLICK
CLICK
CLICK

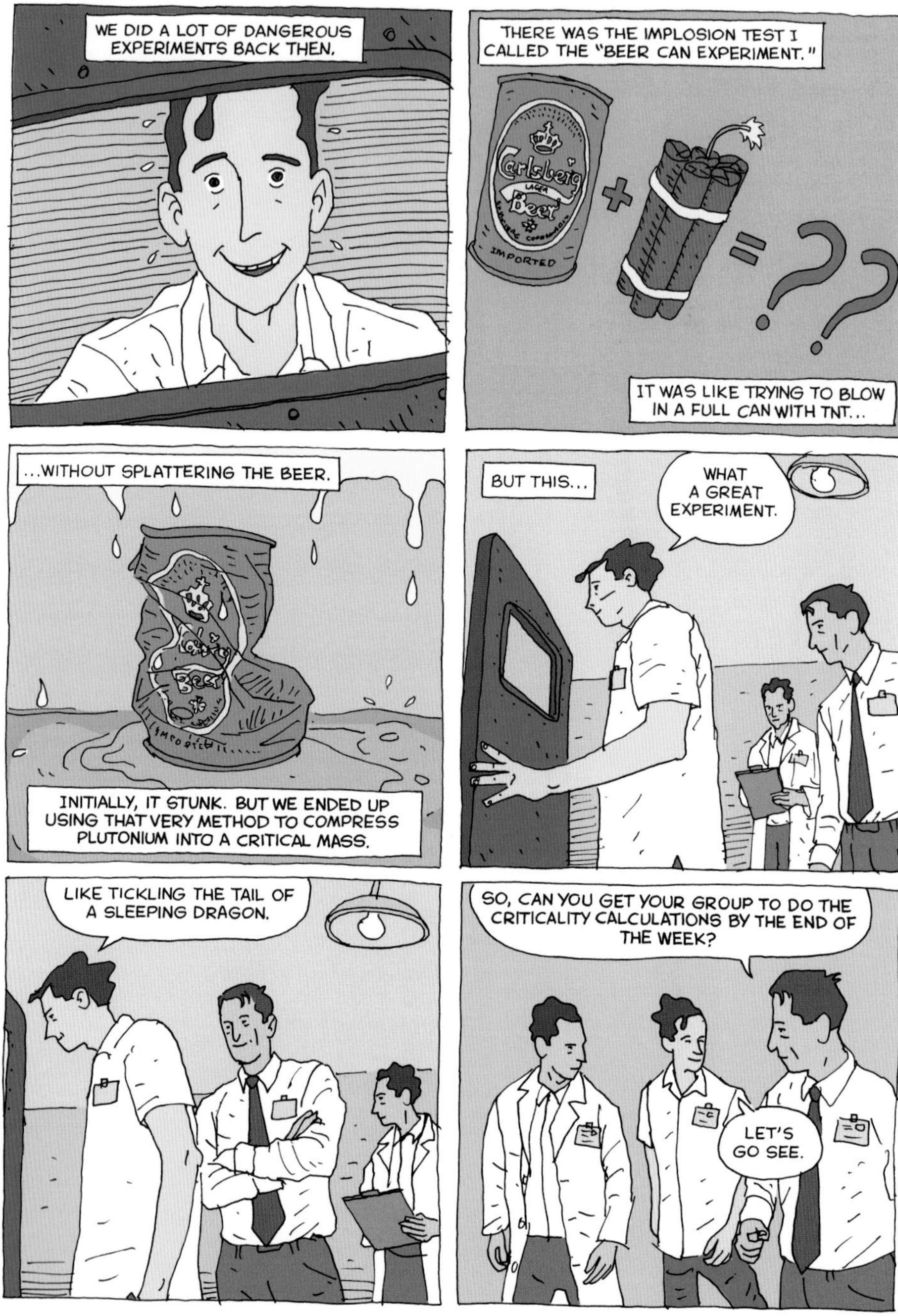
WE DID A LOT OF DANGEROUS EXPERIMENTS BACK THEN.
THERE WAS THE IMPLOSION TEST I CALLED THE "BEER CAN EXPERIMENT."
Carlsberg LAGER Beer
IMPORTED
IT WAS LIKE TRYING TO BLOW IN A FULL CAN WITH TNT...
...WITHOUT SPLATTERING THE BEER.
INITIALLY, IT STUNK. BUT WE ENDED UP USING THAT VERY METHOD TO COMPRESS PLUTONIUM INTO A CRITICAL MASS.
BUT THIS...
WHAT A GREAT EXPERIMENT.
LIKE TICKLING THE TAIL OF A SLEEPING DRAGON.
SO, CAN YOU GET YOUR GROUP TO DO THE CRITICALITY CALCULATIONS BY THE END OF THE WEEK?
LET'S GO SEE.

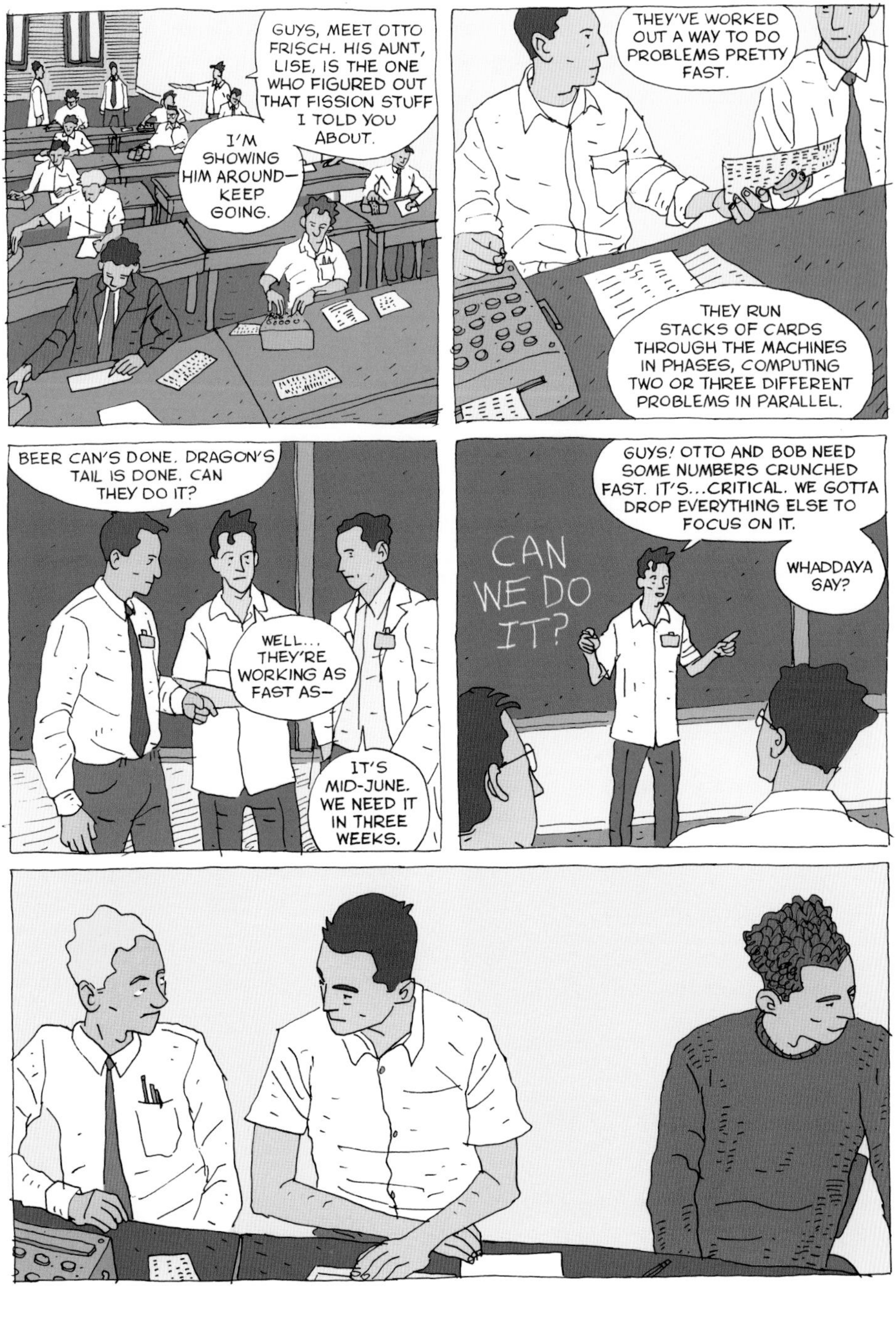
GUYS, MEET OTTO FRISCH. HIS AUNT, LISE, IS THE ONE WHO FIGURED OUT THAT FISSION STUFF I TOLD YOU ABOUT.
I'M SHOWING HIM AROUND— KEEP GOING.
THEY'VE WORKED OUT A WAY TO DO PROBLEMS PRETTY FAST.
THEY RUN STACKS OF CARDS THROUGH THE MACHINES IN PHASES, COMPUTING TWO OR THREE DIFFERENT PROBLEMS IN PARALLEL.
BEER CAN'S DONE. DRAGON'S TAIL IS DONE. CAN THEY DO IT?
WELL... THEY'RE WORKING AS FAST AS—
IT'S MID-JUNE. WE NEED IT IN THREE WEEKS.
GUYS! OTTO AND BOB NEED SOME NUMBERS CRUNCHED FAST. IT'S...CRITICAL. WE GOTTA DROP EVERYTHING ELSE TO FOCUS ON IT.
CAN WE DO IT?
WHADDAYA SAY?

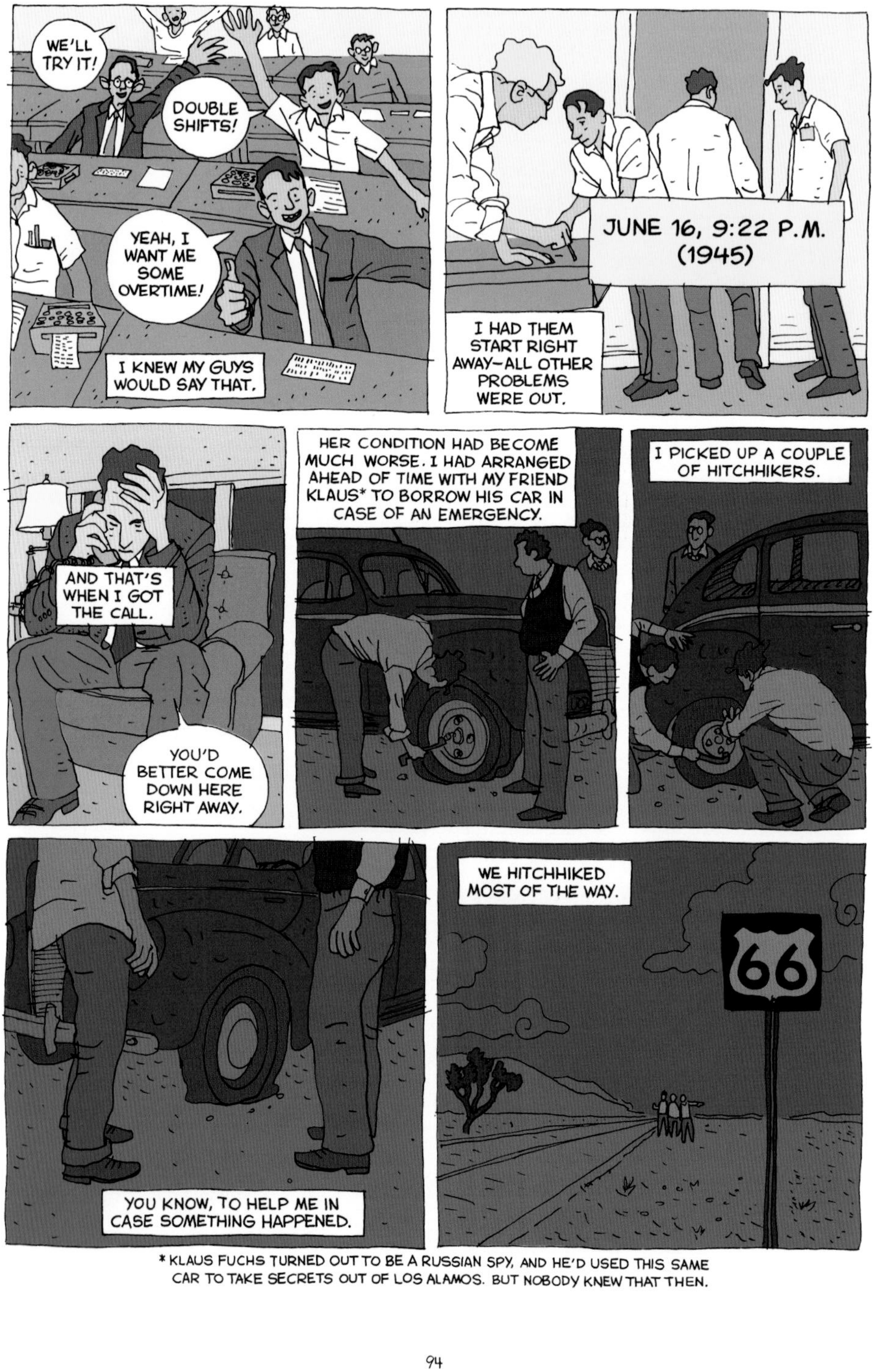
WE'LL TRY IT!
DOUBLE SHIFTS!
YEAH, I WANT ME SOME OVERTIME!
I KNEW MY GUYS WOULD SAY THAT.
JUNE 16, 9:22 P.M. (1945)
I HAD THEM START RIGHT AWAY—ALL OTHER PROBLEMS WERE OUT.
AND THAT'S WHEN I GOT THE CALL.
YOU'D BETTER COME DOWN HERE RIGHT AWAY.
HER CONDITION HAD BECOME MUCH WORSE. I HAD ARRANGED AHEAD OF TIME WITH MY FRIEND KLAUS* TO BORROW HIS CAR IN CASE OF AN EMERGENCY.
I PICKED UP A COUPLE OF HITCHHIKERS.
YOU KNOW, TO HELP ME IN CASE SOMETHING HAPPENED.
WE HITCHHIKED MOST OF THE WAY.
66

* KLAUS FUCHS TURNED OUT TO BE A RUSSIAN SPY, AND HE'D USED THIS SAME CAR TO TAKE SECRETS OUT OF LOS ALAMOS. BUT NOBODY KNEW THAT THEN.

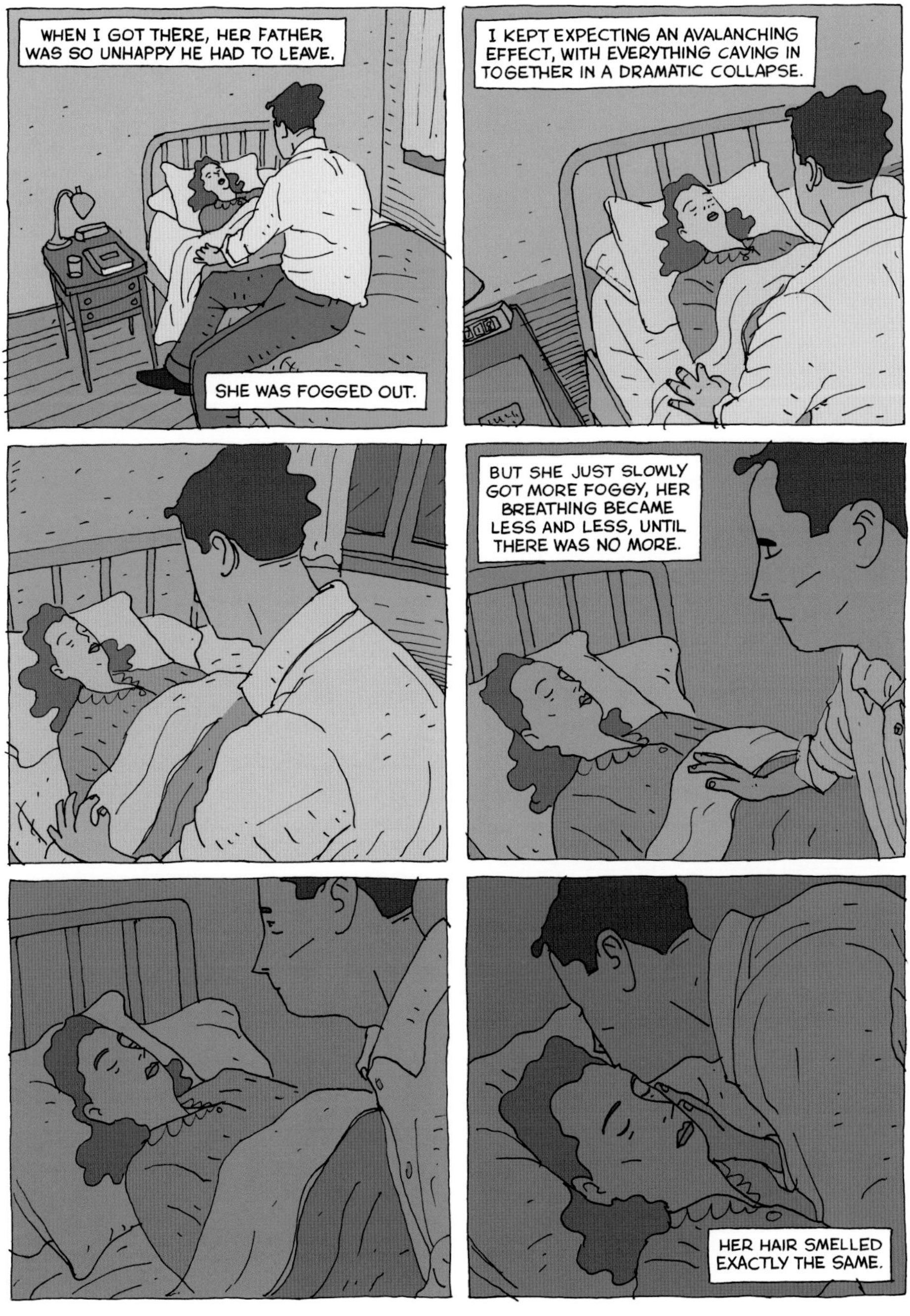
WHEN I GOT THERE, HER FATHER WAS SO UNHAPPY HE HAD TO LEAVE.
SHE WAS FOGGED OUT.
I KEPT EXPECTING AN AVALANCHING EFFECT, WITH EVERYTHING CAVING IN TOGETHER IN A DRAMATIC COLLAPSE.
BUT SHE JUST SLOWLY GOT MORE FOGGY, HER BREATHING BECAME LESS AND LESS, UNTIL THERE WAS NO MORE.
HER HAIR SMELLED EXACTLY THE SAME.

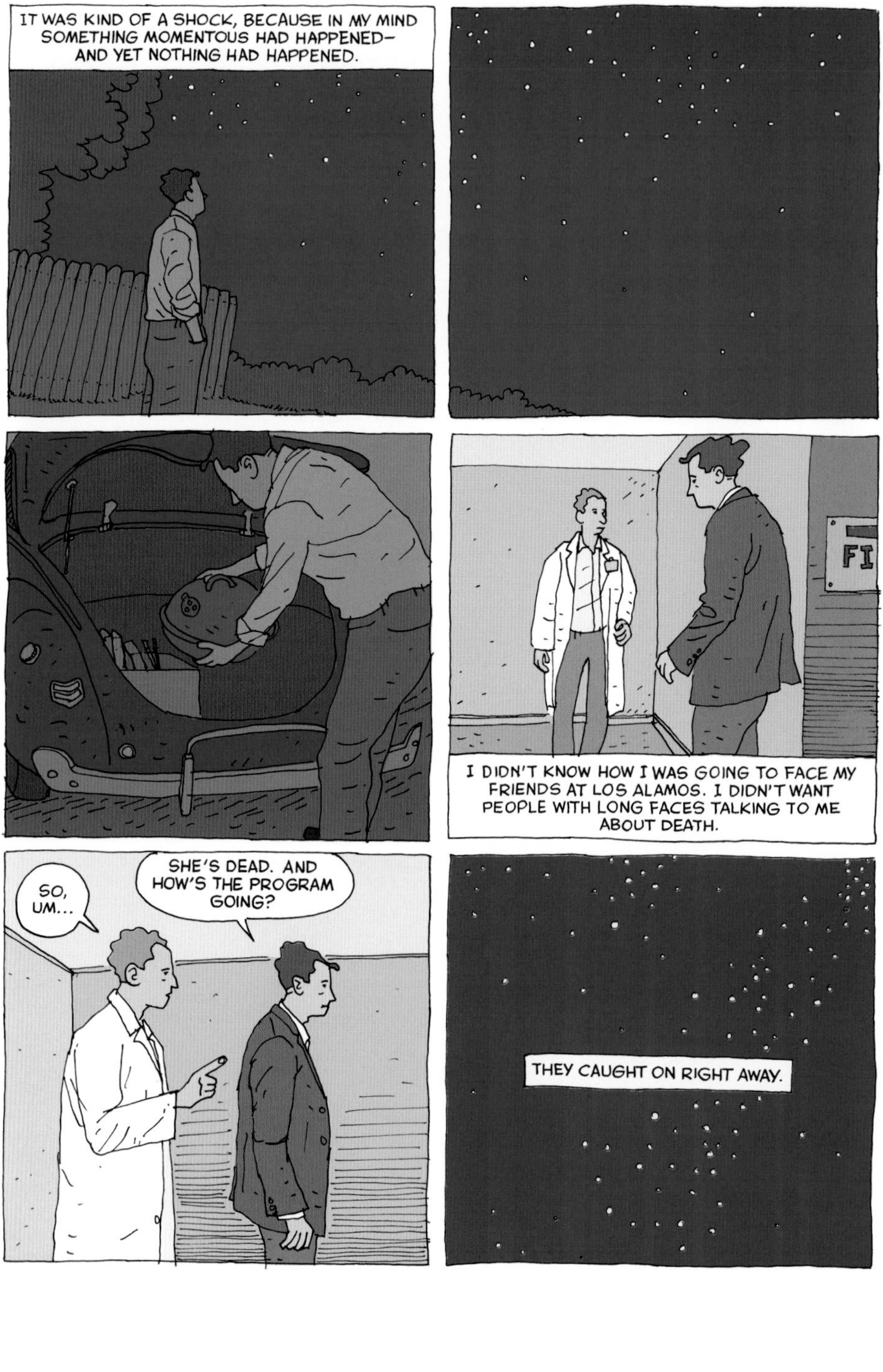
IT WAS KIND OF A SHOCK, BECAUSE IN MY MIND SOMETHING MOMENTOUS HAD HAPPENED— AND YET NOTHING HAD HAPPENED.
FI
I DIDN'T KNOW HOW I WAS GOING TO FACE MY FRIENDS AT LOS ALAMOS. I DIDN'T WANT PEOPLE WITH LONG FACES TALKING TO ME ABOUT DEATH.
SO, UM...
SHE'S DEAD. AND HOW'S THE PROGRAM GOING?
THEY CAUGHT ON RIGHT AWAY.

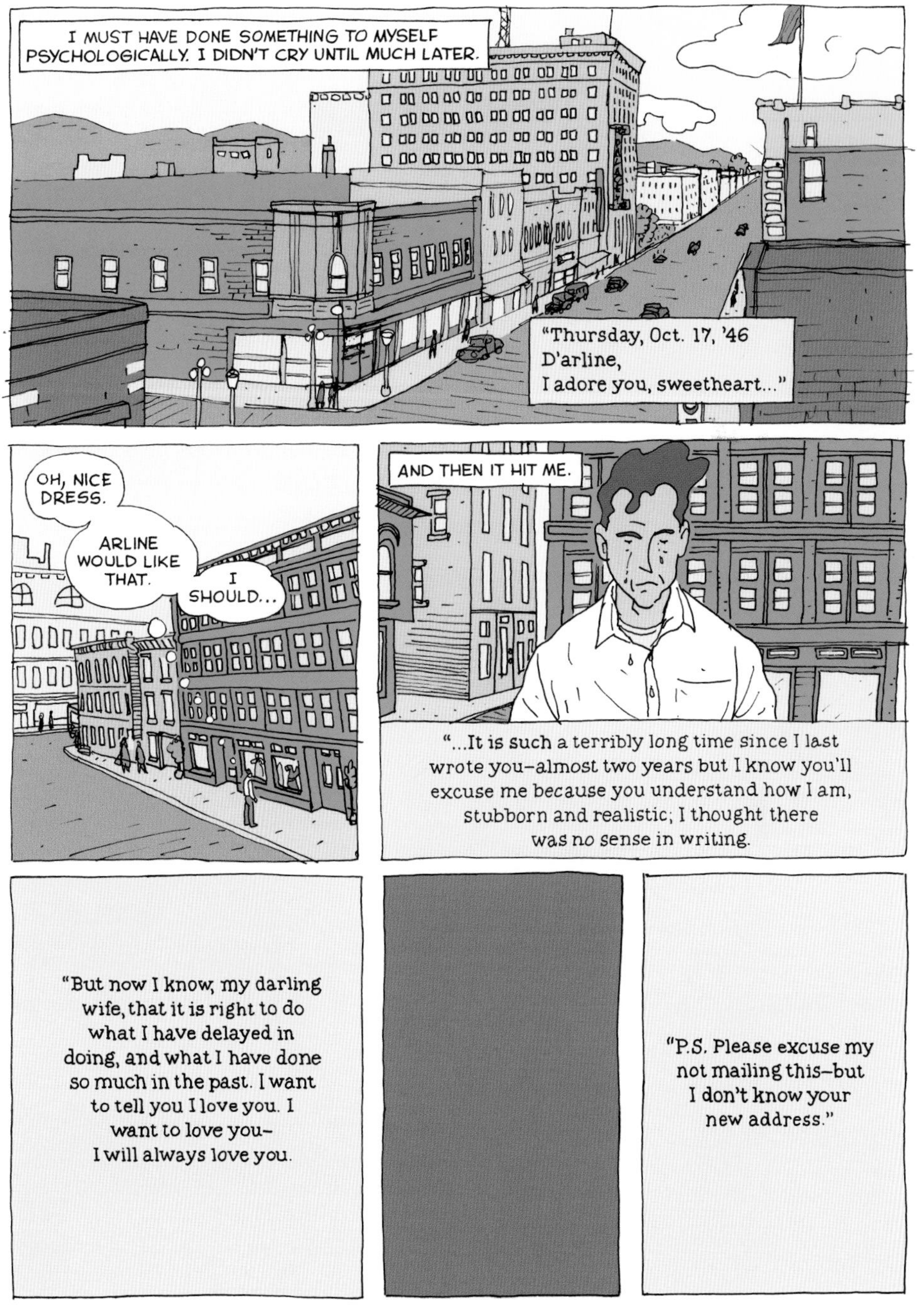
I MUST HAVE DONE SOMETHING TO MYSELF PSYCHOLOGICALLY. I DIDN'T CRY UNTIL MUCH LATER.
"Thursday, Oct. 17, '46
D'arline,
I adore you, sweetheart..."
OH, NICE DRESS.
ARLINE WOULD LIKE THAT.
I SHOULD...
AND THEN IT HIT ME.
"...It is such a terribly long time since I last wrote you–almost two years but I know you'll excuse me because you understand how I am, stubborn and realistic; I thought there was no sense in writing.
"But now I know, my darling wife, that it is right to do what I have delayed in doing, and what I have done so much in the past. I want to tell you I love you. I want to love you– I will always love you.
"P.S. Please excuse my not mailing this–but I don't know your new address."

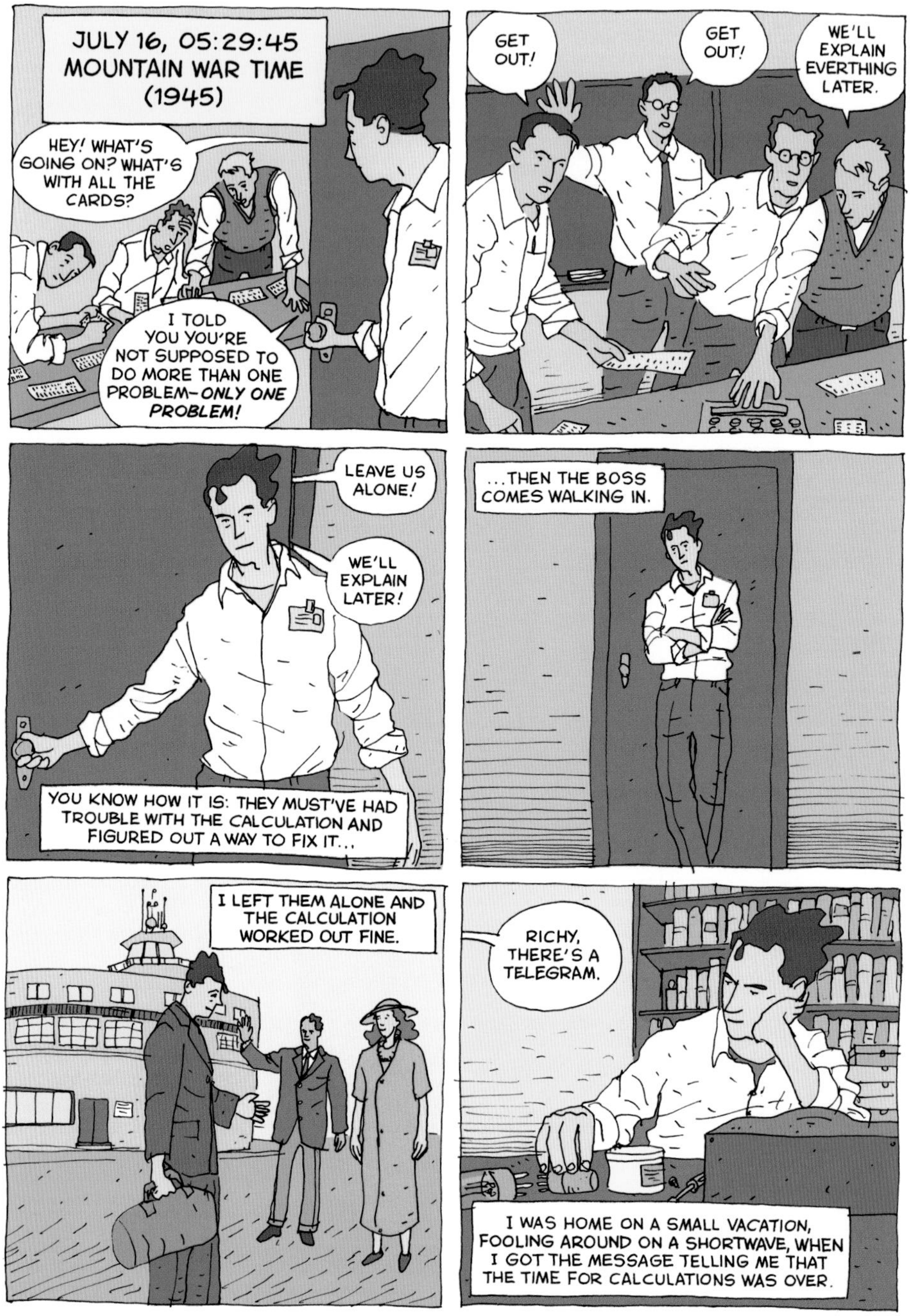
JULY 16, 05:29:45 MOUNTAIN WAR TIME (1945)
HEY! WHAT'S GOING ON? WHAT'S WITH ALL THE CARDS?
I TOLD YOU YOU'RE NOT SUPPOSED TO DO MORE THAN ONE PROBLEM–ONLY ONE PROBLEM!
GET OUT!
GET OUT!
WE'LL EXPLAIN EVERTHING LATER.
LEAVE US ALONE!
WE'LL EXPLAIN LATER!
YOU KNOW HOW IT IS: THEY MUST'VE HAD TROUBLE WITH THE CALCULATION AND FIGURED OUT A WAY TO FIX IT...
...THEN THE BOSS COMES WALKING IN.
I LEFT THEM ALONE AND THE CALCULATION WORKED OUT FINE.
RICHY, THERE'S A TELEGRAM.
I WAS HOME ON A SMALL VACATION, FOOLING AROUND ON A SHORTWAVE, WHEN I GOT THE MESSAGE TELLING ME THAT THE TIME FOR CALCULATIONS WAS OVER.

IT SAID, "THE BABY IS EXPECTED ON JULY 16."

SO I RUSHED BACK, ARRIVING JUST BEFORE THE BUSES LEFT FOR THE TEST.

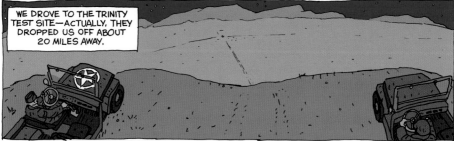

WE DROVE TO THE TRINITY TEST SITE—ACTUALLY, THEY DROPPED US OFF ABOUT 20 MILES AWAY.

WE SETTLED IN FOR THE WAIT.

MUNCH MUNCH

I CHECKED ON THE OTHER GROUP TO MAKE SURE THEIR RADIOS WORKED...

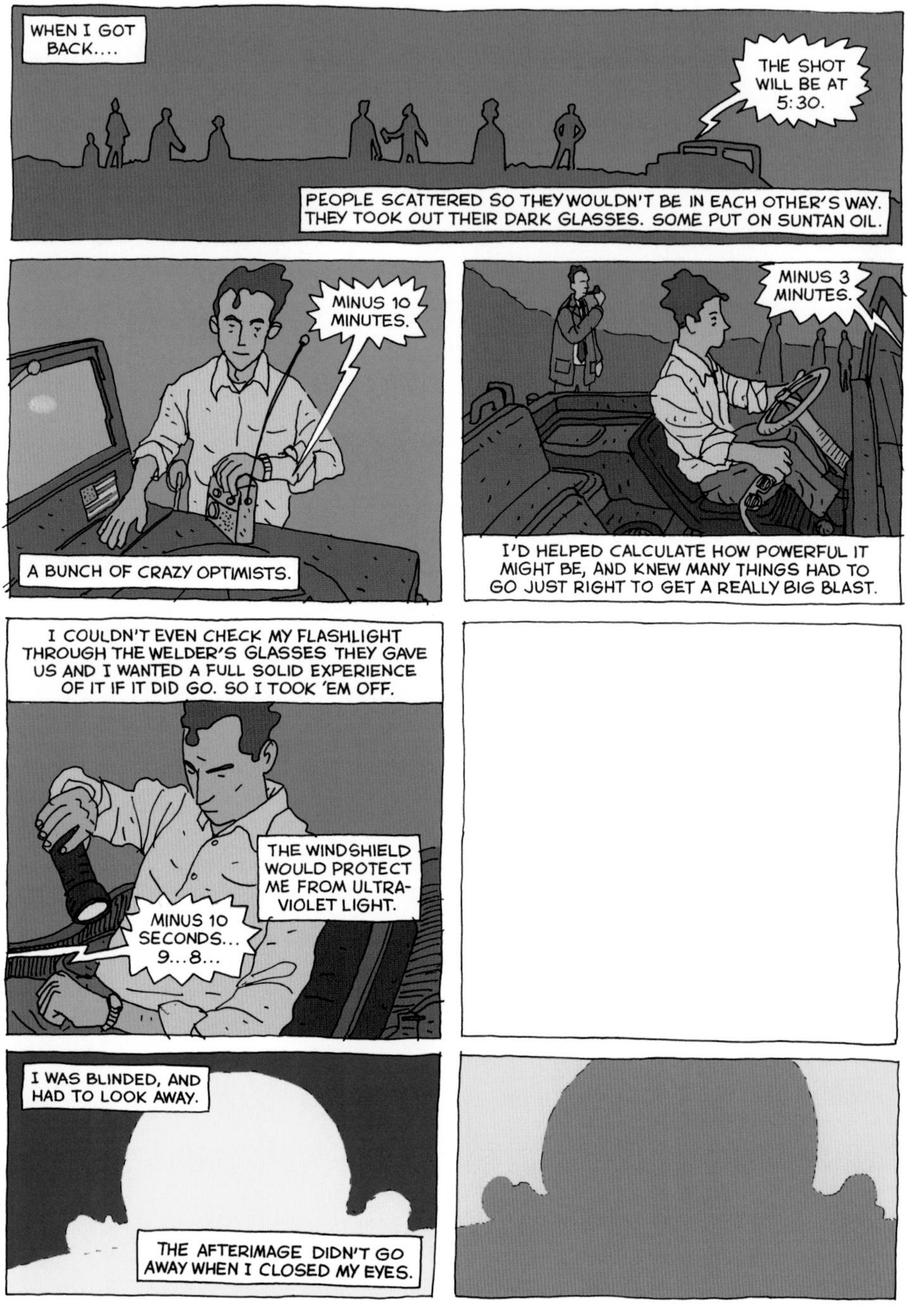
WHEN I GOT BACK....
THE SHOT WILL BE AT 5:30.
PEOPLE SCATTERED SO THEY WOULDN'T BE IN EACH OTHER'S WAY. THEY TOOK OUT THEIR DARK GLASSES. SOME PUT ON SUNTAN OIL.
MINUS 10 MINUTES.
A BUNCH OF CRAZY OPTIMISTS.
MINUS 3 MINUTES.
I'D HELPED CALCULATE HOW POWERFUL IT MIGHT BE, AND KNEW MANY THINGS HAD TO GO JUST RIGHT TO GET A REALLY BIG BLAST.
I COULDN'T EVEN CHECK MY FLASHLIGHT THROUGH THE WELDER'S GLASSES THEY GAVE US AND I WANTED A FULL SOLID EXPERIENCE OF IT IF IT DID GO. SO I TOOK 'EM OFF.
THE WINDSHIELD WOULD PROTECT ME FROM ULTRA-VIOLET LIGHT.
MINUS 10 SECONDS... 9...8...
I WAS BLINDED, AND HAD TO LOOK AWAY.
THE AFTERIMAGE DIDN'T GO AWAY WHEN I CLOSED MY EYES.

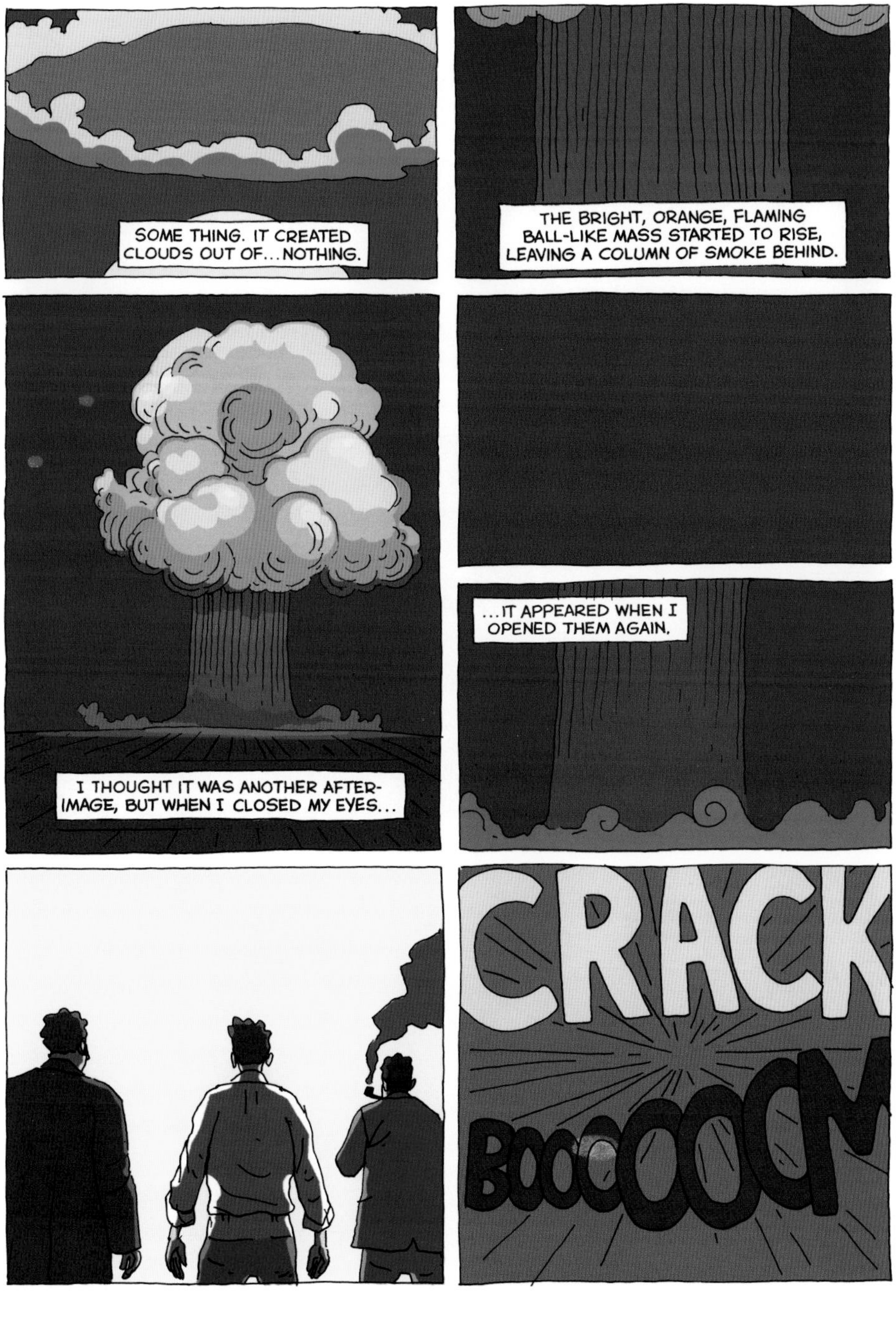
SOME THING. IT CREATED CLOUDS OUT OF... NOTHING.
THE BRIGHT, ORANGE, FLAMING BALL-LIKE MASS STARTED TO RISE, LEAVING A COLUMN OF SMOKE BEHIND.
I THOUGHT IT WAS ANOTHER AFTER-IMAGE, BUT WHEN I CLOSED MY EYES...
...IT APPEARED WHEN I OPENED THEM AGAIN.
CRACK
BOOOOOOM

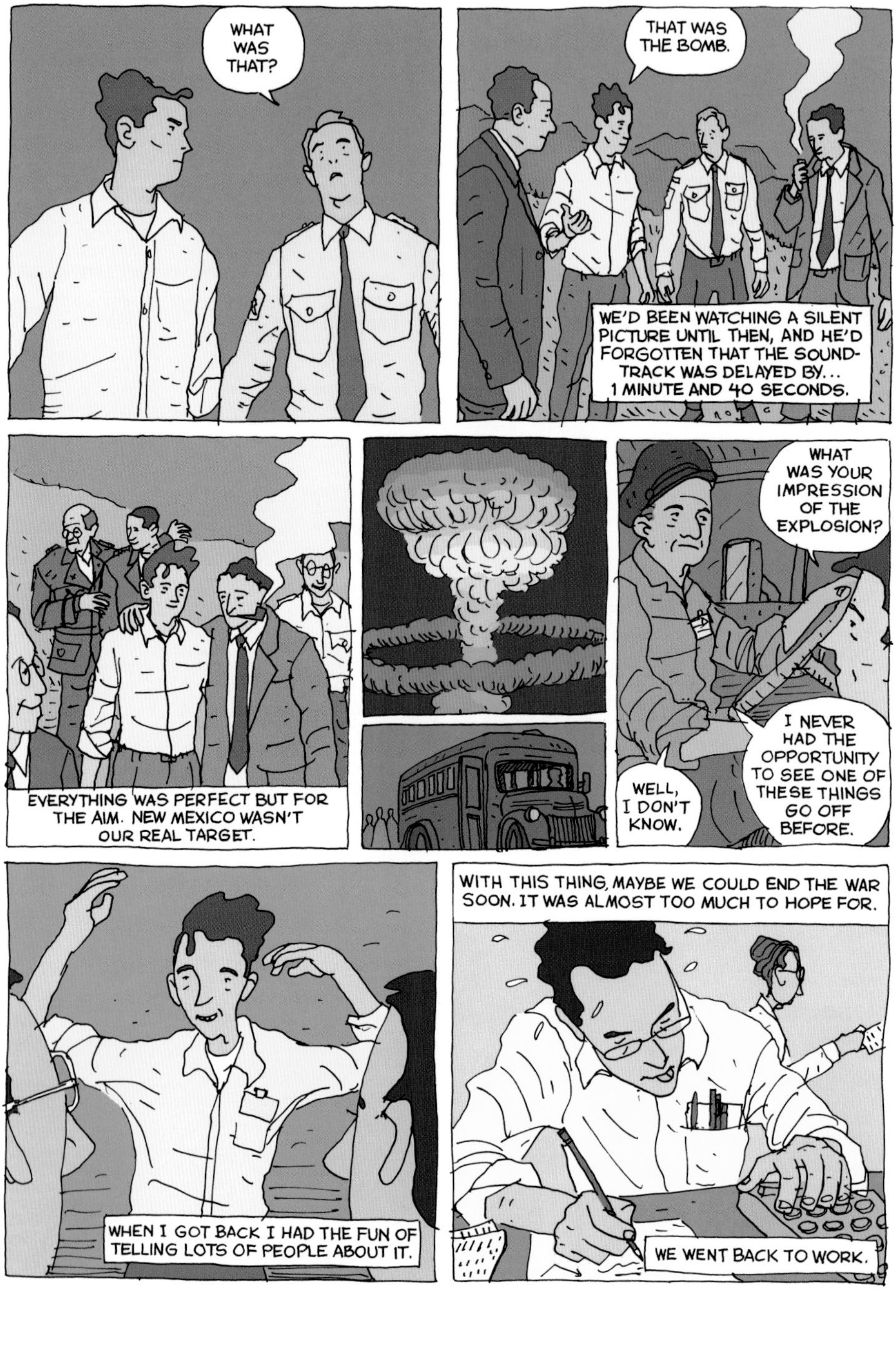
WHAT WAS THAT?
THAT WAS THE BOMB.
WE'D BEEN WATCHING A SILENT PICTURE UNTIL THEN, AND HE'D FORGOTTEN THAT THE SOUND-TRACK WAS DELAYED BY... 1 MINUTE AND 40 SECONDS.
EVERYTHING WAS PERFECT BUT FOR THE AIM. NEW MEXICO WASN'T OUR REAL TARGET.
WHAT WAS YOUR IMPRESSION OF THE EXPLOSION?
WELL, I DON'T KNOW.
I NEVER HAD THE OPPORTUNITY TO SEE ONE OF THESE THINGS GO OFF BEFORE.
WHEN I GOT BACK I HAD THE FUN OF TELLING LOTS OF PEOPLE ABOUT IT.
WITH THIS THING, MAYBE WE COULD END THE WAR SOON. IT WAS ALMOST TOO MUCH TO HOPE FOR.
WE WENT BACK TO WORK.

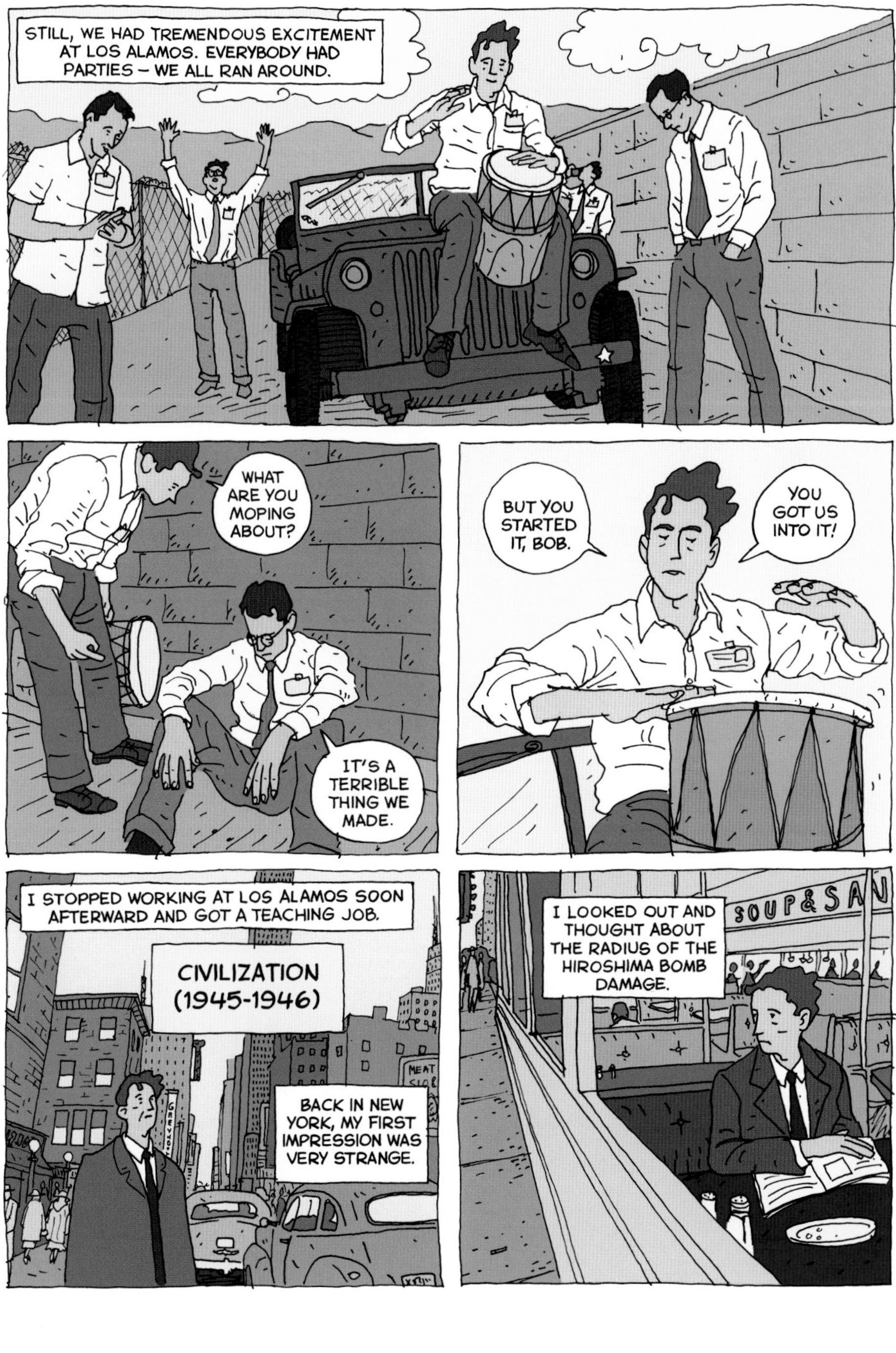
STILL, WE HAD TREMENDOUS EXCITEMENT AT LOS ALAMOS. EVERYBODY HAD PARTIES – WE ALL RAN AROUND.
WHAT ARE YOU MOPING ABOUT?
IT'S A TERRIBLE THING WE MADE.
BUT YOU STARTED IT, BOB.
YOU GOT US INTO IT!
I STOPPED WORKING AT LOS ALAMOS SOON AFTERWARD AND GOT A TEACHING JOB.
CIVILIZATION (1945-1946)
BACK IN NEW YORK, MY FIRST IMPRESSION WAS VERY STRANGE.
I LOOKED OUT AND THOUGHT ABOUT THE RADIUS OF THE HIROSHIMA BOMB DAMAGE.
SOUP&SAN

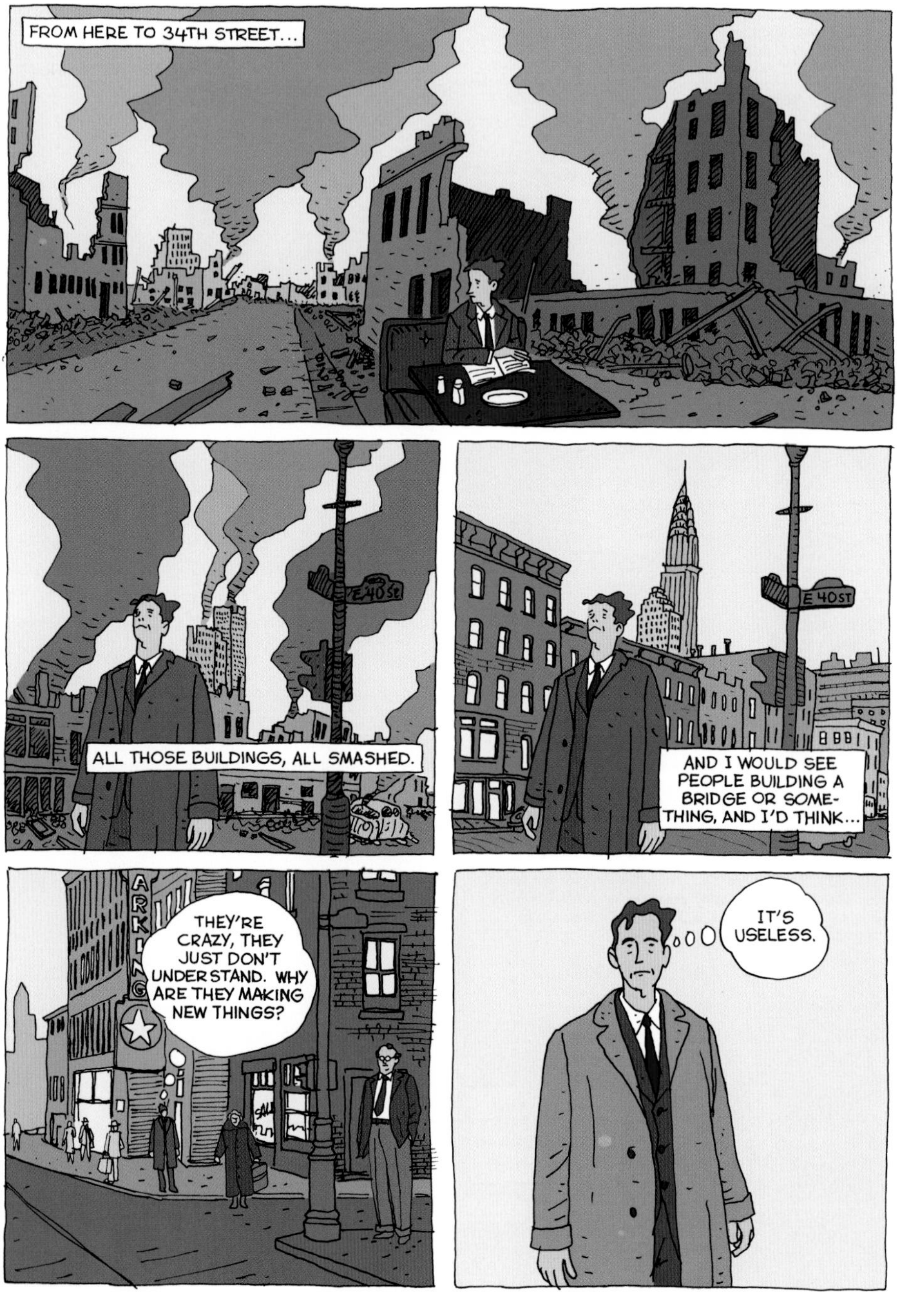
FROM HERE TO 34TH STREET...
E 40 St
ALL THOSE BUILDINGS, ALL SMASHED.
E 40 ST
AND I WOULD SEE PEOPLE BUILDING A BRIDGE OR SOME-THING, AND I'D THINK...
ARKING
THEY'RE CRAZY, THEY JUST DON'T UNDERSTAND. WHY ARE THEY MAKING NEW THINGS?
IT'S USELESS.

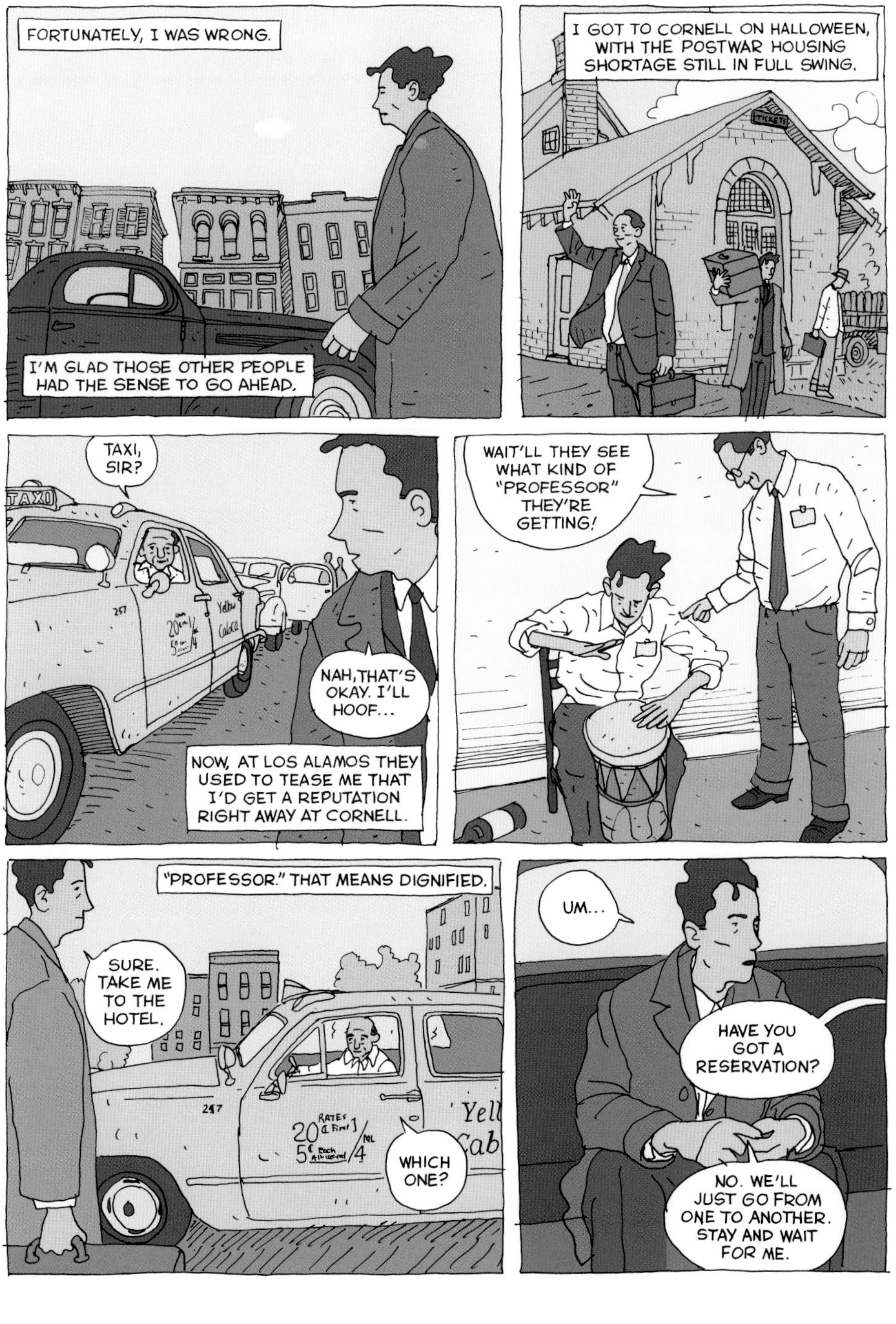
FORTUNATELY, I WAS WRONG.
I'M GLAD THOSE OTHER PEOPLE HAD THE SENSE TO GO AHEAD.
I GOT TO CORNELL ON HALLOWEEN, WITH THE POSTWAR HOUSING SHORTAGE STILL IN FULL SWING.
TAXI, SIR?
NAH, THAT'S OKAY. I'LL HOOF...
NOW, AT LOS ALAMOS THEY USED TO TEASE ME THAT I'D GET A REPUTATION RIGHT AWAY AT CORNELL.
WAIT'LL THEY SEE WHAT KIND OF "PROFESSOR" THEY'RE GETTING!
"PROFESSOR." THAT MEANS DIGNIFIED.
SURE. TAKE ME TO THE HOTEL.
WHICH ONE?
UM...
HAVE YOU GOT A RESERVATION?
NO. WE'LL JUST GO FROM ONE TO ANOTHER. STAY AND WAIT FOR ME.

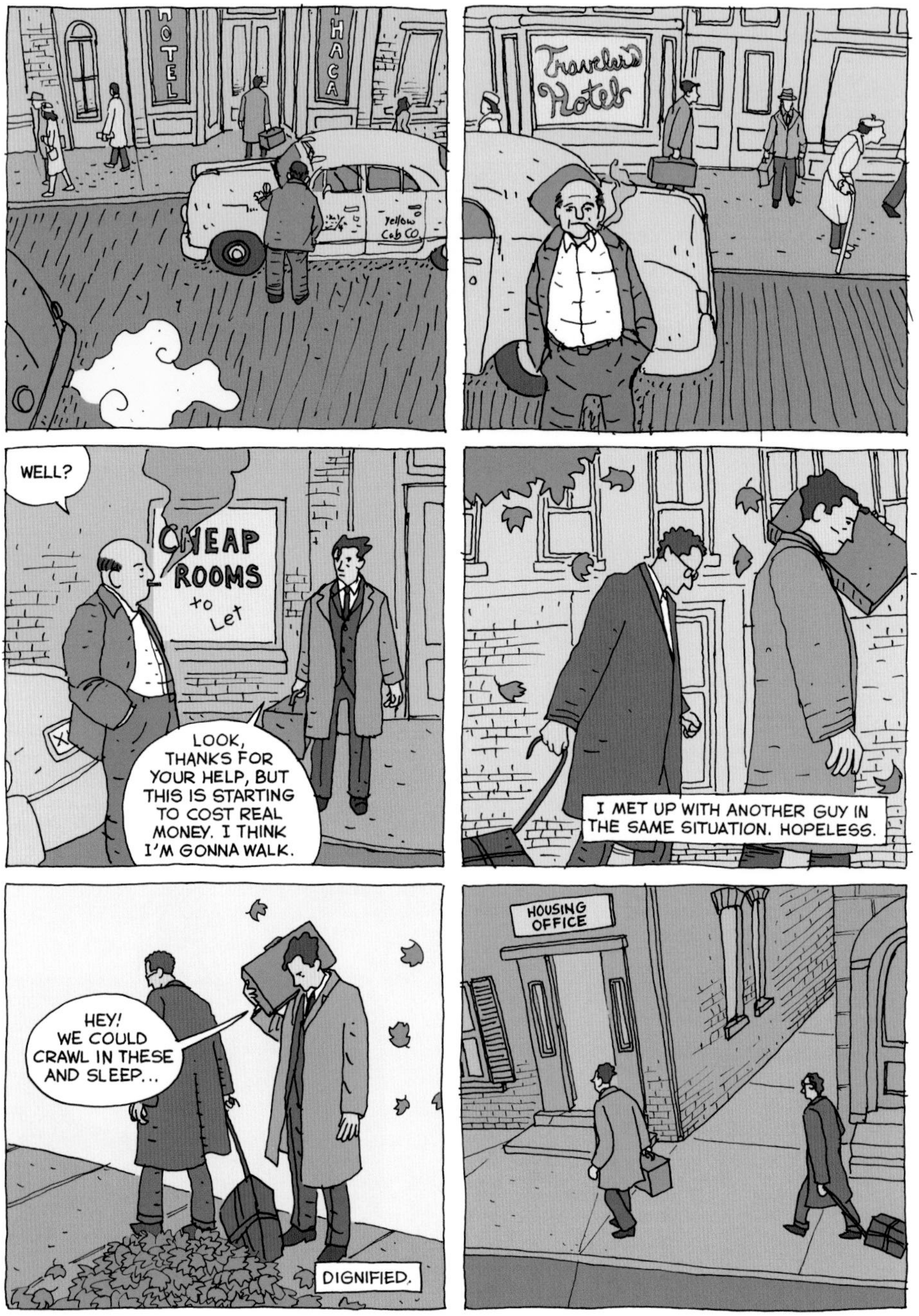
HOTEL
ITHACA
Yellow Cab CO
Traveler's Hotel
WELL?
CHEAP ROOMS to Let
LOOK, THANKS FOR YOUR HELP, BUT THIS IS STARTING TO COST REAL MONEY. I THINK I'M GONNA WALK.
I MET UP WITH ANOTHER GUY IN THE SAME SITUATION. HOPELESS.
HEY! WE COULD CRAWL IN THESE AND SLEEP...
DIGNIFIED.
HOUSING OFFICE

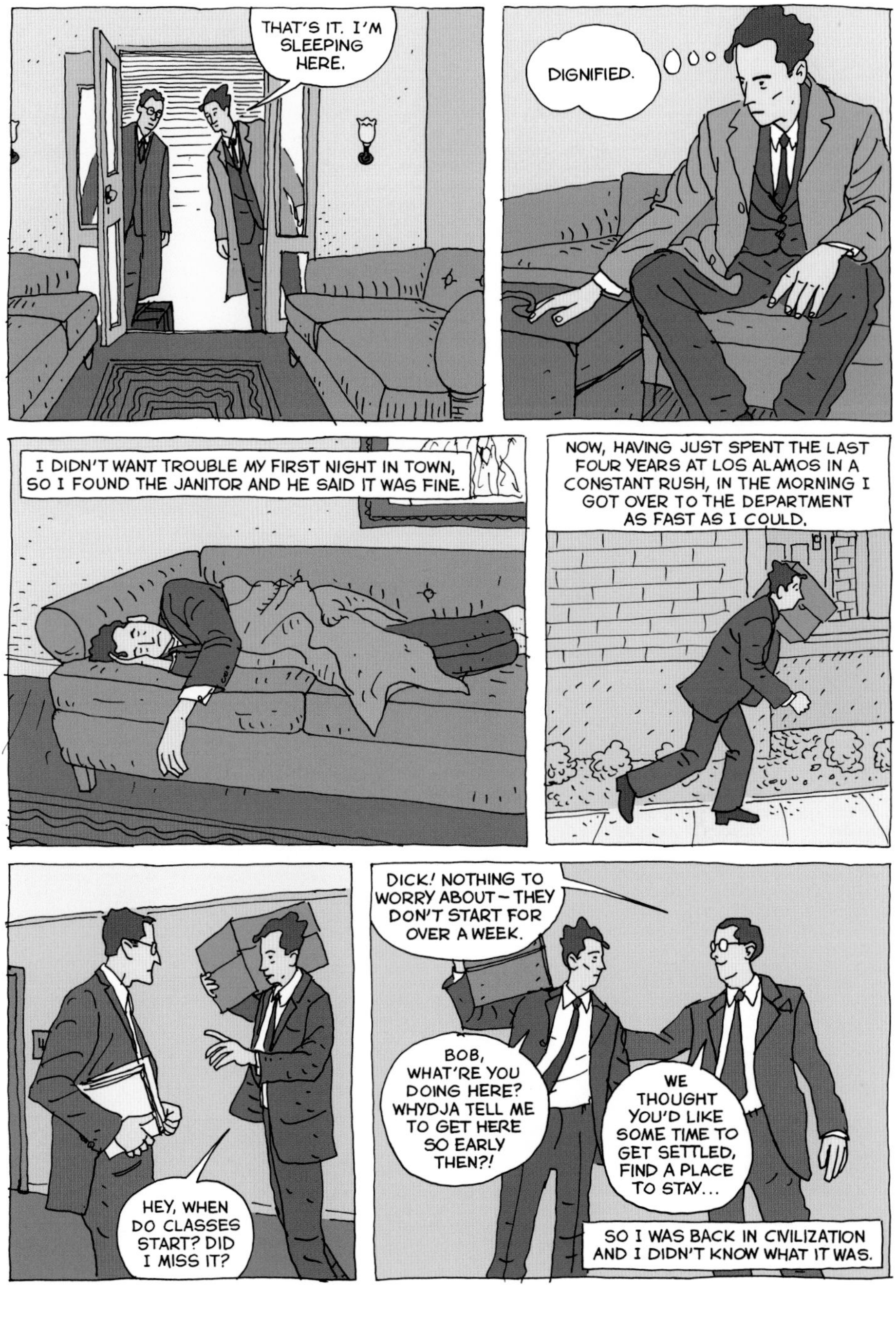
THAT'S IT. I'M SLEEPING HERE.
DIGNIFIED.
I DIDN'T WANT TROUBLE MY FIRST NIGHT IN TOWN, SO I FOUND THE JANITOR AND HE SAID IT WAS FINE.
NOW, HAVING JUST SPENT THE LAST FOUR YEARS AT LOS ALAMOS IN A CONSTANT RUSH, IN THE MORNING I GOT OVER TO THE DEPARTMENT AS FAST AS I COULD.
HEY, WHEN DO CLASSES START? DID I MISS IT?
DICK! NOTHING TO WORRY ABOUT—THEY DON'T START FOR OVER A WEEK.
BOB, WHAT'RE YOU DOING HERE? WHYDJA TELL ME TO GET HERE SO EARLY THEN?!
WE THOUGHT YOU'D LIKE SOME TIME TO GET SETTLED, FIND A PLACE TO STAY...
SO I WAS BACK IN CIVILIZATION AND I DIDN'T KNOW WHAT IT WAS.

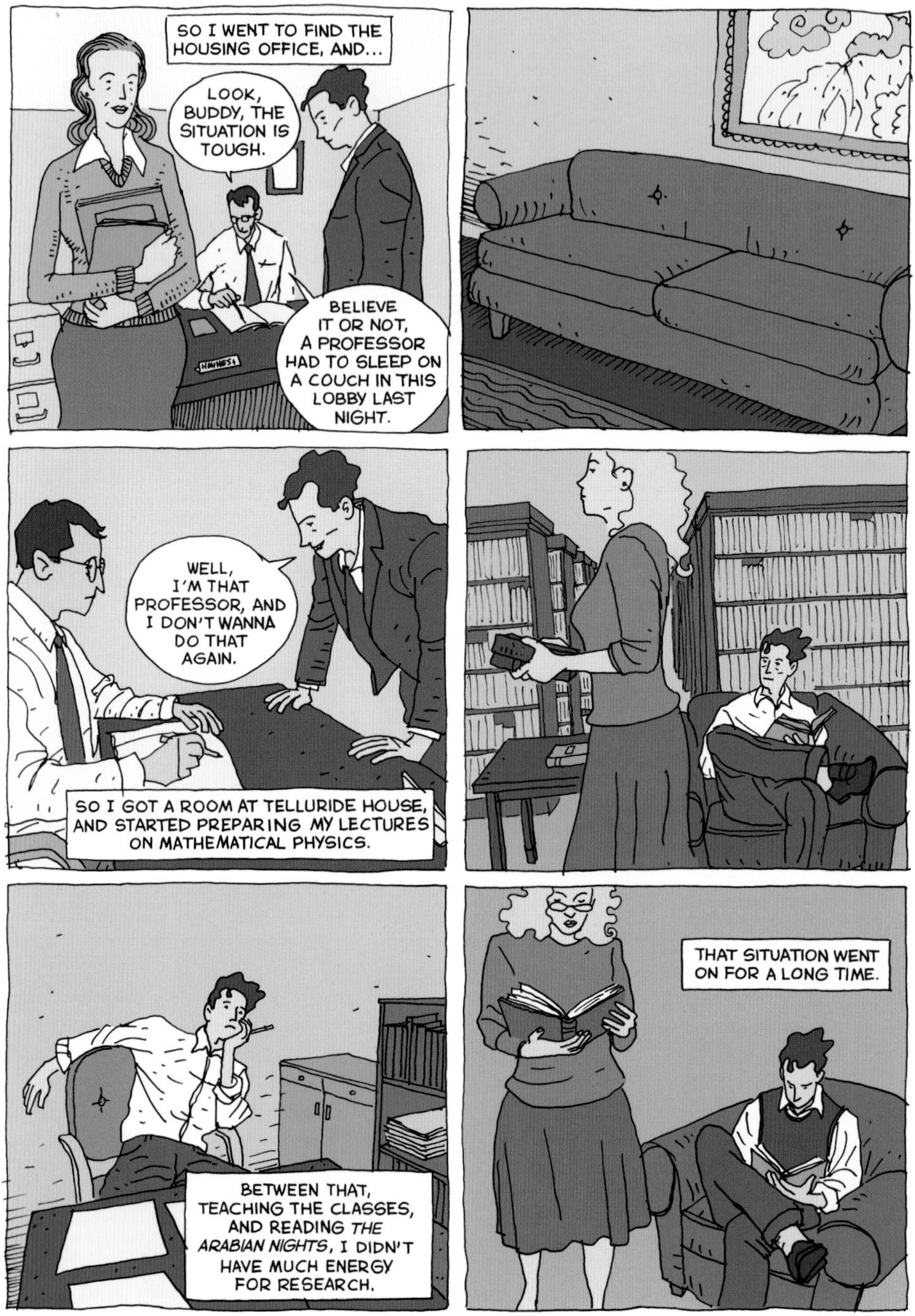
SO I WENT TO FIND THE HOUSING OFFICE, AND...
LOOK, BUDDY, THE SITUATION IS TOUGH.
BELIEVE IT OR NOT, A PROFESSOR HAD TO SLEEP ON A COUCH IN THIS LOBBY LAST NIGHT.
WELL, I'M THAT PROFESSOR, AND I DON'T WANNA DO THAT AGAIN.
SO I GOT A ROOM AT TELLURIDE HOUSE, AND STARTED PREPARING MY LECTURES ON MATHEMATICAL PHYSICS.
BETWEEN THAT, TEACHING THE CLASSES, AND READING *THE ARABIAN NIGHTS*, I DIDN'T HAVE MUCH ENERGY FOR RESEARCH.
THAT SITUATION WENT ON FOR A LONG TIME.

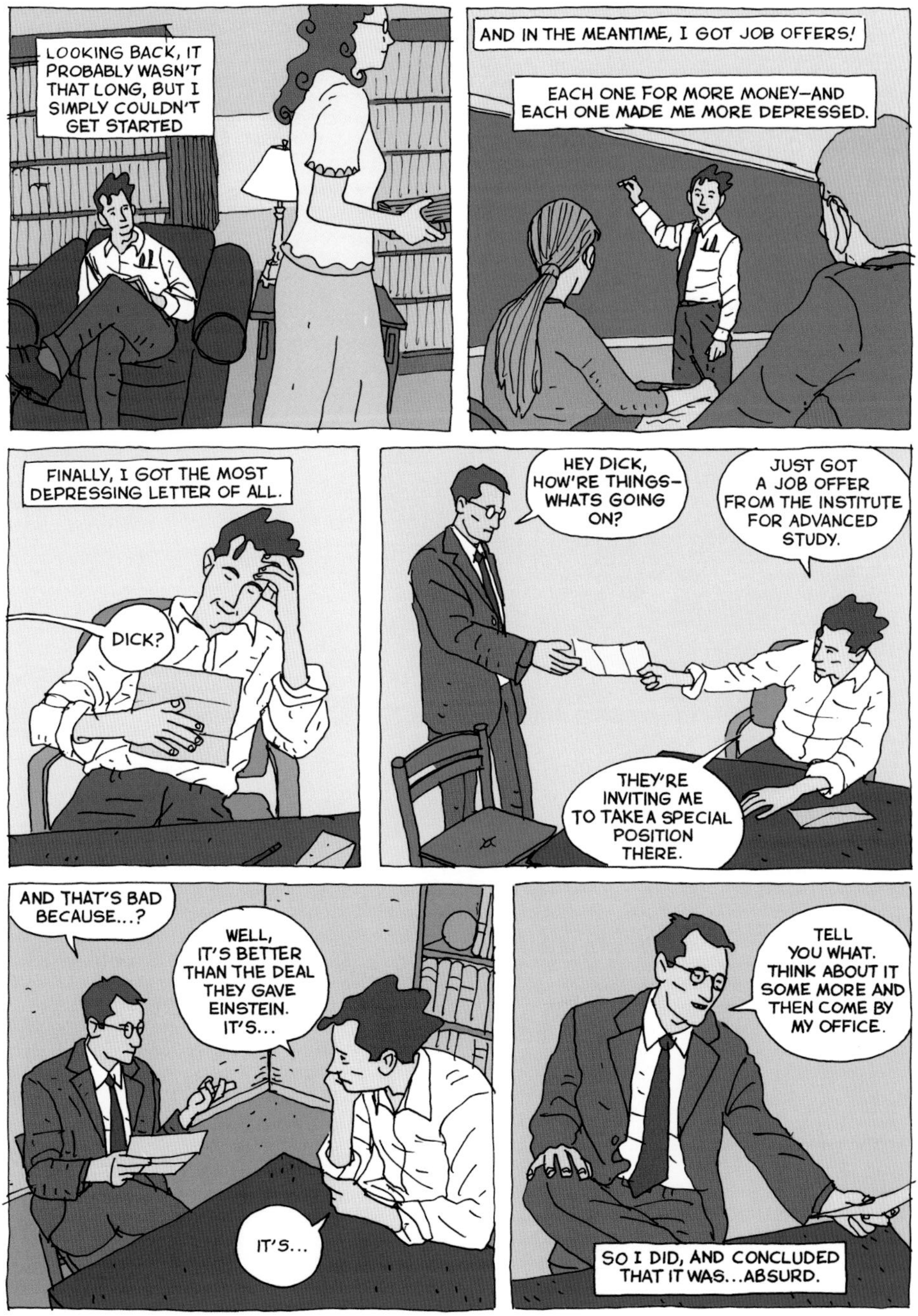
LOOKING BACK, IT PROBABLY WASN'T THAT LONG, BUT I SIMPLY COULDN'T GET STARTED
AND IN THE MEANTIME, I GOT JOB OFFERS!
EACH ONE FOR MORE MONEY—AND EACH ONE MADE ME MORE DEPRESSED.
FINALLY, I GOT THE MOST DEPRESSING LETTER OF ALL.
DICK?
HEY DICK, HOW'RE THINGS—WHATS GOING ON?
JUST GOT A JOB OFFER FROM THE INSTITUTE FOR ADVANCED STUDY.
THEY'RE INVITING ME TO TAKE A SPECIAL POSITION THERE.
AND THAT'S BAD BECAUSE...?
WELL, IT'S BETTER THAN THE DEAL THEY GAVE EINSTEIN. IT'S...
IT'S...
TELL YOU WHAT. THINK ABOUT IT SOME MORE AND THEN COME BY MY OFFICE.
SO I DID, AND CONCLUDED THAT IT WAS...ABSURD.

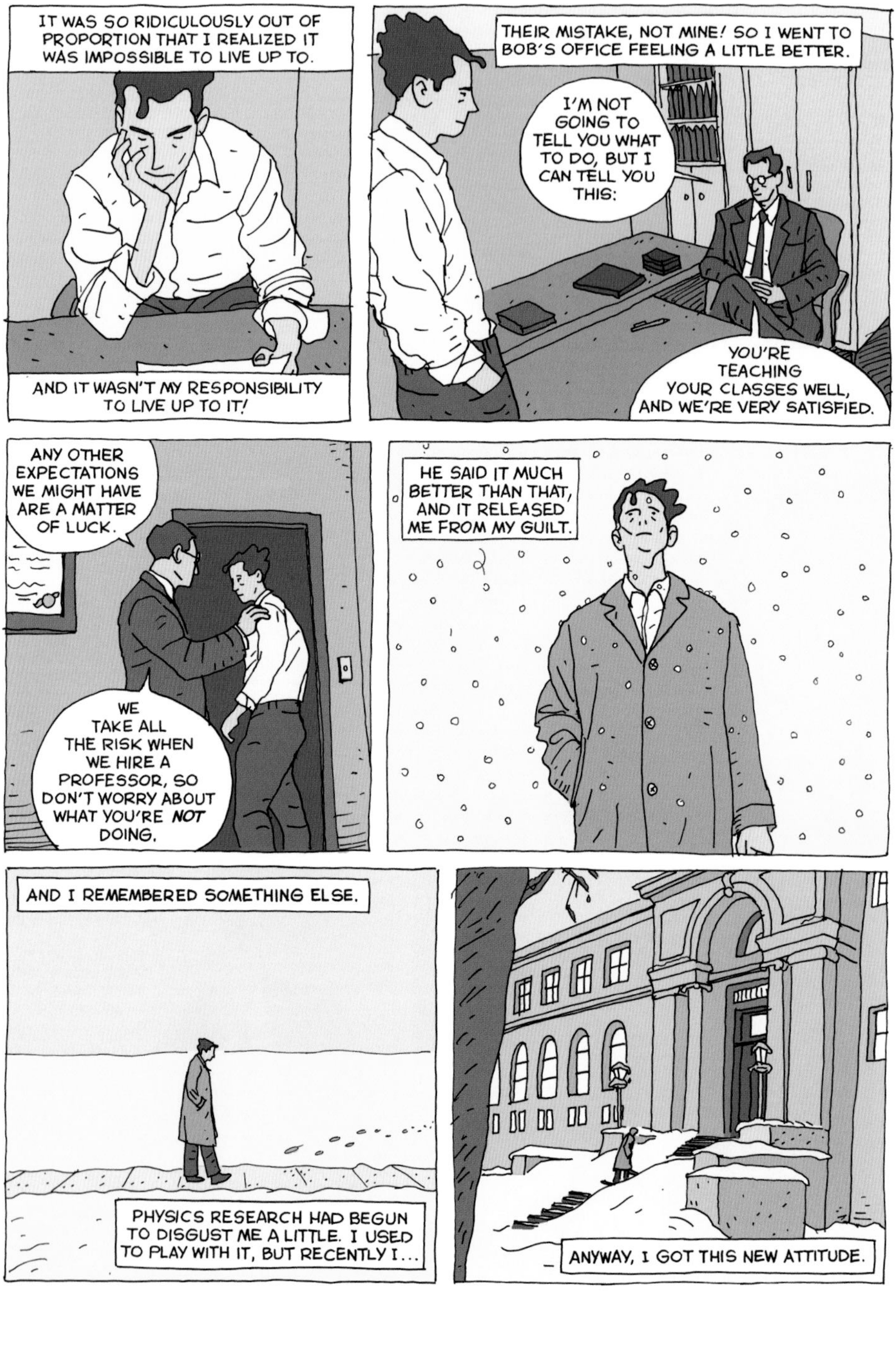

IT WAS SO RIDICULOUSLY OUT OF PROPORTION THAT I REALIZED IT WAS IMPOSSIBLE TO LIVE UP TO.
AND IT WASN'T MY RESPONSIBILITY TO LIVE UP TO IT!
THEIR MISTAKE, NOT MINE! SO I WENT TO BOB'S OFFICE FEELING A LITTLE BETTER.
I'M NOT GOING TO TELL YOU WHAT TO DO, BUT I CAN TELL YOU THIS:
YOU'RE TEACHING YOUR CLASSES WELL, AND WE'RE VERY SATISFIED.
ANY OTHER EXPECTATIONS WE MIGHT HAVE ARE A MATTER OF LUCK.
WE TAKE ALL THE RISK WHEN WE HIRE A PROFESSOR, SO DON'T WORRY ABOUT WHAT YOU'RE *NOT* DOING.
HE SAID IT MUCH BETTER THAN THAT, AND IT RELEASED ME FROM MY GUILT.
AND I REMEMBERED SOMETHING ELSE.
PHYSICS RESEARCH HAD BEGUN TO DISGUST ME A LITTLE. I USED TO PLAY WITH IT, BUT RECENTLY I...
ANYWAY, I GOT THIS NEW ATTITUDE.

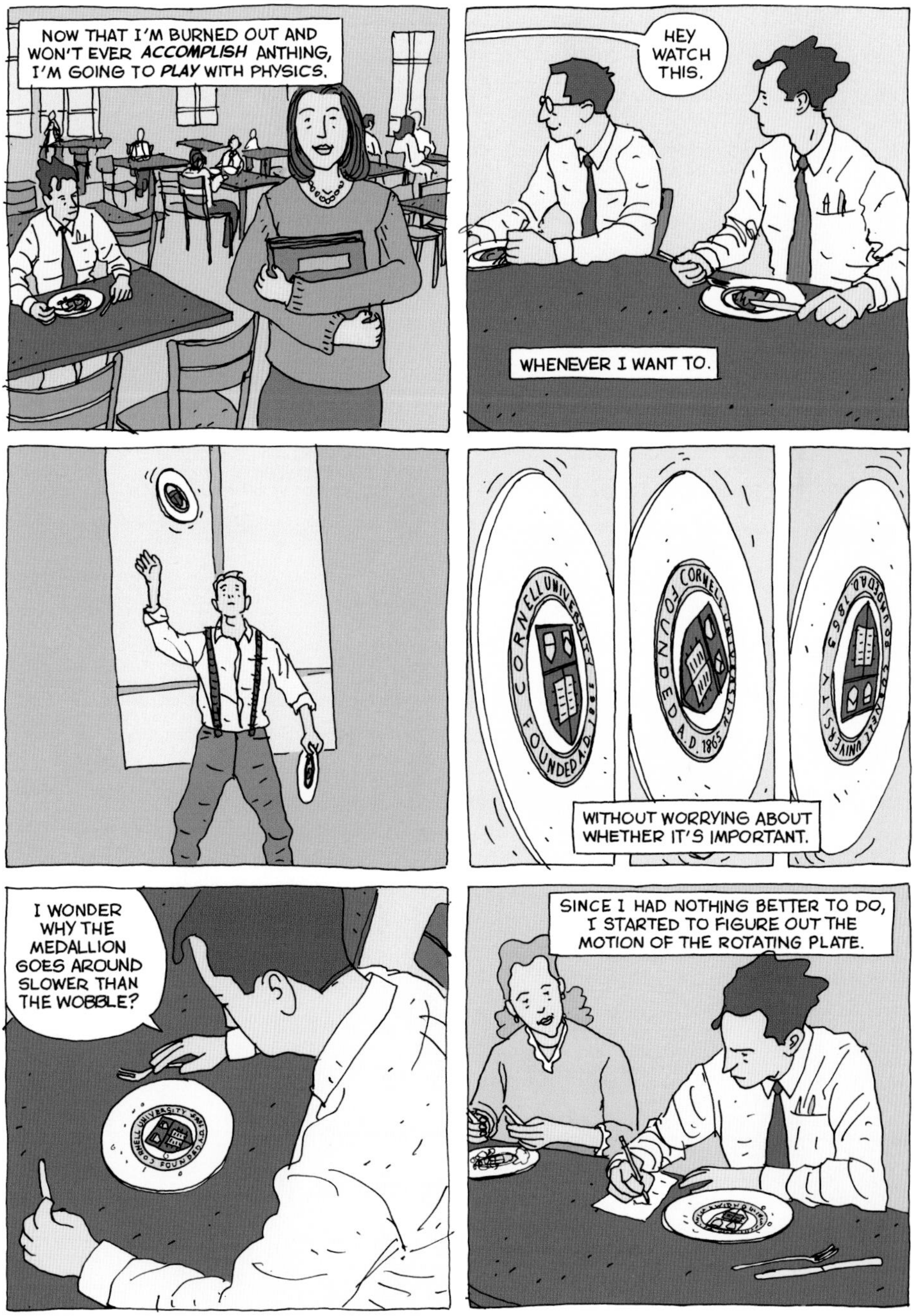
NOW THAT I'M BURNED OUT AND WON'T EVER ACCOMPLISH ANTHING, I'M GOING TO PLAY WITH PHYSICS.
HEY WATCH THIS.
WHENEVER I WANT TO.
WITHOUT WORRYING ABOUT WHETHER IT'S IMPORTANT.
I WONDER WHY THE MEDALLION GOES AROUND SLOWER THAN THE WOBBLE?
SINCE I HAD NOTHING BETTER TO DO, I STARTED TO FIGURE OUT THE MOTION OF THE ROTATING PLATE.

THE "WHY" OF IT BECAME A COMPLICATED EQUATION.

SO I LOOKED FOR A MORE FUNDA-MENTAL WAY TO EXPLAIN THIS 2-WOBBLES/ 1-ROTATION RATIO I SAW IN THE CAFETERIA.

$m_e = 9.11 \times$
$m_n = 1.$

I DON'T REMEMBER HOW, BUT I WORKED OUT HOW ALL THE ACCELERATIONS BALANCED TO MAKE IT COME OUT.

HANS BETHE WAS BACK AT CORNELL TOO.

HEY HANS, WHILE YOU'RE HERE, LET ME SHOW YOU SOMETHING INTERESTING.

THE PLATE GOES AROUND SO...

AND THE REASON IT'S TWO-TO-ONE IS...

LOOK, THAT'S PRETTY INTERESTING, BUT WHY ARE YOU DOING THIS?

WHAT'S THE IMPORTANCE?

HAH!

THERE'S NO IMPORTANCE WHATSOEVER!

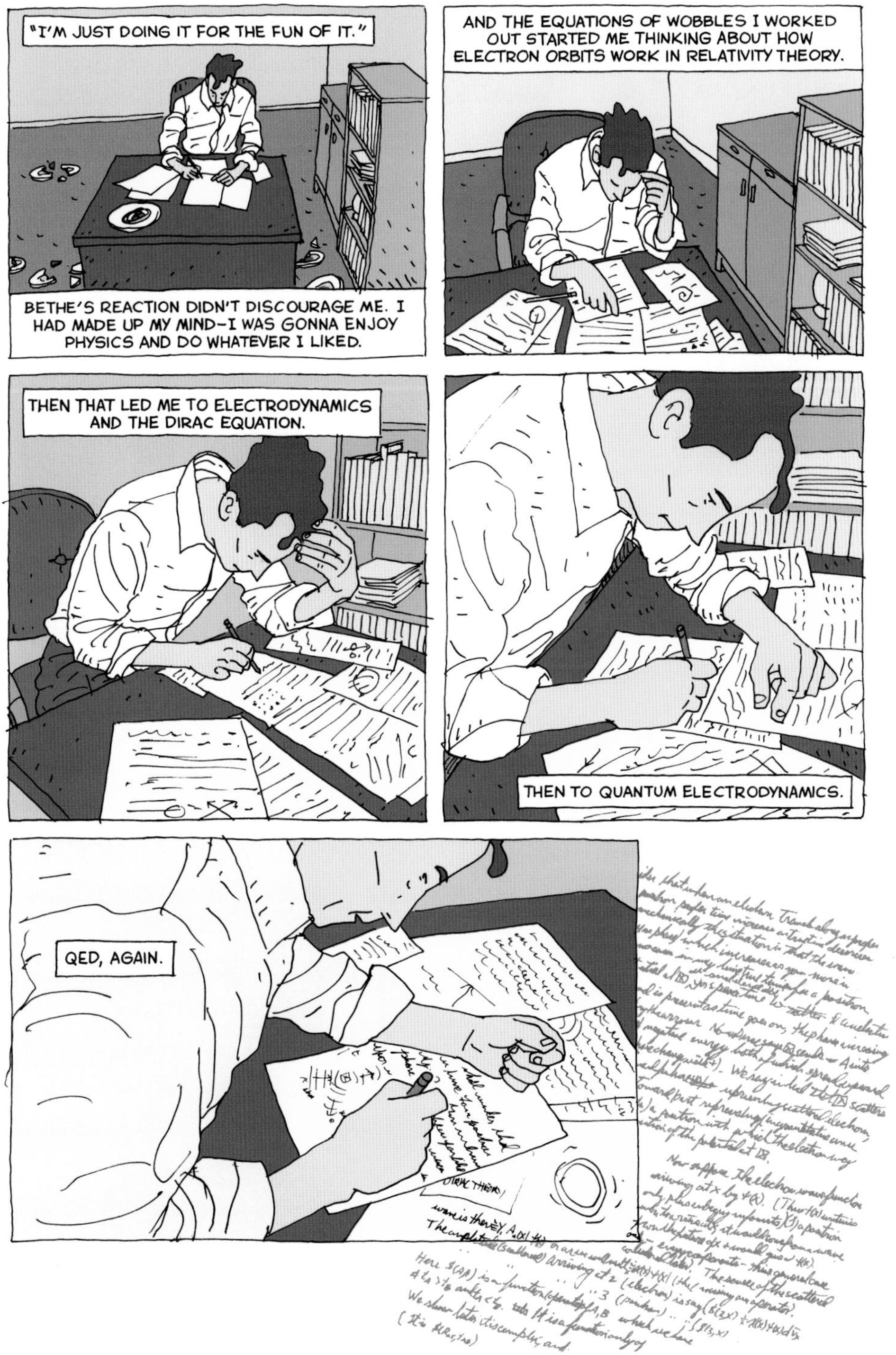
"I'M JUST DOING IT FOR THE FUN OF IT."
BETHE'S REACTION DIDN'T DISCOURAGE ME. I HAD MADE UP MY MIND—I WAS GONNA ENJOY PHYSICS AND DO WHATEVER I LIKED.
AND THE EQUATIONS OF WOBBLES I WORKED OUT STARTED ME THINKING ABOUT HOW ELECTRON ORBITS WORK IN RELATIVITY THEORY.
THEN THAT LED ME TO ELECTRODYNAMICS AND THE DIRAC EQUATION.
THEN TO QUANTUM ELECTRODYNAMICS.
QED, AGAIN.

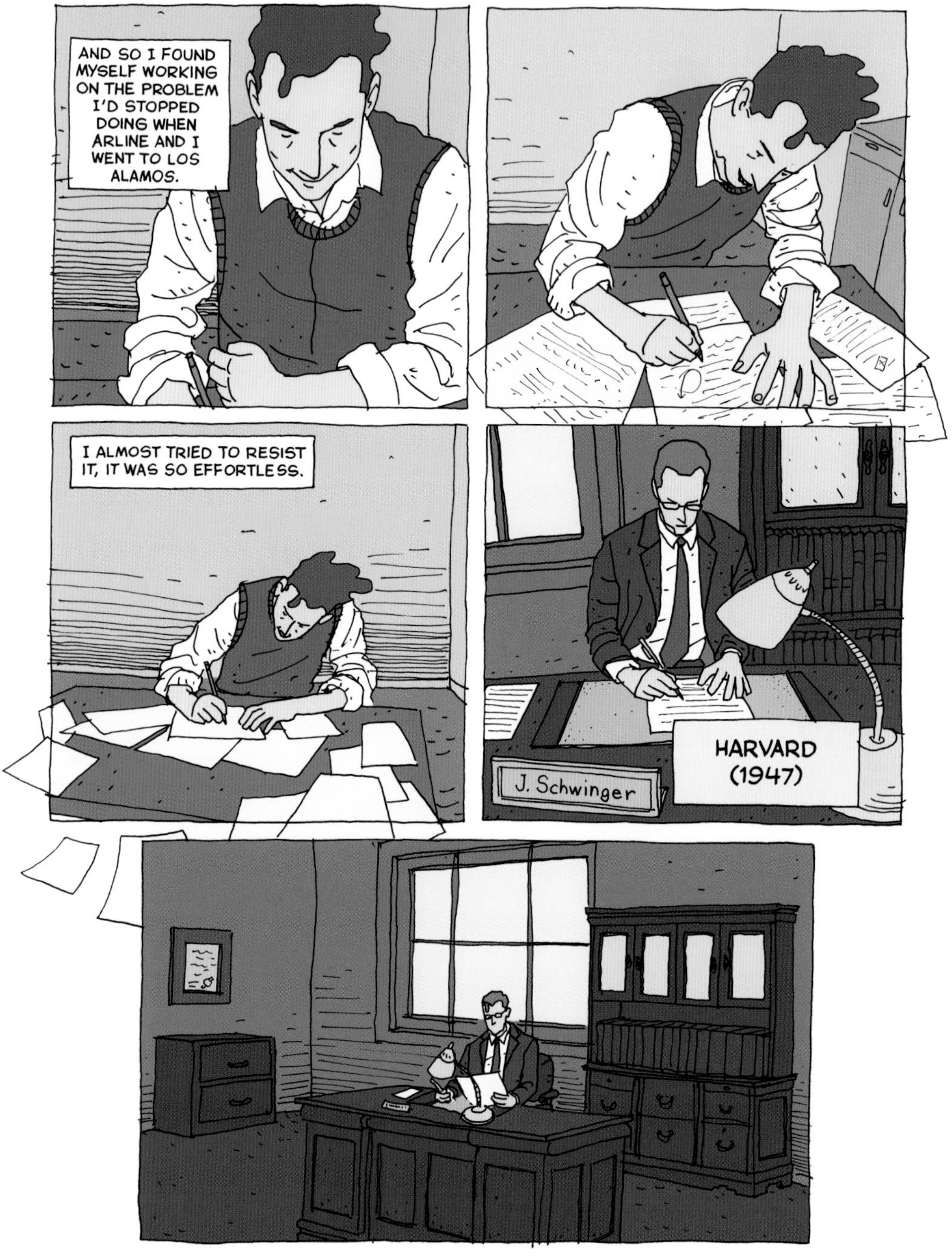
AND SO I FOUND MYSELF WORKING ON THE PROBLEM I'D STOPPED DOING WHEN ARLINE AND I WENT TO LOS ALAMOS.
I ALMOST TRIED TO RESIST IT, IT WAS SO EFFORTLESS.
J. Schwinger
HARVARD (1947)

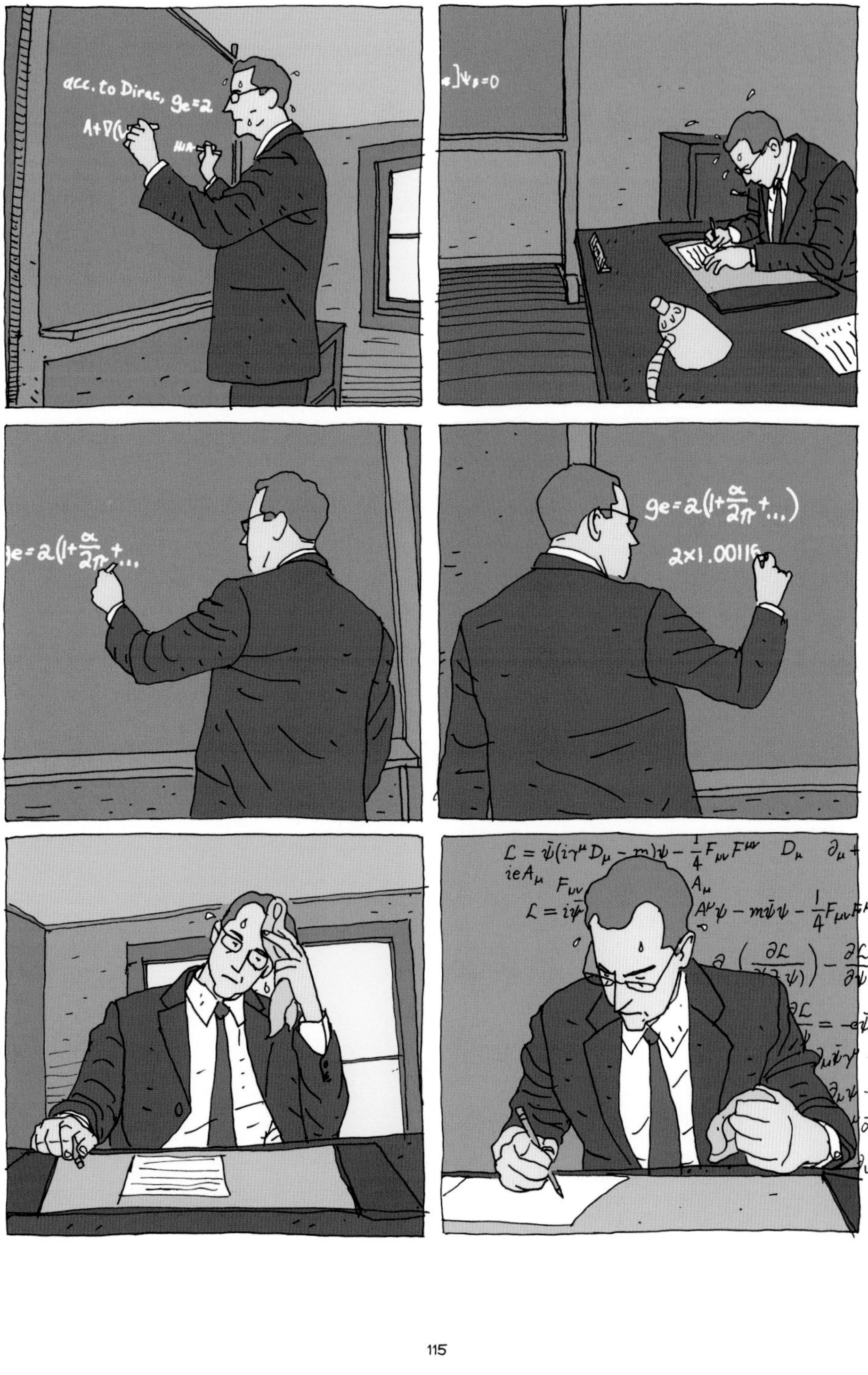
acc. to Dirac, ge=2
ge=2(1+α/2π+...
ge=2(1+α/2π+...)
2×1.00116
L = ψ̄(iγ^μ D_μ − m)ψ − 1/4 F_μν F^μν
D_μ
∂_μ +
ieA_μ
F_μν
A_μ

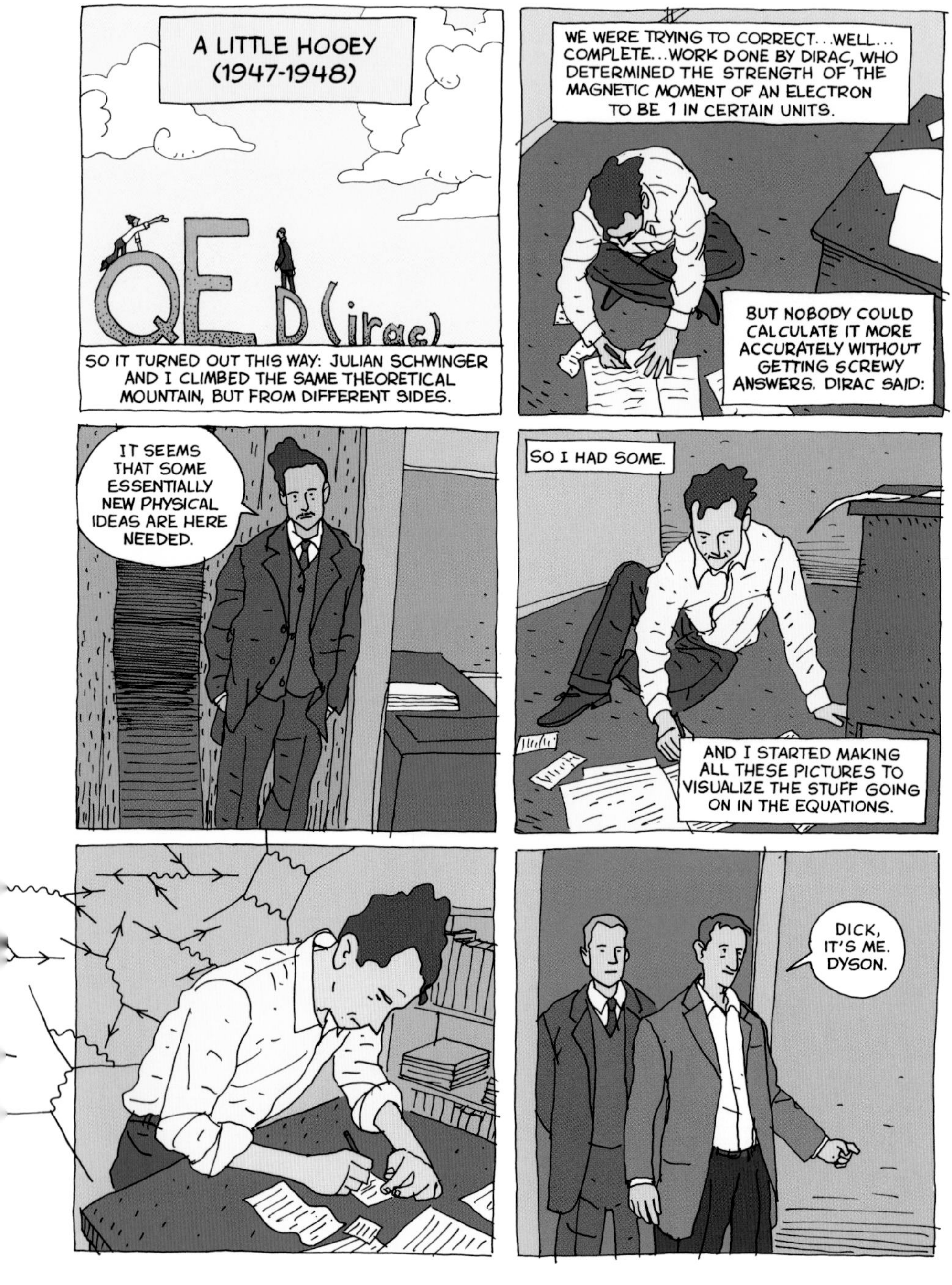

A LITTLE HOOEY (1947-1948)
QED (irac)
SO IT TURNED OUT THIS WAY: JULIAN SCHWINGER AND I CLIMBED THE SAME THEORETICAL MOUNTAIN, BUT FROM DIFFERENT SIDES.
WE WERE TRYING TO CORRECT...WELL... COMPLETE...WORK DONE BY DIRAC, WHO DETERMINED THE STRENGTH OF THE MAGNETIC MOMENT OF AN ELECTRON TO BE 1 IN CERTAIN UNITS.
BUT NOBODY COULD CALCULATE IT MORE ACCURATELY WITHOUT GETTING SCREWY ANSWERS. DIRAC SAID:
IT SEEMS THAT SOME ESSENTIALLY NEW PHYSICAL IDEAS ARE HERE NEEDED.
SO I HAD SOME.
AND I STARTED MAKING ALL THESE PICTURES TO VISUALIZE THE STUFF GOING ON IN THE EQUATIONS.
DICK, IT'S ME. DYSON.

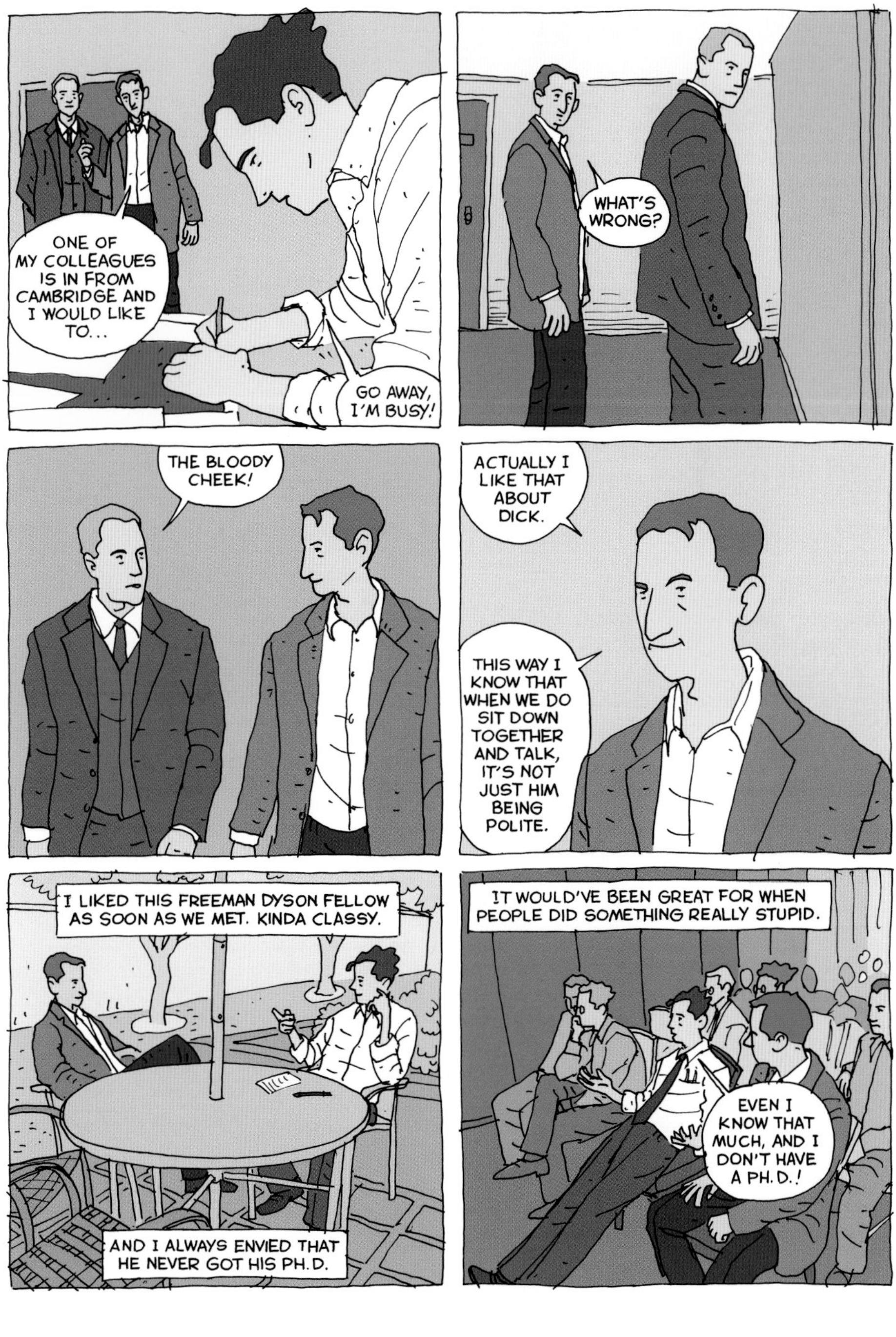
ONE OF MY COLLEAGUES IS IN FROM CAMBRIDGE AND I WOULD LIKE TO...
GO AWAY, I'M BUSY!
WHAT'S WRONG?
THE BLOODY CHEEK!
ACTUALLY I LIKE THAT ABOUT DICK.
THIS WAY I KNOW THAT WHEN WE DO SIT DOWN TOGETHER AND TALK, IT'S NOT JUST HIM BEING POLITE.
I LIKED THIS FREEMAN DYSON FELLOW AS SOON AS WE MET. KINDA CLASSY.
AND I ALWAYS ENVIED THAT HE NEVER GOT HIS PH.D.
IT WOULD'VE BEEN GREAT FOR WHEN PEOPLE DID SOMETHING REALLY STUPID.
EVEN I KNOW THAT MUCH, AND I DON'T HAVE A PH.D.!

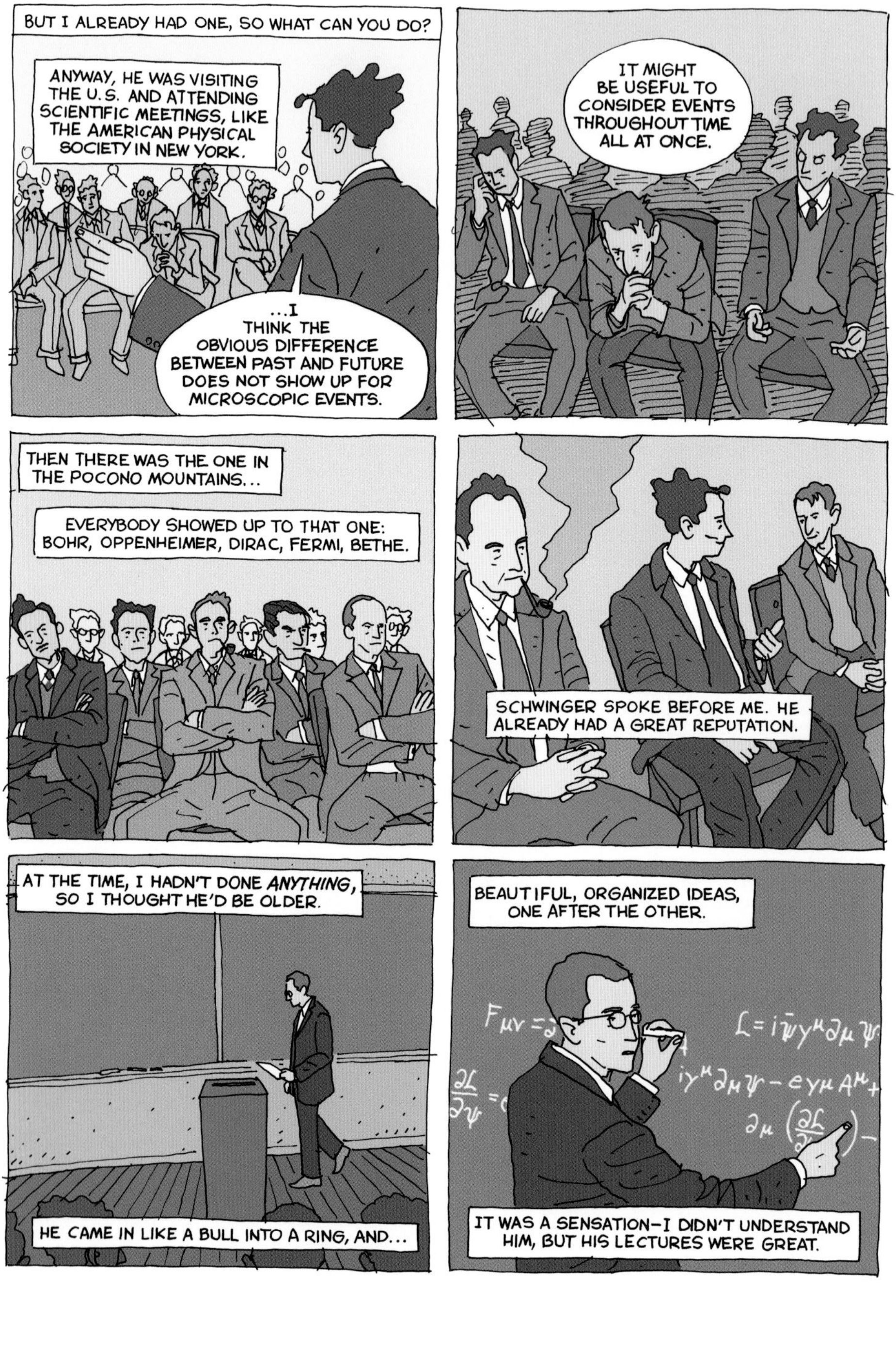
BUT I ALREADY HAD ONE, SO WHAT CAN YOU DO?
ANYWAY, HE WAS VISITING THE U.S. AND ATTENDING SCIENTIFIC MEETINGS, LIKE THE AMERICAN PHYSICAL SOCIETY IN NEW YORK.
...I THINK THE OBVIOUS DIFFERENCE BETWEEN PAST AND FUTURE DOES NOT SHOW UP FOR MICROSCOPIC EVENTS.
IT MIGHT BE USEFUL TO CONSIDER EVENTS THROUGHOUT TIME ALL AT ONCE.
THEN THERE WAS THE ONE IN THE POCONO MOUNTAINS...
EVERYBODY SHOWED UP TO THAT ONE: BOHR, OPPENHEIMER, DIRAC, FERMI, BETHE.
SCHWINGER SPOKE BEFORE ME. HE ALREADY HAD A GREAT REPUTATION.
AT THE TIME, I HADN'T DONE *ANYTHING*, SO I THOUGHT HE'D BE OLDER.
HE CAME IN LIKE A BULL INTO A RING, AND...
BEAUTIFUL, ORGANIZED IDEAS, ONE AFTER THE OTHER.
IT WAS A SENSATION—I DIDN'T UNDERSTAND HIM, BUT HIS LECTURES WERE GREAT.

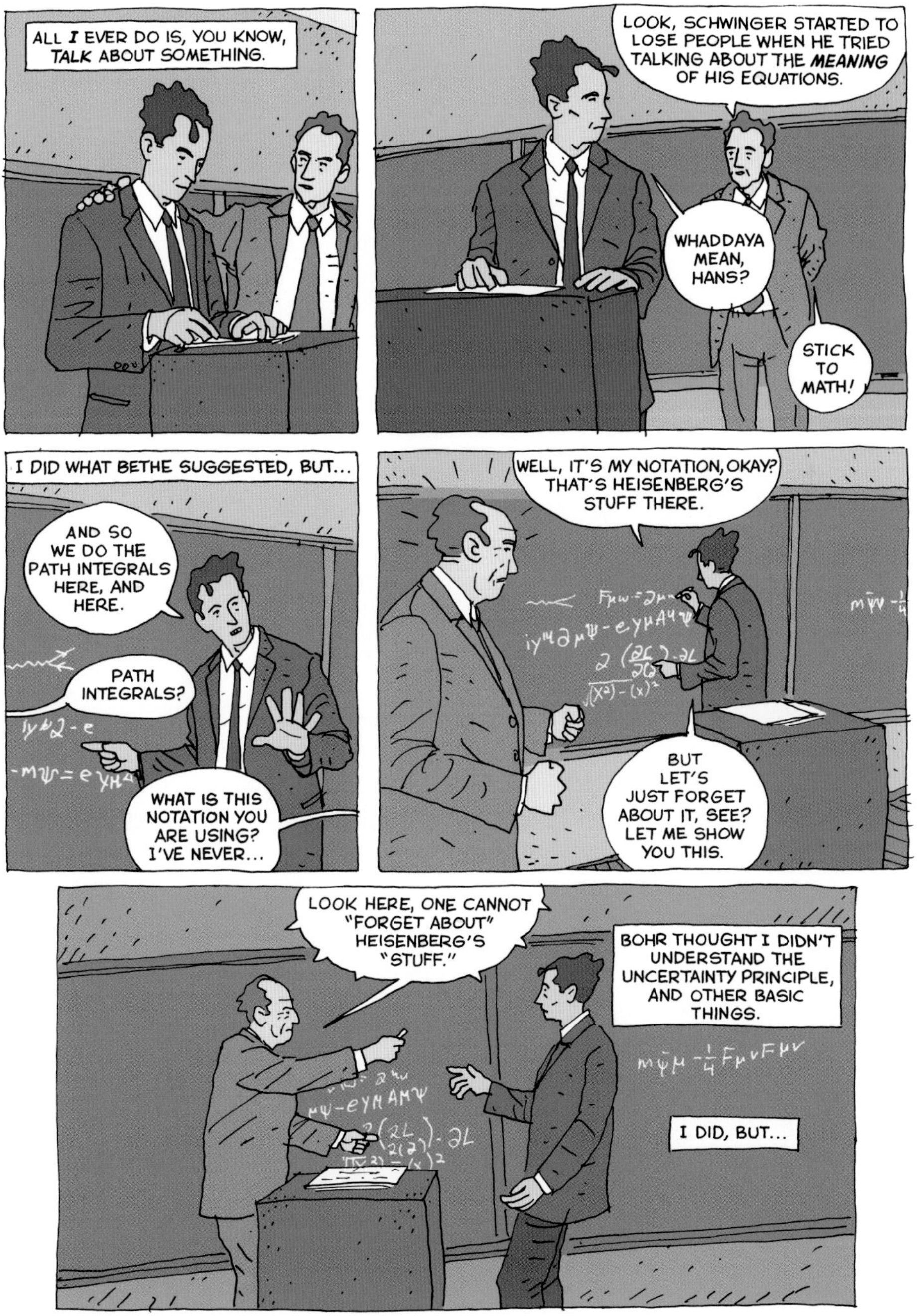
ALL I EVER DO IS, YOU KNOW, TALK ABOUT SOMETHING.
LOOK, SCHWINGER STARTED TO LOSE PEOPLE WHEN HE TRIED TALKING ABOUT THE MEANING OF HIS EQUATIONS.
WHADDAYA MEAN, HANS?
STICK TO MATH!
I DID WHAT BETHE SUGGESTED, BUT...
AND SO WE DO THE PATH INTEGRALS HERE, AND HERE.
PATH INTEGRALS?
WHAT IS THIS NOTATION YOU ARE USING? I'VE NEVER...
WELL, IT'S MY NOTATION, OKAY? THAT'S HEISENBERG'S STUFF THERE.
BUT LET'S JUST FORGET ABOUT IT, SEE? LET ME SHOW YOU THIS.
LOOK HERE, ONE CANNOT "FORGET ABOUT" HEISENBERG'S "STUFF."
BOHR THOUGHT I DIDN'T UNDERSTAND THE UNCERTAINTY PRINCIPLE, AND OTHER BASIC THINGS.
I DID, BUT...

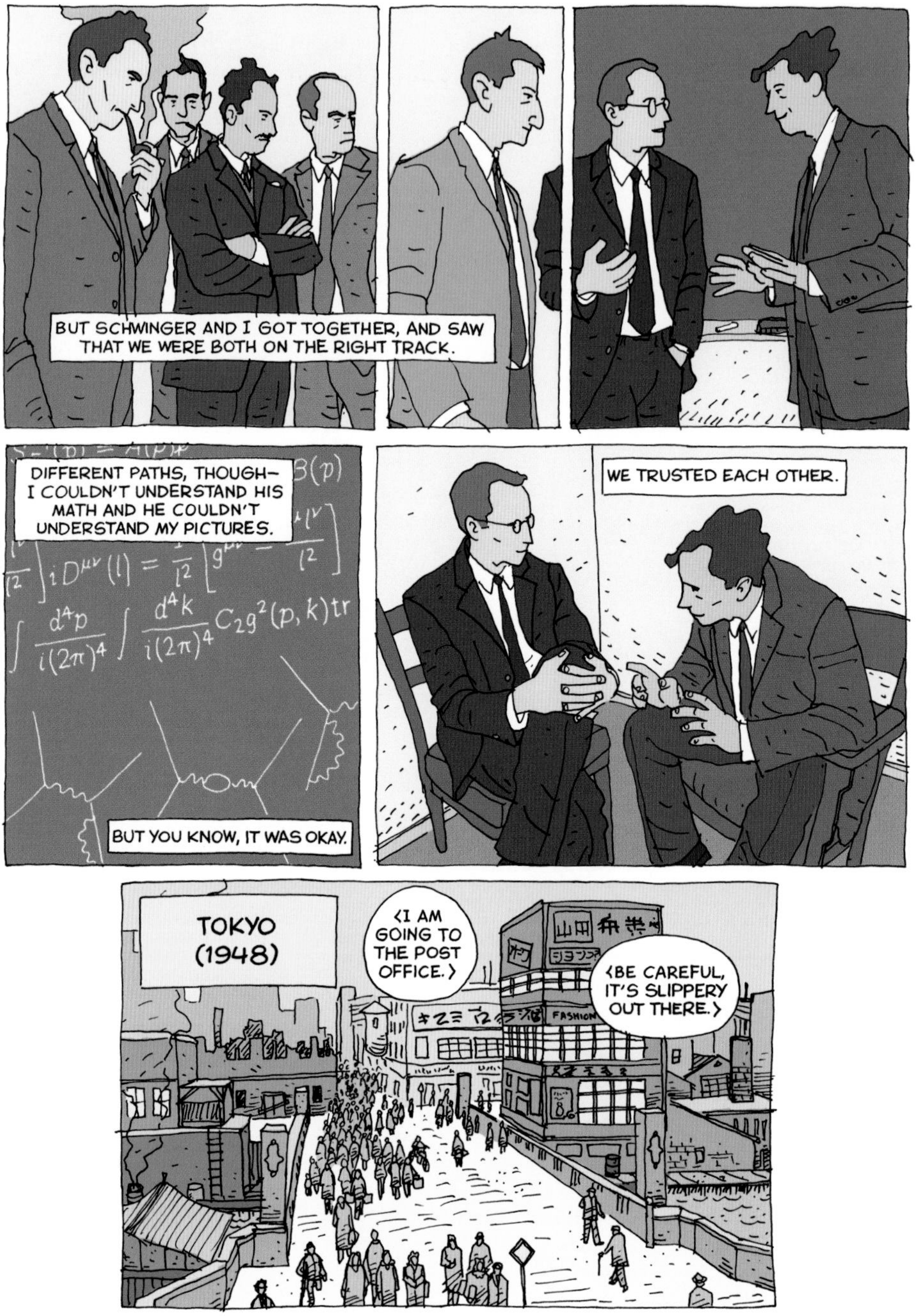
BUT SCHWINGER AND I GOT TOGETHER, AND SAW THAT WE WERE BOTH ON THE RIGHT TRACK.
DIFFERENT PATHS, THOUGH– I COULDN'T UNDERSTAND HIS MATH AND HE COULDN'T UNDERSTAND MY PICTURES.
BUT YOU KNOW, IT WAS OKAY.
WE TRUSTED EACH OTHER.
TOKYO (1948)
‹I AM GOING TO THE POST OFFICE.›
‹BE CAREFUL, IT'S SLIPPERY OUT THERE.›
FASHION

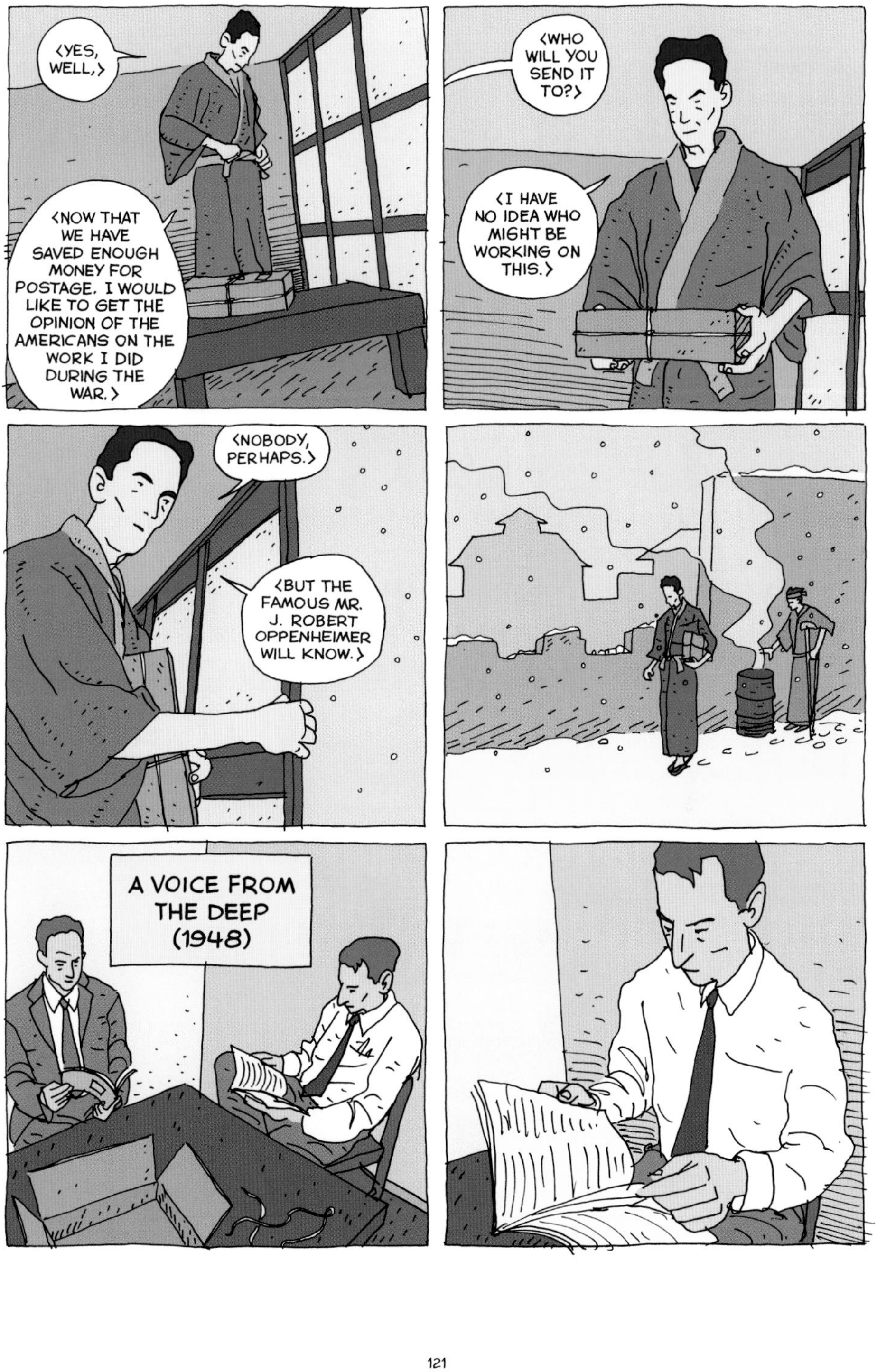
⟨YES, WELL,⟩
⟨NOW THAT WE HAVE SAVED ENOUGH MONEY FOR POSTAGE, I WOULD LIKE TO GET THE OPINION OF THE AMERICANS ON THE WORK I DID DURING THE WAR.⟩
⟨WHO WILL YOU SEND IT TO?⟩
⟨I HAVE NO IDEA WHO MIGHT BE WORKING ON THIS.⟩
⟨NOBODY, PERHAPS.⟩
⟨BUT THE FAMOUS MR. J. ROBERT OPPENHEIMER WILL KNOW.⟩
A VOICE FROM THE DEEP (1948)

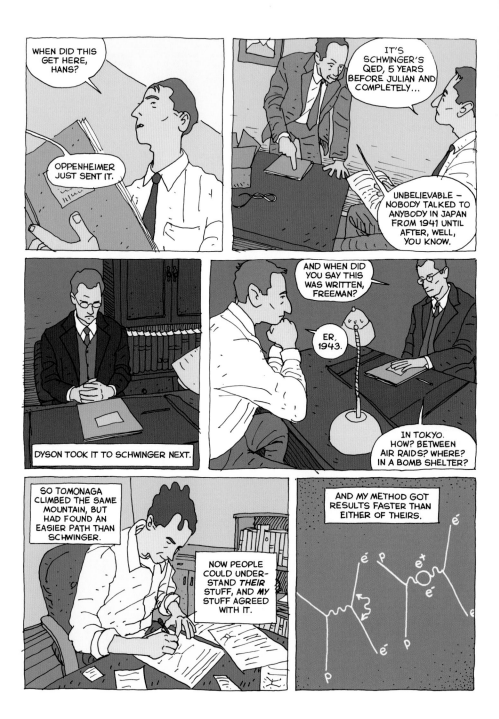

WHEN DID THIS GET HERE, HANS?

OPPENHEIMER JUST SENT IT.

IT'S SCHWINGER'S QED, 5 YEARS BEFORE JULIAN AND COMPLETELY...

UNBELIEVABLE — NOBODY TALKED TO ANYBODY IN JAPAN FROM 1941 UNTIL AFTER, WELL, YOU KNOW,

DYSON TOOK IT TO SCHWINGER NEXT.

AND WHEN DID YOU SAY THIS WAS WRITTEN, FREEMAN?

ER, 1943.

IN TOKYO. HOW? BETWEEN AIR RAIDS? WHERE? IN A BOMB SHELTER?

SO TOMONAGA CLIMBED THE SAME MOUNTAIN, BUT HAD FOUND AN EASIER PATH THAN SCHWINGER.

NOW PEOPLE COULD UNDER-STAND *THEIR* STUFF, AND *MY* STUFF AGREED WITH IT.

AND MY METHOD GOT RESULTS FASTER THAN EITHER OF THEIRS.

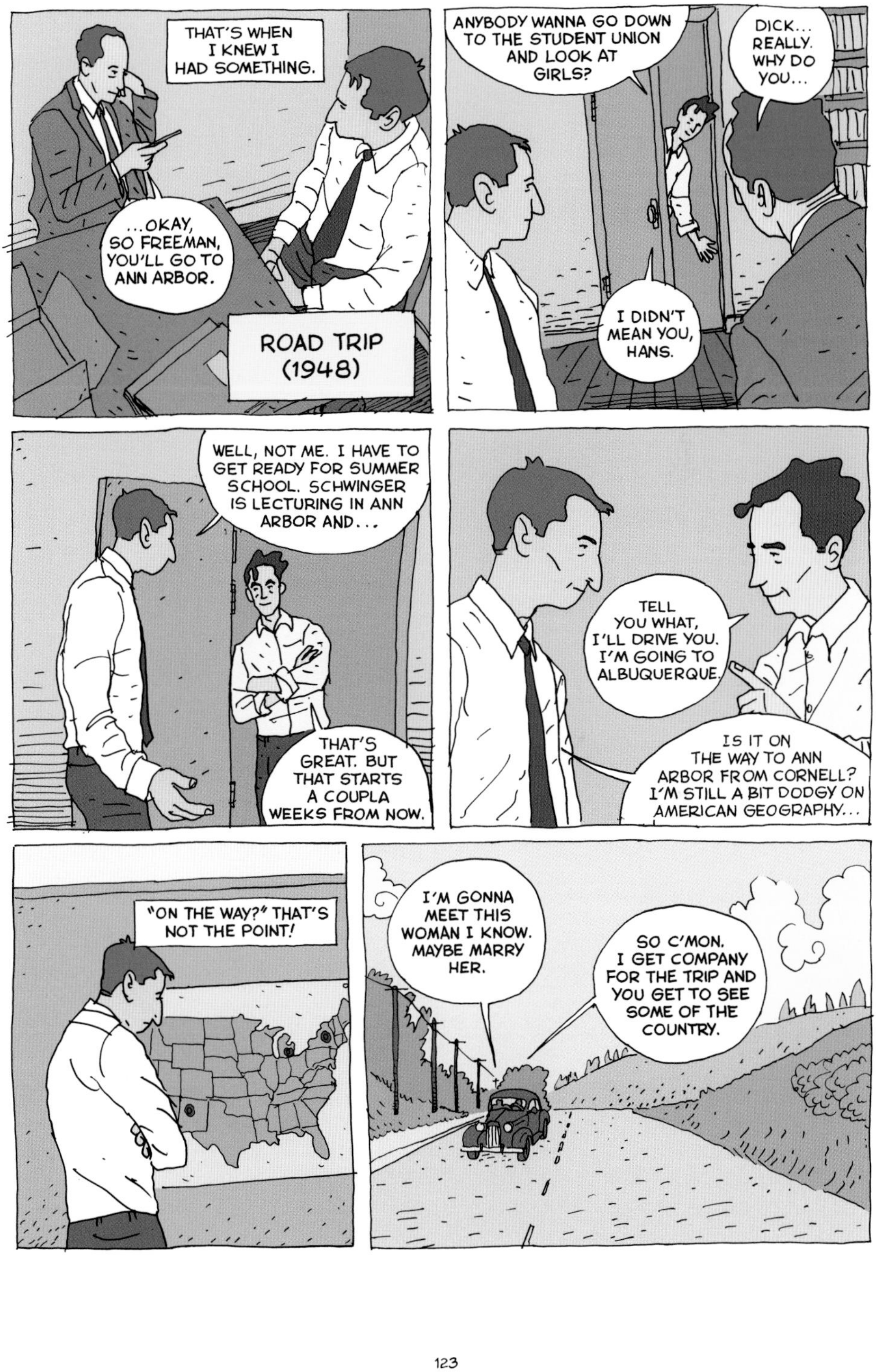
THAT'S WHEN I KNEW I HAD SOMETHING.
...OKAY, SO FREEMAN, YOU'LL GO TO ANN ARBOR.
ROAD TRIP (1948)
ANYBODY WANNA GO DOWN TO THE STUDENT UNION AND LOOK AT GIRLS?
DICK... REALLY. WHY DO YOU...
I DIDN'T MEAN YOU, HANS.
WELL, NOT ME. I HAVE TO GET READY FOR SUMMER SCHOOL. SCHWINGER IS LECTURING IN ANN ARBOR AND...
THAT'S GREAT. BUT THAT STARTS A COUPLA WEEKS FROM NOW.
TELL YOU WHAT, I'LL DRIVE YOU. I'M GOING TO ALBUQUERQUE.
IS IT ON THE WAY TO ANN ARBOR FROM CORNELL? I'M STILL A BIT DODGY ON AMERICAN GEOGRAPHY...
"ON THE WAY?" THAT'S NOT THE POINT!
I'M GONNA MEET THIS WOMAN I KNOW. MAYBE MARRY HER.
SO C'MON. I GET COMPANY FOR THE TRIP AND YOU GET TO SEE SOME OF THE COUNTRY.

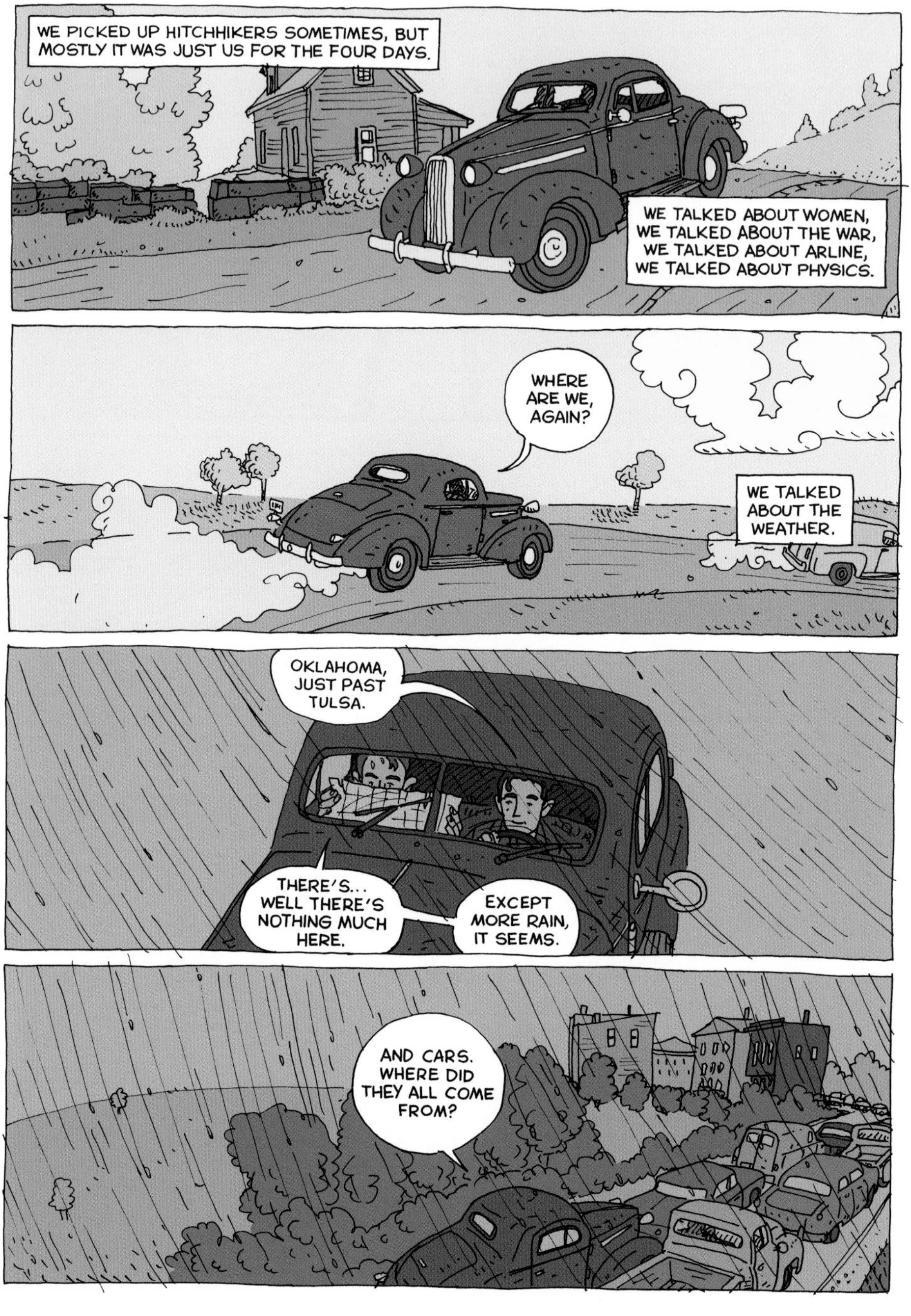
WE PICKED UP HITCHHIKERS SOMETIMES, BUT MOSTLY IT WAS JUST US FOR THE FOUR DAYS.
WE TALKED ABOUT WOMEN, WE TALKED ABOUT THE WAR, WE TALKED ABOUT ARLINE, WE TALKED ABOUT PHYSICS.
WHERE ARE WE, AGAIN?
WE TALKED ABOUT THE WEATHER.
OKLAHOMA, JUST PAST TULSA.
THERE'S... WELL THERE'S NOTHING MUCH HERE.
EXCEPT MORE RAIN, IT SEEMS.
AND CARS. WHERE DID THEY ALL COME FROM?

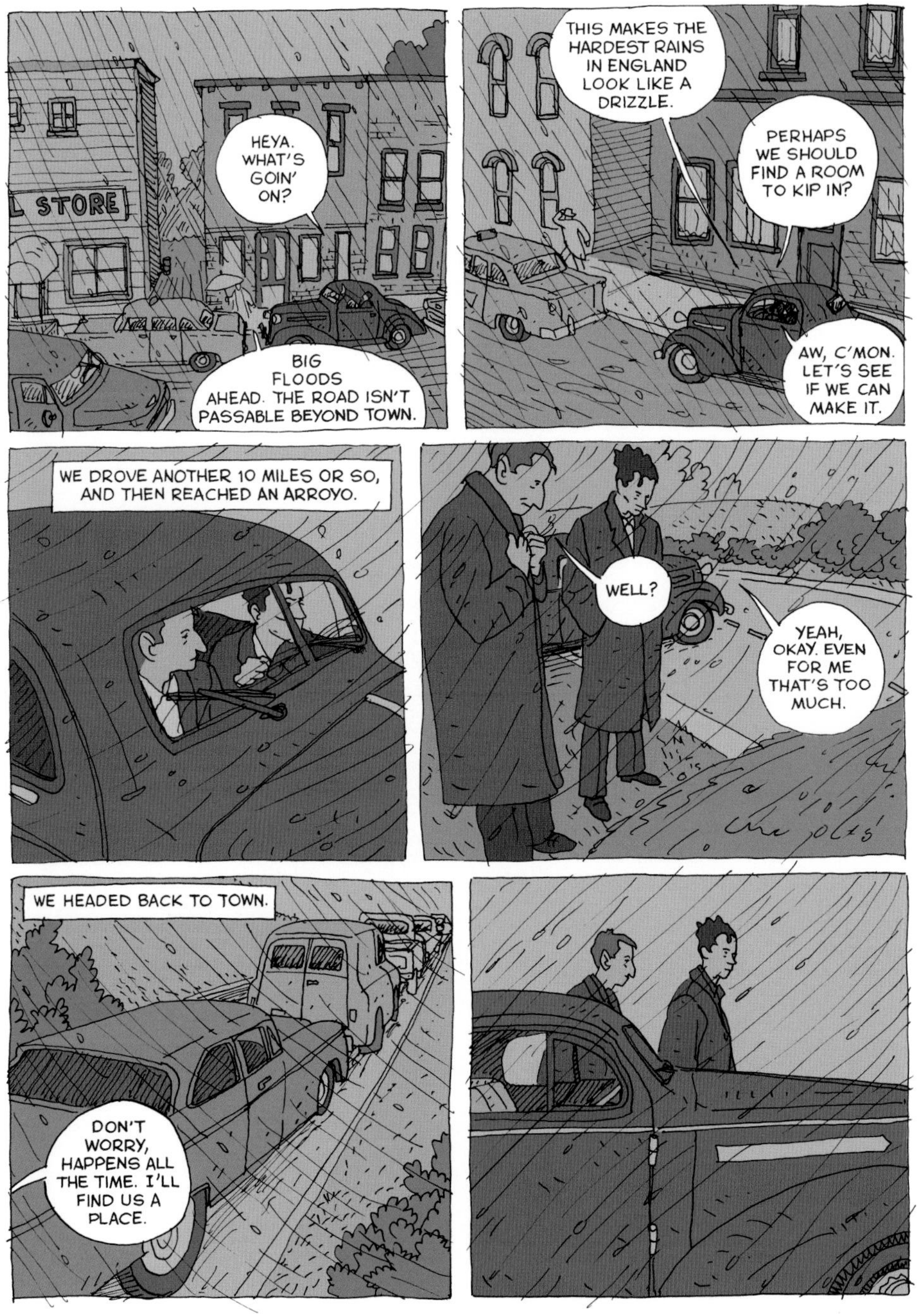

L STORE
HEYA. WHAT'S GOIN' ON?
BIG FLOODS AHEAD. THE ROAD ISN'T PASSABLE BEYOND TOWN.
THIS MAKES THE HARDEST RAINS IN ENGLAND LOOK LIKE A DRIZZLE.
PERHAPS WE SHOULD FIND A ROOM TO KIP IN?
AW, C'MON. LET'S SEE IF WE CAN MAKE IT.
WE DROVE ANOTHER 10 MILES OR SO, AND THEN REACHED AN ARROYO.
WELL?
YEAH, OKAY. EVEN FOR ME THAT'S TOO MUCH.
WE HEADED BACK TO TOWN.
DON'T WORRY, HAPPENS ALL THE TIME. I'LL FIND US A PLACE.

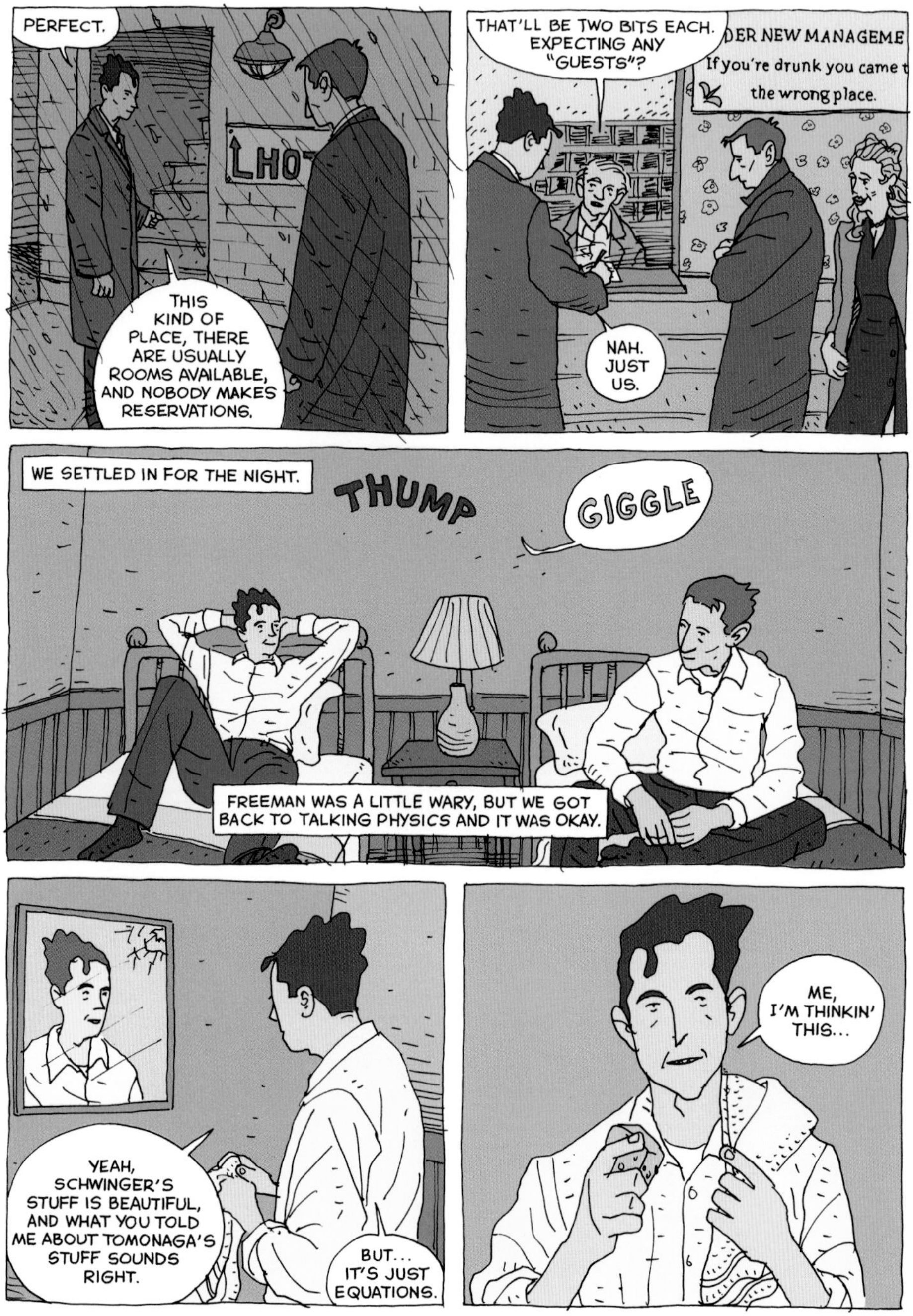
PERFECT.
THIS KIND OF PLACE, THERE ARE USUALLY ROOMS AVAILABLE, AND NOBODY MAKES RESERVATIONS.
LHOT
THAT'LL BE TWO BITS EACH. EXPECTING ANY "GUESTS"?
DER NEW MANAGEME
If you're drunk you came t
the wrong place.
NAH. JUST US.
WE SETTLED IN FOR THE NIGHT.
THUMP
GIGGLE
FREEMAN WAS A LITTLE WARY, BUT WE GOT BACK TO TALKING PHYSICS AND IT WAS OKAY.
YEAH, SCHWINGER'S STUFF IS BEAUTIFUL, AND WHAT YOU TOLD ME ABOUT TOMONAGA'S STUFF SOUNDS RIGHT.
BUT... IT'S JUST EQUATIONS.
ME, I'M THINKIN' THIS...

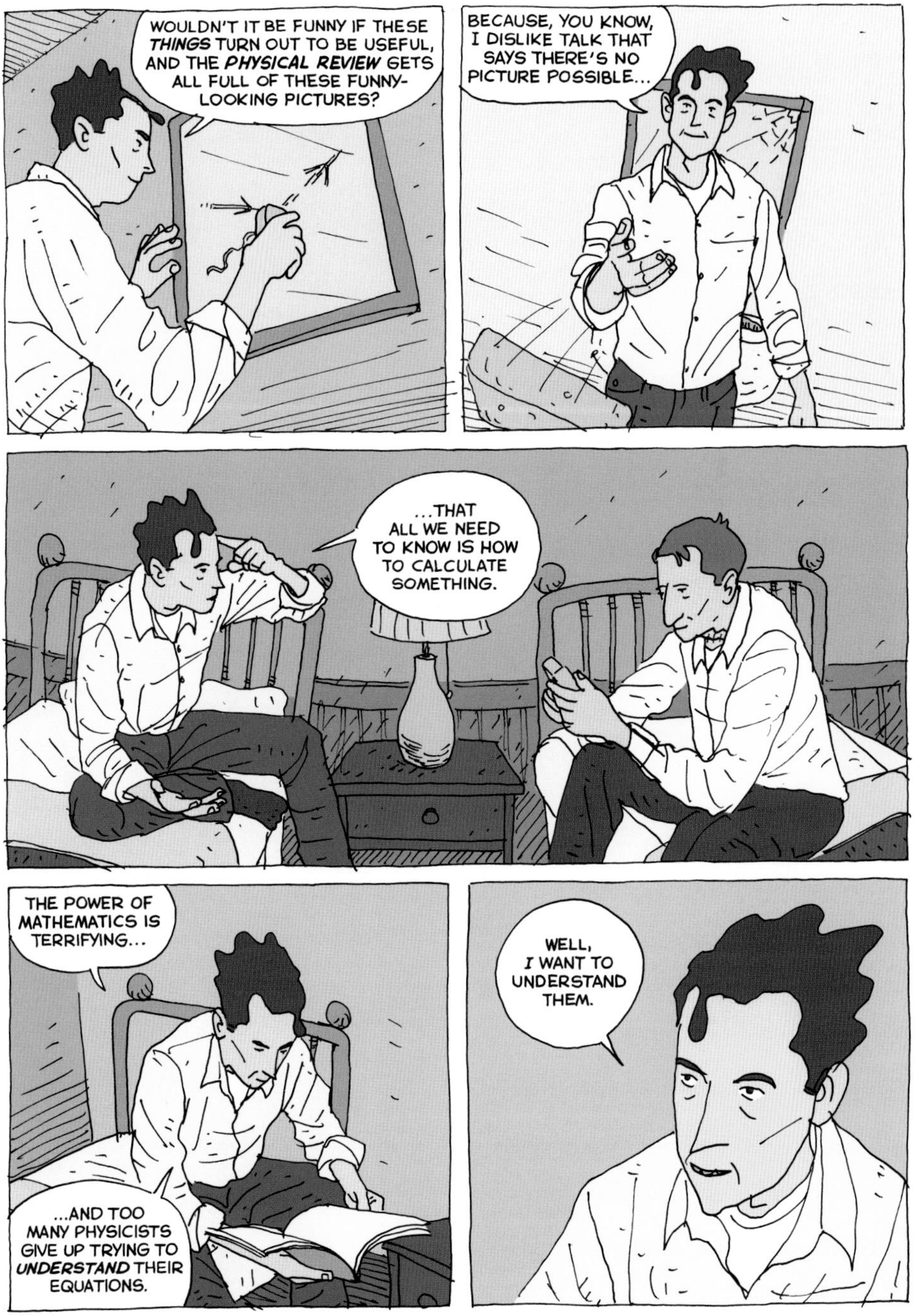
WOULDN'T IT BE FUNNY IF THESE *THINGS* TURN OUT TO BE USEFUL, AND THE ***PHYSICAL REVIEW*** GETS ALL FULL OF THESE FUNNY-LOOKING PICTURES?
BECAUSE, YOU KNOW, I DISLIKE TALK THAT SAYS THERE'S NO PICTURE POSSIBLE...
...THAT ALL WE NEED TO KNOW IS HOW TO CALCULATE SOMETHING.
THE POWER OF MATHEMATICS IS TERRIFYING...
...AND TOO MANY PHYSICISTS GIVE UP TRYING TO ***UNDERSTAND*** THEIR EQUATIONS.
WELL, *I* WANT TO UNDERSTAND THEM.

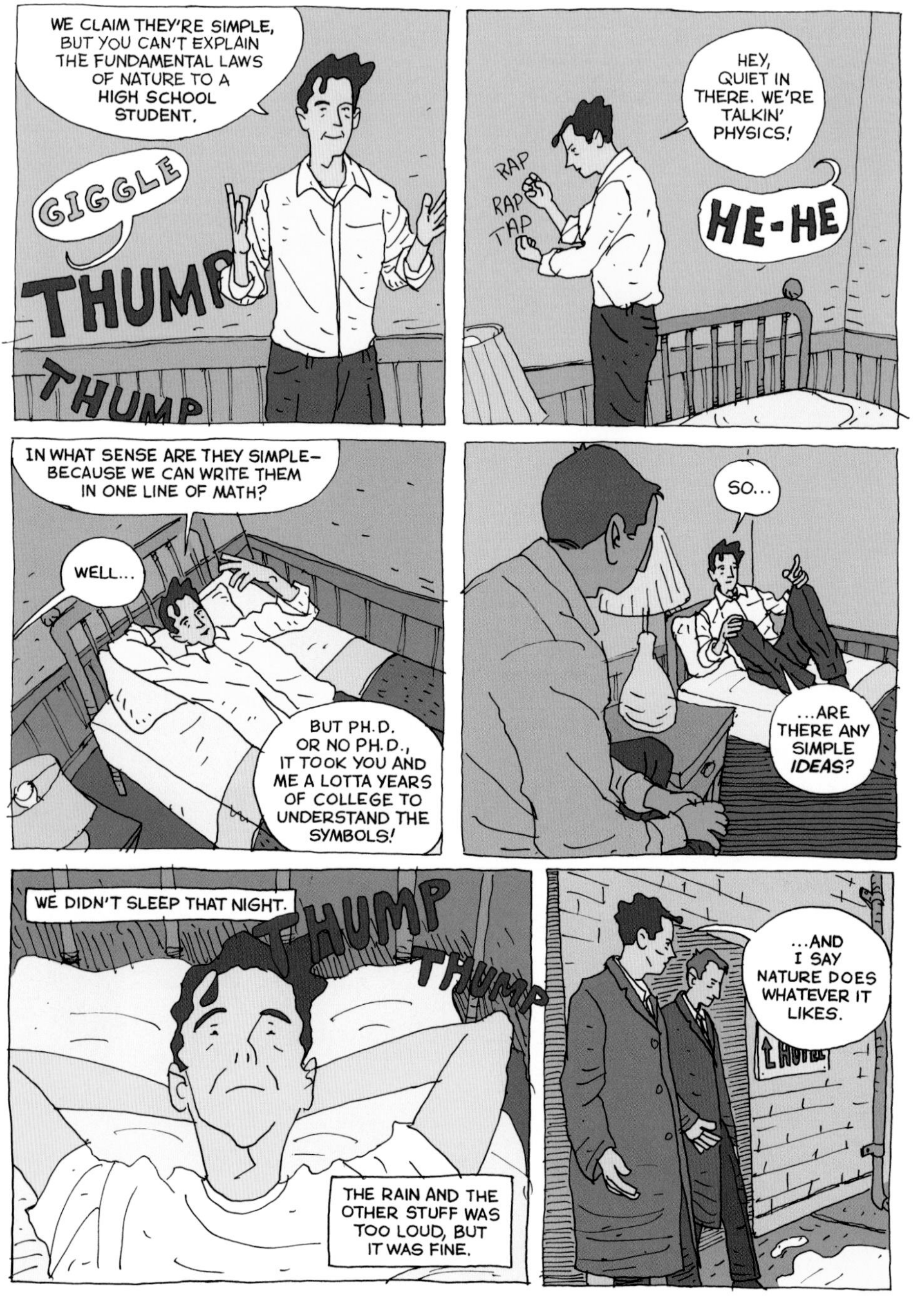
WE CLAIM THEY'RE SIMPLE, BUT YOU CAN'T EXPLAIN THE FUNDAMENTAL LAWS OF NATURE TO A HIGH SCHOOL STUDENT.
GIGGLE
THUMP
THUMP
HEY, QUIET IN THERE. WE'RE TALKIN' PHYSICS!
RAP
RAP
TAP
HE-HE
IN WHAT SENSE ARE THEY SIMPLE– BECAUSE WE CAN WRITE THEM IN ONE LINE OF MATH?
WELL...
BUT PH.D. OR NO PH.D., IT TOOK YOU AND ME A LOTTA YEARS OF COLLEGE TO UNDERSTAND THE SYMBOLS!
SO...
...ARE THERE ANY SIMPLE *IDEAS*?
WE DIDN'T SLEEP THAT NIGHT.
THUMP
THUMP
THE RAIN AND THE OTHER STUFF WAS TOO LOUD, BUT IT WAS FINE.
...AND I SAY NATURE DOES WHATEVER IT LIKES.

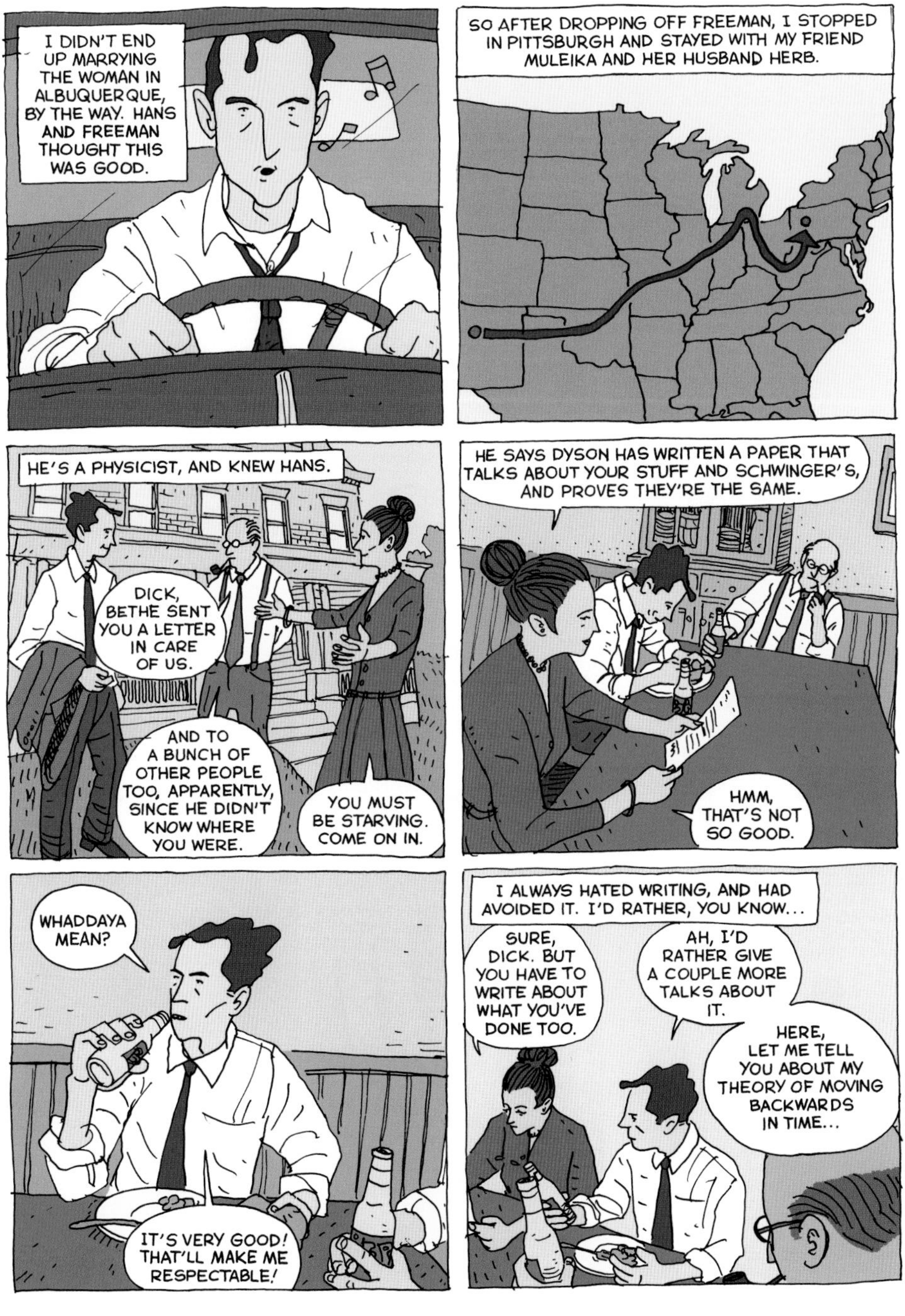
I DIDN'T END UP MARRYING THE WOMAN IN ALBUQUERQUE, BY THE WAY. HANS AND FREEMAN THOUGHT THIS WAS GOOD.
SO AFTER DROPPING OFF FREEMAN, I STOPPED IN PITTSBURGH AND STAYED WITH MY FRIEND MULEIKA AND HER HUSBAND HERB.
HE'S A PHYSICIST, AND KNEW HANS.
DICK, BETHE SENT YOU A LETTER IN CARE OF US.
AND TO A BUNCH OF OTHER PEOPLE TOO, APPARENTLY, SINCE HE DIDN'T KNOW WHERE YOU WERE.
YOU MUST BE STARVING. COME ON IN.
HE SAYS DYSON HAS WRITTEN A PAPER THAT TALKS ABOUT YOUR STUFF AND SCHWINGER'S, AND PROVES THEY'RE THE SAME.
HMM, THAT'S NOT SO GOOD.
WHADDAYA MEAN?
IT'S VERY GOOD! THAT'LL MAKE ME RESPECTABLE!
I ALWAYS HATED WRITING, AND HAD AVOIDED IT. I'D RATHER, YOU KNOW...
SURE, DICK. BUT YOU HAVE TO WRITE ABOUT WHAT YOU'VE DONE TOO.
AH, I'D RATHER GIVE A COUPLE MORE TALKS ABOUT IT.
HERE, LET ME TELL YOU ABOUT MY THEORY OF MOVING BACKWARDS IN TIME...

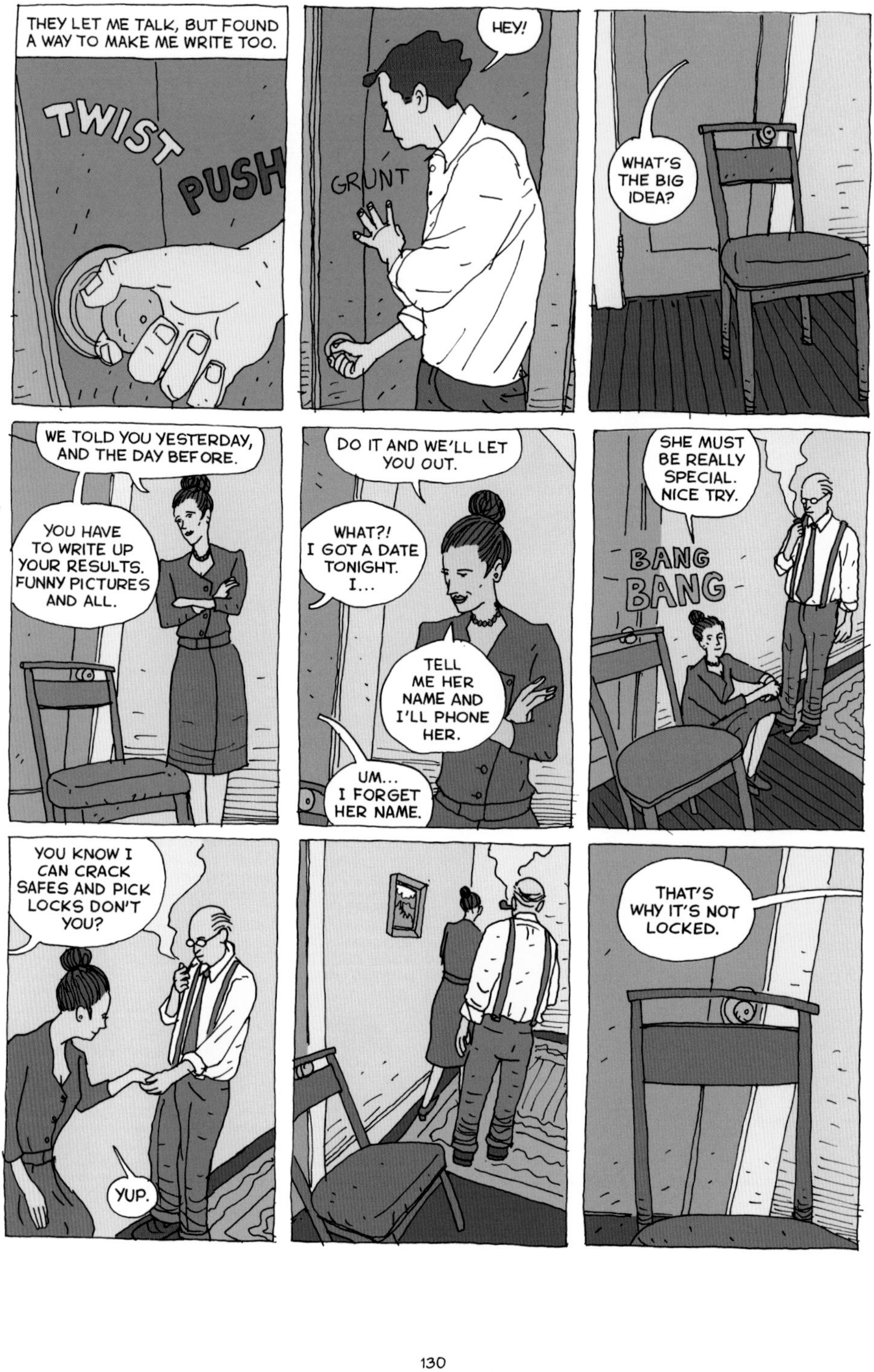
THEY LET ME TALK, BUT FOUND A WAY TO MAKE ME WRITE TOO.
TWIST
PUSH
HEY!
GRUNT
WHAT'S THE BIG IDEA?
WE TOLD YOU YESTERDAY, AND THE DAY BEFORE.
YOU HAVE TO WRITE UP YOUR RESULTS. FUNNY PICTURES AND ALL.
DO IT AND WE'LL LET YOU OUT.
WHAT?! I GOT A DATE TONIGHT. I...
TELL ME HER NAME AND I'LL PHONE HER.
UM... I FORGET HER NAME.
SHE MUST BE REALLY SPECIAL. NICE TRY.
BANG
BANG
YOU KNOW I CAN CRACK SAFES AND PICK LOCKS DON'T YOU?
YUP.
THAT'S WHY IT'S NOT LOCKED.

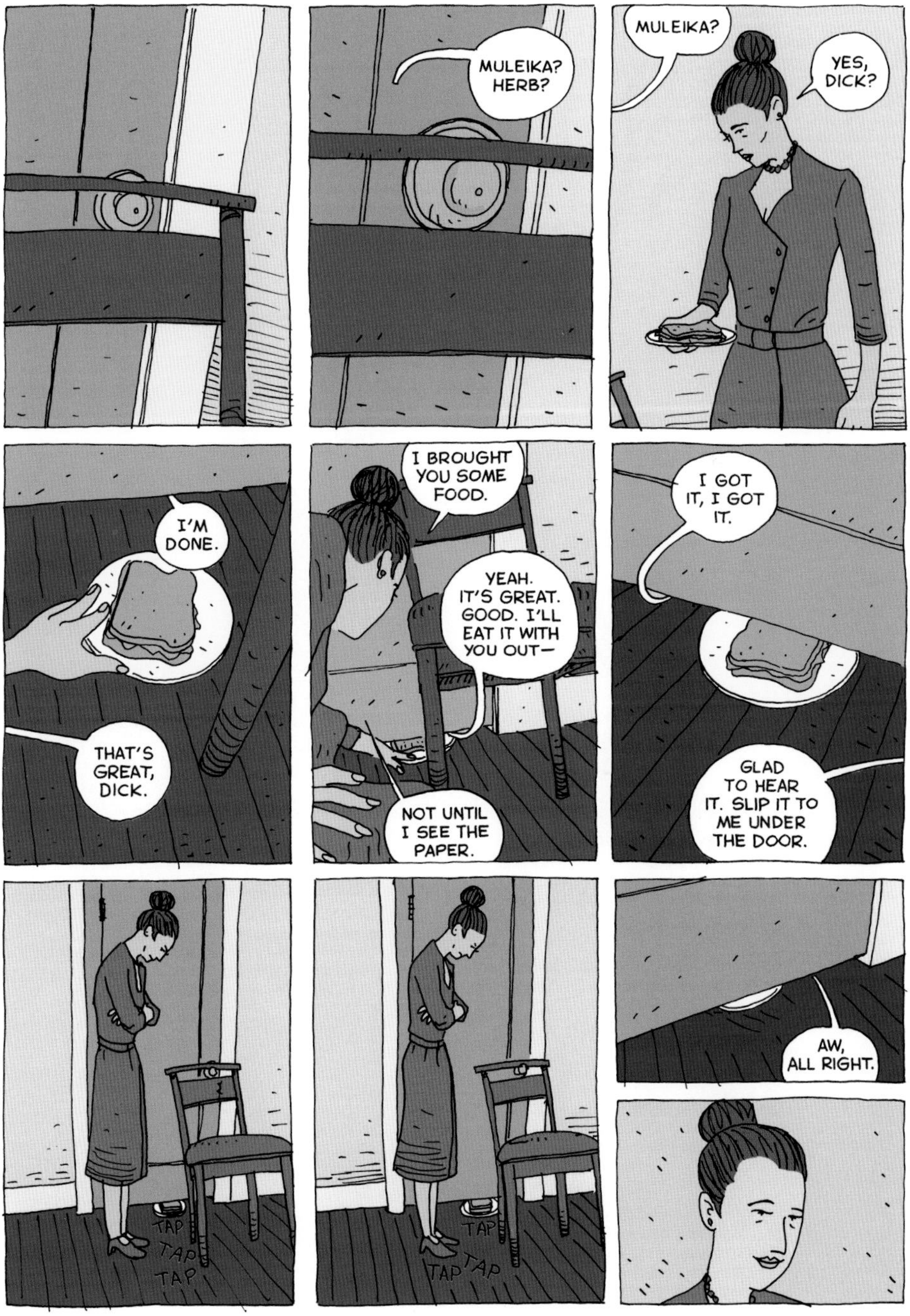
MULEIKA? HERB?
MULEIKA?
YES, DICK?
I'M DONE.
THAT'S GREAT, DICK.
I BROUGHT YOU SOME FOOD.
YEAH. IT'S GREAT. GOOD. I'LL EAT IT WITH YOU OUT—
NOT UNTIL I SEE THE PAPER.
I GOT IT, I GOT IT.
GLAD TO HEAR IT. SLIP IT TO ME UNDER THE DOOR.
TAP TAP TAP
TAP TAP TAP
AW, ALL RIGHT.

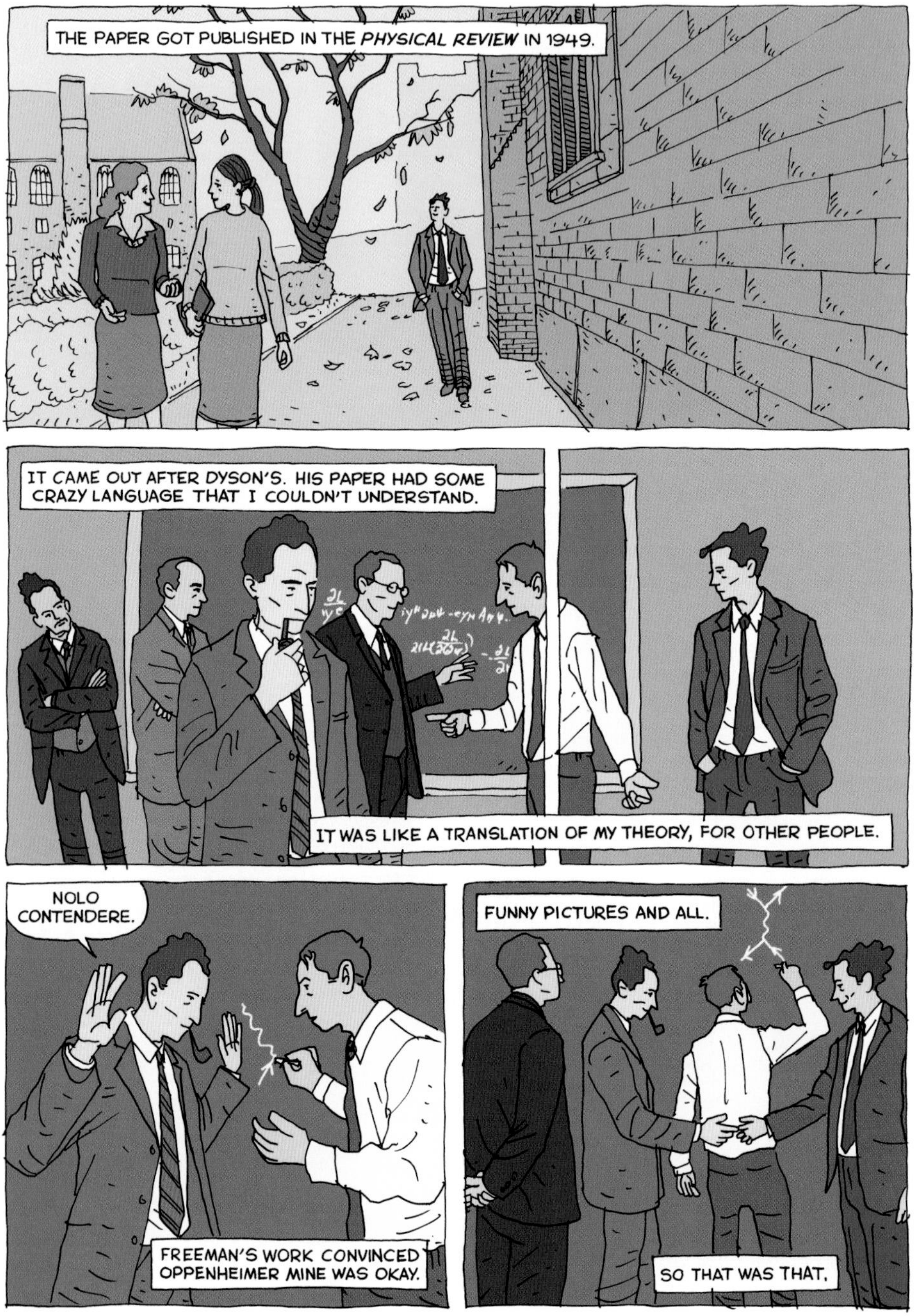
THE PAPER GOT PUBLISHED IN THE *PHYSICAL REVIEW* IN 1949.
IT CAME OUT AFTER DYSON'S. HIS PAPER HAD SOME CRAZY LANGUAGE THAT I COULDN'T UNDERSTAND.
IT WAS LIKE A TRANSLATION OF MY THEORY, FOR OTHER PEOPLE.
NOLO CONTENDERE.
FREEMAN'S WORK CONVINCED OPPENHEIMER MINE WAS OKAY.
FUNNY PICTURES AND ALL.
SO THAT WAS THAT.

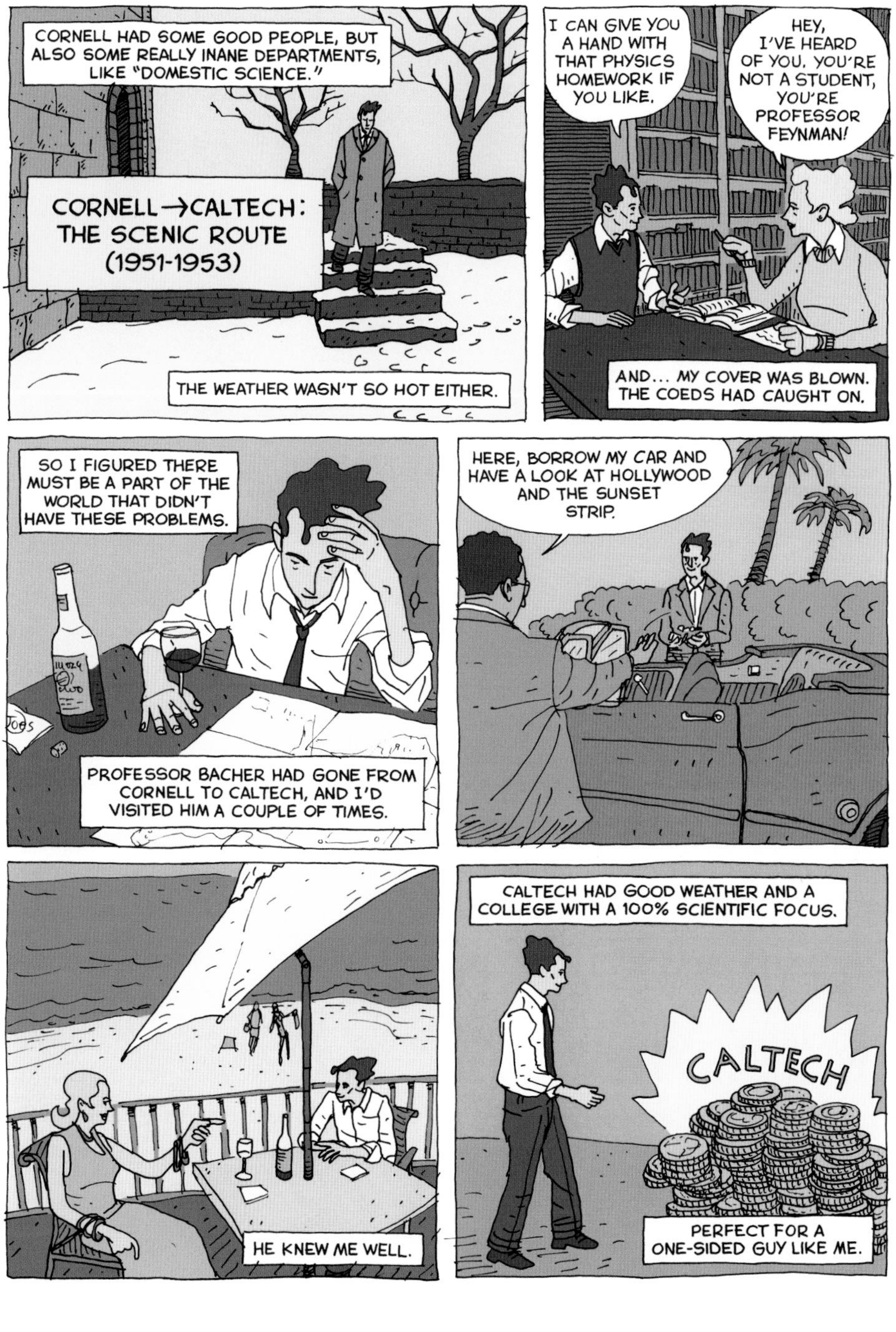
CORNELL HAD SOME GOOD PEOPLE, BUT ALSO SOME REALLY INANE DEPARTMENTS, LIKE "DOMESTIC SCIENCE."
CORNELL→CALTECH: THE SCENIC ROUTE (1951-1953)
THE WEATHER WASN'T SO HOT EITHER.
I CAN GIVE YOU A HAND WITH THAT PHYSICS HOMEWORK IF YOU LIKE.
HEY, I'VE HEARD OF YOU. YOU'RE NOT A STUDENT, YOU'RE PROFESSOR FEYNMAN!
AND... MY COVER WAS BLOWN. THE COEDS HAD CAUGHT ON.
SO I FIGURED THERE MUST BE A PART OF THE WORLD THAT DIDN'T HAVE THESE PROBLEMS.
PROFESSOR BACHER HAD GONE FROM CORNELL TO CALTECH, AND I'D VISITED HIM A COUPLE OF TIMES.
HERE, BORROW MY CAR AND HAVE A LOOK AT HOLLYWOOD AND THE SUNSET STRIP.
HE KNEW ME WELL.
CALTECH HAD GOOD WEATHER AND A COLLEGE WITH A 100% SCIENTIFIC FOCUS.
CALTECH
PERFECT FOR A ONE-SIDED GUY LIKE ME.

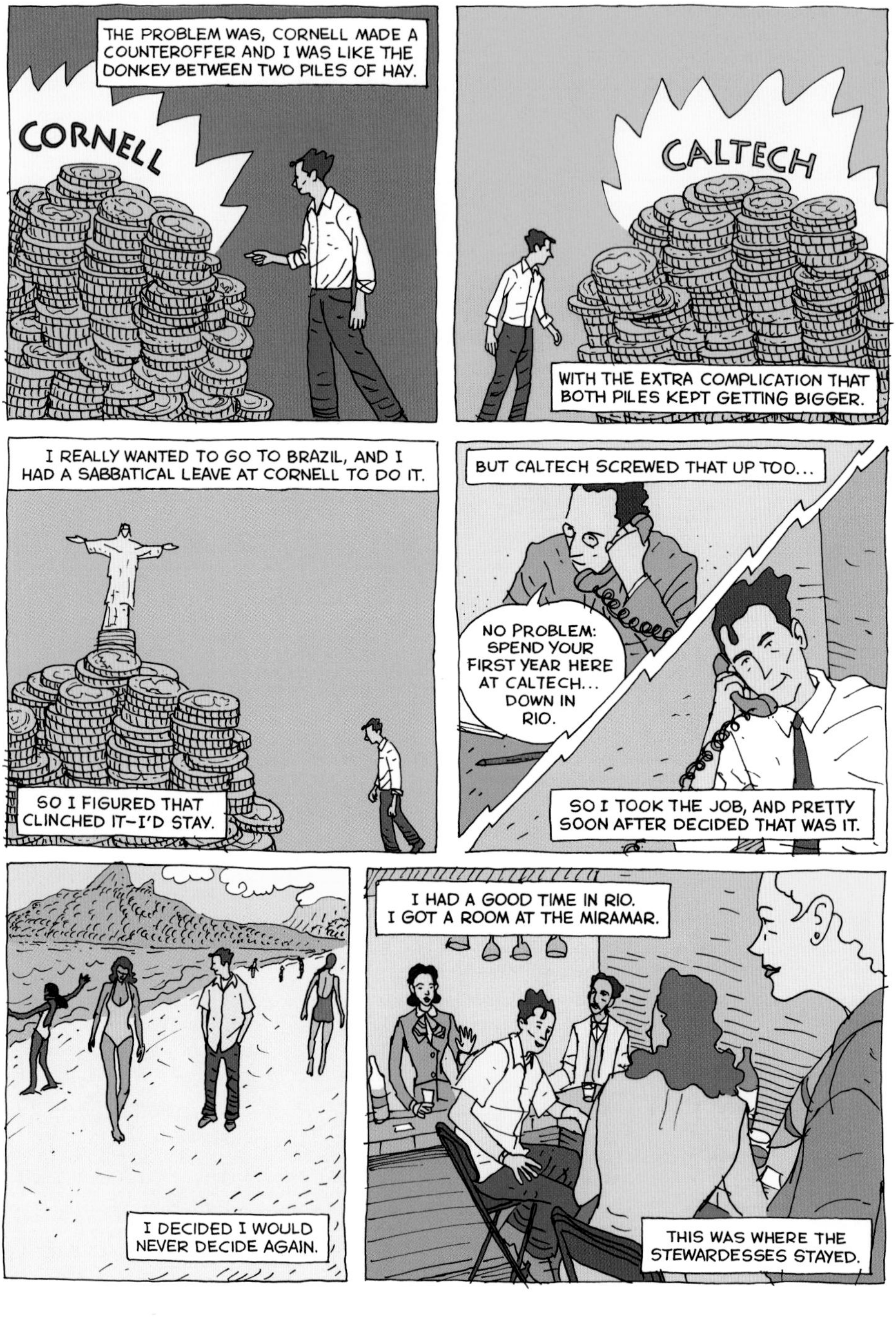
THE PROBLEM WAS, CORNELL MADE A COUNTEROFFER AND I WAS LIKE THE DONKEY BETWEEN TWO PILES OF HAY.
CORNELL
CALTECH
WITH THE EXTRA COMPLICATION THAT BOTH PILES KEPT GETTING BIGGER.
I REALLY WANTED TO GO TO BRAZIL, AND I HAD A SABBATICAL LEAVE AT CORNELL TO DO IT.
SO I FIGURED THAT CLINCHED IT—I'D STAY.
BUT CALTECH SCREWED THAT UP TOO...
NO PROBLEM: SPEND YOUR FIRST YEAR HERE AT CALTECH... DOWN IN RIO.
SO I TOOK THE JOB, AND PRETTY SOON AFTER DECIDED THAT WAS IT.
I DECIDED I WOULD NEVER DECIDE AGAIN.
I HAD A GOOD TIME IN RIO. I GOT A ROOM AT THE MIRAMAR.
THIS WAS WHERE THE STEWARDESSES STAYED.

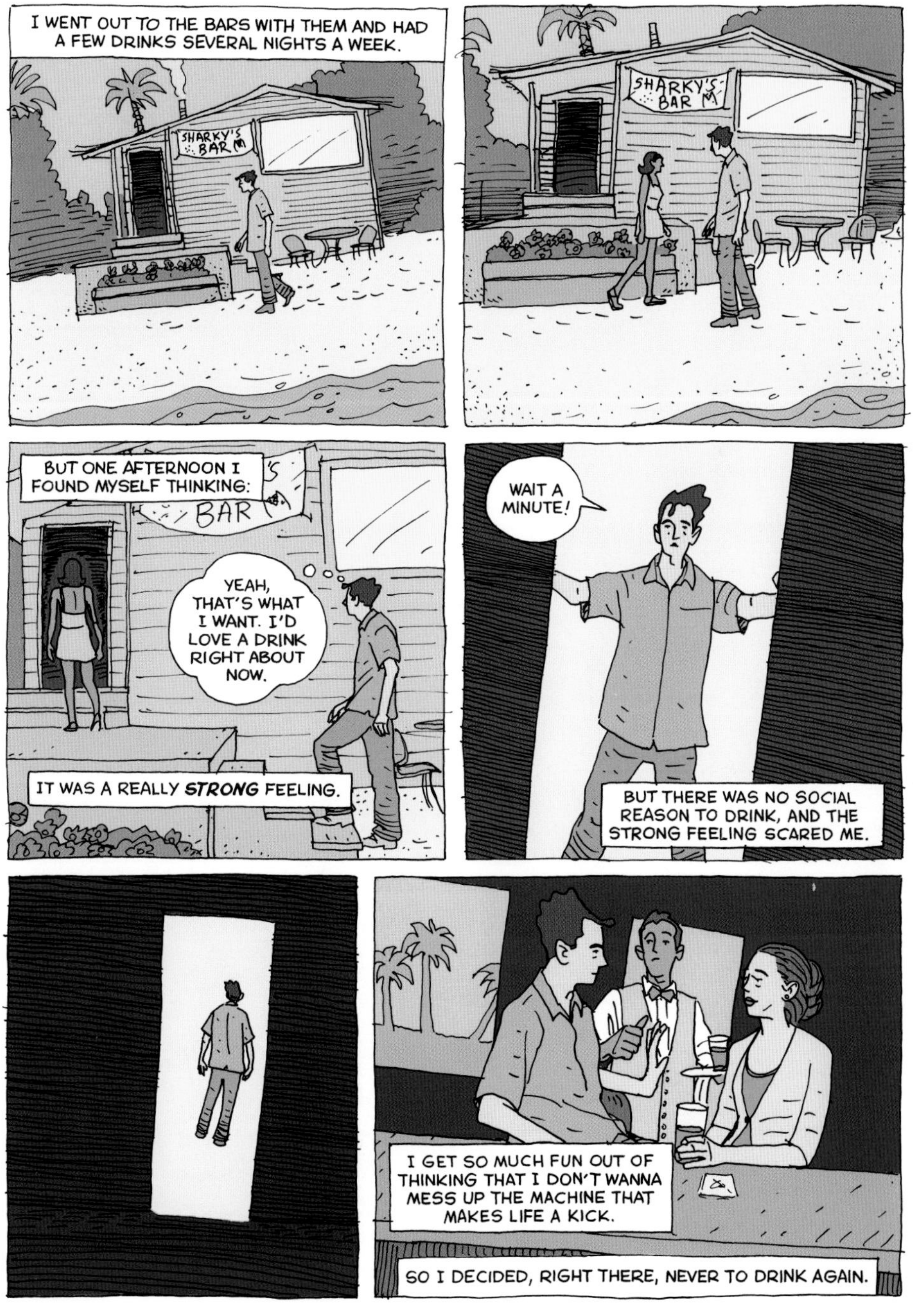
I WENT OUT TO THE BARS WITH THEM AND HAD A FEW DRINKS SEVERAL NIGHTS A WEEK.
SHARKY'S BAR
SHARKY'S BAR
BUT ONE AFTERNOON I FOUND MYSELF THINKING:
BAR
YEAH, THAT'S WHAT I WANT. I'D LOVE A DRINK RIGHT ABOUT NOW.
IT WAS A REALLY **STRONG** FEELING.
WAIT A MINUTE!
BUT THERE WAS NO SOCIAL REASON TO DRINK, AND THE STRONG FEELING SCARED ME.
I GET SO MUCH FUN OUT OF THINKING THAT I DON'T WANNA MESS UP THE MACHINE THAT MAKES LIFE A KICK.
SO I DECIDED, RIGHT THERE, NEVER TO DRINK AGAIN.

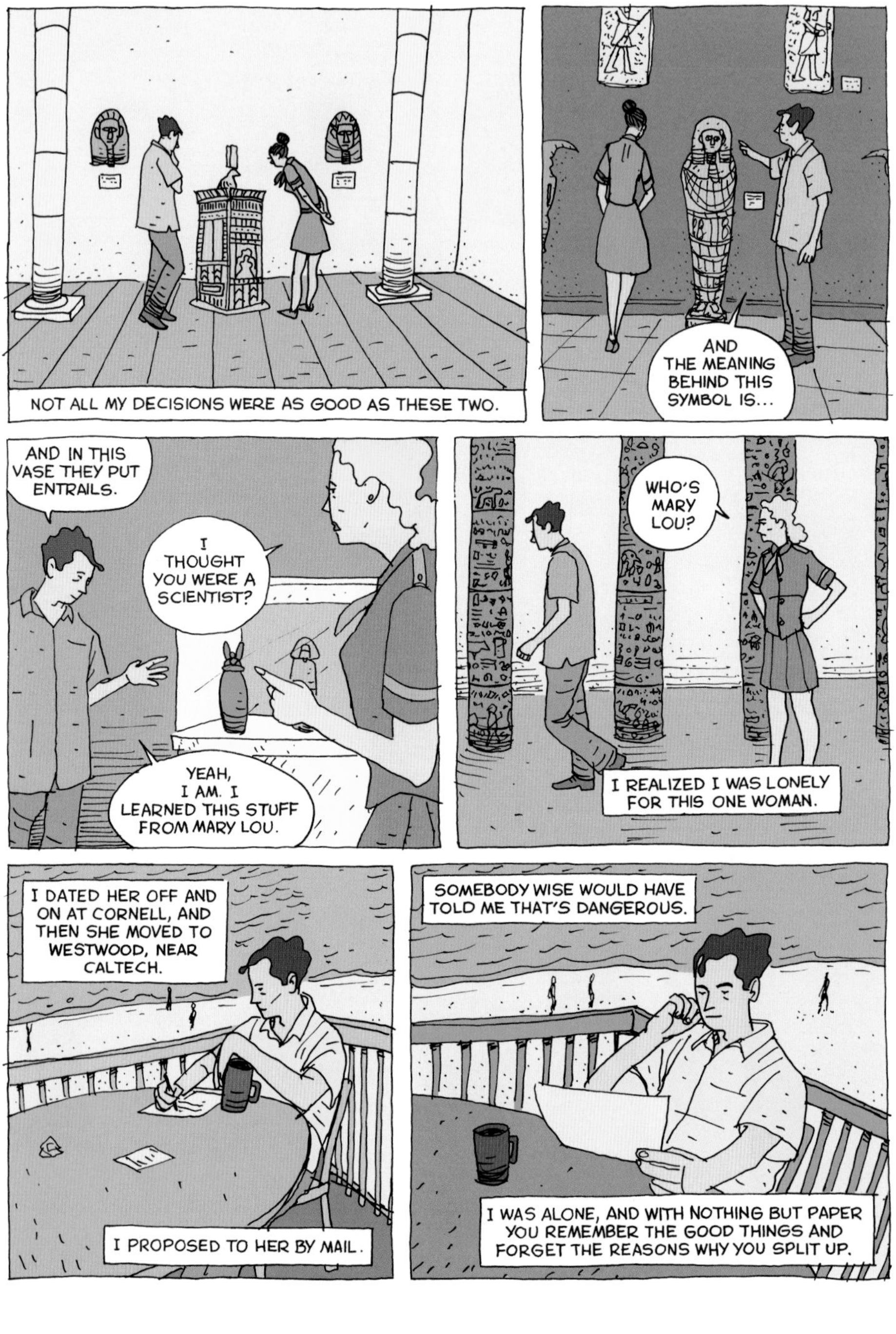
NOT ALL MY DECISIONS WERE AS GOOD AS THESE TWO.
AND THE MEANING BEHIND THIS SYMBOL IS...
AND IN THIS VASE THEY PUT ENTRAILS.
I THOUGHT YOU WERE A SCIENTIST?
YEAH, I AM. I LEARNED THIS STUFF FROM MARY LOU.
WHO'S MARY LOU?
I REALIZED I WAS LONELY FOR THIS ONE WOMAN.
I DATED HER OFF AND ON AT CORNELL, AND THEN SHE MOVED TO WESTWOOD, NEAR CALTECH.
I PROPOSED TO HER BY MAIL.
SOMEBODY WISE WOULD HAVE TOLD ME THAT'S DANGEROUS.
I WAS ALONE, AND WITH NOTHING BUT PAPER YOU REMEMBER THE GOOD THINGS AND FORGET THE REASONS WHY YOU SPLIT UP.

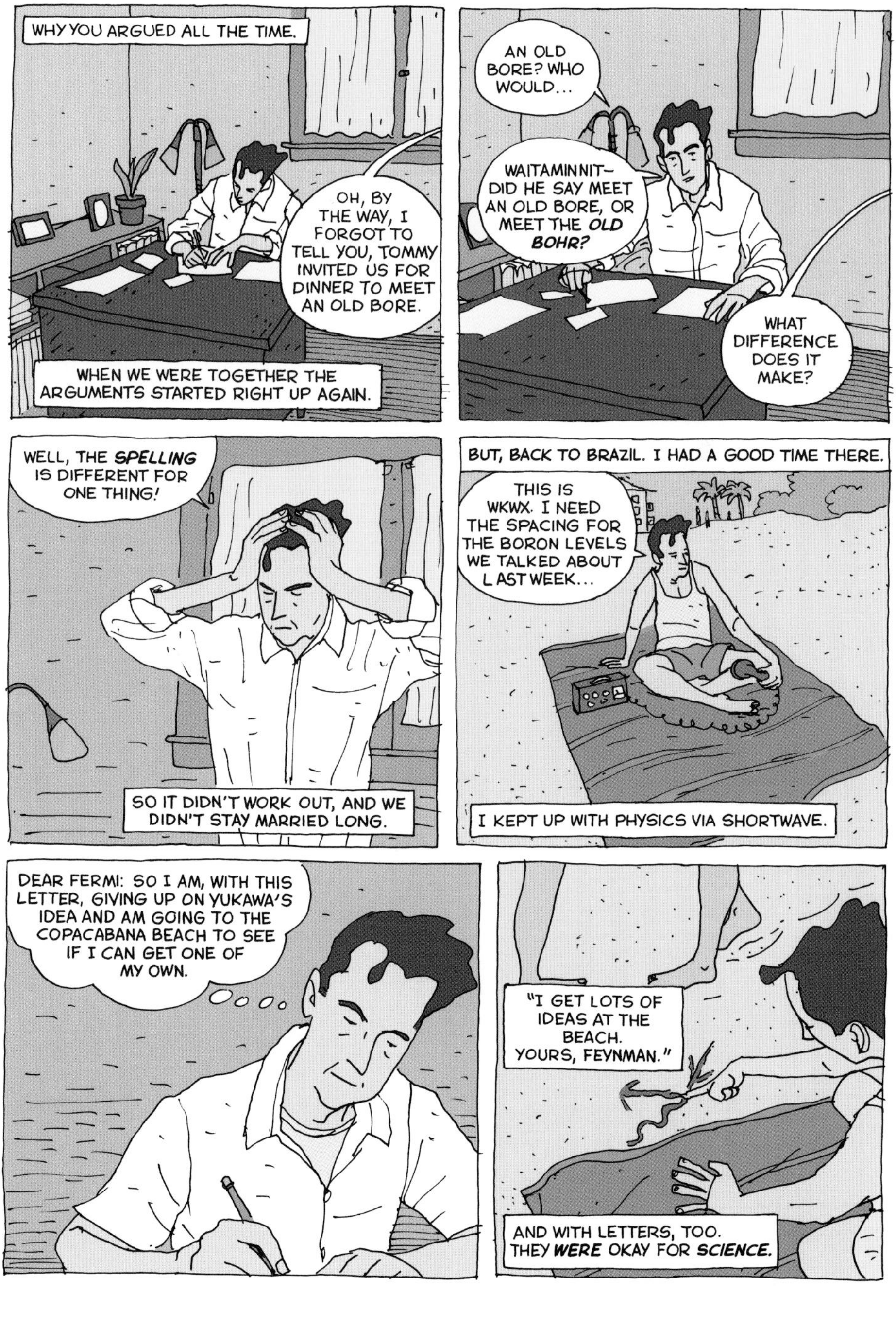
WHY YOU ARGUED ALL THE TIME.
OH, BY THE WAY, I FORGOT TO TELL YOU, TOMMY INVITED US FOR DINNER TO MEET AN OLD BORE.
WHEN WE WERE TOGETHER THE ARGUMENTS STARTED RIGHT UP AGAIN.
AN OLD BORE? WHO WOULD...
WAITAMINNIT— DID HE SAY MEET AN OLD BORE, OR MEET THE ***OLD BOHR?***
WHAT DIFFERENCE DOES IT MAKE?
WELL, THE ***SPELLING*** IS DIFFERENT FOR ONE THING!
SO IT DIDN'T WORK OUT, AND WE DIDN'T STAY MARRIED LONG.
BUT, BACK TO BRAZIL. I HAD A GOOD TIME THERE.
THIS IS WKWX. I NEED THE SPACING FOR THE BORON LEVELS WE TALKED ABOUT LAST WEEK...
I KEPT UP WITH PHYSICS VIA SHORTWAVE.
DEAR FERMI: SO I AM, WITH THIS LETTER, GIVING UP ON YUKAWA'S IDEA AND AM GOING TO THE COPACABANA BEACH TO SEE IF I CAN GET ONE OF MY OWN.
"I GET LOTS OF IDEAS AT THE BEACH. YOURS, FEYNMAN."
AND WITH LETTERS, TOO. THEY ***WERE*** OKAY FOR ***SCIENCE.***

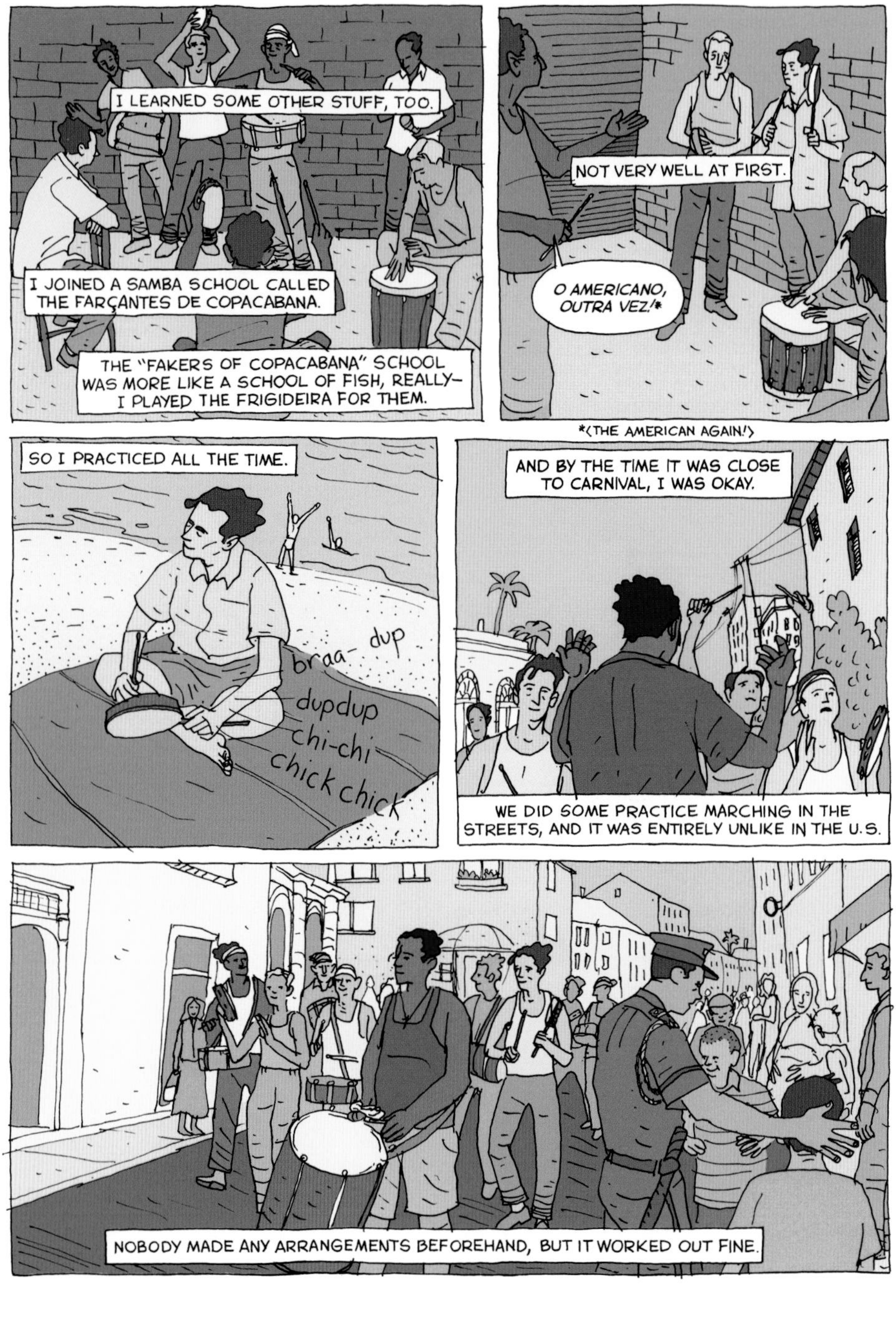
I LEARNED SOME OTHER STUFF, TOO.
I JOINED A SAMBA SCHOOL CALLED THE FARÇANTES DE COPACABANA.
THE "FAKERS OF COPACABANA" SCHOOL WAS MORE LIKE A SCHOOL OF FISH, REALLY— I PLAYED THE FRIGIDEIRA FOR THEM.
NOT VERY WELL AT FIRST.
O AMERICANO, OUTRA VEZ!*
*(THE AMERICAN AGAIN!)
SO I PRACTICED ALL THE TIME.
braa- dup
dupdup
chi-chi
chick chick
AND BY THE TIME IT WAS CLOSE TO CARNIVAL, I WAS OKAY.
WE DID SOME PRACTICE MARCHING IN THE STREETS, AND IT WAS ENTIRELY UNLIKE IN THE U.S.
NOBODY MADE ANY ARRANGEMENTS BEFOREHAND, BUT IT WORKED OUT FINE.

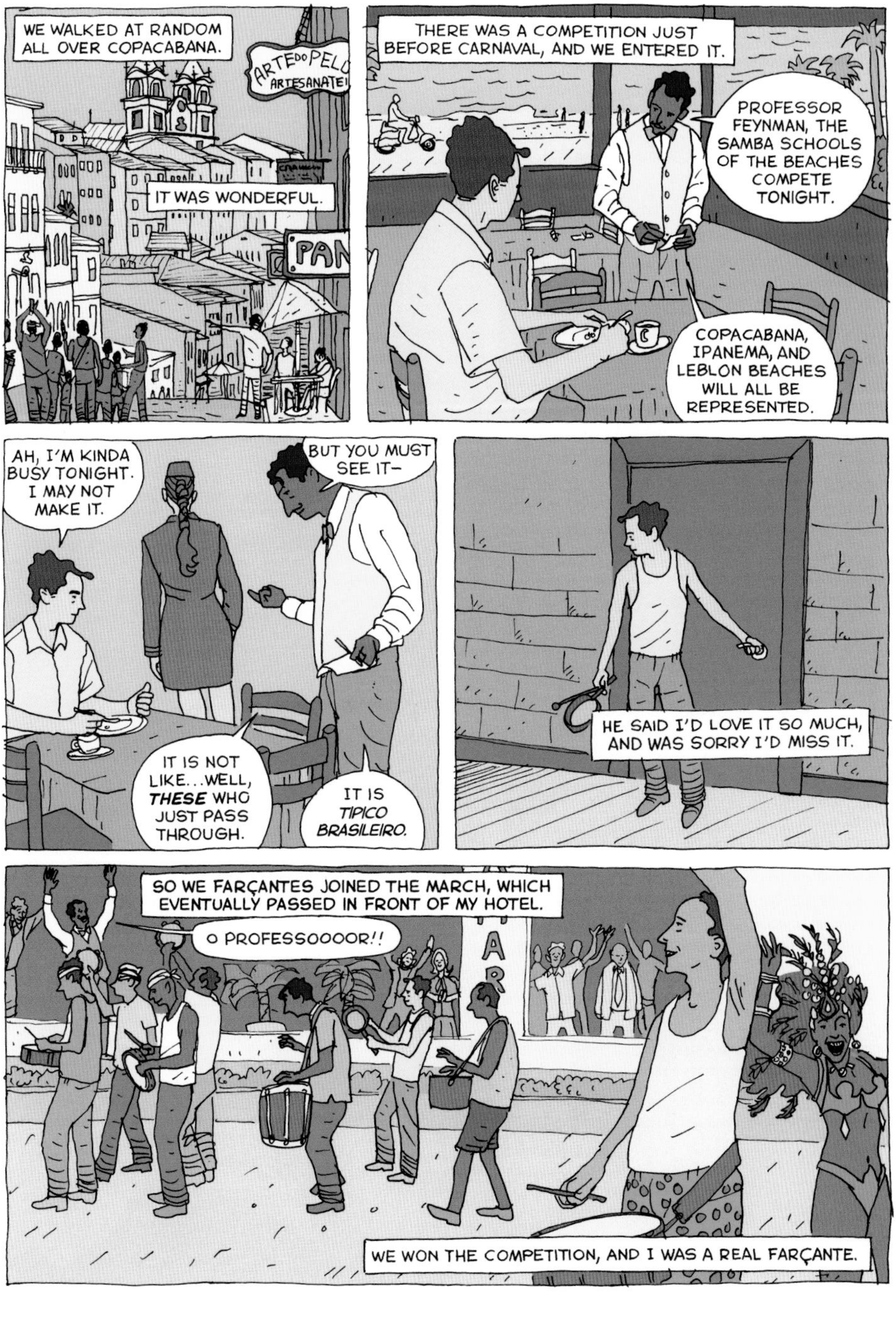
WE WALKED AT RANDOM ALL OVER COPACABANA.
IT WAS WONDERFUL.
THERE WAS A COMPETITION JUST BEFORE CARNAVAL, AND WE ENTERED IT.
PROFESSOR FEYNMAN, THE SAMBA SCHOOLS OF THE BEACHES COMPETE TONIGHT.
COPACABANA, IPANEMA, AND LEBLON BEACHES WILL ALL BE REPRESENTED.
AH, I'M KINDA BUSY TONIGHT. I MAY NOT MAKE IT.
BUT YOU MUST SEE IT—
IT IS NOT LIKE...WELL, ***THESE*** WHO JUST PASS THROUGH.
IT IS *TIPICO BRASILEIRO.*
HE SAID I'D LOVE IT SO MUCH, AND WAS SORRY I'D MISS IT.
SO WE FARÇANTES JOINED THE MARCH, WHICH EVENTUALLY PASSED IN FRONT OF MY HOTEL.
O PROFESSOOOOR!!
WE WON THE COMPETITION, AND I WAS A REAL FARÇANTE.

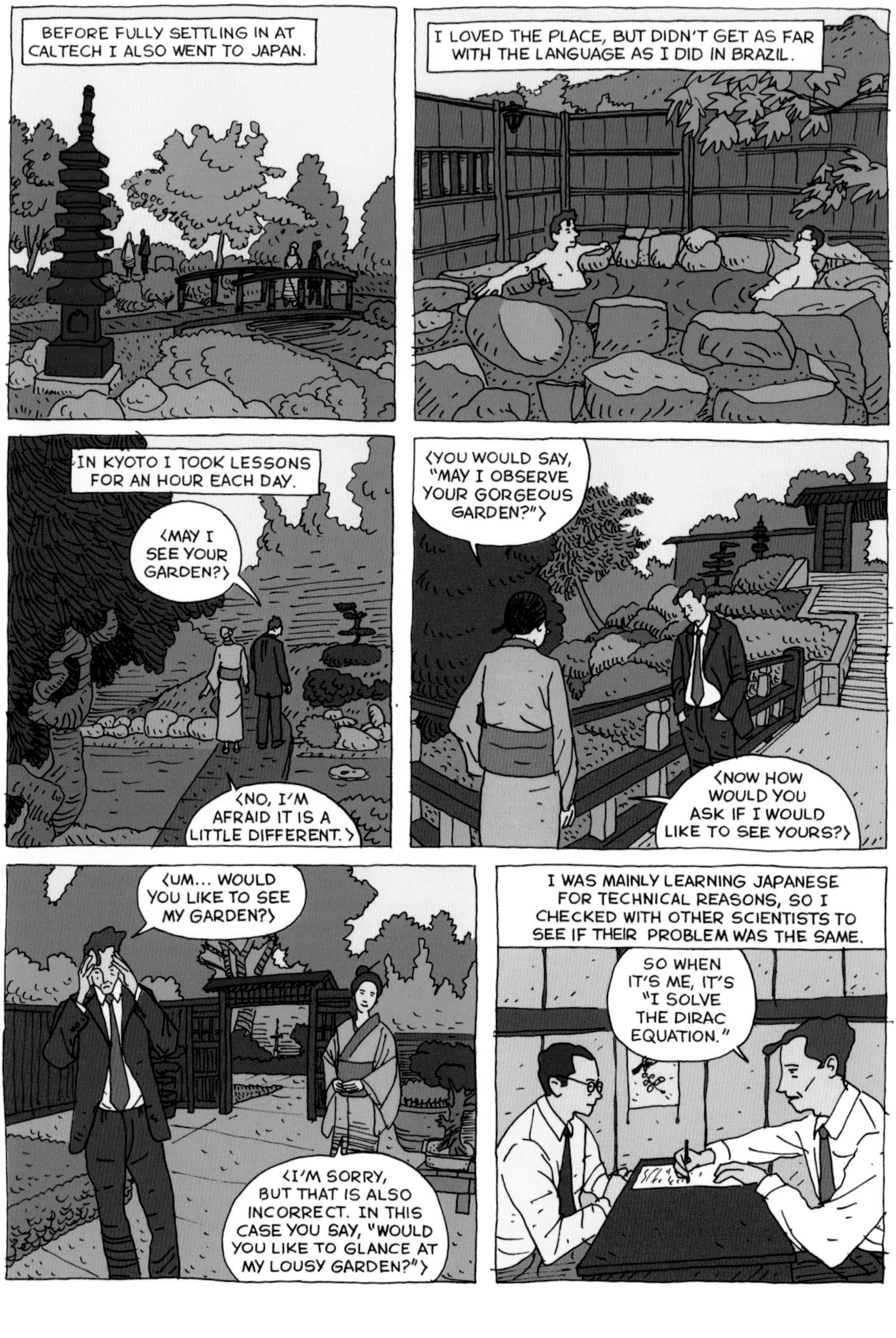
BEFORE FULLY SETTLING IN AT CALTECH I ALSO WENT TO JAPAN.
I LOVED THE PLACE, BUT DIDN'T GET AS FAR WITH THE LANGUAGE AS I DID IN BRAZIL.
IN KYOTO I TOOK LESSONS FOR AN HOUR EACH DAY.
‹MAY I SEE YOUR GARDEN?›
‹NO, I'M AFRAID IT IS A LITTLE DIFFERENT.›
‹YOU WOULD SAY, "MAY I OBSERVE YOUR GORGEOUS GARDEN?"›
‹NOW HOW WOULD YOU ASK IF I WOULD LIKE TO SEE YOURS?›
‹UM... WOULD YOU LIKE TO SEE MY GARDEN?›
‹I'M SORRY, BUT THAT IS ALSO INCORRECT. IN THIS CASE YOU SAY, "WOULD YOU LIKE TO GLANCE AT MY LOUSY GARDEN?"›
I WAS MAINLY LEARNING JAPANESE FOR TECHNICAL REASONS, SO I CHECKED WITH OTHER SCIENTISTS TO SEE IF THEIR PROBLEM WAS THE SAME.
SO WHEN IT'S ME, IT'S "I SOLVE THE DIRAC EQUATION."

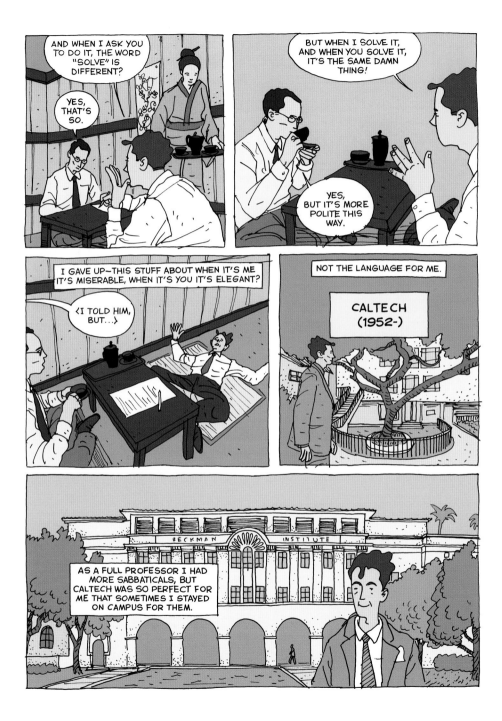

AND WHEN I ASK YOU TO DO IT, THE WORD "SOLVE" IS DIFFERENT?

YES, THAT'S SO.

BUT WHEN I SOLVE IT, AND WHEN YOU SOLVE IT, IT'S THE SAME DAMN THING!

YES, BUT IT'S MORE POLITE THIS WAY.

I GAVE UP—THIS STUFF ABOUT WHEN IT'S ME IT'S MISERABLE, WHEN IT'S YOU IT'S ELEGANT?

⟨I TOLD HIM, BUT...⟩

NOT THE LANGUAGE FOR ME.

CALTECH
(1952–)

AS A FULL PROFESSOR I HAD MORE SABBATICALS, BUT CALTECH WAS SO PERFECT FOR ME THAT SOMETIMES I STAYED ON CAMPUS FOR THEM.

BECKMAN INSTITUTE

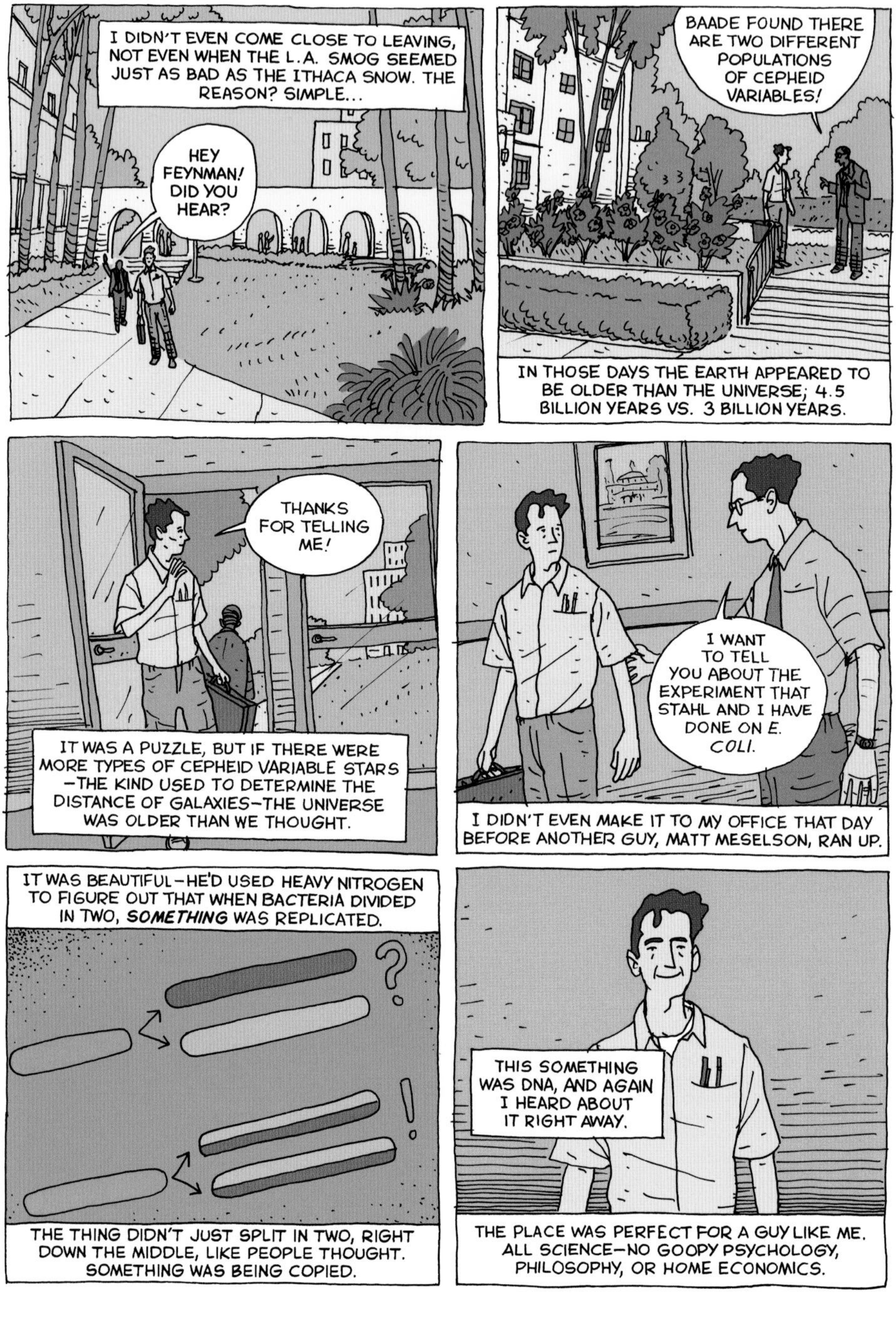
I DIDN'T EVEN COME CLOSE TO LEAVING, NOT EVEN WHEN THE L.A. SMOG SEEMED JUST AS BAD AS THE ITHACA SNOW. THE REASON? SIMPLE...
HEY FEYNMAN! DID YOU HEAR?
BAADE FOUND THERE ARE TWO DIFFERENT POPULATIONS OF CEPHEID VARIABLES!
IN THOSE DAYS THE EARTH APPEARED TO BE OLDER THAN THE UNIVERSE; 4.5 BILLION YEARS VS. 3 BILLION YEARS.
THANKS FOR TELLING ME!
IT WAS A PUZZLE, BUT IF THERE WERE MORE TYPES OF CEPHEID VARIABLE STARS—THE KIND USED TO DETERMINE THE DISTANCE OF GALAXIES—THE UNIVERSE WAS OLDER THAN WE THOUGHT.
I WANT TO TELL YOU ABOUT THE EXPERIMENT THAT STAHL AND I HAVE DONE ON *E. COLI*.
I DIDN'T EVEN MAKE IT TO MY OFFICE THAT DAY BEFORE ANOTHER GUY, MATT MESELSON, RAN UP.
IT WAS BEAUTIFUL—HE'D USED HEAVY NITROGEN TO FIGURE OUT THAT WHEN BACTERIA DIVIDED IN TWO, *SOMETHING* WAS REPLICATED.
?
!
THE THING DIDN'T JUST SPLIT IN TWO, RIGHT DOWN THE MIDDLE, LIKE PEOPLE THOUGHT. SOMETHING WAS BEING COPIED.
THIS SOMETHING WAS DNA, AND AGAIN I HEARD ABOUT IT RIGHT AWAY.
THE PLACE WAS PERFECT FOR A GUY LIKE ME. ALL SCIENCE—NO GOOPY PSYCHOLOGY, PHILOSOPHY, OR HOME ECONOMICS.

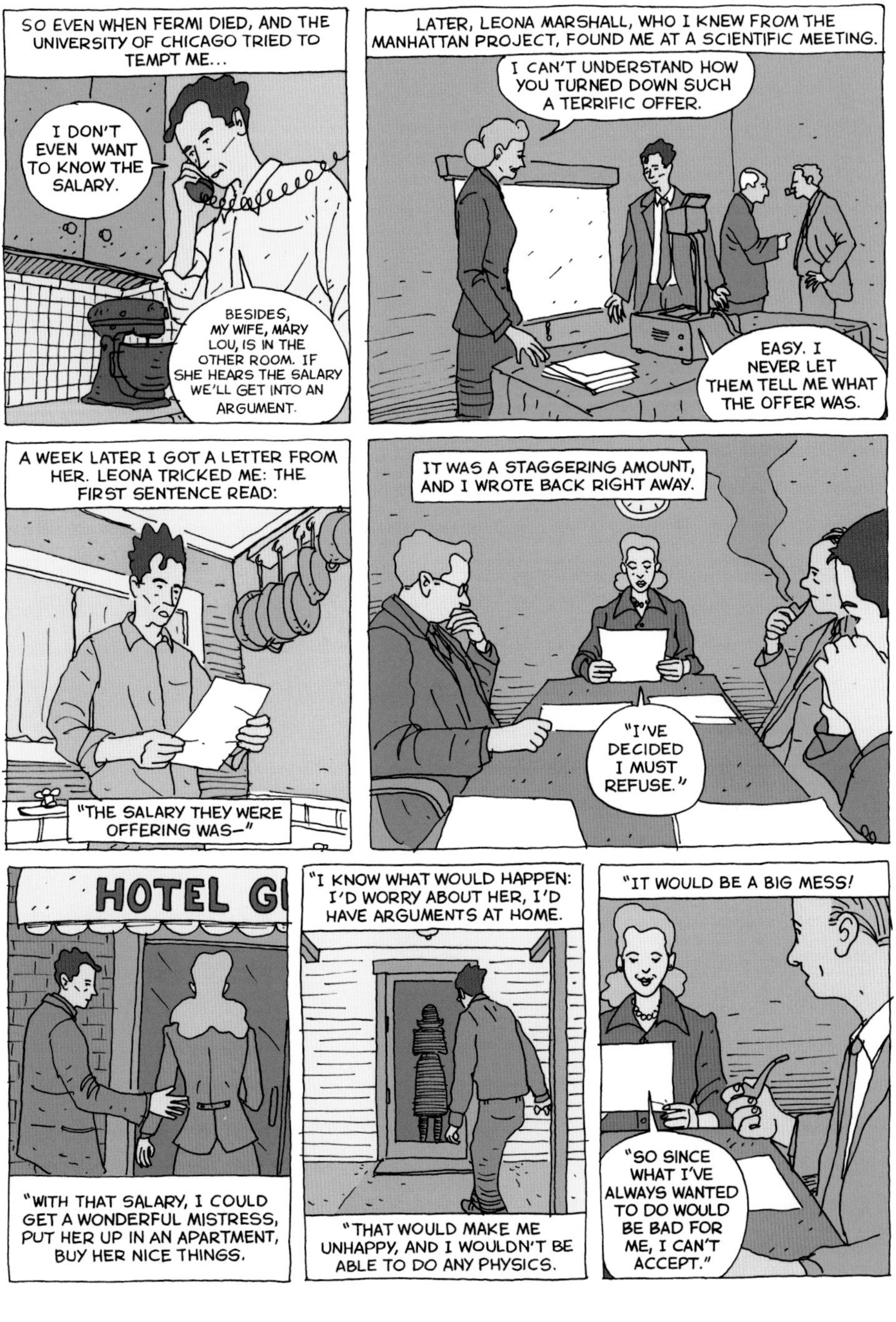
SO EVEN WHEN FERMI DIED, AND THE UNIVERSITY OF CHICAGO TRIED TO TEMPT ME...
I DON'T EVEN WANT TO KNOW THE SALARY.
BESIDES, MY WIFE, MARY LOU, IS IN THE OTHER ROOM. IF SHE HEARS THE SALARY WE'LL GET INTO AN ARGUMENT.
LATER, LEONA MARSHALL, WHO I KNEW FROM THE MANHATTAN PROJECT, FOUND ME AT A SCIENTIFIC MEETING.
I CAN'T UNDERSTAND HOW YOU TURNED DOWN SUCH A TERRIFIC OFFER.
EASY. I NEVER LET THEM TELL ME WHAT THE OFFER WAS.
A WEEK LATER I GOT A LETTER FROM HER. LEONA TRICKED ME: THE FIRST SENTENCE READ:
"THE SALARY THEY WERE OFFERING WAS—"
IT WAS A STAGGERING AMOUNT, AND I WROTE BACK RIGHT AWAY.
"I'VE DECIDED I MUST REFUSE."
HOTEL G
"WITH THAT SALARY, I COULD GET A WONDERFUL MISTRESS, PUT HER UP IN AN APARTMENT, BUY HER NICE THINGS.
"I KNOW WHAT WOULD HAPPEN: I'D WORRY ABOUT HER, I'D HAVE ARGUMENTS AT HOME.
"THAT WOULD MAKE ME UNHAPPY, AND I WOULDN'T BE ABLE TO DO ANY PHYSICS.
"IT WOULD BE A BIG MESS!
"SO SINCE WHAT I'VE ALWAYS WANTED TO DO WOULD BE BAD FOR ME, I CAN'T ACCEPT."

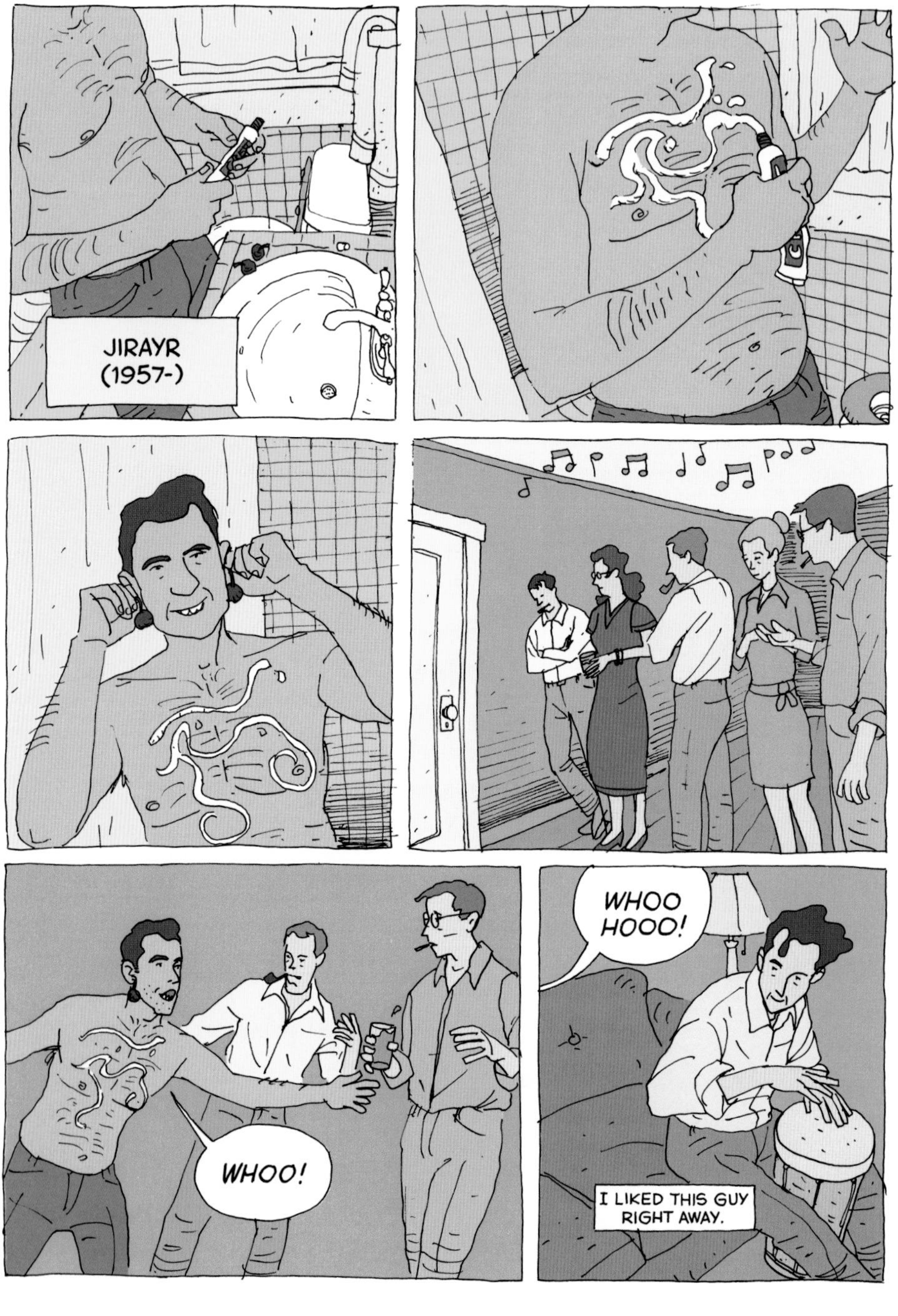
JIRAYR
(1957-)
WHOO!
WHOO
HOOO!
I LIKED THIS GUY
RIGHT AWAY.

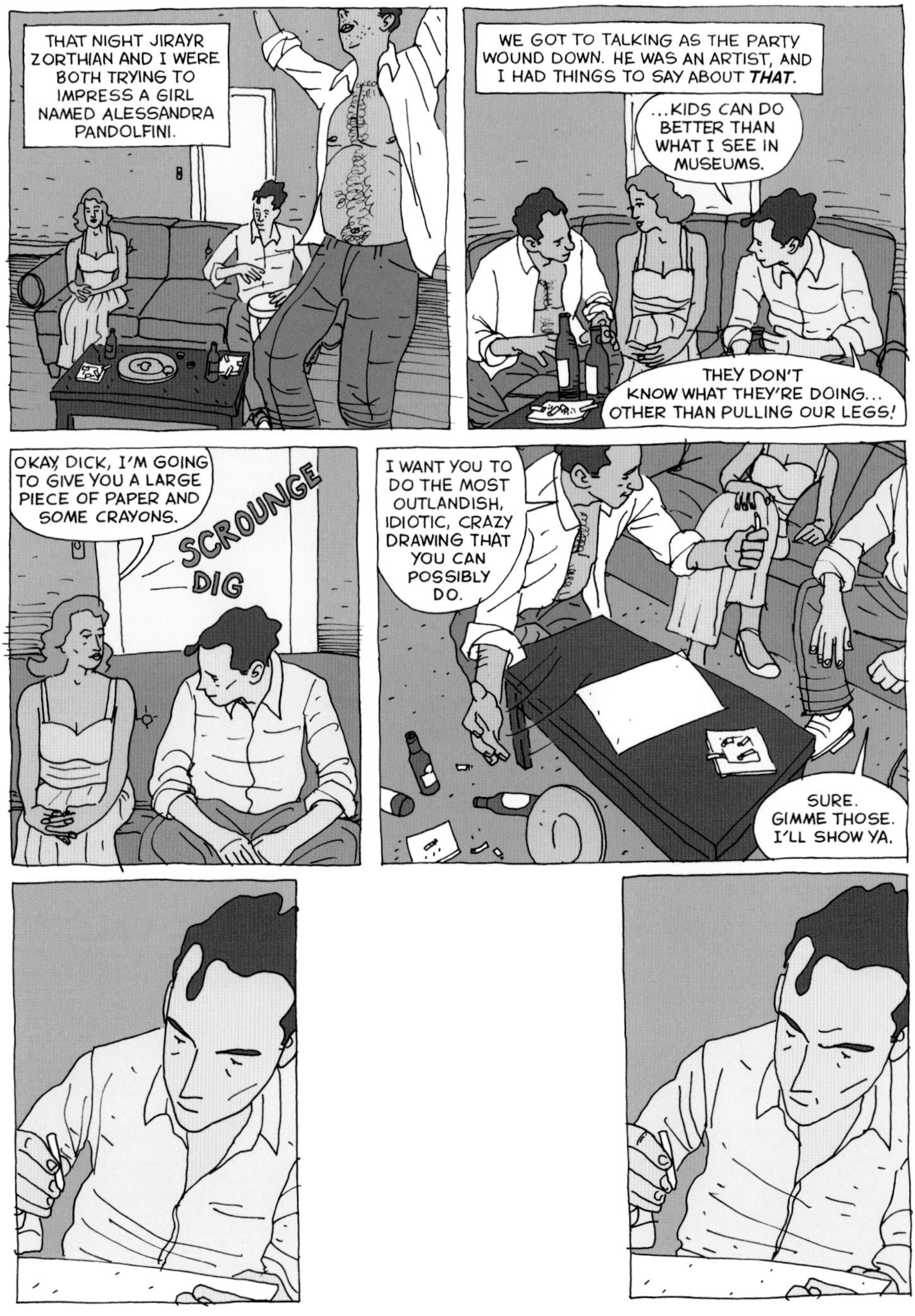
THAT NIGHT JIRAYR ZORTHIAN AND I WERE BOTH TRYING TO IMPRESS A GIRL NAMED ALESSANDRA PANDOLFINI.
WE GOT TO TALKING AS THE PARTY WOUND DOWN. HE WAS AN ARTIST, AND I HAD THINGS TO SAY ABOUT THAT.
...KIDS CAN DO BETTER THAN WHAT I SEE IN MUSEUMS.
THEY DON'T KNOW WHAT THEY'RE DOING... OTHER THAN PULLING OUR LEGS!
OKAY, DICK, I'M GOING TO GIVE YOU A LARGE PIECE OF PAPER AND SOME CRAYONS.
SCROUNGE
DIG
I WANT YOU TO DO THE MOST OUTLANDISH, IDIOTIC, CRAZY DRAWING THAT YOU CAN POSSIBLY DO.
SURE. GIMME THOSE. I'LL SHOW YA.

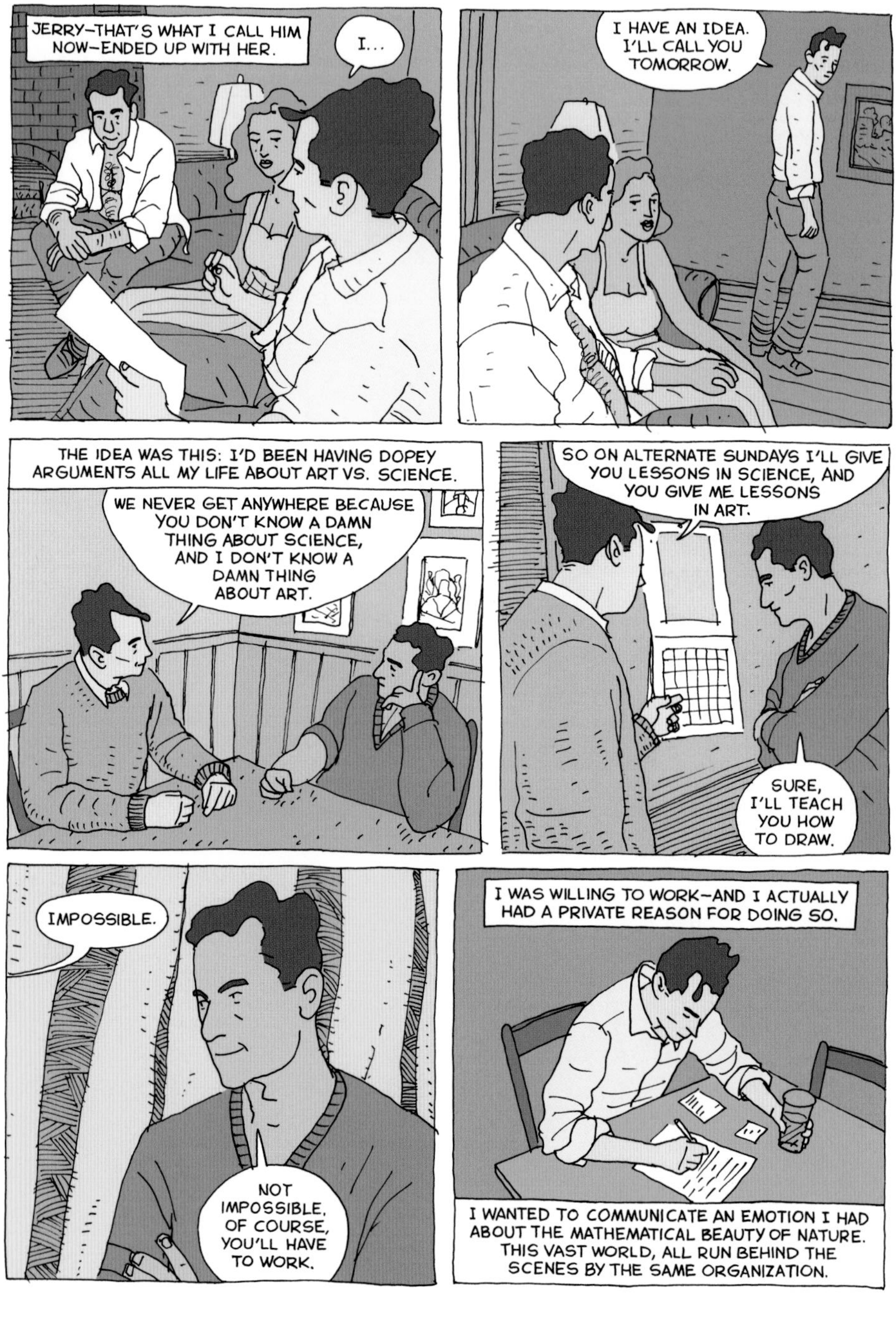
JERRY—THAT'S WHAT I CALL HIM NOW—ENDED UP WITH HER.
I...
I HAVE AN IDEA. I'LL CALL YOU TOMORROW.
THE IDEA WAS THIS: I'D BEEN HAVING DOPEY ARGUMENTS ALL MY LIFE ABOUT ART VS. SCIENCE.
WE NEVER GET ANYWHERE BECAUSE YOU DON'T KNOW A DAMN THING ABOUT SCIENCE, AND I DON'T KNOW A DAMN THING ABOUT ART.
SO ON ALTERNATE SUNDAYS I'LL GIVE YOU LESSONS IN SCIENCE, AND YOU GIVE ME LESSONS IN ART.
SURE, I'LL TEACH YOU HOW TO DRAW.
IMPOSSIBLE.
NOT IMPOSSIBLE. OF COURSE, YOU'LL HAVE TO WORK.
I WAS WILLING TO WORK—AND I ACTUALLY HAD A PRIVATE REASON FOR DOING SO.
I WANTED TO COMMUNICATE AN EMOTION I HAD ABOUT THE MATHEMATICAL BEAUTY OF NATURE. THIS VAST WORLD, ALL RUN BEHIND THE SCENES BY THE SAME ORGANIZATION.

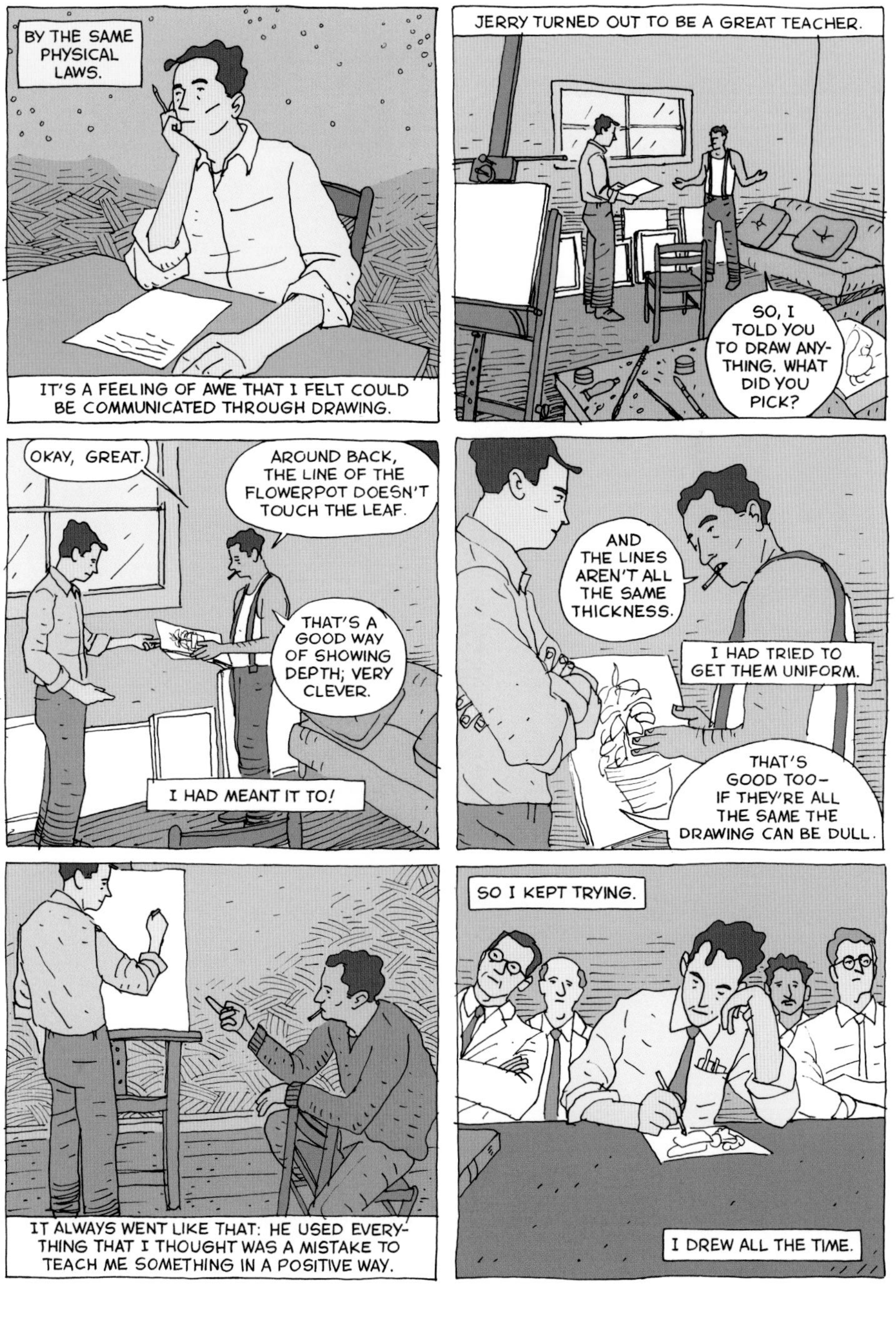
BY THE SAME PHYSICAL LAWS.
IT'S A FEELING OF AWE THAT I FELT COULD BE COMMUNICATED THROUGH DRAWING.
JERRY TURNED OUT TO BE A GREAT TEACHER.
SO, I TOLD YOU TO DRAW ANYTHING. WHAT DID YOU PICK?
OKAY, GREAT.
AROUND BACK, THE LINE OF THE FLOWERPOT DOESN'T TOUCH THE LEAF.
THAT'S A GOOD WAY OF SHOWING DEPTH; VERY CLEVER.
I HAD MEANT IT TO!
AND THE LINES AREN'T ALL THE SAME THICKNESS.
I HAD TRIED TO GET THEM UNIFORM.
THAT'S GOOD TOO– IF THEY'RE ALL THE SAME THE DRAWING CAN BE DULL.
IT ALWAYS WENT LIKE THAT: HE USED EVERYTHING THAT I THOUGHT WAS A MISTAKE TO TEACH ME SOMETHING IN A POSITIVE WAY.
SO I KEPT TRYING.
I DREW ALL THE TIME.

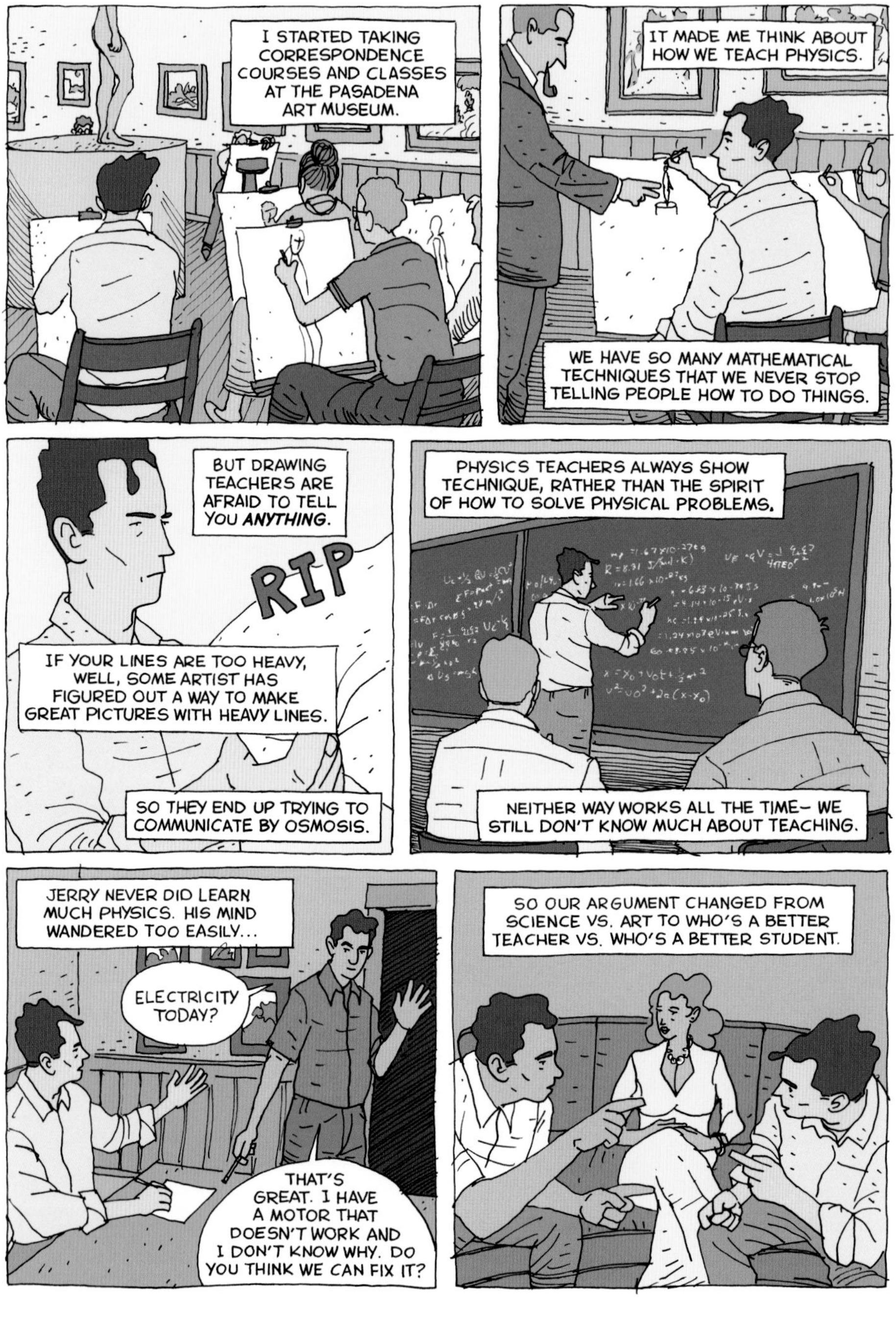
I STARTED TAKING CORRESPONDENCE COURSES AND CLASSES AT THE PASADENA ART MUSEUM.
IT MADE ME THINK ABOUT HOW WE TEACH PHYSICS.
WE HAVE SO MANY MATHEMATICAL TECHNIQUES THAT WE NEVER STOP TELLING PEOPLE HOW TO DO THINGS.
BUT DRAWING TEACHERS ARE AFRAID TO TELL YOU *ANYTHING*.
RIP
IF YOUR LINES ARE TOO HEAVY, WELL, SOME ARTIST HAS FIGURED OUT A WAY TO MAKE GREAT PICTURES WITH HEAVY LINES.
SO THEY END UP TRYING TO COMMUNICATE BY OSMOSIS.
PHYSICS TEACHERS ALWAYS SHOW TECHNIQUE, RATHER THAN THE SPIRIT OF HOW TO SOLVE PHYSICAL PROBLEMS.
NEITHER WAY WORKS ALL THE TIME– WE STILL DON'T KNOW MUCH ABOUT TEACHING.
JERRY NEVER DID LEARN MUCH PHYSICS. HIS MIND WANDERED TOO EASILY...
ELECTRICITY TODAY?
THAT'S GREAT. I HAVE A MOTOR THAT DOESN'T WORK AND I DON'T KNOW WHY. DO YOU THINK WE CAN FIX IT?
SO OUR ARGUMENT CHANGED FROM SCIENCE VS. ART TO WHO'S A BETTER TEACHER VS. WHO'S A BETTER STUDENT.

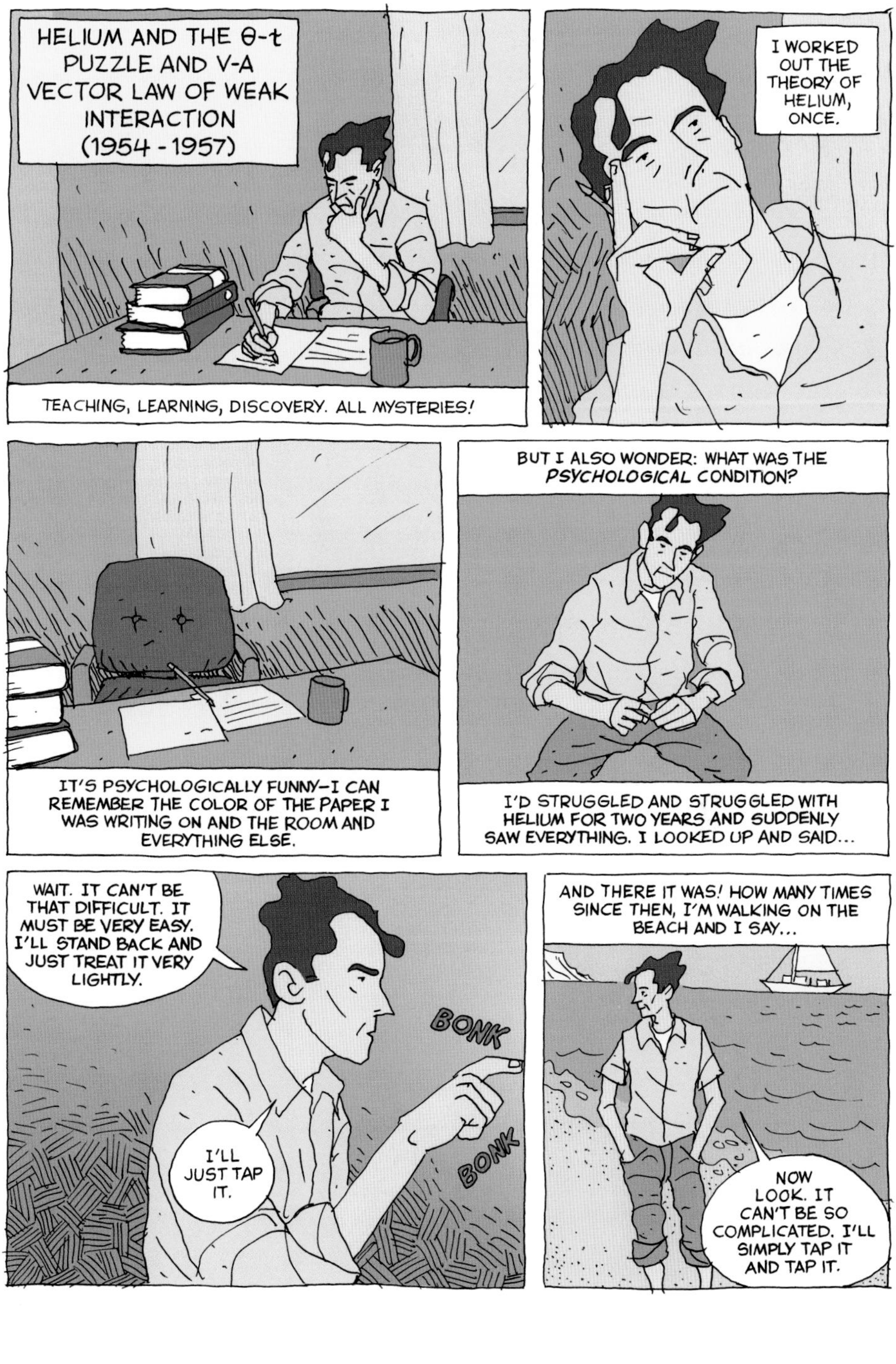
HELIUM AND THE θ-τ PUZZLE AND V-A VECTOR LAW OF WEAK INTERACTION (1954 - 1957)
TEACHING, LEARNING, DISCOVERY. ALL MYSTERIES!
I WORKED OUT THE THEORY OF HELIUM, ONCE.
IT'S PSYCHOLOGICALLY FUNNY—I CAN REMEMBER THE COLOR OF THE PAPER I WAS WRITING ON AND THE ROOM AND EVERYTHING ELSE.
BUT I ALSO WONDER: WHAT WAS THE *PSYCHOLOGICAL* CONDITION?
I'D STRUGGLED AND STRUGGLED WITH HELIUM FOR TWO YEARS AND SUDDENLY SAW EVERYTHING. I LOOKED UP AND SAID...
WAIT. IT CAN'T BE THAT DIFFICULT. IT MUST BE VERY EASY. I'LL STAND BACK AND JUST TREAT IT VERY LIGHTLY.
BONK
I'LL JUST TAP IT.
BONK
AND THERE IT WAS! HOW MANY TIMES SINCE THEN, I'M WALKING ON THE BEACH AND I SAY...
NOW LOOK. IT CAN'T BE SO COMPLICATED. I'LL SIMPLY TAP IT AND TAP IT.

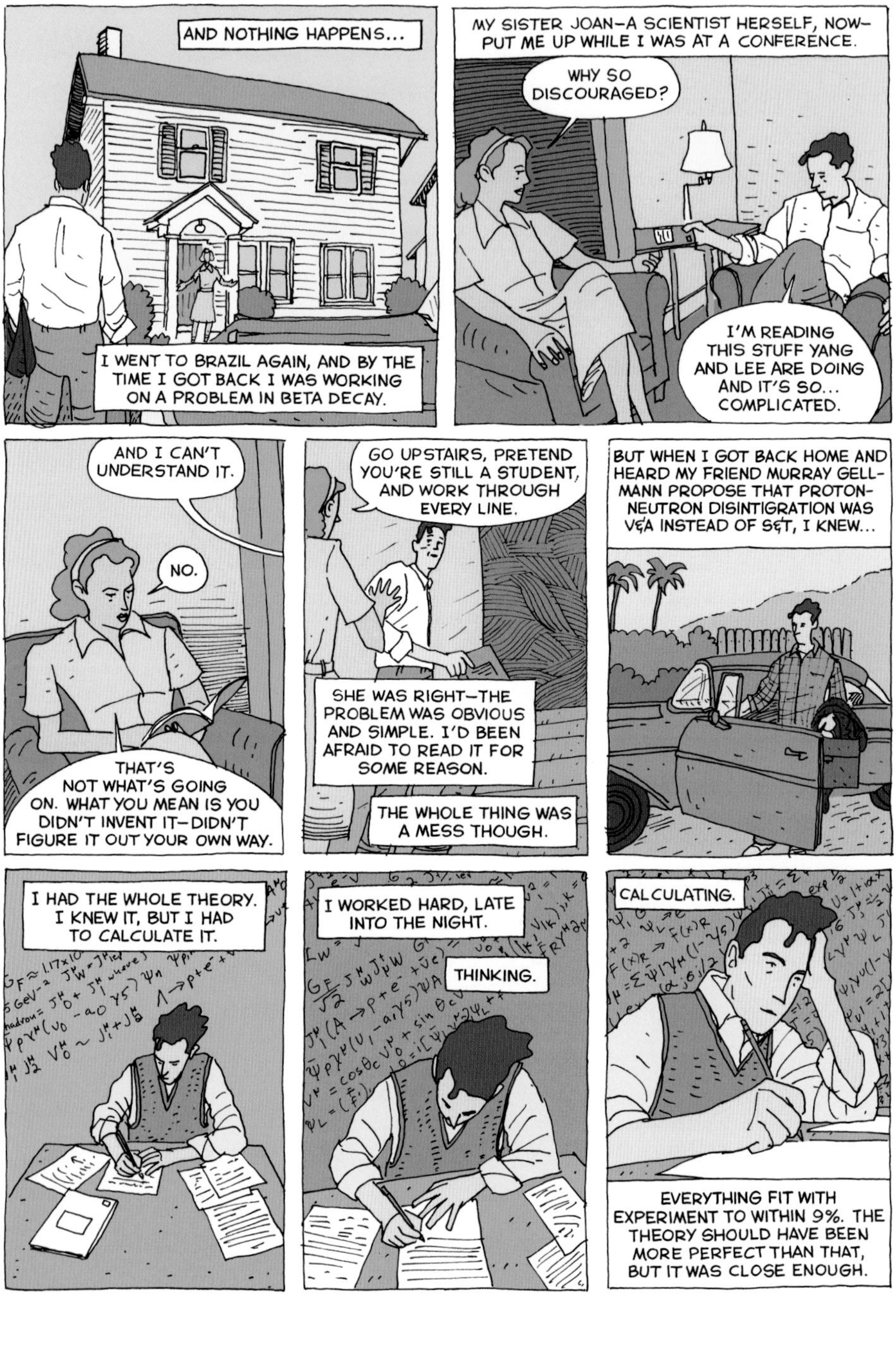
AND NOTHING HAPPENS...
I WENT TO BRAZIL AGAIN, AND BY THE TIME I GOT BACK I WAS WORKING ON A PROBLEM IN BETA DECAY.
MY SISTER JOAN—A SCIENTIST HERSELF, NOW—PUT ME UP WHILE I WAS AT A CONFERENCE.
WHY SO DISCOURAGED?
I'M READING THIS STUFF YANG AND LEE ARE DOING AND IT'S SO... COMPLICATED.
AND I CAN'T UNDERSTAND IT.
NO.
THAT'S NOT WHAT'S GOING ON. WHAT YOU MEAN IS YOU DIDN'T INVENT IT—DIDN'T FIGURE IT OUT YOUR OWN WAY.
GO UPSTAIRS, PRETEND YOU'RE STILL A STUDENT, AND WORK THROUGH EVERY LINE.
SHE WAS RIGHT—THE PROBLEM WAS OBVIOUS AND SIMPLE. I'D BEEN AFRAID TO READ IT FOR SOME REASON.
THE WHOLE THING WAS A MESS THOUGH.
BUT WHEN I GOT BACK HOME AND HEARD MY FRIEND MURRAY GELL-MANN PROPOSE THAT PROTON-NEUTRON DISINTIGRATION WAS V&A INSTEAD OF S&T, I KNEW...
I HAD THE WHOLE THEORY. I KNEW IT, BUT I HAD TO CALCULATE IT.
I WORKED HARD, LATE INTO THE NIGHT.
THINKING.
CALCULATING.
EVERYTHING FIT WITH EXPERIMENT TO WITHIN 9%. THE THEORY SHOULD HAVE BEEN MORE PERFECT THAN THAT, BUT IT WAS CLOSE ENOUGH.

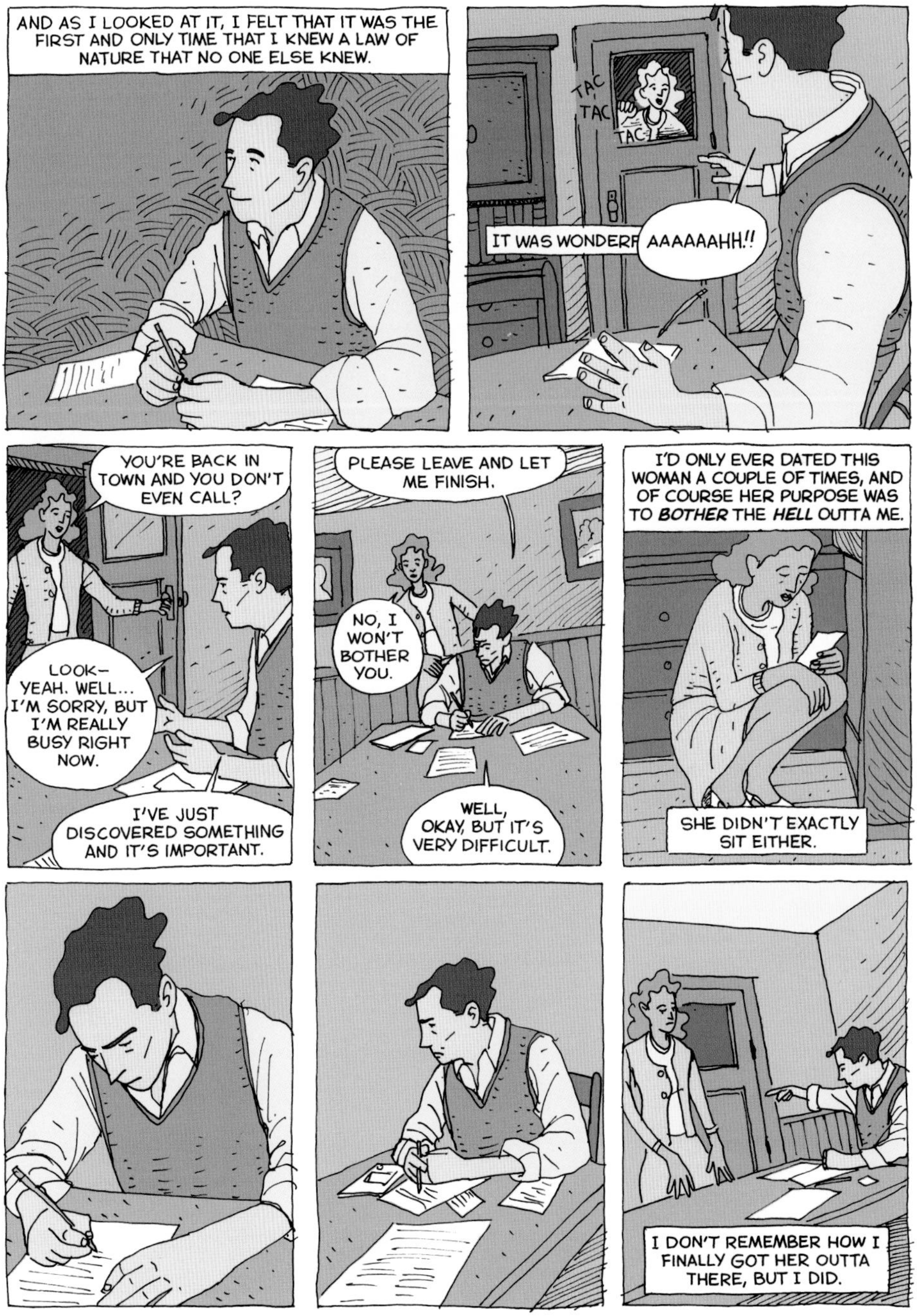
AND AS I LOOKED AT IT, I FELT THAT IT WAS THE FIRST AND ONLY TIME THAT I KNEW A LAW OF NATURE THAT NO ONE ELSE KNEW.
TAC
TAC
TAC
IT WAS WONDERF
AAAAAAHH!!
YOU'RE BACK IN TOWN AND YOU DON'T EVEN CALL?
LOOK— YEAH. WELL... I'M SORRY, BUT I'M REALLY BUSY RIGHT NOW.
I'VE JUST DISCOVERED SOMETHING AND IT'S IMPORTANT.
PLEASE LEAVE AND LET ME FINISH.
NO, I WON'T BOTHER YOU.
WELL, OKAY, BUT IT'S VERY DIFFICULT.
I'D ONLY EVER DATED THIS WOMAN A COUPLE OF TIMES, AND OF COURSE HER PURPOSE WAS TO BOTHER THE HELL OUTTA ME.
SHE DIDN'T EXACTLY SIT EITHER.
I DON'T REMEMBER HOW I FINALLY GOT HER OUTTA THERE, BUT I DID.

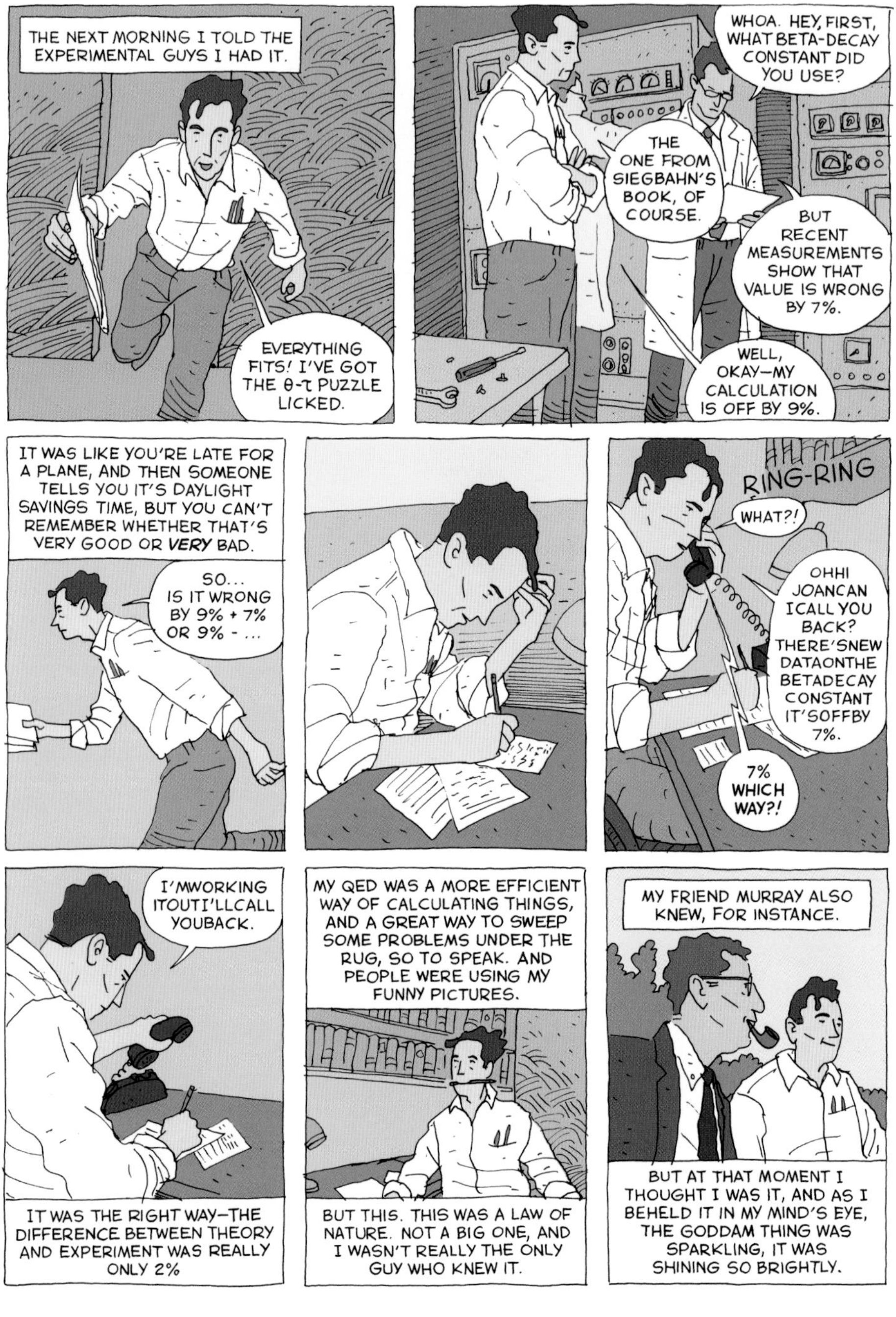
THE NEXT MORNING I TOLD THE EXPERIMENTAL GUYS I HAD IT.
EVERYTHING FITS! I'VE GOT THE θ-τ PUZZLE LICKED.
WHOA. HEY, FIRST, WHAT BETA-DECAY CONSTANT DID YOU USE?
THE ONE FROM SIEGBAHN'S BOOK, OF COURSE.
BUT RECENT MEASUREMENTS SHOW THAT VALUE IS WRONG BY 7%.
WELL, OKAY—MY CALCULATION IS OFF BY 9%.
IT WAS LIKE YOU'RE LATE FOR A PLANE, AND THEN SOMEONE TELLS YOU IT'S DAYLIGHT SAVINGS TIME, BUT YOU CAN'T REMEMBER WHETHER THAT'S VERY GOOD OR ***VERY*** BAD.
SO... IS IT WRONG BY 9% + 7% OR 9% - ...
RING-RING
WHAT?!
OHHI JOANCAN ICALL YOU BACK? THERE'SNEW DATAONTHE BETADECAY CONSTANT IT'SOFFBY 7%.
7% **WHICH WAY?!**
I'MWORKING ITOUTI'LLCALL YOUBACK.
IT WAS THE RIGHT WAY—THE DIFFERENCE BETWEEN THEORY AND EXPERIMENT WAS REALLY ONLY 2%
MY QED WAS A MORE EFFICIENT WAY OF CALCULATING THINGS, AND A GREAT WAY TO SWEEP SOME PROBLEMS UNDER THE RUG, SO TO SPEAK. AND PEOPLE WERE USING MY FUNNY PICTURES.
BUT THIS. THIS WAS A LAW OF NATURE. NOT A BIG ONE, AND I WASN'T REALLY THE ONLY GUY WHO KNEW IT.
MY FRIEND MURRAY ALSO KNEW, FOR INSTANCE.
BUT AT THAT MOMENT I THOUGHT I WAS IT, AND AS I BEHELD IT IN MY MIND'S EYE, THE GODDAM THING WAS SPARKLING, IT WAS SHINING SO BRIGHTLY.

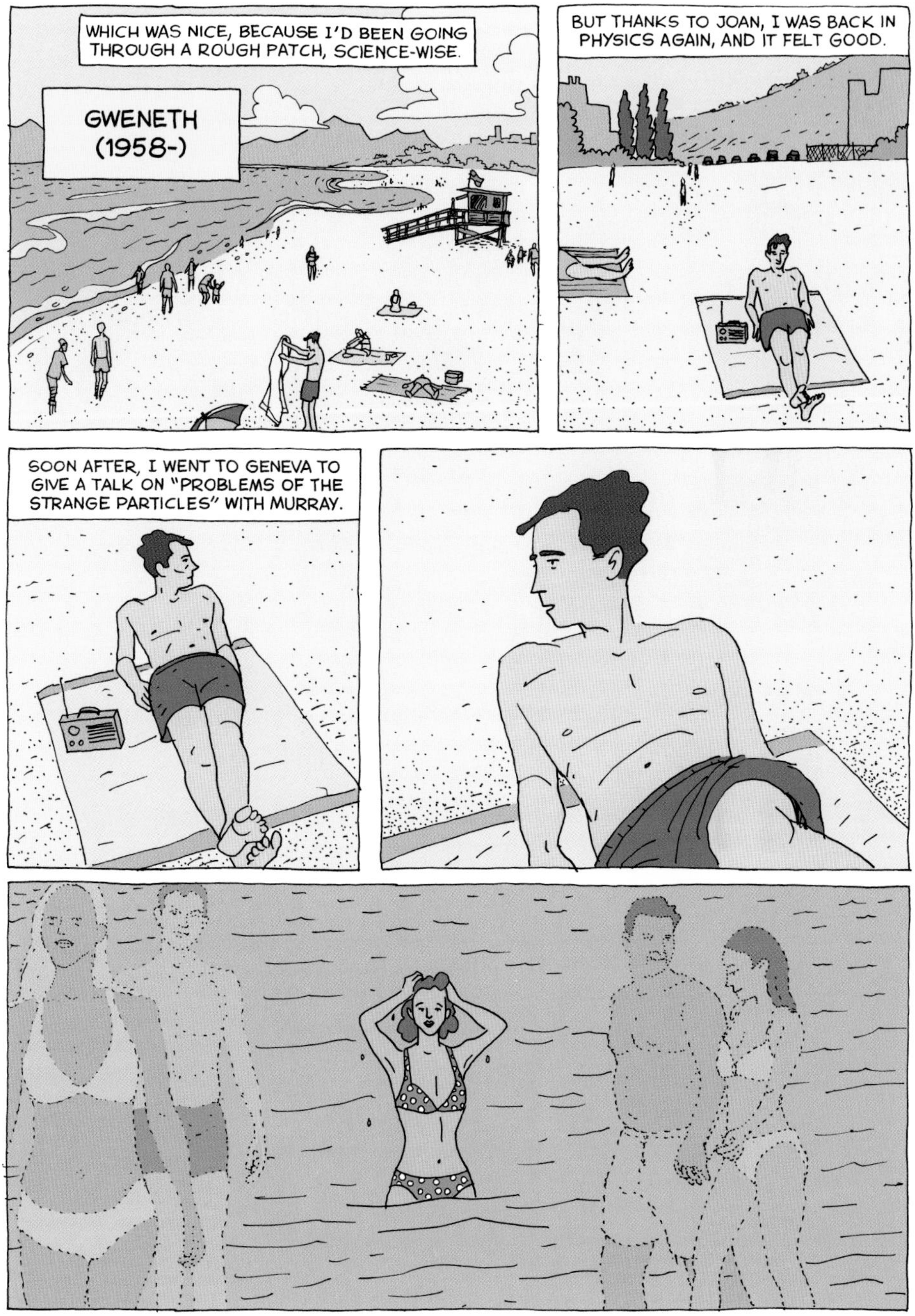
WHICH WAS NICE, BECAUSE I'D BEEN GOING THROUGH A ROUGH PATCH, SCIENCE-WISE.
GWENETH (1958-)
BUT THANKS TO JOAN, I WAS BACK IN PHYSICS AGAIN, AND IT FELT GOOD.
SOON AFTER, I WENT TO GENEVA TO GIVE A TALK ON "PROBLEMS OF THE STRANGE PARTICLES" WITH MURRAY.

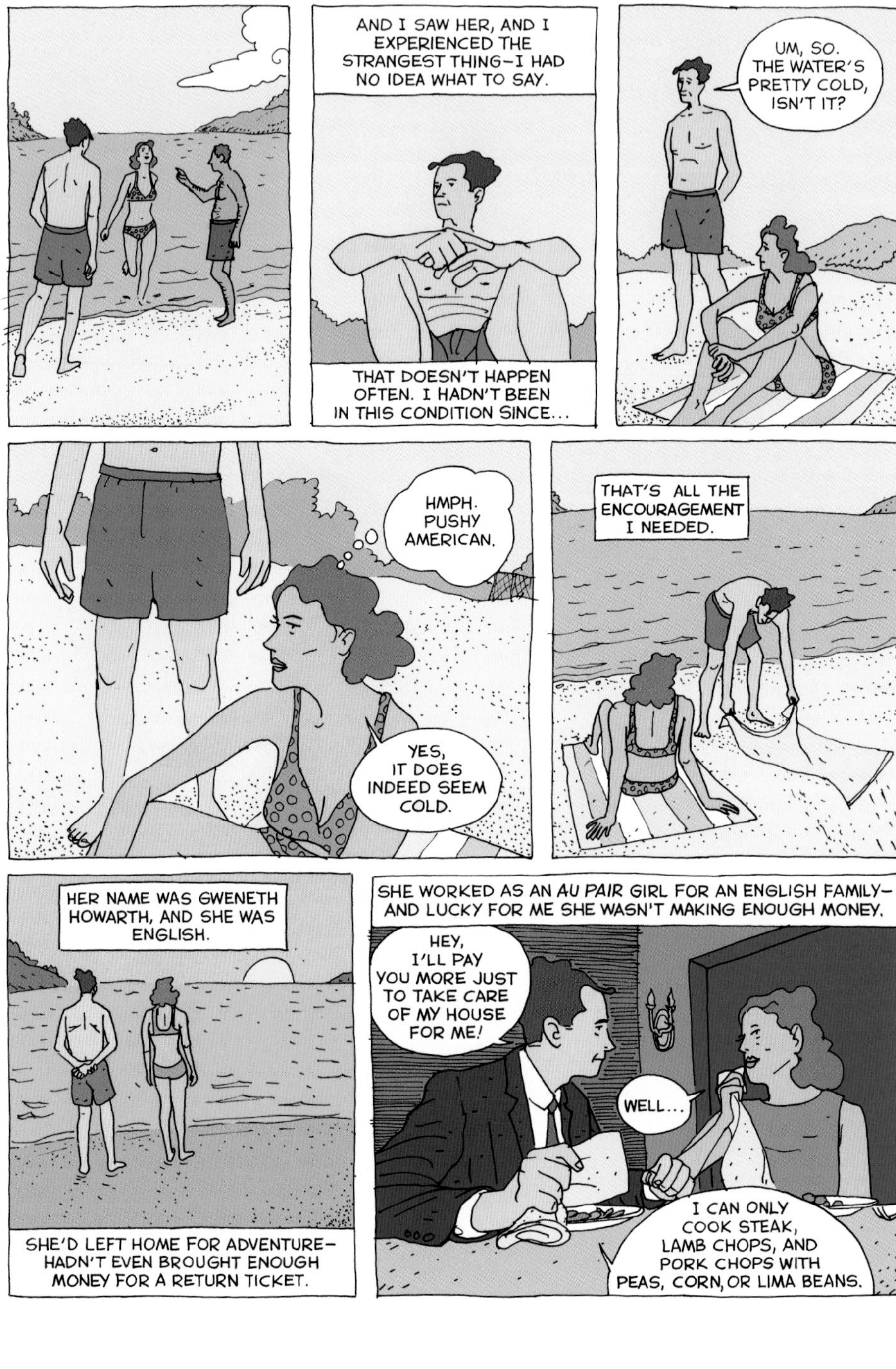
AND I SAW HER, AND I EXPERIENCED THE STRANGEST THING—I HAD NO IDEA WHAT TO SAY.
THAT DOESN'T HAPPEN OFTEN. I HADN'T BEEN IN THIS CONDITION SINCE...
UM, SO. THE WATER'S PRETTY COLD, ISN'T IT?
HMPH. PUSHY AMERICAN.
YES, IT DOES INDEED SEEM COLD.
THAT'S ALL THE ENCOURAGEMENT I NEEDED.
HER NAME WAS GWENETH HOWARTH, AND SHE WAS ENGLISH.
SHE'D LEFT HOME FOR ADVENTURE—HADN'T EVEN BROUGHT ENOUGH MONEY FOR A RETURN TICKET.
SHE WORKED AS AN *AU PAIR* GIRL FOR AN ENGLISH FAMILY—AND LUCKY FOR ME SHE WASN'T MAKING ENOUGH MONEY.
HEY, I'LL PAY YOU MORE JUST TO TAKE CARE OF MY HOUSE FOR ME!
WELL...
I CAN ONLY COOK STEAK, LAMB CHOPS, AND PORK CHOPS WITH PEAS, CORN, OR LIMA BEANS.

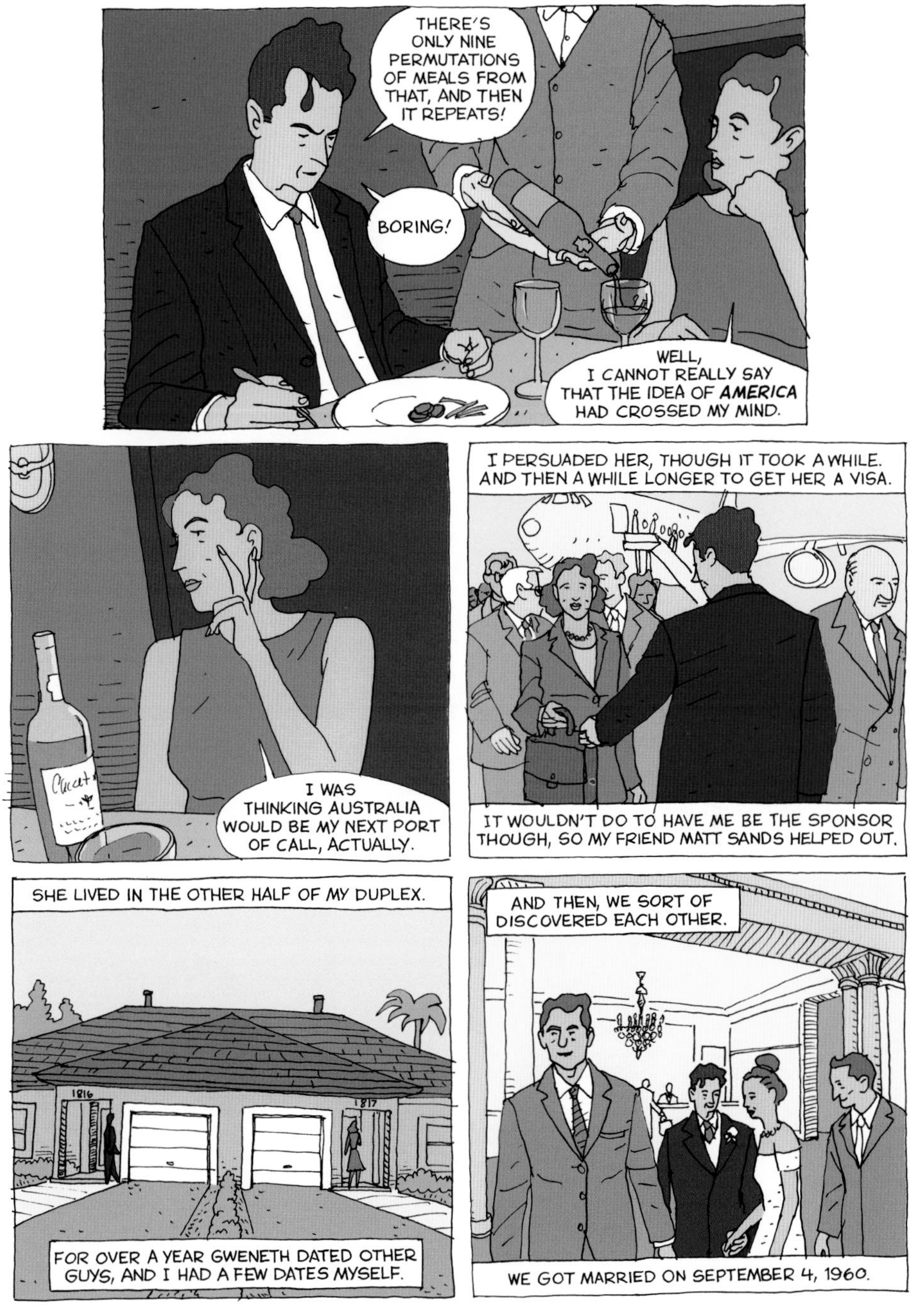
THERE'S ONLY NINE PERMUTATIONS OF MEALS FROM THAT, AND THEN IT REPEATS!
BORING!
WELL, I CANNOT REALLY SAY THAT THE IDEA OF **AMERICA** HAD CROSSED MY MIND.
I WAS THINKING AUSTRALIA WOULD BE MY NEXT PORT OF CALL, ACTUALLY.
I PERSUADED HER, THOUGH IT TOOK A WHILE. AND THEN A WHILE LONGER TO GET HER A VISA.
IT WOULDN'T DO TO HAVE ME BE THE SPONSOR THOUGH, SO MY FRIEND MATT SANDS HELPED OUT.
SHE LIVED IN THE OTHER HALF OF MY DUPLEX.
1816
1817
FOR OVER A YEAR GWENETH DATED OTHER GUYS, AND I HAD A FEW DATES MYSELF.
AND THEN, WE SORT OF DISCOVERED EACH OTHER.
WE GOT MARRIED ON SEPTEMBER 4, 1960.

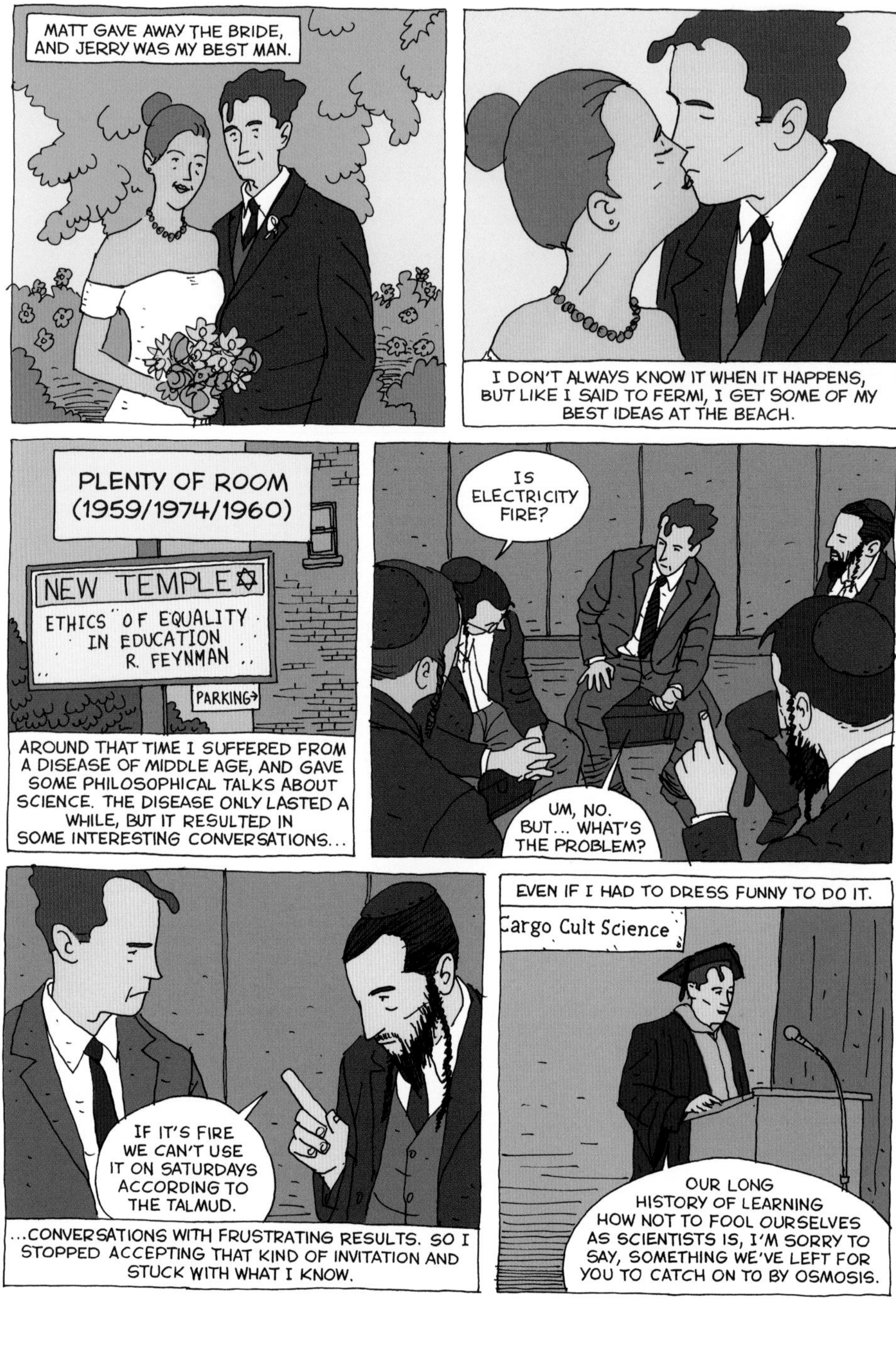

MATT GAVE AWAY THE BRIDE, AND JERRY WAS MY BEST MAN.
I DON'T ALWAYS KNOW IT WHEN IT HAPPENS, BUT LIKE I SAID TO FERMI, I GET SOME OF MY BEST IDEAS AT THE BEACH.
PLENTY OF ROOM (1959/1974/1960)
NEW TEMPLE
ETHICS OF EQUALITY IN EDUCATION R. FEYNMAN
PARKING→
AROUND THAT TIME I SUFFERED FROM A DISEASE OF MIDDLE AGE, AND GAVE SOME PHILOSOPHICAL TALKS ABOUT SCIENCE. THE DISEASE ONLY LASTED A WHILE, BUT IT RESULTED IN SOME INTERESTING CONVERSATIONS...
IS ELECTRICITY FIRE?
UM, NO. BUT... WHAT'S THE PROBLEM?
IF IT'S FIRE WE CAN'T USE IT ON SATURDAYS ACCORDING TO THE TALMUD.
...CONVERSATIONS WITH FRUSTRATING RESULTS. SO I STOPPED ACCEPTING THAT KIND OF INVITATION AND STUCK WITH WHAT I KNOW.
EVEN IF I HAD TO DRESS FUNNY TO DO IT.
Cargo Cult Science
OUR LONG HISTORY OF LEARNING HOW NOT TO FOOL OURSELVES AS SCIENTISTS IS, I'M SORRY TO SAY, SOMETHING WE'VE LEFT FOR YOU TO CATCH ON TO BY OSMOSIS.

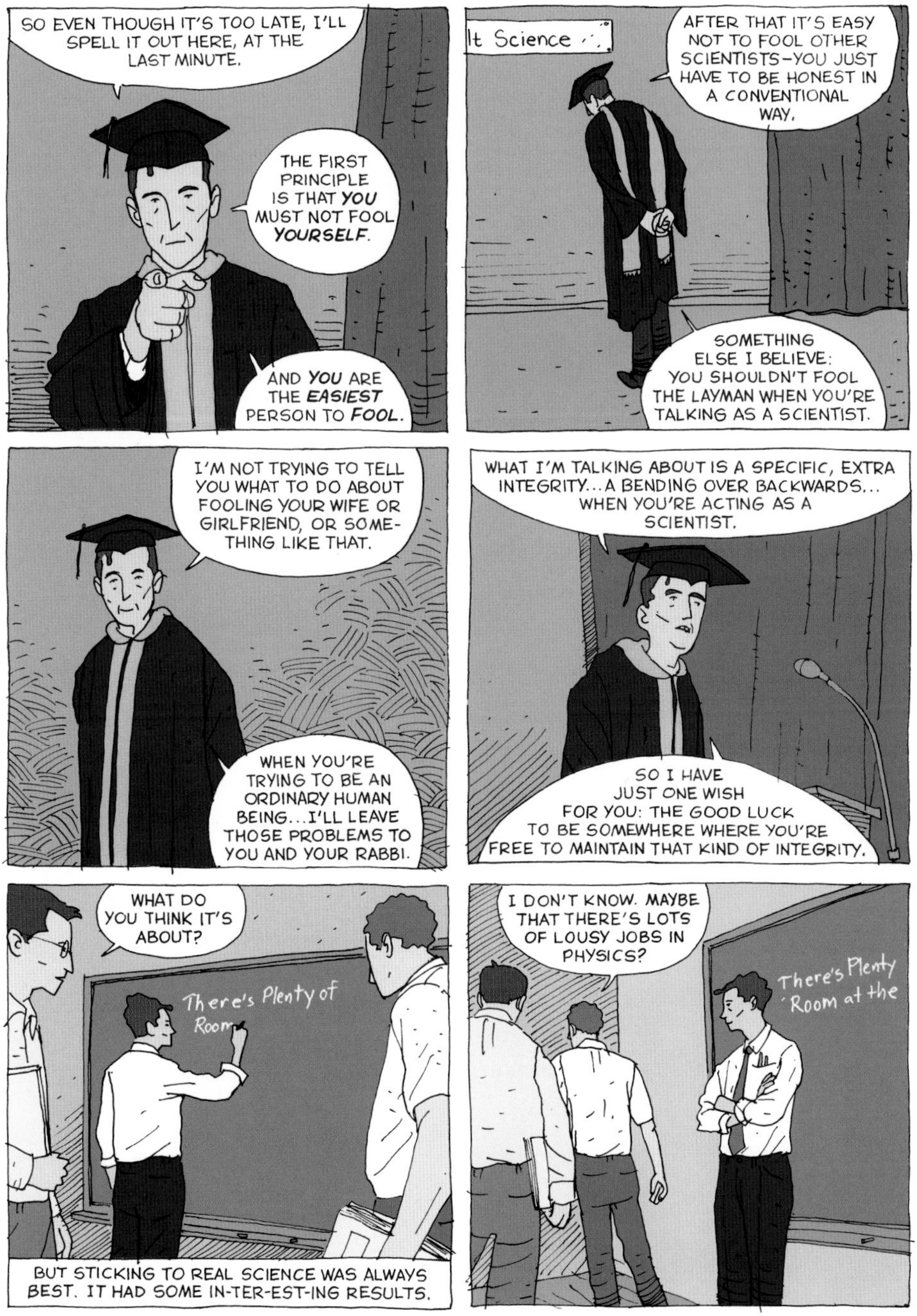

SO EVEN THOUGH IT'S TOO LATE, I'LL SPELL IT OUT HERE, AT THE LAST MINUTE.
THE FIRST PRINCIPLE IS THAT YOU MUST NOT FOOL YOURSELF.
AND YOU ARE THE EASIEST PERSON TO FOOL.
It Science
AFTER THAT IT'S EASY NOT TO FOOL OTHER SCIENTISTS—YOU JUST HAVE TO BE HONEST IN A CONVENTIONAL WAY.
SOMETHING ELSE I BELIEVE: YOU SHOULDN'T FOOL THE LAYMAN WHEN YOU'RE TALKING AS A SCIENTIST.
I'M NOT TRYING TO TELL YOU WHAT TO DO ABOUT FOOLING YOUR WIFE OR GIRLFRIEND, OR SOMETHING LIKE THAT.
WHEN YOU'RE TRYING TO BE AN ORDINARY HUMAN BEING...I'LL LEAVE THOSE PROBLEMS TO YOU AND YOUR RABBI.
WHAT I'M TALKING ABOUT IS A SPECIFIC, EXTRA INTEGRITY...A BENDING OVER BACKWARDS... WHEN YOU'RE ACTING AS A SCIENTIST.
SO I HAVE JUST ONE WISH FOR YOU: THE GOOD LUCK TO BE SOMEWHERE WHERE YOU'RE FREE TO MAINTAIN THAT KIND OF INTEGRITY.
WHAT DO YOU THINK IT'S ABOUT?
There's Plenty of Room
BUT STICKING TO REAL SCIENCE WAS ALWAYS BEST. IT HAD SOME IN-TER-EST-ING RESULTS.
I DON'T KNOW. MAYBE THAT THERE'S LOTS OF LOUSY JOBS IN PHYSICS?
There's Plenty Room at the

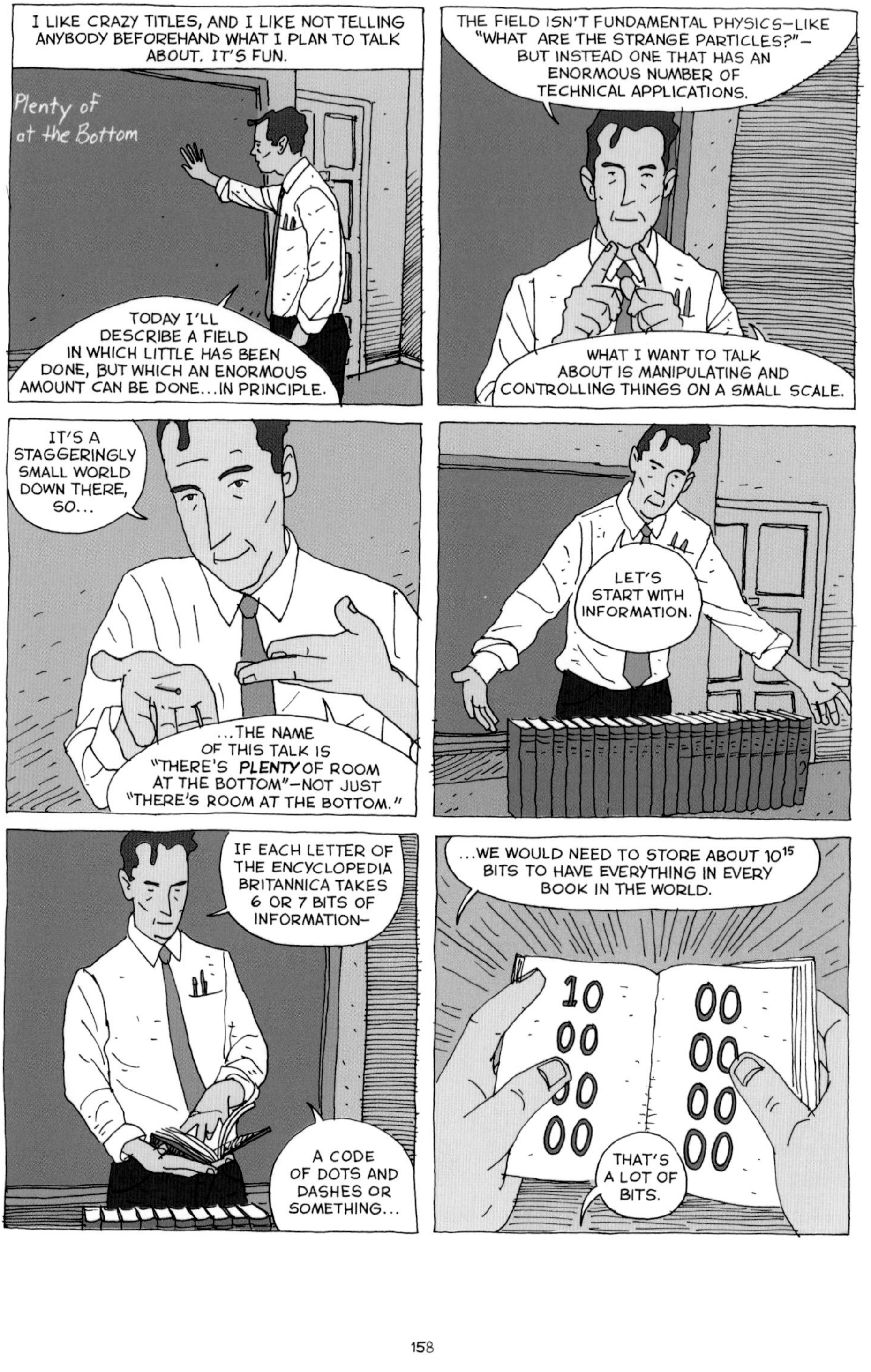
I LIKE CRAZY TITLES, AND I LIKE NOT TELLING ANYBODY BEFOREHAND WHAT I PLAN TO TALK ABOUT. IT'S FUN.
Plenty of at the Bottom
TODAY I'LL DESCRIBE A FIELD IN WHICH LITTLE HAS BEEN DONE, BUT WHICH AN ENORMOUS AMOUNT CAN BE DONE...IN PRINCIPLE.
THE FIELD ISN'T FUNDAMENTAL PHYSICS—LIKE "WHAT ARE THE STRANGE PARTICLES?"—BUT INSTEAD ONE THAT HAS AN ENORMOUS NUMBER OF TECHNICAL APPLICATIONS.
WHAT I WANT TO TALK ABOUT IS MANIPULATING AND CONTROLLING THINGS ON A SMALL SCALE.
IT'S A STAGGERINGLY SMALL WORLD DOWN THERE, SO...
...THE NAME OF THIS TALK IS "THERE'S **PLENTY** OF ROOM AT THE BOTTOM"—NOT JUST "THERE'S ROOM AT THE BOTTOM."
LET'S START WITH INFORMATION.
IF EACH LETTER OF THE ENCYCLOPEDIA BRITANNICA TAKES 6 OR 7 BITS OF INFORMATION—
A CODE OF DOTS AND DASHES OR SOMETHING...
...WE WOULD NEED TO STORE ABOUT 10^{15} BITS TO HAVE EVERYTHING IN EVERY BOOK IN THE WORLD.
10 00 00 00 00 00 00 00
THAT'S A LOT OF BITS.

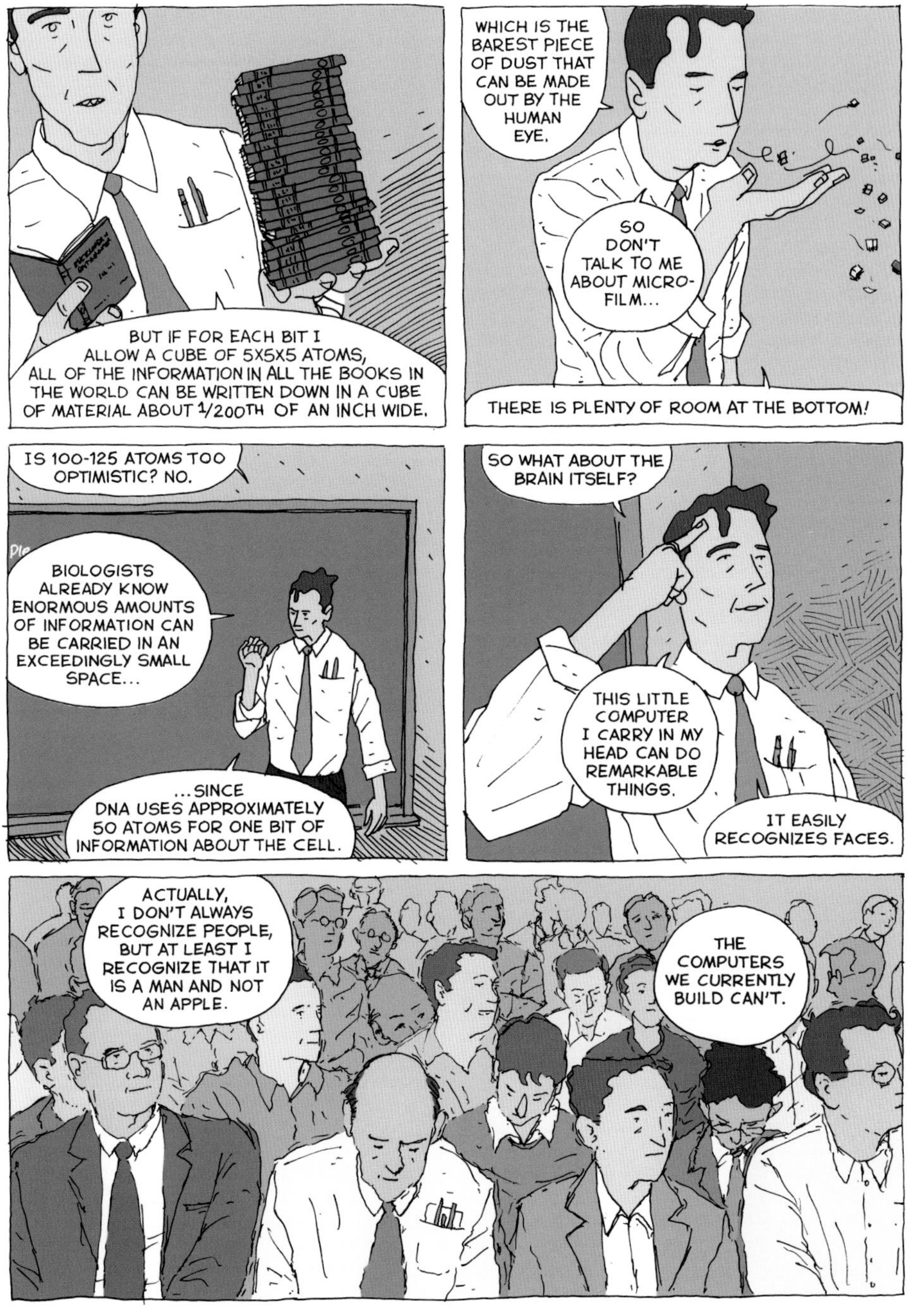
BUT IF FOR EACH BIT I ALLOW A CUBE OF 5X5X5 ATOMS, ALL OF THE INFORMATION IN ALL THE BOOKS IN THE WORLD CAN BE WRITTEN DOWN IN A CUBE OF MATERIAL ABOUT 1/200TH OF AN INCH WIDE.
WHICH IS THE BAREST PIECE OF DUST THAT CAN BE MADE OUT BY THE HUMAN EYE.
SO DON'T TALK TO ME ABOUT MICRO-FILM...
THERE IS PLENTY OF ROOM AT THE BOTTOM!
IS 100-125 ATOMS TOO OPTIMISTIC? NO.
BIOLOGISTS ALREADY KNOW ENORMOUS AMOUNTS OF INFORMATION CAN BE CARRIED IN AN EXCEEDINGLY SMALL SPACE...
...SINCE DNA USES APPROXIMATELY 50 ATOMS FOR ONE BIT OF INFORMATION ABOUT THE CELL.
SO WHAT ABOUT THE BRAIN ITSELF?
THIS LITTLE COMPUTER I CARRY IN MY HEAD CAN DO REMARKABLE THINGS.
IT EASILY RECOGNIZES FACES.
ACTUALLY, I DON'T ALWAYS RECOGNIZE PEOPLE, BUT AT LEAST I RECOGNIZE THAT IT IS A MAN AND NOT AN APPLE.
THE COMPUTERS WE CURRENTLY BUILD CAN'T.

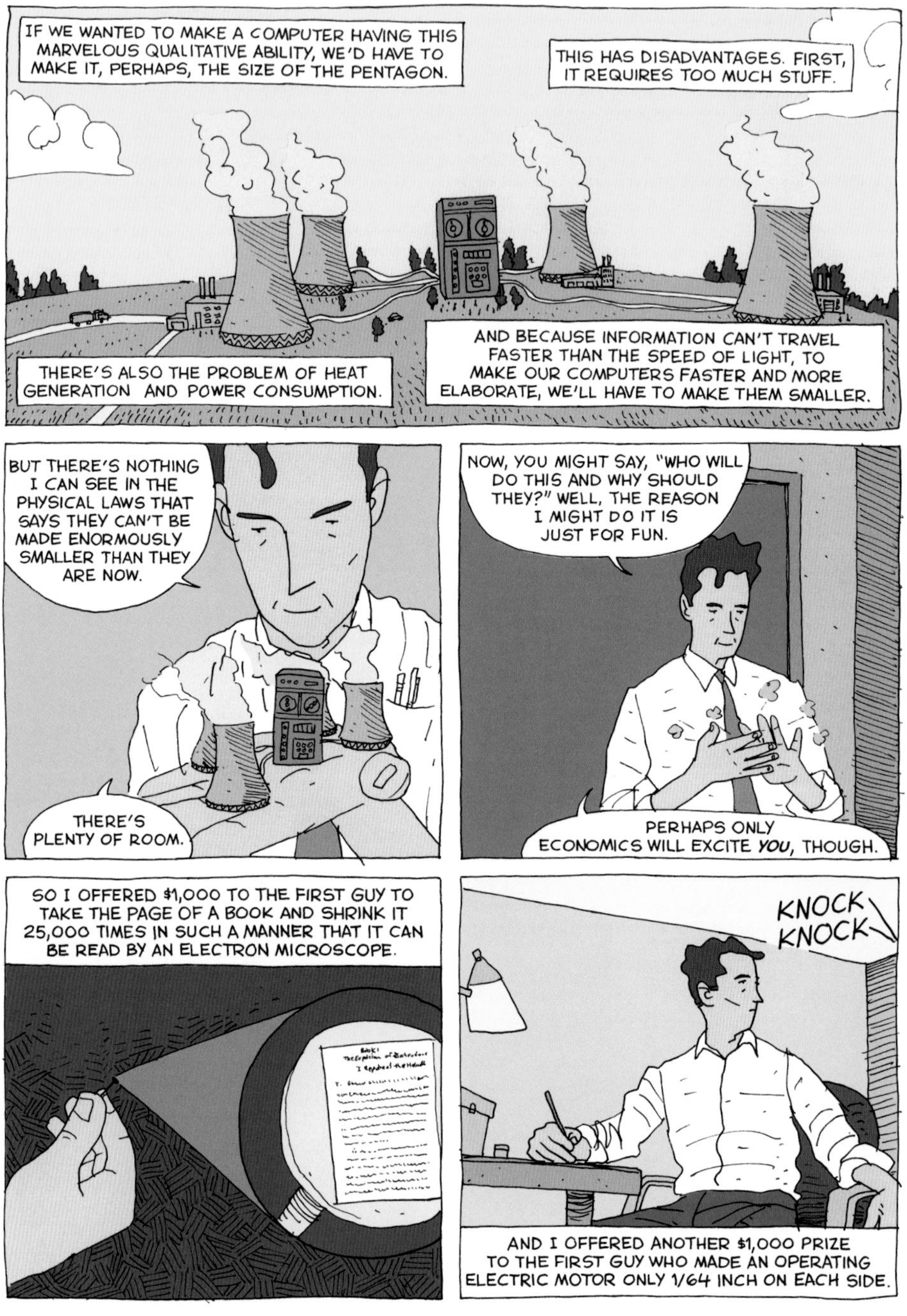
IF WE WANTED TO MAKE A COMPUTER HAVING THIS MARVELOUS QUALITATIVE ABILITY, WE'D HAVE TO MAKE IT, PERHAPS, THE SIZE OF THE PENTAGON.
THIS HAS DISADVANTAGES. FIRST, IT REQUIRES TOO MUCH STUFF.
THERE'S ALSO THE PROBLEM OF HEAT GENERATION AND POWER CONSUMPTION.
AND BECAUSE INFORMATION CAN'T TRAVEL FASTER THAN THE SPEED OF LIGHT, TO MAKE OUR COMPUTERS FASTER AND MORE ELABORATE, WE'LL HAVE TO MAKE THEM SMALLER.
BUT THERE'S NOTHING I CAN SEE IN THE PHYSICAL LAWS THAT SAYS THEY CAN'T BE MADE ENORMOUSLY SMALLER THAN THEY ARE NOW.
THERE'S PLENTY OF ROOM.
NOW, YOU MIGHT SAY, "WHO WILL DO THIS AND WHY SHOULD THEY?" WELL, THE REASON I MIGHT DO IT IS JUST FOR FUN.
PERHAPS ONLY ECONOMICS WILL EXCITE *YOU*, THOUGH.
SO I OFFERED $1,000 TO THE FIRST GUY TO TAKE THE PAGE OF A BOOK AND SHRINK IT 25,000 TIMES IN SUCH A MANNER THAT IT CAN BE READ BY AN ELECTRON MICROSCOPE.
KNOCK KNOCK
AND I OFFERED ANOTHER $1,000 PRIZE TO THE FIRST GUY WHO MADE AN OPERATING ELECTRIC MOTOR ONLY 1/64 INCH ON EACH SIDE.

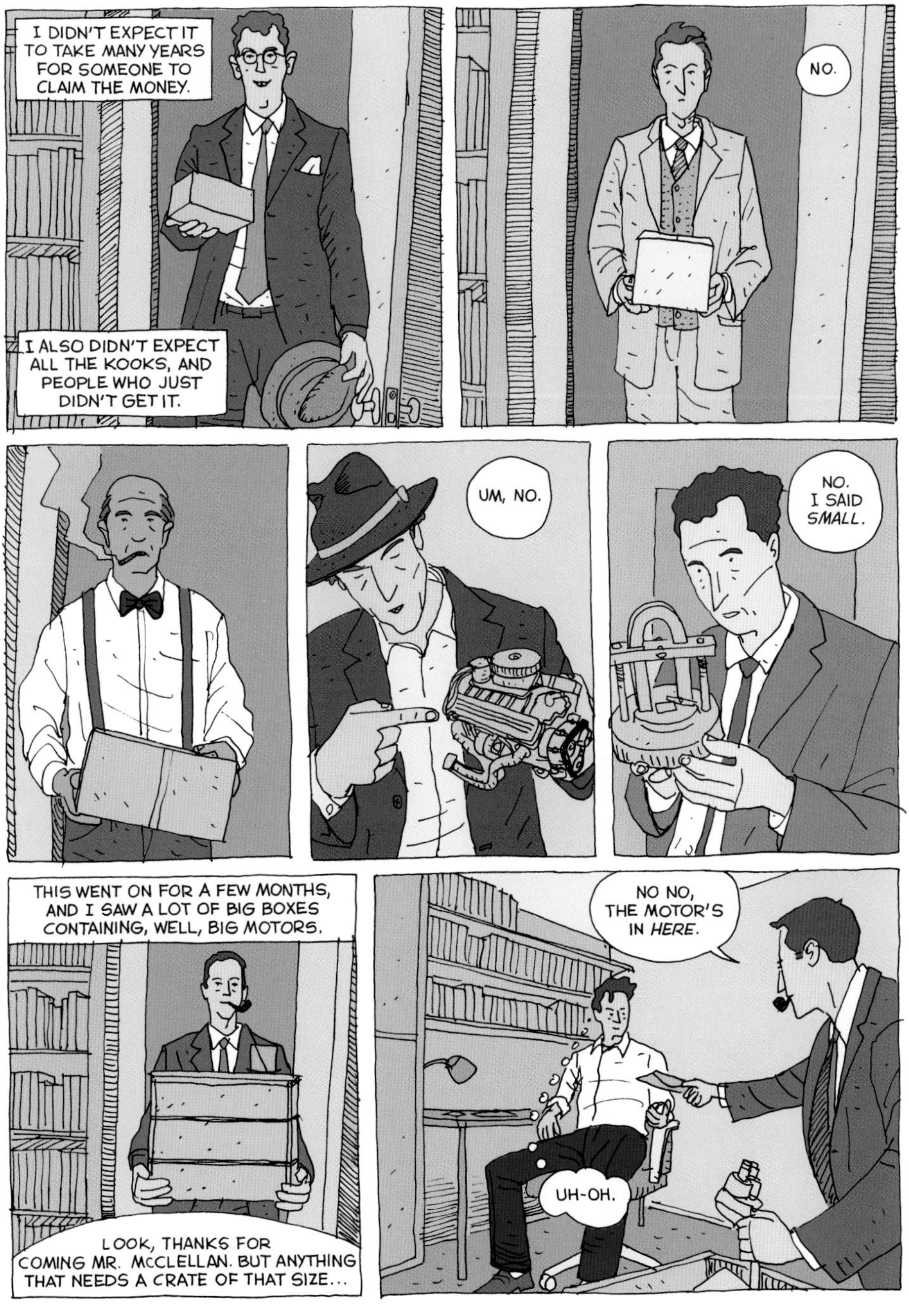
I DIDN'T EXPECT IT TO TAKE MANY YEARS FOR SOMEONE TO CLAIM THE MONEY.
I ALSO DIDN'T EXPECT ALL THE KOOKS, AND PEOPLE WHO JUST DIDN'T GET IT.
NO.
UM, NO.
NO. I SAID *SMALL*.
THIS WENT ON FOR A FEW MONTHS, AND I SAW A LOT OF BIG BOXES CONTAINING, WELL, BIG MOTORS.
LOOK, THANKS FOR COMING MR. McCLELLAN. BUT ANYTHING THAT NEEDS A CRATE OF THAT SIZE...
NO NO, THE MOTOR'S IN *HERE*.
UH-OH.

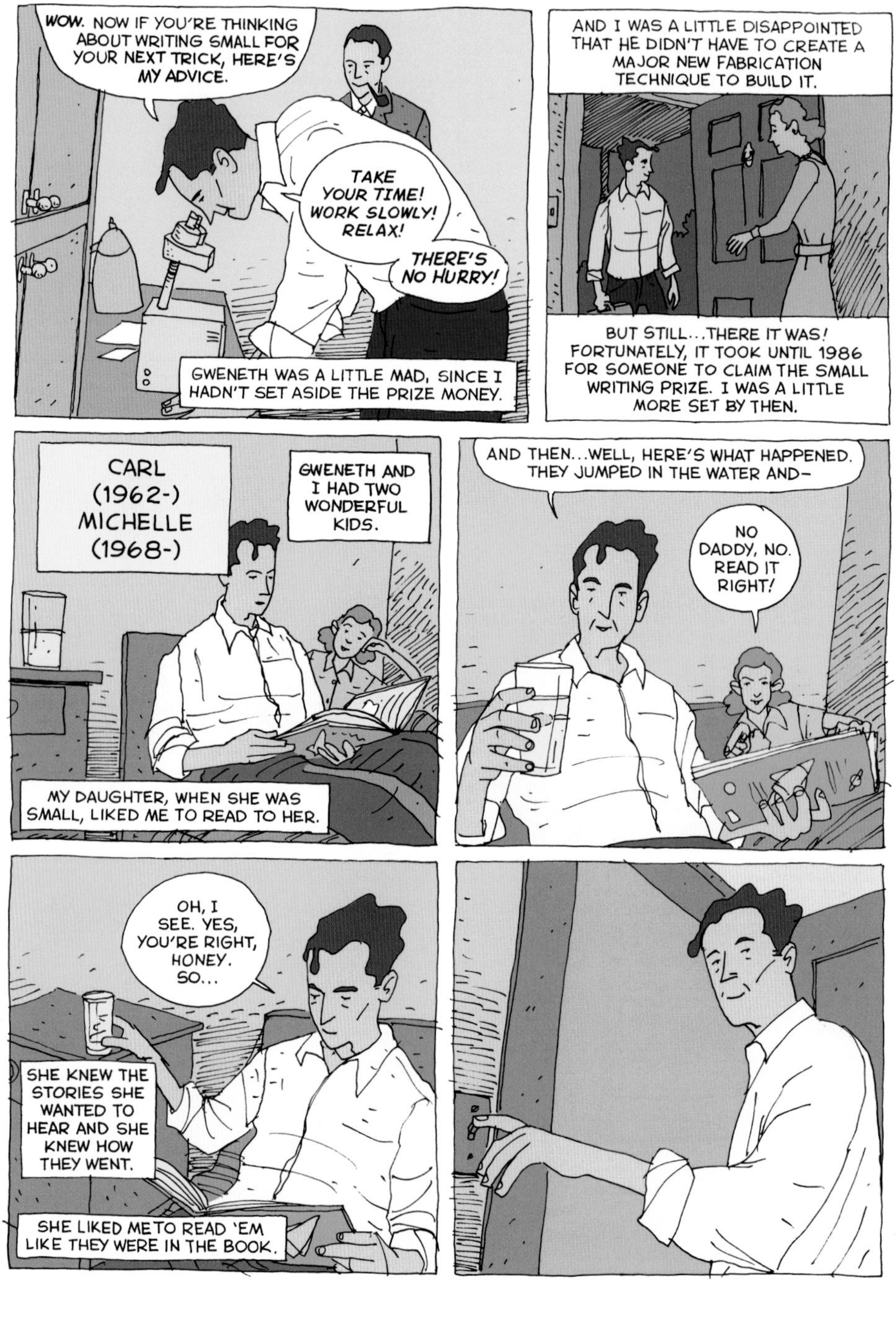
WOW. NOW IF YOU'RE THINKING ABOUT WRITING SMALL FOR YOUR NEXT TRICK, HERE'S MY ADVICE.
TAKE YOUR TIME! WORK SLOWLY! RELAX!
THERE'S NO HURRY!
GWENETH WAS A LITTLE MAD, SINCE I HADN'T SET ASIDE THE PRIZE MONEY.
AND I WAS A LITTLE DISAPPOINTED THAT HE DIDN'T HAVE TO CREATE A MAJOR NEW FABRICATION TECHNIQUE TO BUILD IT.
BUT STILL...THERE IT WAS! FORTUNATELY, IT TOOK UNTIL 1986 FOR SOMEONE TO CLAIM THE SMALL WRITING PRIZE. I WAS A LITTLE MORE SET BY THEN.
CARL (1962-) MICHELLE (1968-)
GWENETH AND I HAD TWO WONDERFUL KIDS.
MY DAUGHTER, WHEN SHE WAS SMALL, LIKED ME TO READ TO HER.
AND THEN...WELL, HERE'S WHAT HAPPENED. THEY JUMPED IN THE WATER AND–
NO DADDY, NO. READ IT RIGHT!
OH, I SEE. YES, YOU'RE RIGHT, HONEY. SO...
SHE KNEW THE STORIES SHE WANTED TO HEAR AND SHE KNEW HOW THEY WENT.
SHE LIKED ME TO READ 'EM LIKE THEY WERE IN THE BOOK.

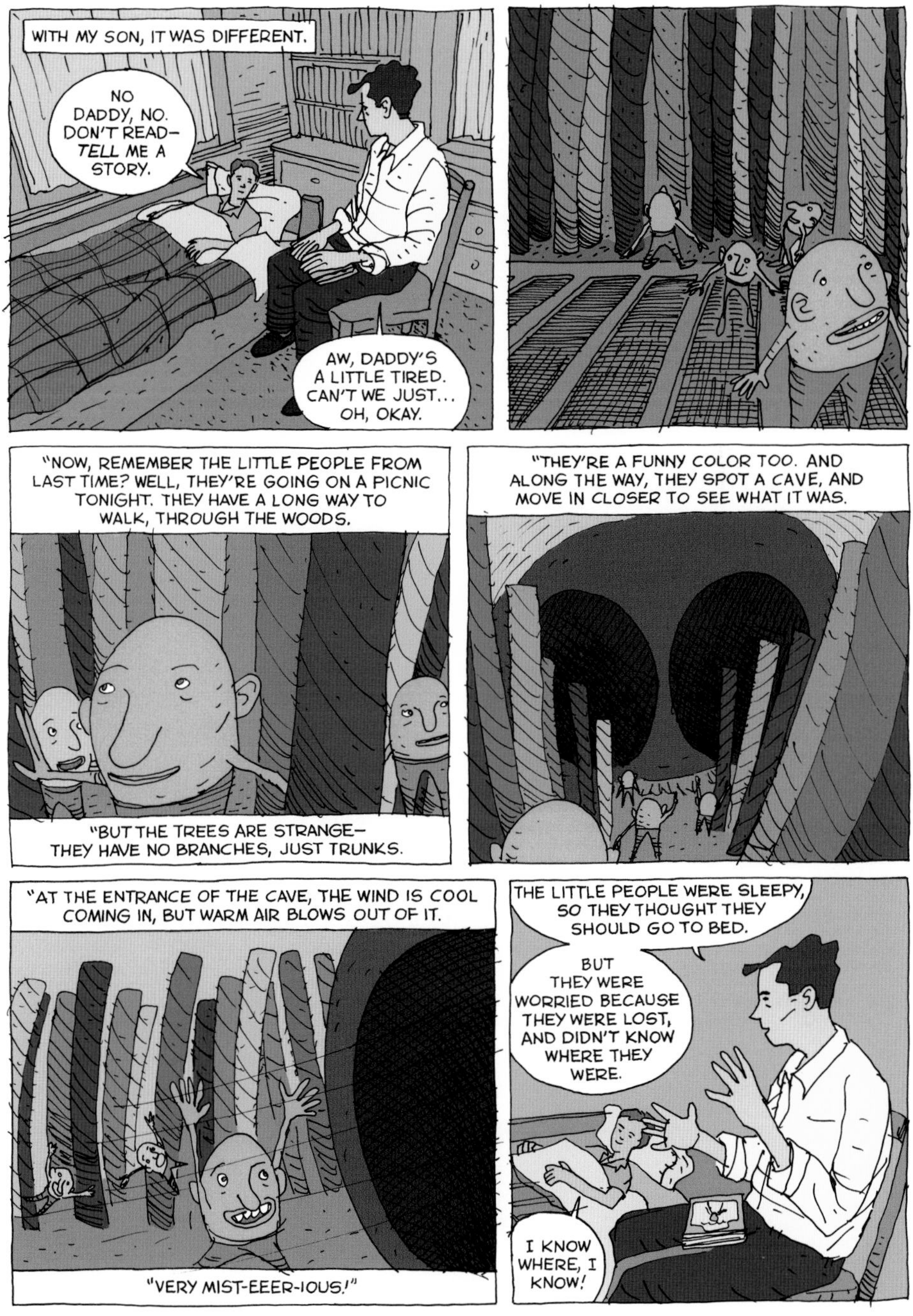

WITH MY SON, IT WAS DIFFERENT.
NO DADDY, NO. DON'T READ— *TELL* ME A STORY.
AW, DADDY'S A LITTLE TIRED. CAN'T WE JUST... OH, OKAY.
"NOW, REMEMBER THE LITTLE PEOPLE FROM LAST TIME? WELL, THEY'RE GOING ON A PICNIC TONIGHT. THEY HAVE A LONG WAY TO WALK, THROUGH THE WOODS.
"BUT THE TREES ARE STRANGE— THEY HAVE NO BRANCHES, JUST TRUNKS.
"THEY'RE A FUNNY COLOR TOO. AND ALONG THE WAY, THEY SPOT A CAVE, AND MOVE IN CLOSER TO SEE WHAT IT WAS.
"AT THE ENTRANCE OF THE CAVE, THE WIND IS COOL COMING IN, BUT WARM AIR BLOWS OUT OF IT.
"VERY MIST-EEER-IOUS!"
THE LITTLE PEOPLE WERE SLEEPY, SO THEY THOUGHT THEY SHOULD GO TO BED.
BUT THEY WERE WORRIED BECAUSE THEY WERE LOST, AND DIDN'T KNOW WHERE THEY WERE.
I KNOW WHERE, I KNOW!

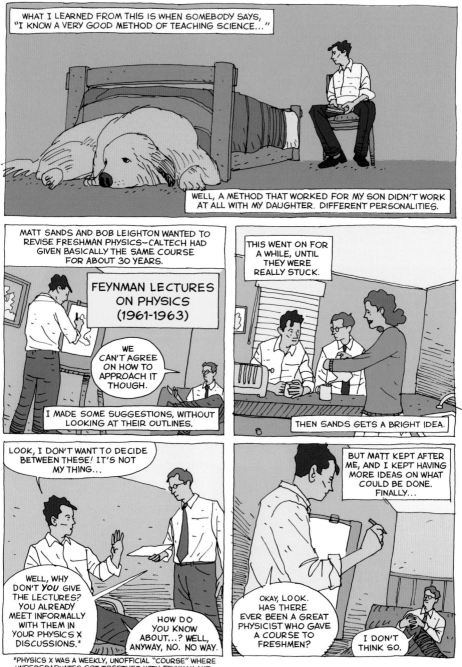

WHAT I LEARNED FROM THIS IS WHEN SOMEBODY SAYS, "I KNOW A VERY GOOD METHOD OF TEACHING SCIENCE..."

WELL, A METHOD THAT WORKED FOR MY SON DIDN'T WORK AT ALL WITH MY DAUGHTER. DIFFERENT PERSONALITIES.

MATT SANDS AND BOB LEIGHTON WANTED TO REVISE FRESHMAN PHYSICS—CALTECH HAD GIVEN BASICALLY THE SAME COURSE FOR ABOUT 30 YEARS.

FEYNMAN LECTURES ON PHYSICS (1961-1963)

WE CAN'T AGREE ON HOW TO APPROACH IT THOUGH.

I MADE SOME SUGGESTIONS, WITHOUT LOOKING AT THEIR OUTLINES.

THIS WENT ON FOR A WHILE, UNTIL THEY WERE REALLY STUCK.

THEN SANDS GETS A BRIGHT IDEA.

LOOK, I DON'T WANT TO DECIDE BETWEEN THESE! IT'S NOT MY THING...

WELL, WHY DON'T YOU GIVE THE LECTURES? YOU ALREADY MEET INFORMALLY WITH THEM IN YOUR PHYSICS X DISCUSSIONS.*

HOW DO YOU KNOW ABOUT...? WELL, ANYWAY, NO. NO WAY.

*PHYSICS X WAS A WEEKLY, UNOFFICIAL "COURSE" WHERE UNDERGRADUATES GOT TOGETHER WITH FEYNMAN AND ASKED HIM ANY (PHYSICS) QUESTIONS THEY WANTED.

BUT MATT KEPT AFTER ME, AND I KEPT HAVING MORE IDEAS ON WHAT COULD BE DONE. FINALLY...

OKAY, LOOK. HAS THERE EVER BEEN A GREAT PHYSICIST WHO GAVE A COURSE TO FRESHMEN?

I DON'T THINK SO.

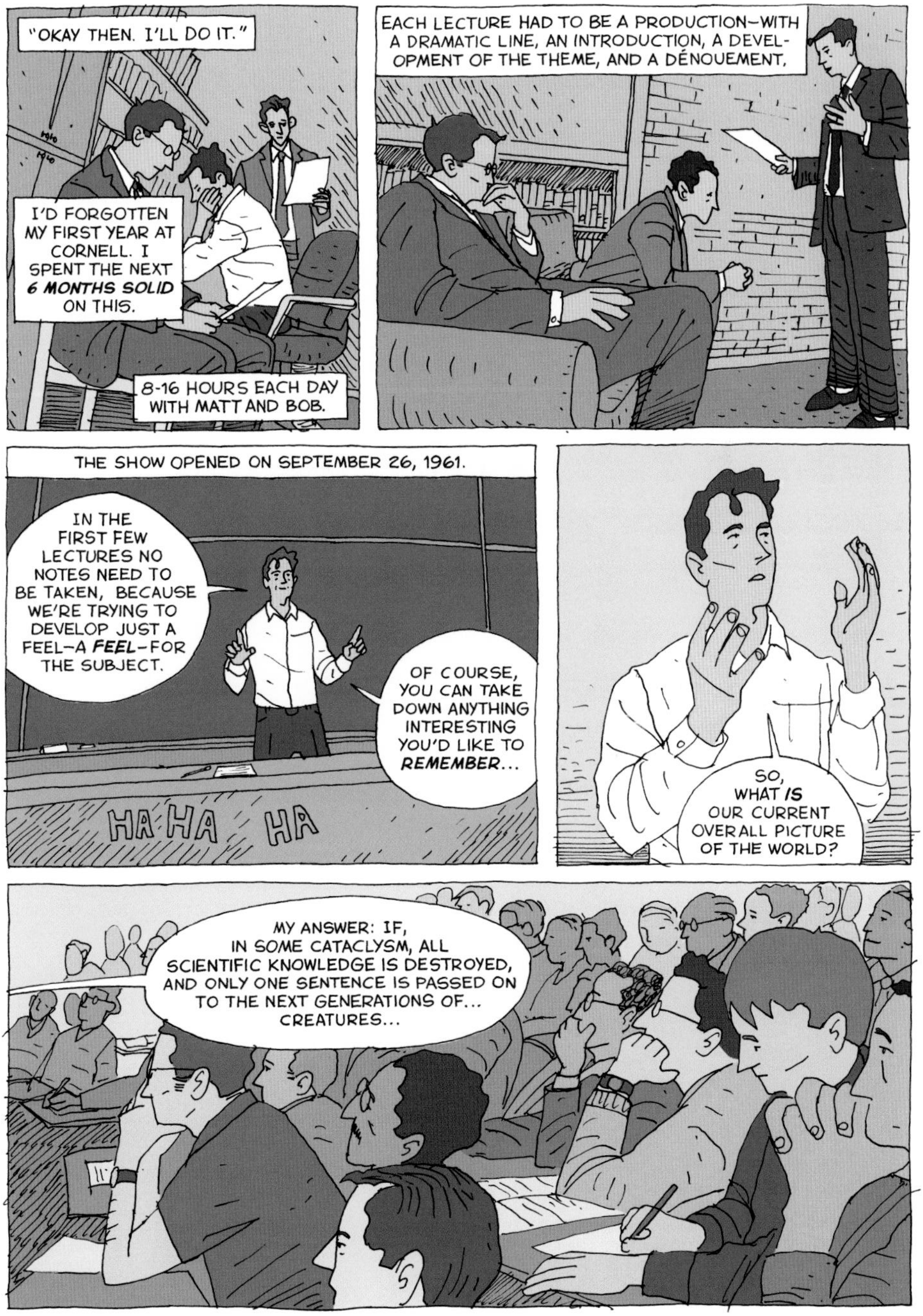
"OKAY THEN. I'LL DO IT."
I'D FORGOTTEN MY FIRST YEAR AT CORNELL. I SPENT THE NEXT **6 MONTHS SOLID** ON THIS.
8-16 HOURS EACH DAY WITH MATT AND BOB.
EACH LECTURE HAD TO BE A PRODUCTION–WITH A DRAMATIC LINE, AN INTRODUCTION, A DEVELOPMENT OF THE THEME, AND A DÉNOUEMENT.
THE SHOW OPENED ON SEPTEMBER 26, 1961.
IN THE FIRST FEW LECTURES NO NOTES NEED TO BE TAKEN, BECAUSE WE'RE TRYING TO DEVELOP JUST A FEEL–A **FEEL**–FOR THE SUBJECT.
OF COURSE, YOU CAN TAKE DOWN ANYTHING INTERESTING YOU'D LIKE TO **REMEMBER**...
HA HA HA
SO, WHAT ***IS*** OUR CURRENT OVERALL PICTURE OF THE WORLD?
MY ANSWER: IF, IN SOME CATACLYSM, ALL SCIENTIFIC KNOWLEDGE IS DESTROYED, AND ONLY ONE SENTENCE IS PASSED ON TO THE NEXT GENERATIONS OF... CREATURES...

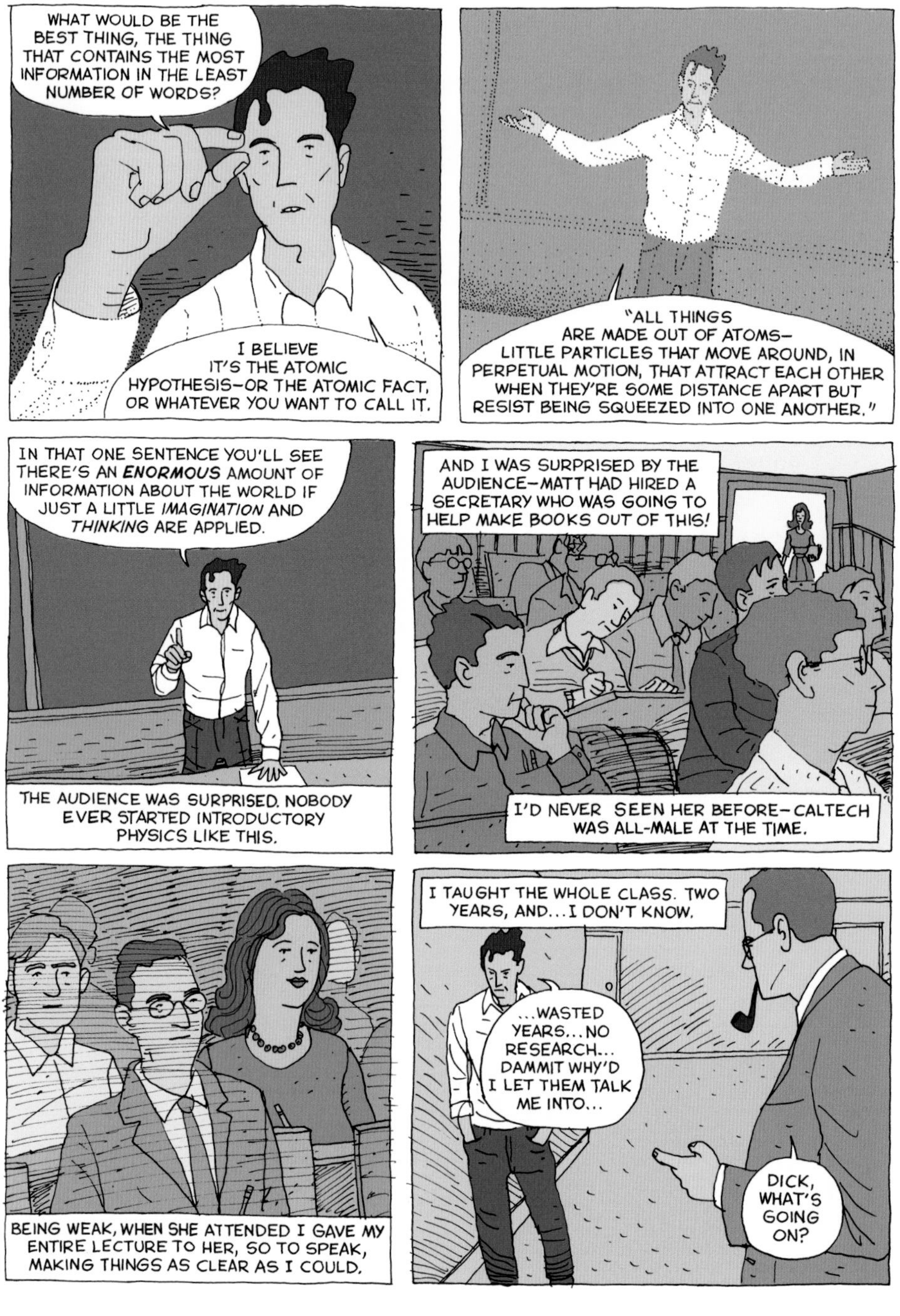
WHAT WOULD BE THE BEST THING, THE THING THAT CONTAINS THE MOST INFORMATION IN THE LEAST NUMBER OF WORDS?
I BELIEVE IT'S THE ATOMIC HYPOTHESIS–OR THE ATOMIC FACT, OR WHATEVER YOU WANT TO CALL IT.
"ALL THINGS ARE MADE OUT OF ATOMS– LITTLE PARTICLES THAT MOVE AROUND, IN PERPETUAL MOTION, THAT ATTRACT EACH OTHER WHEN THEY'RE SOME DISTANCE APART BUT RESIST BEING SQUEEZED INTO ONE ANOTHER."
IN THAT ONE SENTENCE YOU'LL SEE THERE'S AN ENORMOUS AMOUNT OF INFORMATION ABOUT THE WORLD IF JUST A LITTLE IMAGINATION AND THINKING ARE APPLIED.
THE AUDIENCE WAS SURPRISED. NOBODY EVER STARTED INTRODUCTORY PHYSICS LIKE THIS.
AND I WAS SURPRISED BY THE AUDIENCE–MATT HAD HIRED A SECRETARY WHO WAS GOING TO HELP MAKE BOOKS OUT OF THIS!
I'D NEVER SEEN HER BEFORE–CALTECH WAS ALL-MALE AT THE TIME.
BEING WEAK, WHEN SHE ATTENDED I GAVE MY ENTIRE LECTURE TO HER, SO TO SPEAK, MAKING THINGS AS CLEAR AS I COULD.
I TAUGHT THE WHOLE CLASS. TWO YEARS, AND... I DON'T KNOW.
...WASTED YEARS... NO RESEARCH... DAMMIT WHY'D I LET THEM TALK ME INTO...
DICK, WHAT'S GOING ON?

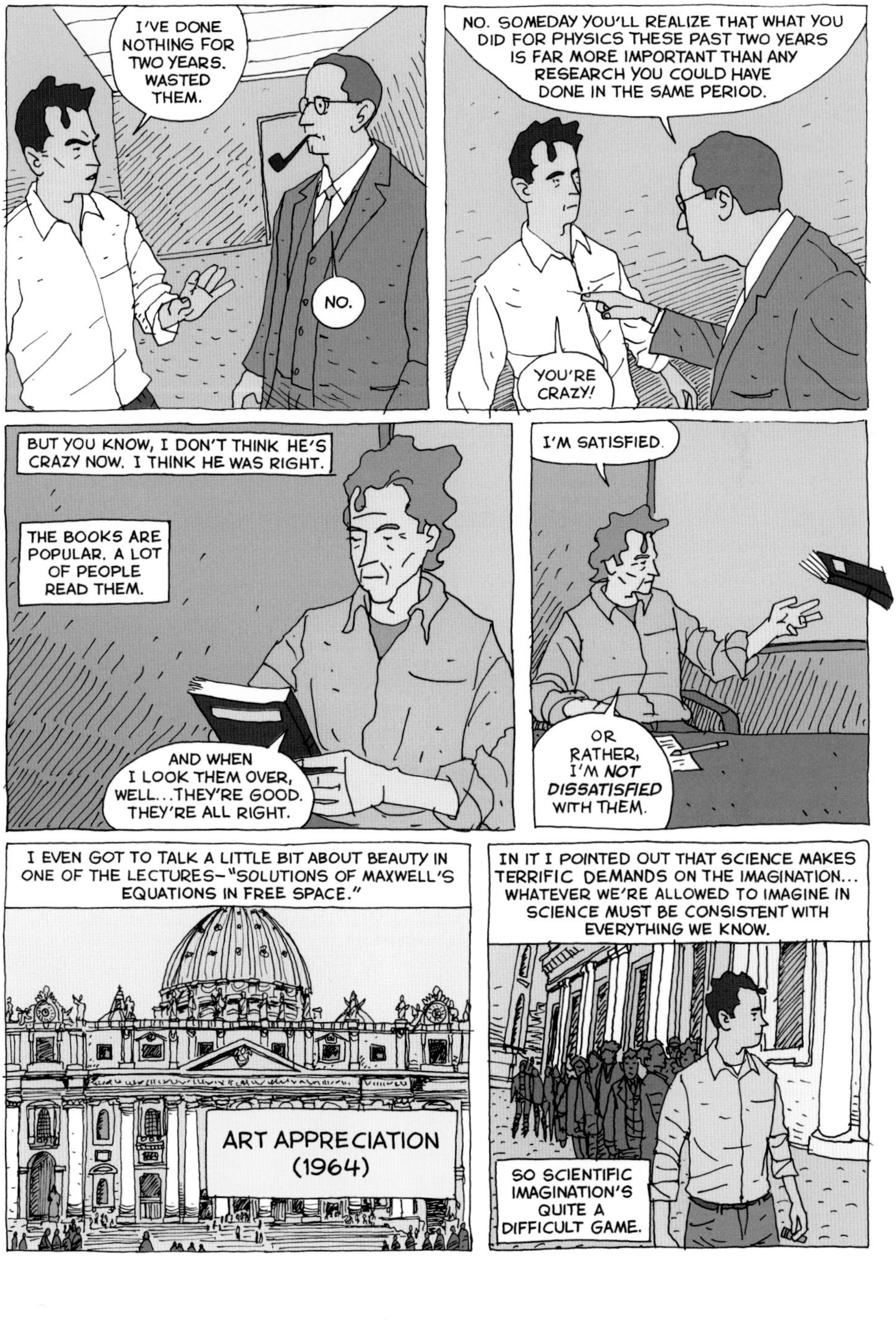
I'VE DONE NOTHING FOR TWO YEARS. WASTED THEM.
NO.
NO. SOMEDAY YOU'LL REALIZE THAT WHAT YOU DID FOR PHYSICS THESE PAST TWO YEARS IS FAR MORE IMPORTANT THAN ANY RESEARCH YOU COULD HAVE DONE IN THE SAME PERIOD.
YOU'RE CRAZY!
BUT YOU KNOW, I DON'T THINK HE'S CRAZY NOW. I THINK HE WAS RIGHT.
THE BOOKS ARE POPULAR. A LOT OF PEOPLE READ THEM.
AND WHEN I LOOK THEM OVER, WELL... THEY'RE GOOD. THEY'RE ALL RIGHT.
I'M SATISFIED.
OR RATHER, I'M *NOT DISSATISFIED* WITH THEM.
I EVEN GOT TO TALK A LITTLE BIT ABOUT BEAUTY IN ONE OF THE LECTURES–"SOLUTIONS OF MAXWELL'S EQUATIONS IN FREE SPACE."
ART APPRECIATION (1964)
IN IT I POINTED OUT THAT SCIENCE MAKES TERRIFIC DEMANDS ON THE IMAGINATION... WHATEVER WE'RE ALLOWED TO IMAGINE IN SCIENCE MUST BE CONSISTENT WITH EVERYTHING WE KNOW.
SO SCIENTIFIC IMAGINATION'S QUITE A DIFFICULT GAME.

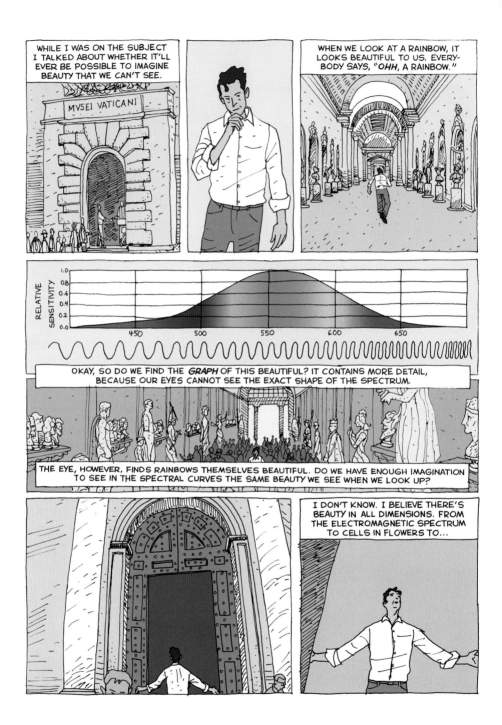

WHILE I WAS ON THE SUBJECT I TALKED ABOUT WHETHER IT'LL EVER BE POSSIBLE TO IMAGINE BEAUTY THAT WE *CAN'T* SEE.

MVSEI VATICANI

WHEN WE LOOK AT A RAINBOW, IT LOOKS BEAUTIFUL TO US. EVERY-BODY SAYS, *"OHH, A RAINBOW."*

OKAY, SO DO WE FIND THE *GRAPH* OF THIS BEAUTIFUL? IT CONTAINS MORE DETAIL, BECAUSE OUR EYES CANNOT SEE THE EXACT SHAPE OF THE SPECTRUM.

THE EYE, HOWEVER, FINDS RAINBOWS THEMSELVES BEAUTIFUL. DO WE HAVE ENOUGH IMAGINATION TO SEE IN THE SPECTRAL CURVES THE SAME BEAUTY WE SEE WHEN WE LOOK UP?

I DON'T KNOW. I BELIEVE THERE'S BEAUTY IN ALL DIMENSIONS. FROM THE ELECTROMAGNETIC SPECTRUM TO CELLS IN FLOWERS TO...

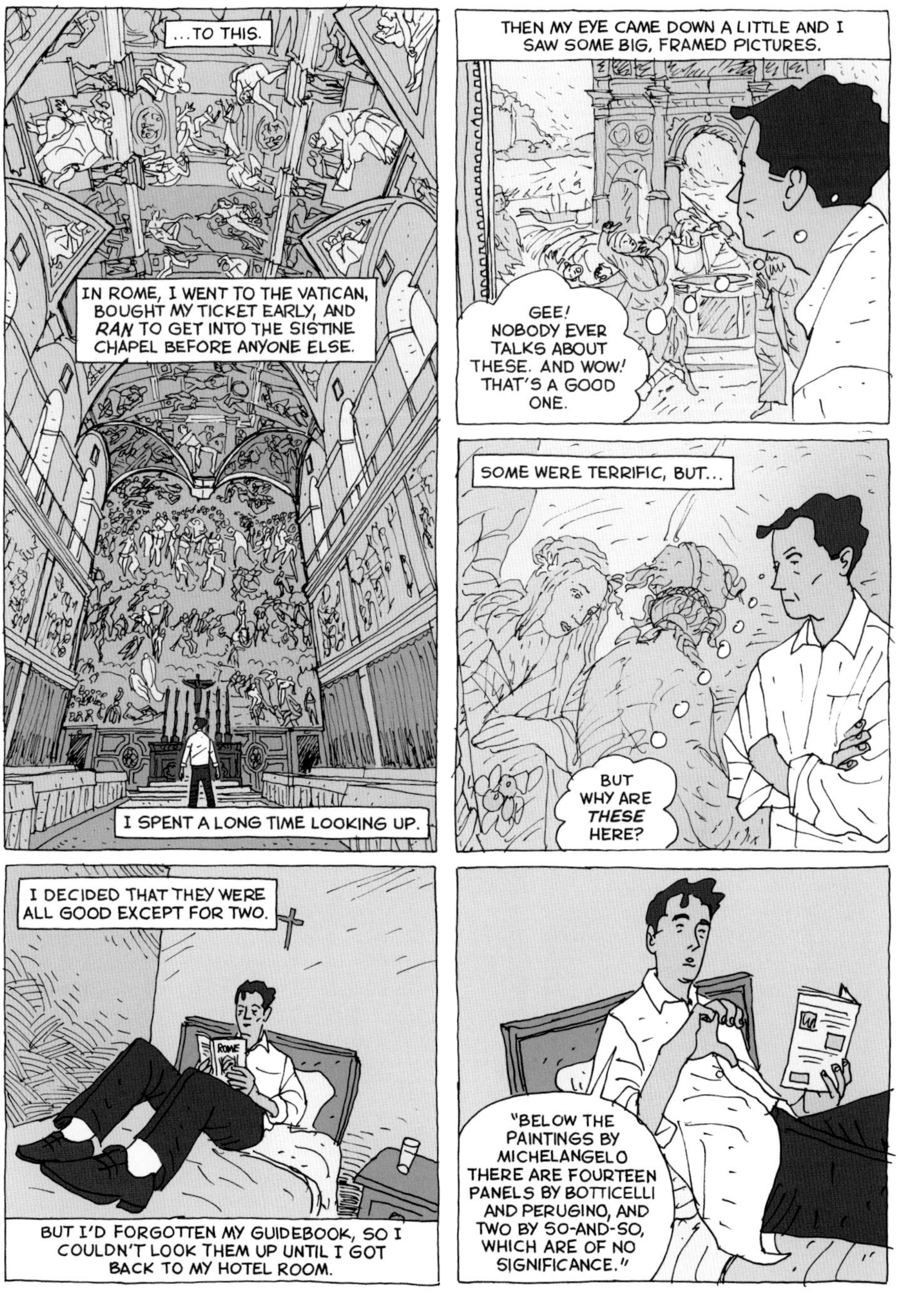

...TO THIS.
IN ROME, I WENT TO THE VATICAN, BOUGHT MY TICKET EARLY, AND RAN TO GET INTO THE SISTINE CHAPEL BEFORE ANYONE ELSE.
I SPENT A LONG TIME LOOKING UP.
THEN MY EYE CAME DOWN A LITTLE AND I SAW SOME BIG, FRAMED PICTURES.
GEE! NOBODY EVER TALKS ABOUT THESE. AND WOW! THAT'S A GOOD ONE.
SOME WERE TERRIFIC, BUT...
BUT WHY ARE THESE HERE?
I DECIDED THAT THEY WERE ALL GOOD EXCEPT FOR TWO.
ROME
BUT I'D FORGOTTEN MY GUIDEBOOK, SO I COULDN'T LOOK THEM UP UNTIL I GOT BACK TO MY HOTEL ROOM.
"BELOW THE PAINTINGS BY MICHELANGELO THERE ARE FOURTEEN PANELS BY BOTTICELLI AND PERUGINO, AND TWO BY SO-AND-SO, WHICH ARE OF NO SIGNIFICANCE."

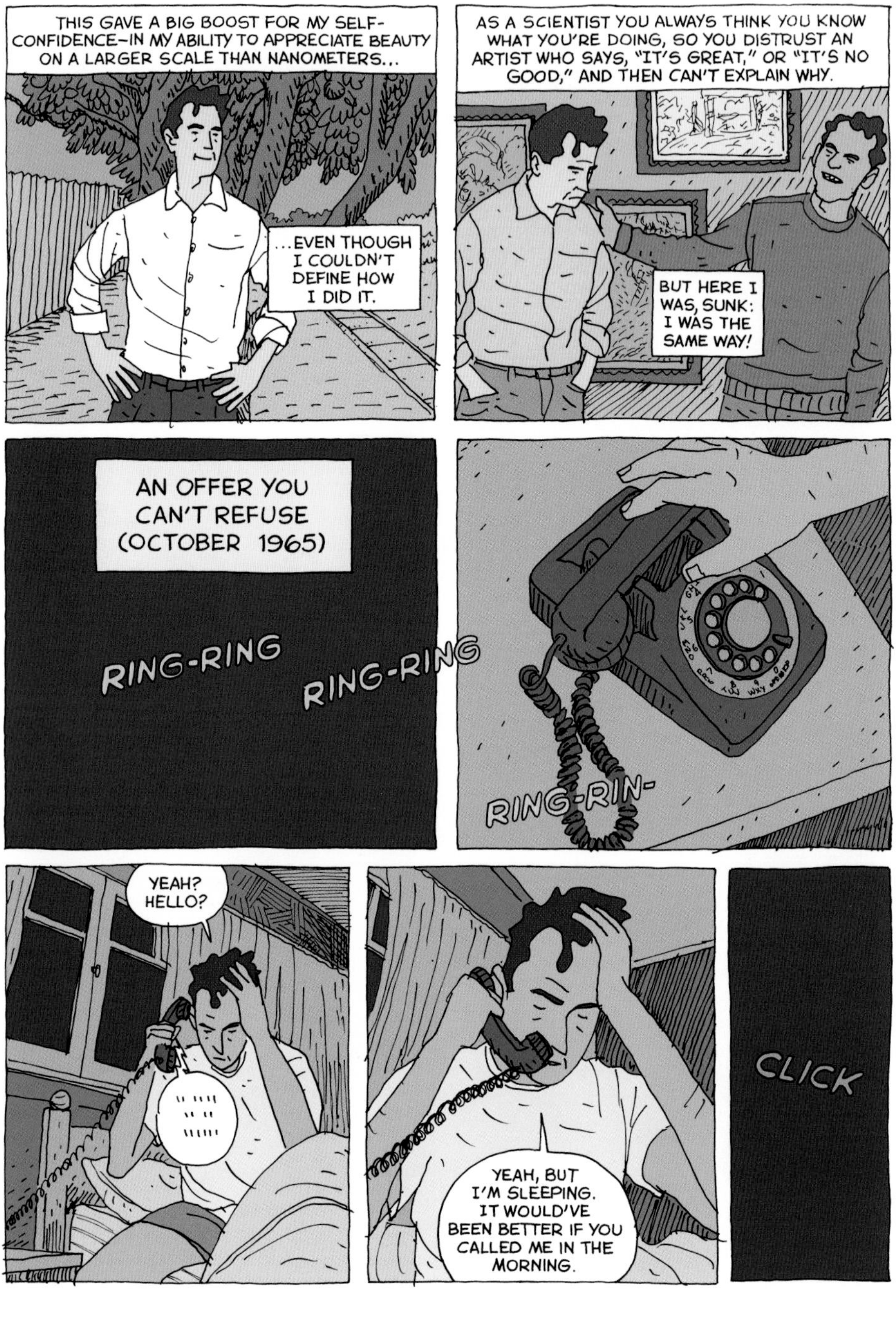

THIS GAVE A BIG BOOST FOR MY SELF-CONFIDENCE—IN MY ABILITY TO APPRECIATE BEAUTY ON A LARGER SCALE THAN NANOMETERS...
...EVEN THOUGH I COULDN'T DEFINE HOW I DID IT.
AS A SCIENTIST YOU ALWAYS THINK YOU KNOW WHAT YOU'RE DOING, SO YOU DISTRUST AN ARTIST WHO SAYS, "IT'S GREAT," OR "IT'S NO GOOD," AND THEN CAN'T EXPLAIN WHY.
BUT HERE I WAS, SUNK: I WAS THE SAME WAY!
AN OFFER YOU CAN'T REFUSE (OCTOBER 1965)
RING-RING
RING-RING
RING-RIN-
YEAH? HELLO?
YEAH, BUT I'M SLEEPING. IT WOULD'VE BEEN BETTER IF YOU CALLED ME IN THE MORNING.
CLICK

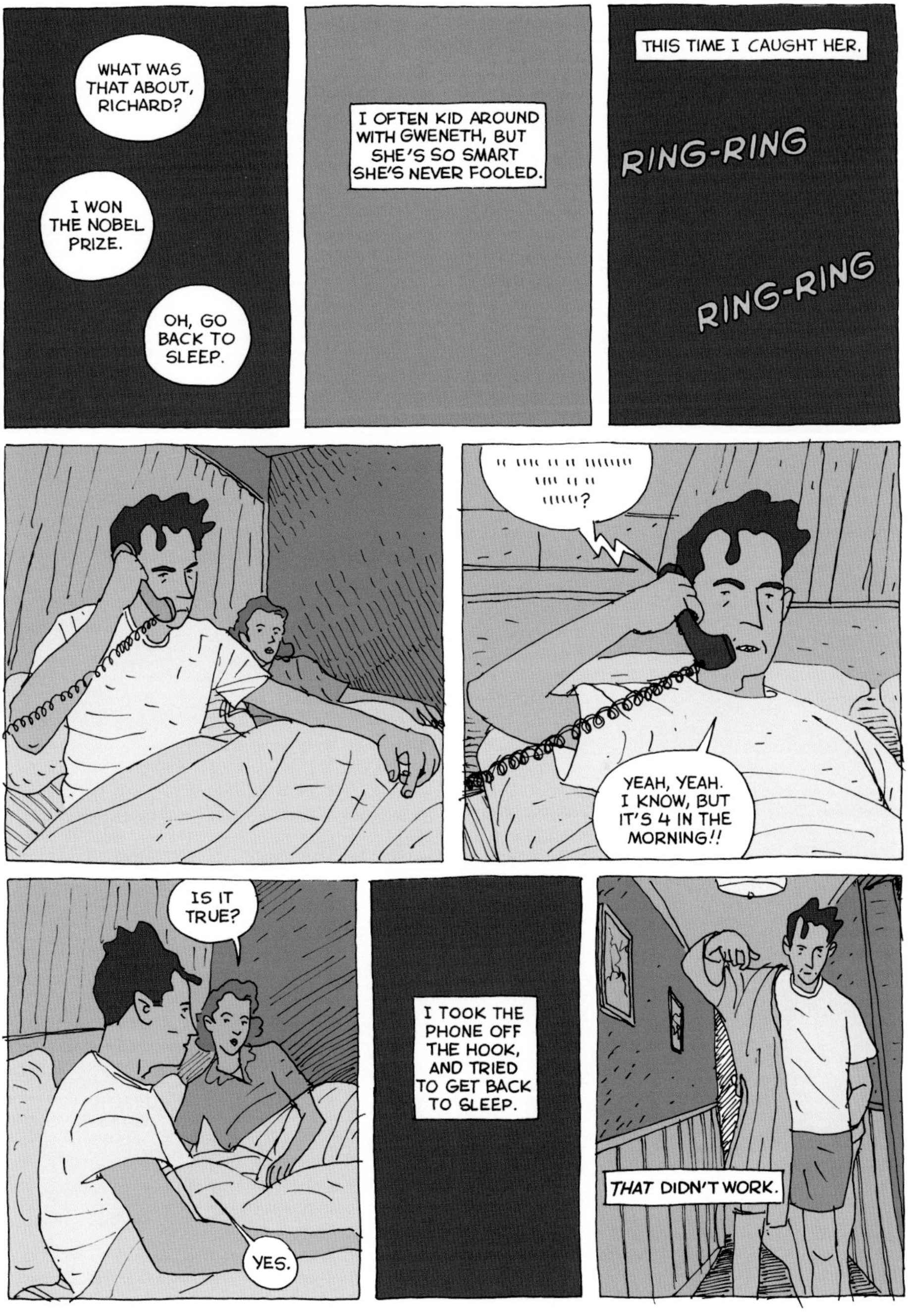
WHAT WAS THAT ABOUT, RICHARD?
I WON THE NOBEL PRIZE.
OH, GO BACK TO SLEEP.
I OFTEN KID AROUND WITH GWENETH, BUT SHE'S SO SMART SHE'S NEVER FOOLED.
THIS TIME I CAUGHT HER.
RING-RING
RING-RING
YEAH, YEAH. I KNOW, BUT IT'S 4 IN THE MORNING!!
IS IT TRUE?
YES.
I TOOK THE PHONE OFF THE HOOK, AND TRIED TO GET BACK TO SLEEP.
THAT DIDN'T WORK.

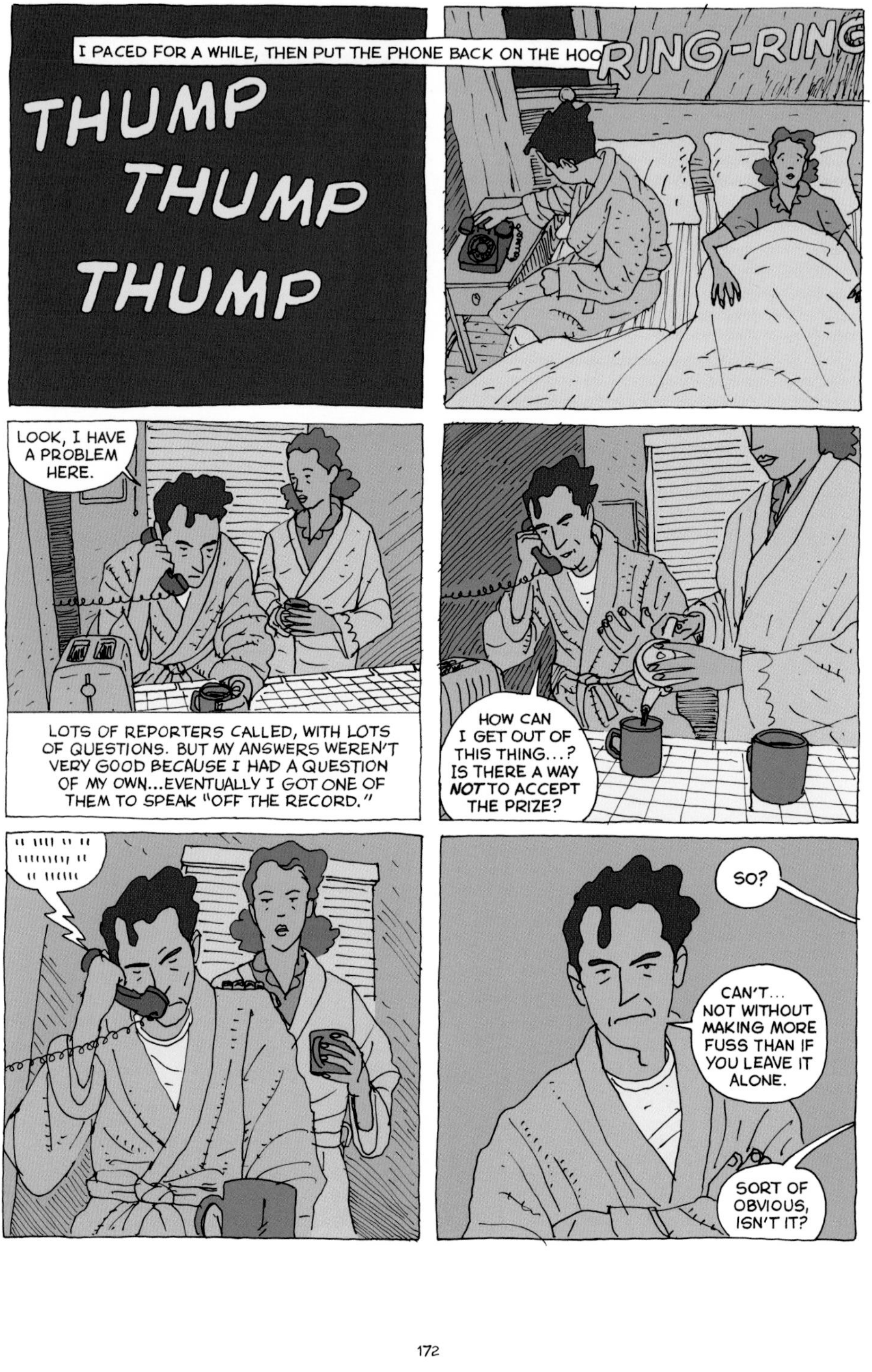
I PACED FOR A WHILE, THEN PUT THE PHONE BACK ON THE HOO
THUMP
THUMP
THUMP
RING-RING
LOOK, I HAVE A PROBLEM HERE.
LOTS OF REPORTERS CALLED, WITH LOTS OF QUESTIONS. BUT MY ANSWERS WEREN'T VERY GOOD BECAUSE I HAD A QUESTION OF MY OWN...EVENTUALLY I GOT ONE OF THEM TO SPEAK "OFF THE RECORD."
HOW CAN I GET OUT OF THIS THING...? IS THERE A WAY ***NOT*** TO ACCEPT THE PRIZE?
SO?
CAN'T... NOT WITHOUT MAKING MORE FUSS THAN IF YOU LEAVE IT ALONE.
SORT OF OBVIOUS, ISN'T IT?

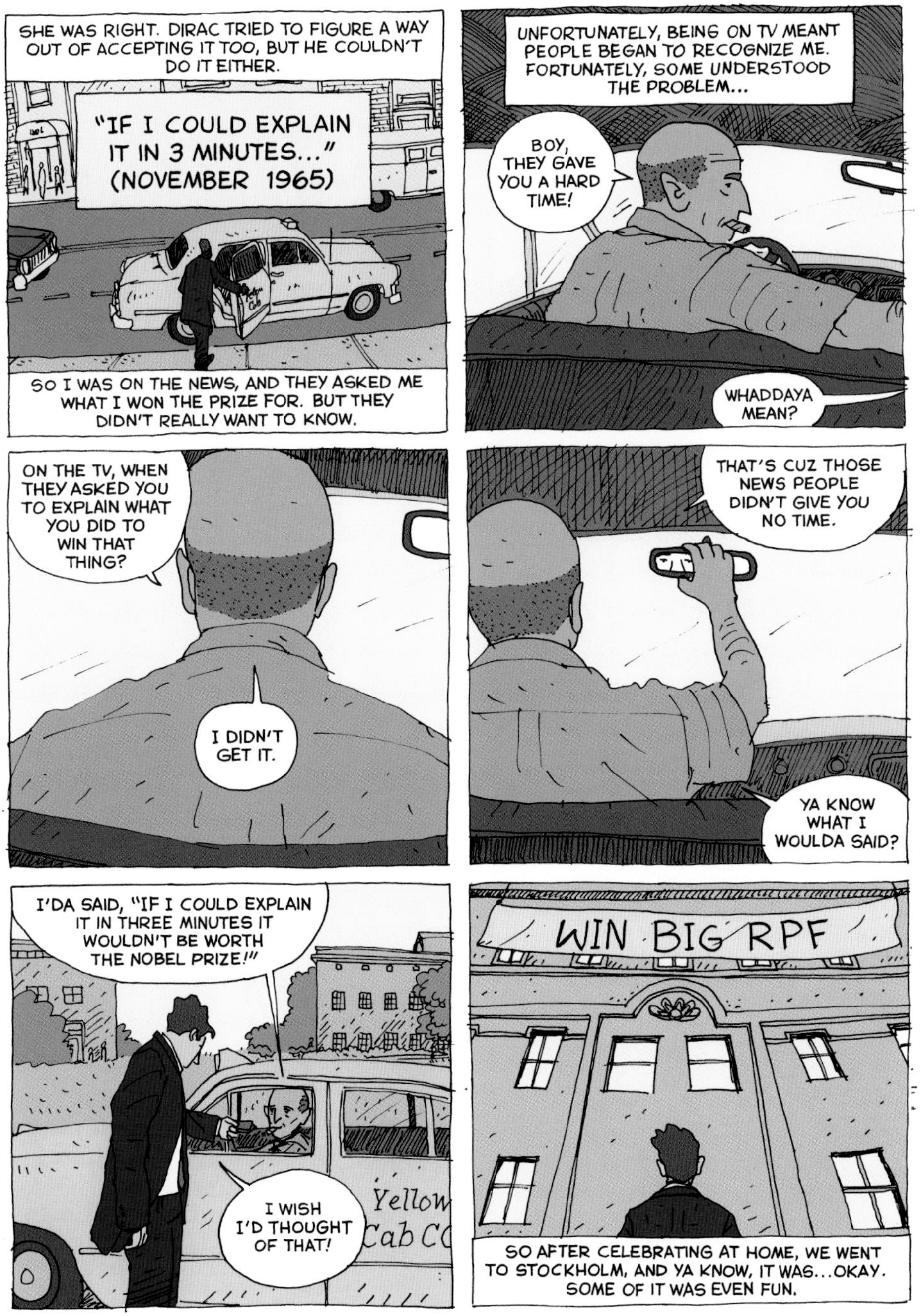
SHE WAS RIGHT. DIRAC TRIED TO FIGURE A WAY OUT OF ACCEPTING IT TOO, BUT HE COULDN'T DO IT EITHER.
"IF I COULD EXPLAIN IT IN 3 MINUTES..." (NOVEMBER 1965)
SO I WAS ON THE NEWS, AND THEY ASKED ME WHAT I WON THE PRIZE FOR. BUT THEY DIDN'T REALLY WANT TO KNOW.
UNFORTUNATELY, BEING ON TV MEANT PEOPLE BEGAN TO RECOGNIZE ME. FORTUNATELY, SOME UNDERSTOOD THE PROBLEM...
BOY, THEY GAVE YOU A HARD TIME!
WHADDAYA MEAN?
ON THE TV, WHEN THEY ASKED YOU TO EXPLAIN WHAT YOU DID TO WIN THAT THING?
I DIDN'T GET IT.
THAT'S CUZ THOSE NEWS PEOPLE DIDN'T GIVE YOU NO TIME.
YA KNOW WHAT I WOULDA SAID?
I'DA SAID, "IF I COULD EXPLAIN IT IN THREE MINUTES IT WOULDN'T BE WORTH THE NOBEL PRIZE!"
I WISH I'D THOUGHT OF THAT!
Yellow Cab Co
WIN BIG RPF
SO AFTER CELEBRATING AT HOME, WE WENT TO STOCKHOLM, AND YA KNOW, IT WAS...OKAY. SOME OF IT WAS EVEN FUN.

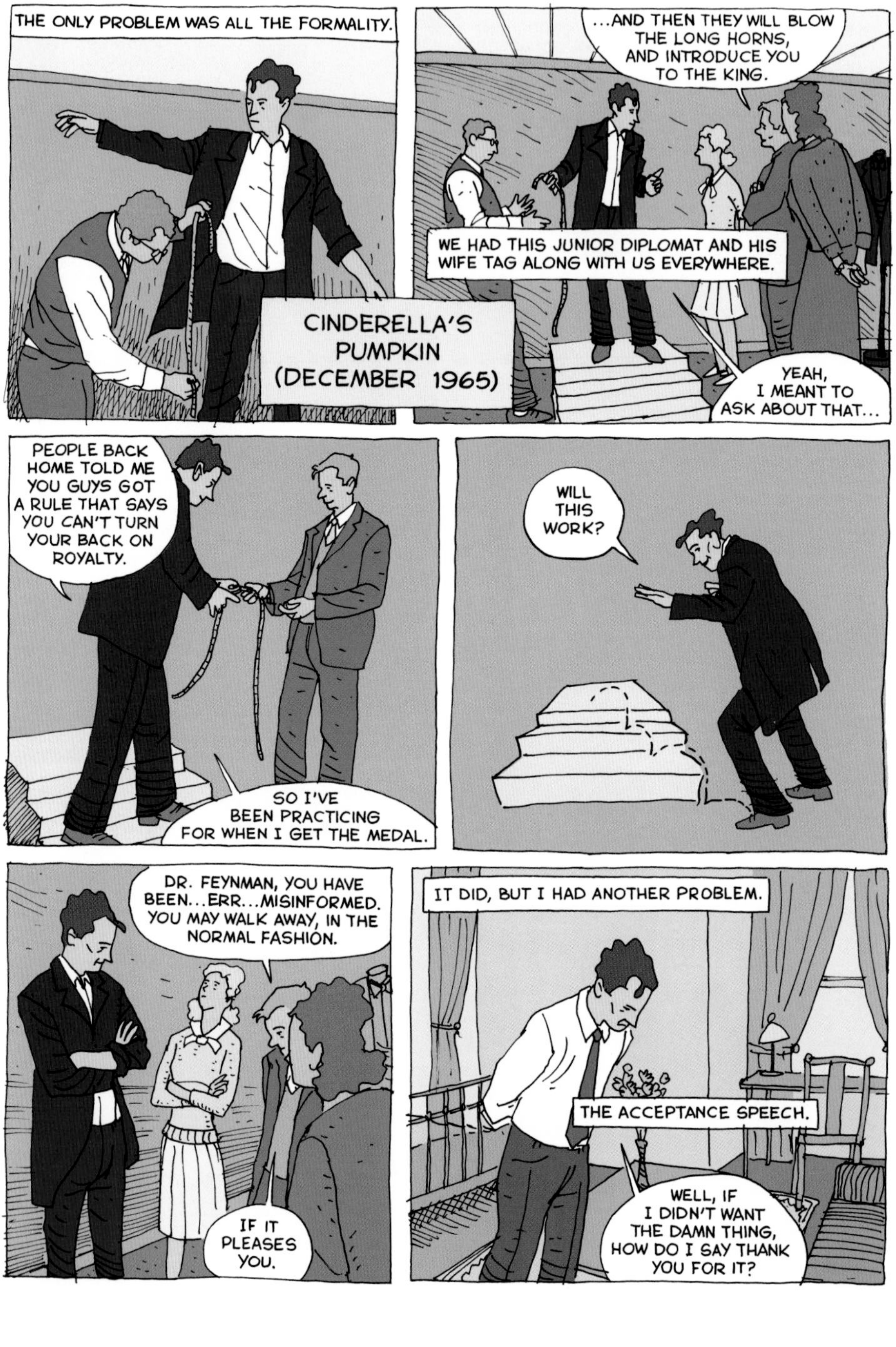
THE ONLY PROBLEM WAS ALL THE FORMALITY.
CINDERELLA'S PUMPKIN (DECEMBER 1965)
...AND THEN THEY WILL BLOW THE LONG HORNS, AND INTRODUCE YOU TO THE KING.
WE HAD THIS JUNIOR DIPLOMAT AND HIS WIFE TAG ALONG WITH US EVERYWHERE.
YEAH, I MEANT TO ASK ABOUT THAT...
PEOPLE BACK HOME TOLD ME YOU GUYS GOT A RULE THAT SAYS YOU CAN'T TURN YOUR BACK ON ROYALTY.
SO I'VE BEEN PRACTICING FOR WHEN I GET THE MEDAL.
WILL THIS WORK?
DR. FEYNMAN, YOU HAVE BEEN...ERR...MISINFORMED. YOU MAY WALK AWAY, IN THE NORMAL FASHION.
IF IT PLEASES YOU.
IT DID, BUT I HAD ANOTHER PROBLEM.
THE ACCEPTANCE SPEECH.
WELL, IF I DIDN'T WANT THE DAMN THING, HOW DO I SAY THANK YOU FOR IT?

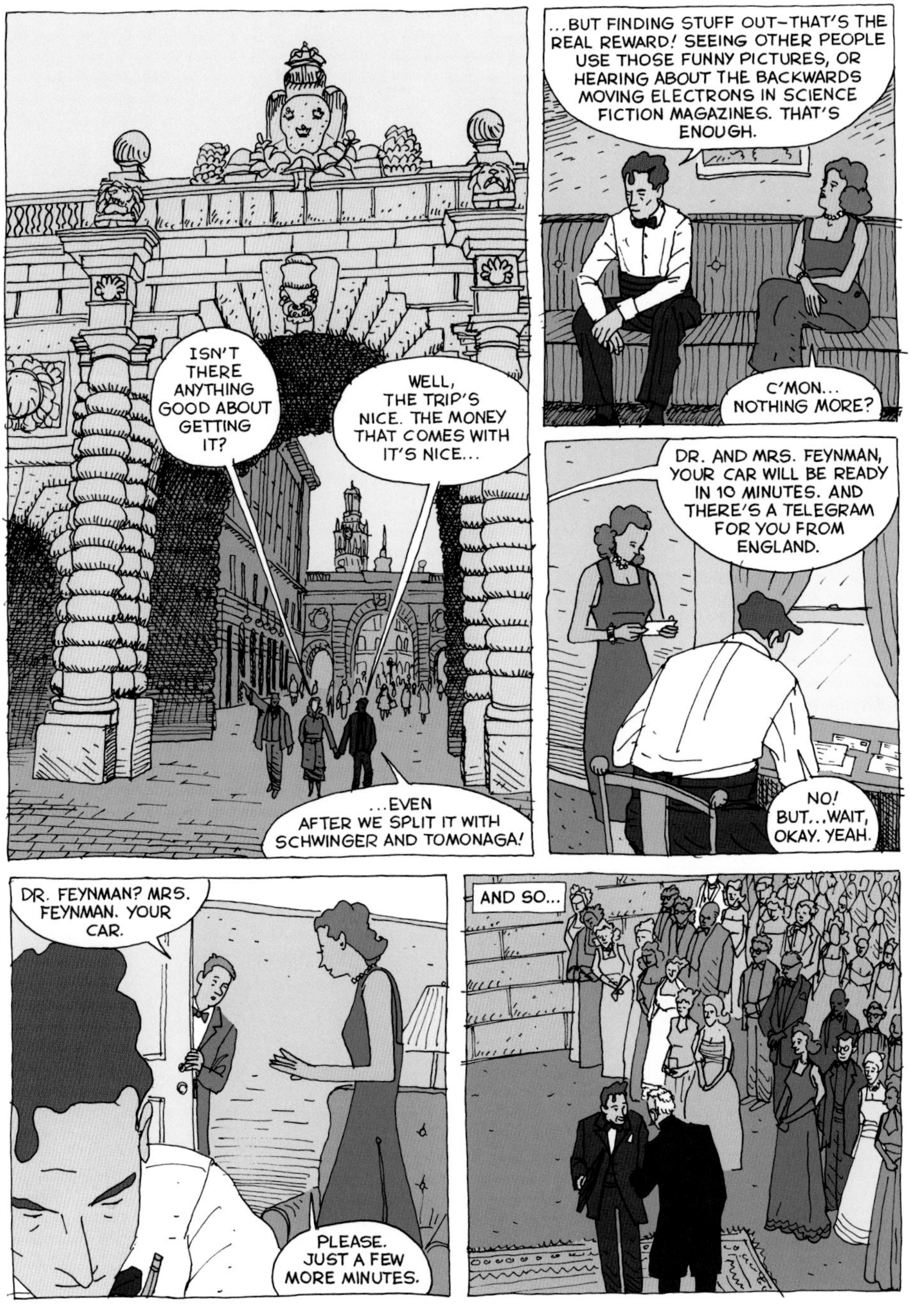
ISN'T THERE ANYTHING GOOD ABOUT GETTING IT?
WELL, THE TRIP'S NICE. THE MONEY THAT COMES WITH IT'S NICE...
...EVEN AFTER WE SPLIT IT WITH SCHWINGER AND TOMONAGA!
...BUT FINDING STUFF OUT–THAT'S THE REAL REWARD! SEEING OTHER PEOPLE USE THOSE FUNNY PICTURES, OR HEARING ABOUT THE BACKWARDS MOVING ELECTRONS IN SCIENCE FICTION MAGAZINES. THAT'S ENOUGH.
C'MON... NOTHING MORE?
DR. AND MRS. FEYNMAN, YOUR CAR WILL BE READY IN 10 MINUTES. AND THERE'S A TELEGRAM FOR YOU FROM ENGLAND.
NO! BUT...WAIT, OKAY. YEAH.
DR. FEYNMAN? MRS. FEYNMAN. YOUR CAR.
PLEASE. JUST A FEW MORE MINUTES.
AND SO...

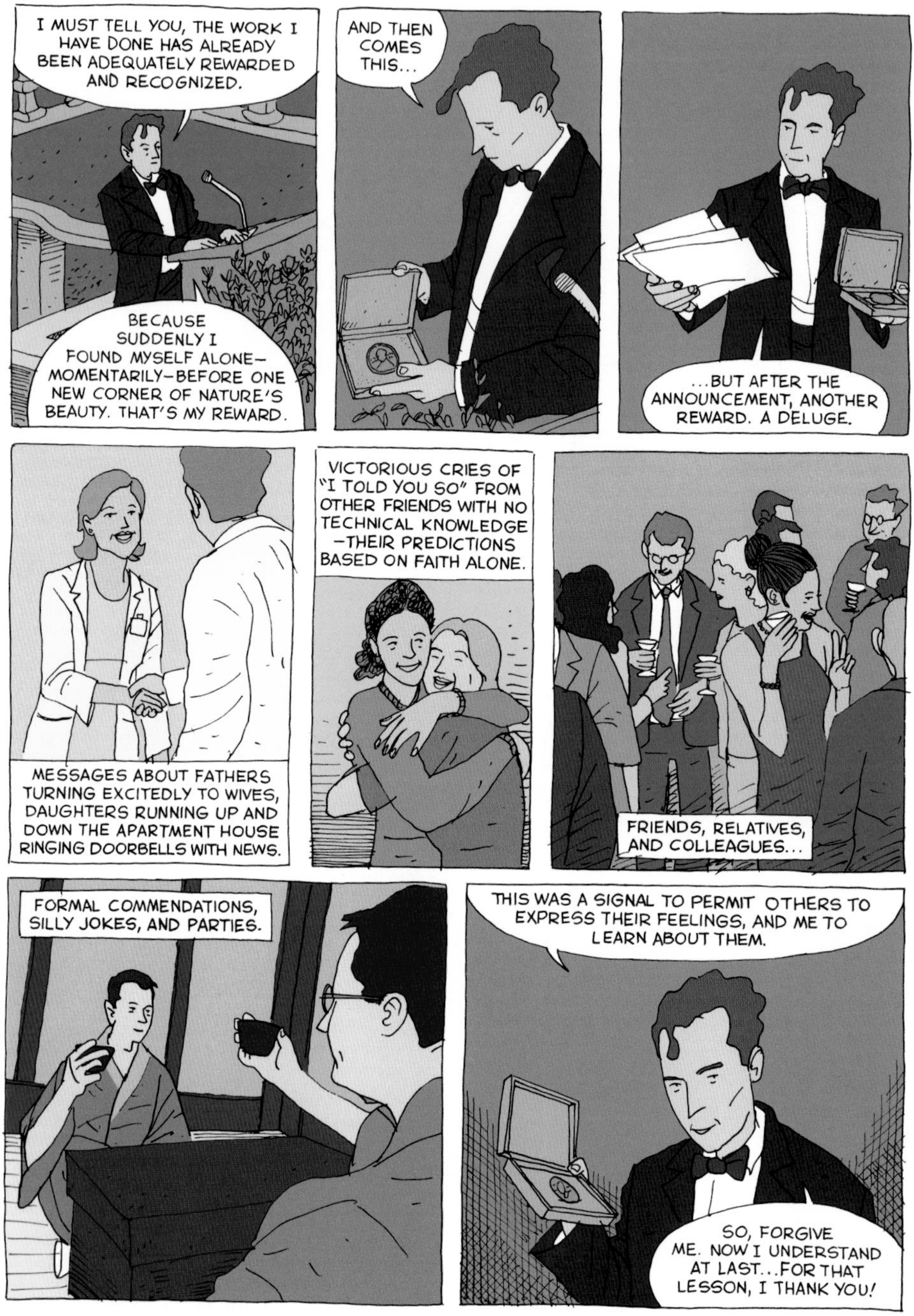
I MUST TELL YOU, THE WORK I HAVE DONE HAS ALREADY BEEN ADEQUATELY REWARDED AND RECOGNIZED.
BECAUSE SUDDENLY I FOUND MYSELF ALONE—MOMENTARILY—BEFORE ONE NEW CORNER OF NATURE'S BEAUTY. THAT'S MY REWARD.
AND THEN COMES THIS...
...BUT AFTER THE ANNOUNCEMENT, ANOTHER REWARD. A DELUGE.
MESSAGES ABOUT FATHERS TURNING EXCITEDLY TO WIVES, DAUGHTERS RUNNING UP AND DOWN THE APARTMENT HOUSE RINGING DOORBELLS WITH NEWS.
VICTORIOUS CRIES OF "I TOLD YOU SO" FROM OTHER FRIENDS WITH NO TECHNICAL KNOWLEDGE —THEIR PREDICTIONS BASED ON FAITH ALONE.
FRIENDS, RELATIVES, AND COLLEAGUES...
FORMAL COMMENDATIONS, SILLY JOKES, AND PARTIES.
THIS WAS A SIGNAL TO PERMIT OTHERS TO EXPRESS THEIR FEELINGS, AND ME TO LEARN ABOUT THEM.
SO, FORGIVE ME. NOW I UNDERSTAND AT LAST...FOR THAT LESSON, I THANK YOU!

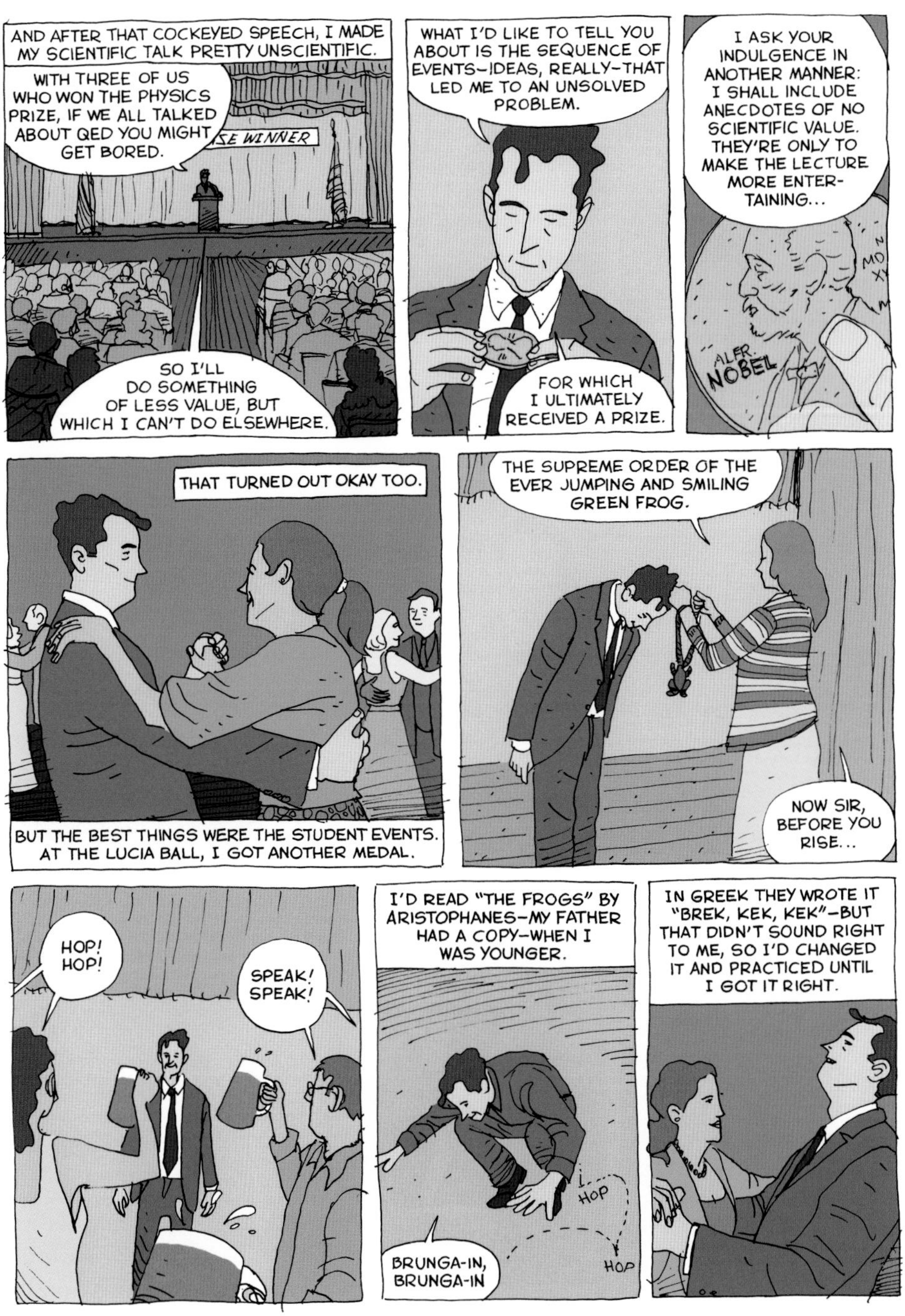
AND AFTER THAT COCKEYED SPEECH, I MADE MY SCIENTIFIC TALK PRETTY UNSCIENTIFIC.
WITH THREE OF US WHO WON THE PHYSICS PRIZE, IF WE ALL TALKED ABOUT QED YOU MIGHT GET BORED.
SE WINNER
SO I'LL DO SOMETHING OF LESS VALUE, BUT WHICH I CAN'T DO ELSEWHERE.
WHAT I'D LIKE TO TELL YOU ABOUT IS THE SEQUENCE OF EVENTS–IDEAS, REALLY–THAT LED ME TO AN UNSOLVED PROBLEM.
FOR WHICH I ULTIMATELY RECEIVED A PRIZE.
I ASK YOUR INDULGENCE IN ANOTHER MANNER: I SHALL INCLUDE ANECDOTES OF NO SCIENTIFIC VALUE. THEY'RE ONLY TO MAKE THE LECTURE MORE ENTERTAINING...
ALFR. NOBEL
THAT TURNED OUT OKAY TOO.
BUT THE BEST THINGS WERE THE STUDENT EVENTS. AT THE LUCIA BALL, I GOT ANOTHER MEDAL.
THE SUPREME ORDER OF THE EVER JUMPING AND SMILING GREEN FROG.
NOW SIR, BEFORE YOU RISE...
HOP! HOP!
SPEAK! SPEAK!
I'D READ "THE FROGS" BY ARISTOPHANES–MY FATHER HAD A COPY–WHEN I WAS YOUNGER.
HOP
HOP
BRUNGA-IN, BRUNGA-IN
IN GREEK THEY WROTE IT "BREK, KEK, KEK"–BUT THAT DIDN'T SOUND RIGHT TO ME, SO I'D CHANGED IT AND PRACTICED UNTIL I GOT IT RIGHT.

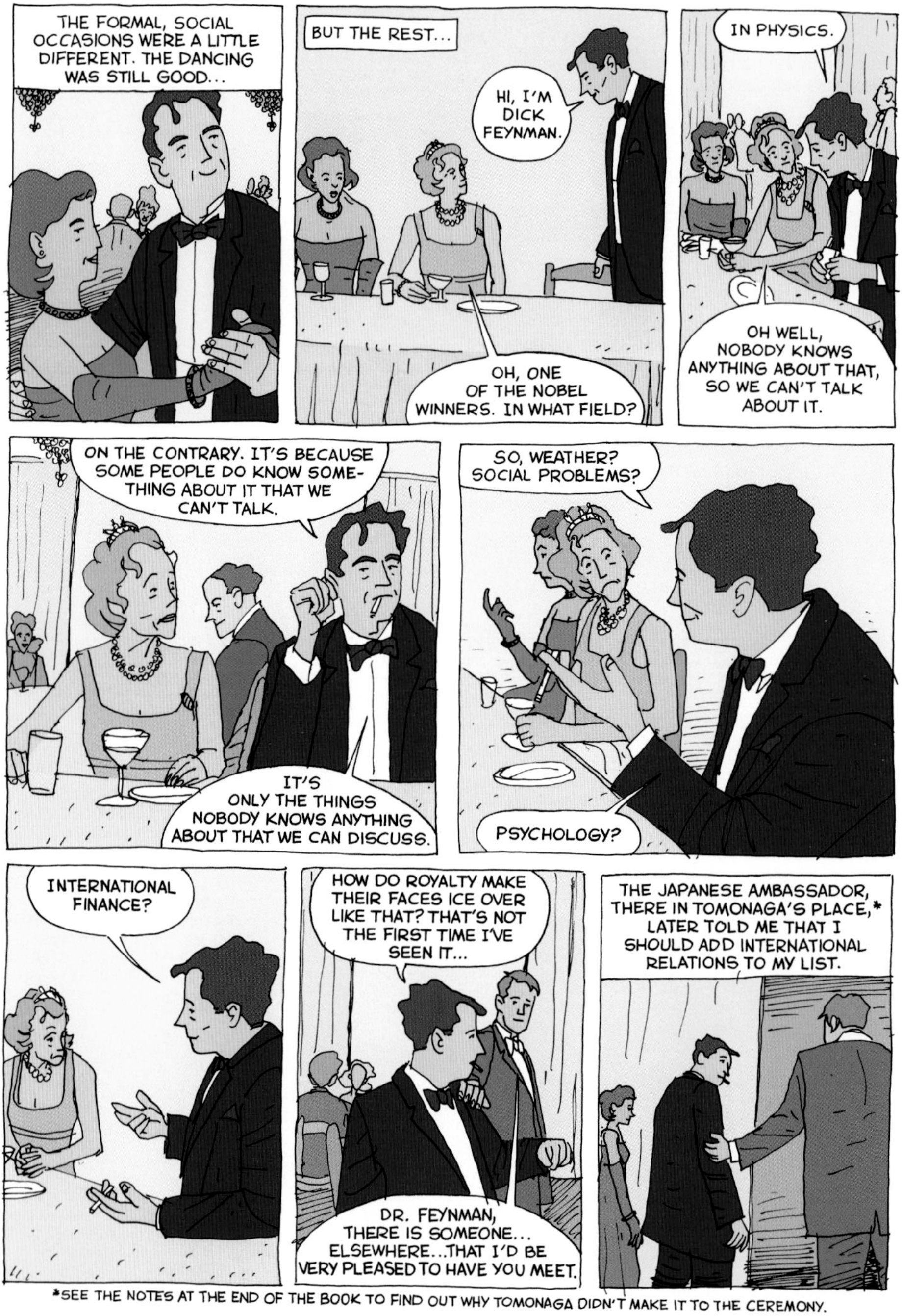

THE FORMAL, SOCIAL OCCASIONS WERE A LITTLE DIFFERENT. THE DANCING WAS STILL GOOD...
BUT THE REST...
HI, I'M DICK FEYNMAN.
OH, ONE OF THE NOBEL WINNERS. IN WHAT FIELD?
IN PHYSICS.
OH WELL, NOBODY KNOWS ANYTHING ABOUT THAT, SO WE CAN'T TALK ABOUT IT.
ON THE CONTRARY. IT'S BECAUSE SOME PEOPLE DO KNOW SOME-THING ABOUT IT THAT WE CAN'T TALK.
IT'S ONLY THE THINGS NOBODY KNOWS ANYTHING ABOUT THAT WE CAN DISCUSS.
SO, WEATHER? SOCIAL PROBLEMS?
PSYCHOLOGY?
INTERNATIONAL FINANCE?
HOW DO ROYALTY MAKE THEIR FACES ICE OVER LIKE THAT? THAT'S NOT THE FIRST TIME I'VE SEEN IT...
DR. FEYNMAN, THERE IS SOMEONE... ELSEWHERE...THAT I'D BE VERY PLEASED TO HAVE YOU MEET.
THE JAPANESE AMBASSADOR, THERE IN TOMONAGA'S PLACE,* LATER TOLD ME THAT I SHOULD ADD INTERNATIONAL RELATIONS TO MY LIST.

*SEE THE NOTES AT THE END OF THE BOOK TO FIND OUT WHY TOMONAGA DIDN'T MAKE IT TO THE CEREMONY.

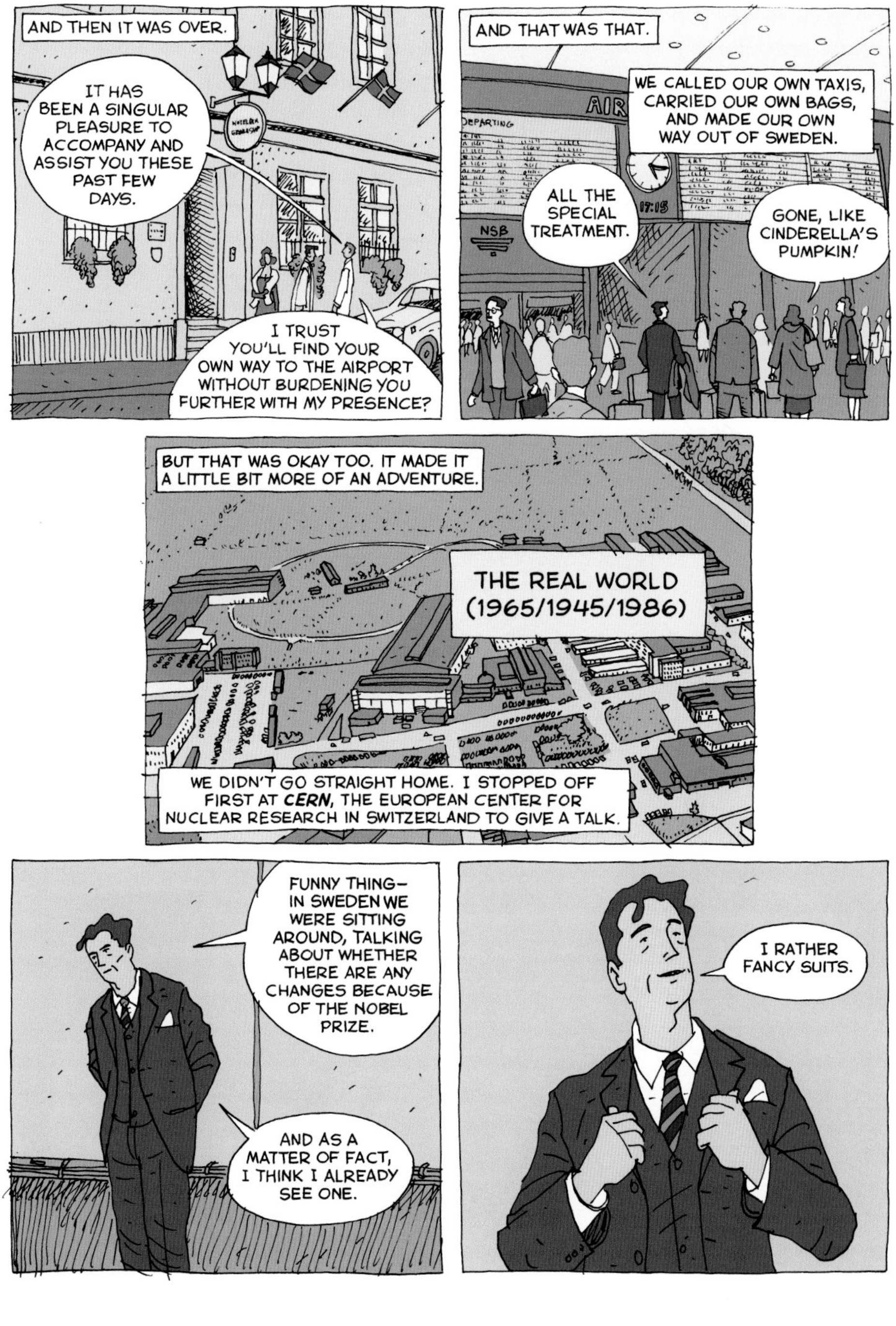

AND THEN IT WAS OVER.
IT HAS BEEN A SINGULAR PLEASURE TO ACCOMPANY AND ASSIST YOU THESE PAST FEW DAYS.
I TRUST YOU'LL FIND YOUR OWN WAY TO THE AIRPORT WITHOUT BURDENING YOU FURTHER WITH MY PRESENCE?
AND THAT WAS THAT.
WE CALLED OUR OWN TAXIS, CARRIED OUR OWN BAGS, AND MADE OUR OWN WAY OUT OF SWEDEN.
AIR
DEPARTING
17:15
NSB
ALL THE SPECIAL TREATMENT.
GONE, LIKE CINDERELLA'S PUMPKIN!
BUT THAT WAS OKAY TOO. IT MADE IT A LITTLE BIT MORE OF AN ADVENTURE.
THE REAL WORLD (1965/1945/1986)
WE DIDN'T GO STRAIGHT HOME. I STOPPED OFF FIRST AT **CERN**, THE EUROPEAN CENTER FOR NUCLEAR RESEARCH IN SWITZERLAND TO GIVE A TALK.
FUNNY THING— IN SWEDEN WE WERE SITTING AROUND, TALKING ABOUT WHETHER THERE ARE ANY CHANGES BECAUSE OF THE NOBEL PRIZE.
AND AS A MATTER OF FACT, I THINK I ALREADY SEE ONE.
I RATHER FANCY SUITS.

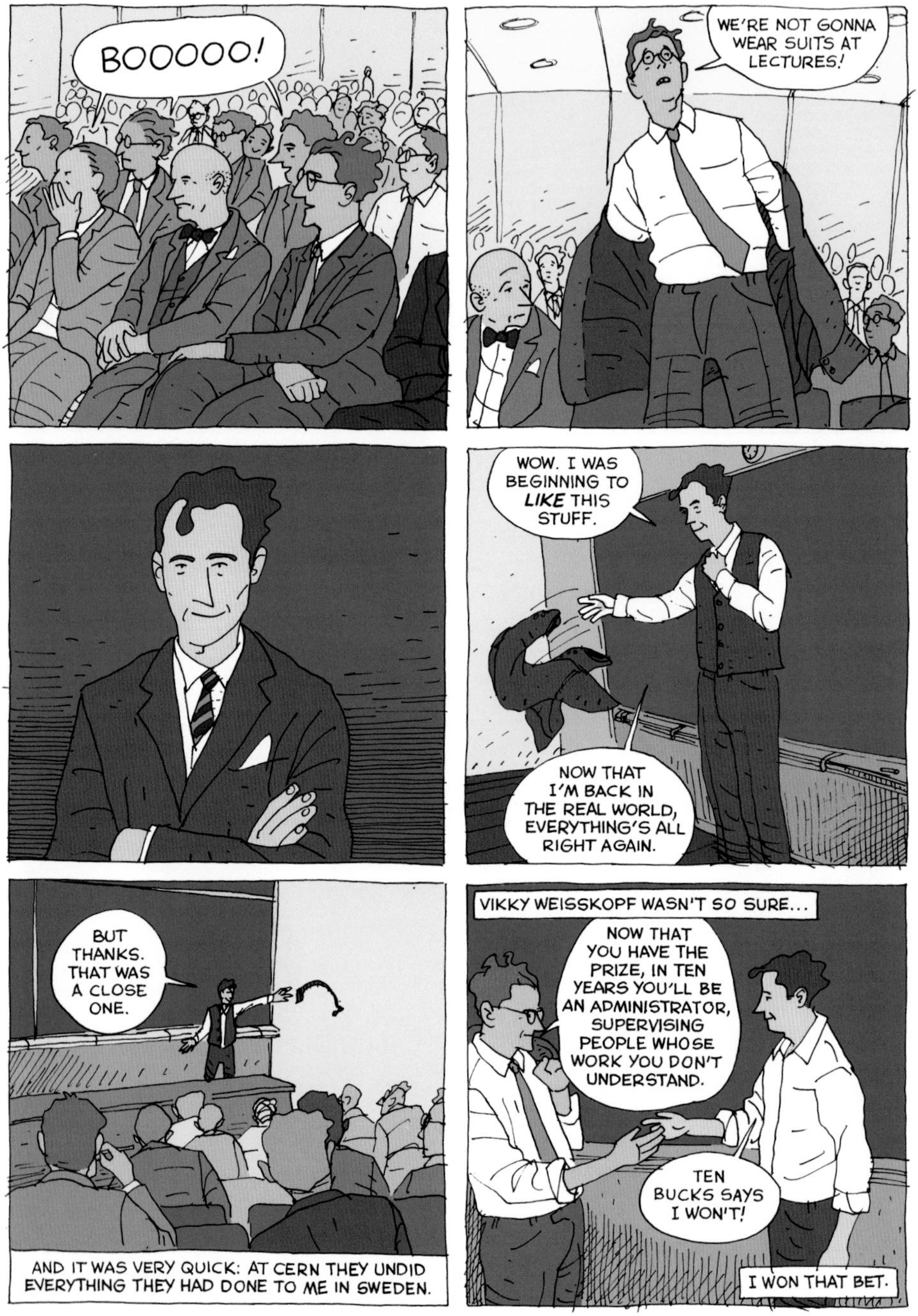
BOOOOO!
WE'RE NOT GONNA WEAR SUITS AT LECTURES!
WOW. I WAS BEGINNING TO *LIKE* THIS STUFF.
NOW THAT I'M BACK IN THE REAL WORLD, EVERYTHING'S ALL RIGHT AGAIN.
BUT THANKS. THAT WAS A CLOSE ONE.
AND IT WAS VERY QUICK: AT CERN THEY UNDID EVERYTHING THEY HAD DONE TO ME IN SWEDEN.
VIKKY WEISSKOPF WASN'T SO SURE...
NOW THAT YOU HAVE THE PRIZE, IN TEN YEARS YOU'LL BE AN ADMINISTRATOR, SUPERVISING PEOPLE WHOSE WORK YOU DON'T UNDERSTAND.
TEN BUCKS SAYS I WON'T!
I WON THAT BET.

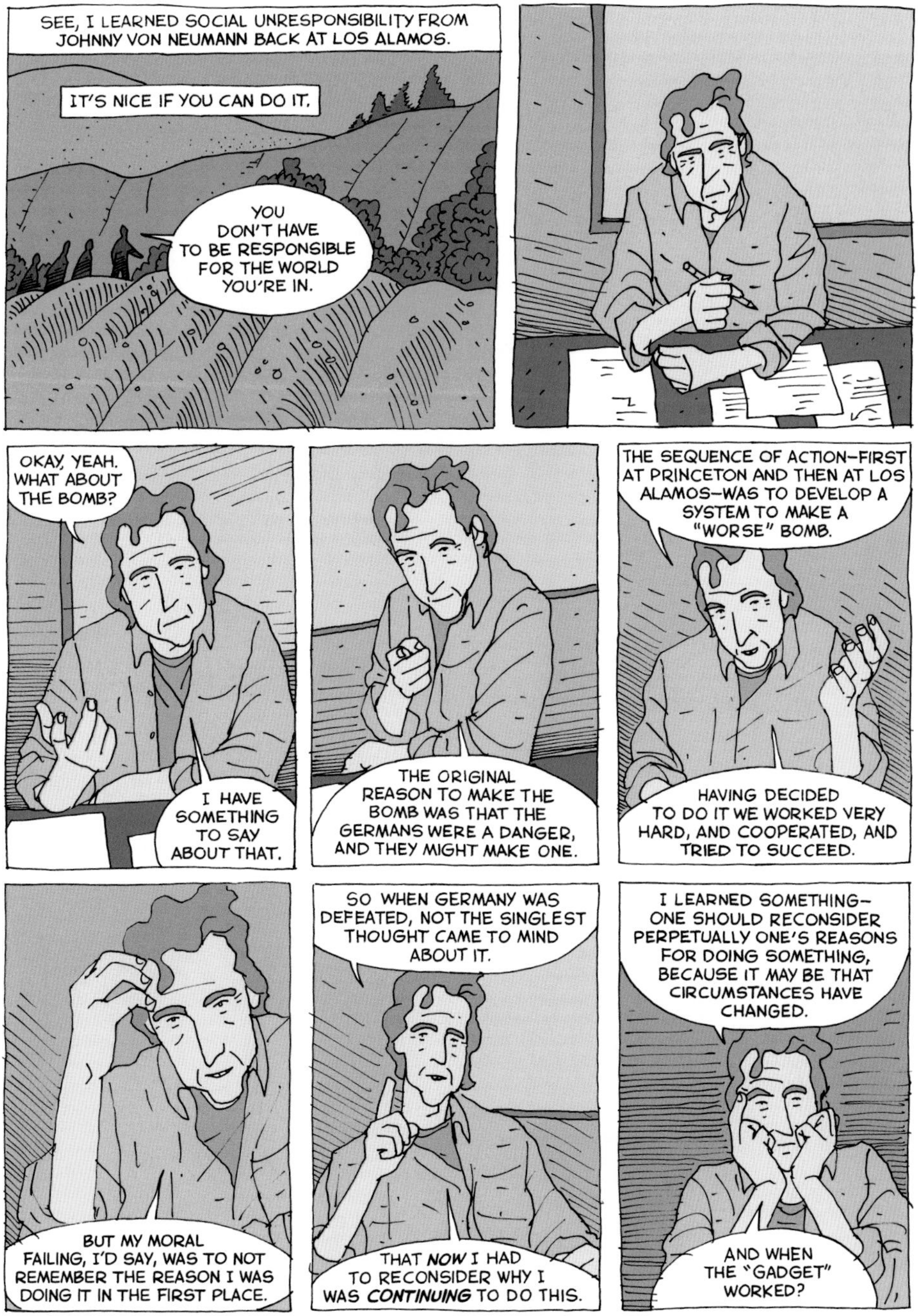
SEE, I LEARNED SOCIAL UNRESPONSIBILITY FROM JOHNNY VON NEUMANN BACK AT LOS ALAMOS.
IT'S NICE IF YOU CAN DO IT.
YOU DON'T HAVE TO BE RESPONSIBLE FOR THE WORLD YOU'RE IN.
OKAY, YEAH. WHAT ABOUT THE BOMB?
I HAVE SOMETHING TO SAY ABOUT THAT.
THE ORIGINAL REASON TO MAKE THE BOMB WAS THAT THE GERMANS WERE A DANGER, AND THEY MIGHT MAKE ONE.
THE SEQUENCE OF ACTION—FIRST AT PRINCETON AND THEN AT LOS ALAMOS—WAS TO DEVELOP A SYSTEM TO MAKE A "WORSE" BOMB.
HAVING DECIDED TO DO IT WE WORKED VERY HARD, AND COOPERATED, AND TRIED TO SUCCEED.
BUT MY MORAL FAILING, I'D SAY, WAS TO NOT REMEMBER THE REASON I WAS DOING IT IN THE FIRST PLACE.
SO WHEN GERMANY WAS DEFEATED, NOT THE SINGLEST THOUGHT CAME TO MIND ABOUT IT.
THAT *NOW* I HAD TO RECONSIDER WHY I WAS ***CONTINUING*** TO DO THIS.
I LEARNED SOMETHING— ONE SHOULD RECONSIDER PERPETUALLY ONE'S REASONS FOR DOING SOMETHING, BECAUSE IT MAY BE THAT CIRCUMSTANCES HAVE CHANGED.
AND WHEN THE "GADGET" WORKED?

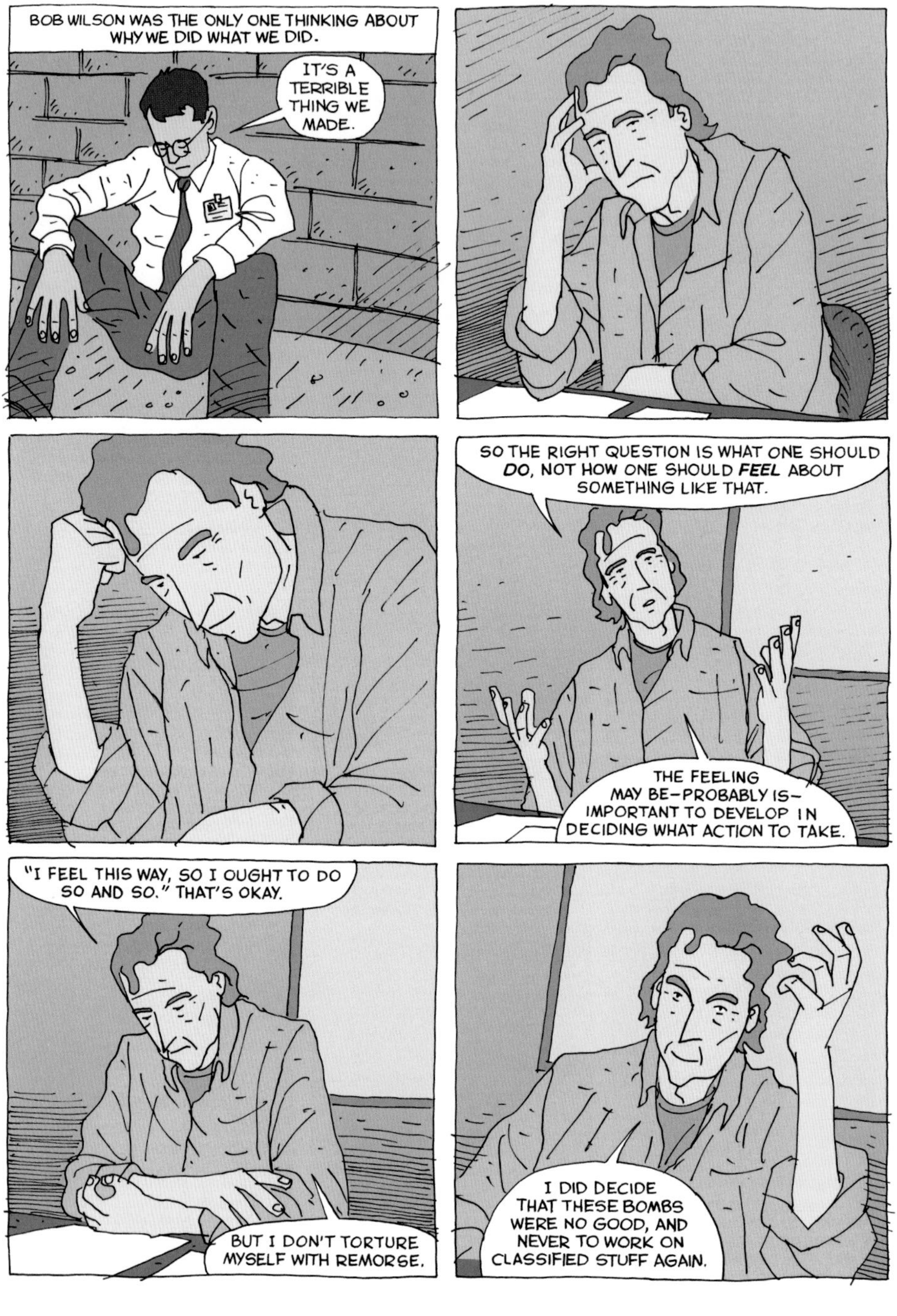
BOB WILSON WAS THE ONLY ONE THINKING ABOUT WHY WE DID WHAT WE DID.
IT'S A TERRIBLE THING WE MADE.
SO THE RIGHT QUESTION IS WHAT ONE SHOULD *DO*, NOT HOW ONE SHOULD *FEEL* ABOUT SOMETHING LIKE THAT.
THE FEELING MAY BE–PROBABLY IS–IMPORTANT TO DEVELOP IN DECIDING WHAT ACTION TO TAKE.
"I FEEL THIS WAY, SO I OUGHT TO DO SO AND SO." THAT'S OKAY.
BUT I DON'T TORTURE MYSELF WITH REMORSE.
I DID DECIDE THAT THESE BOMBS WERE NO GOOD, AND NEVER TO WORK ON CLASSIFIED STUFF AGAIN.

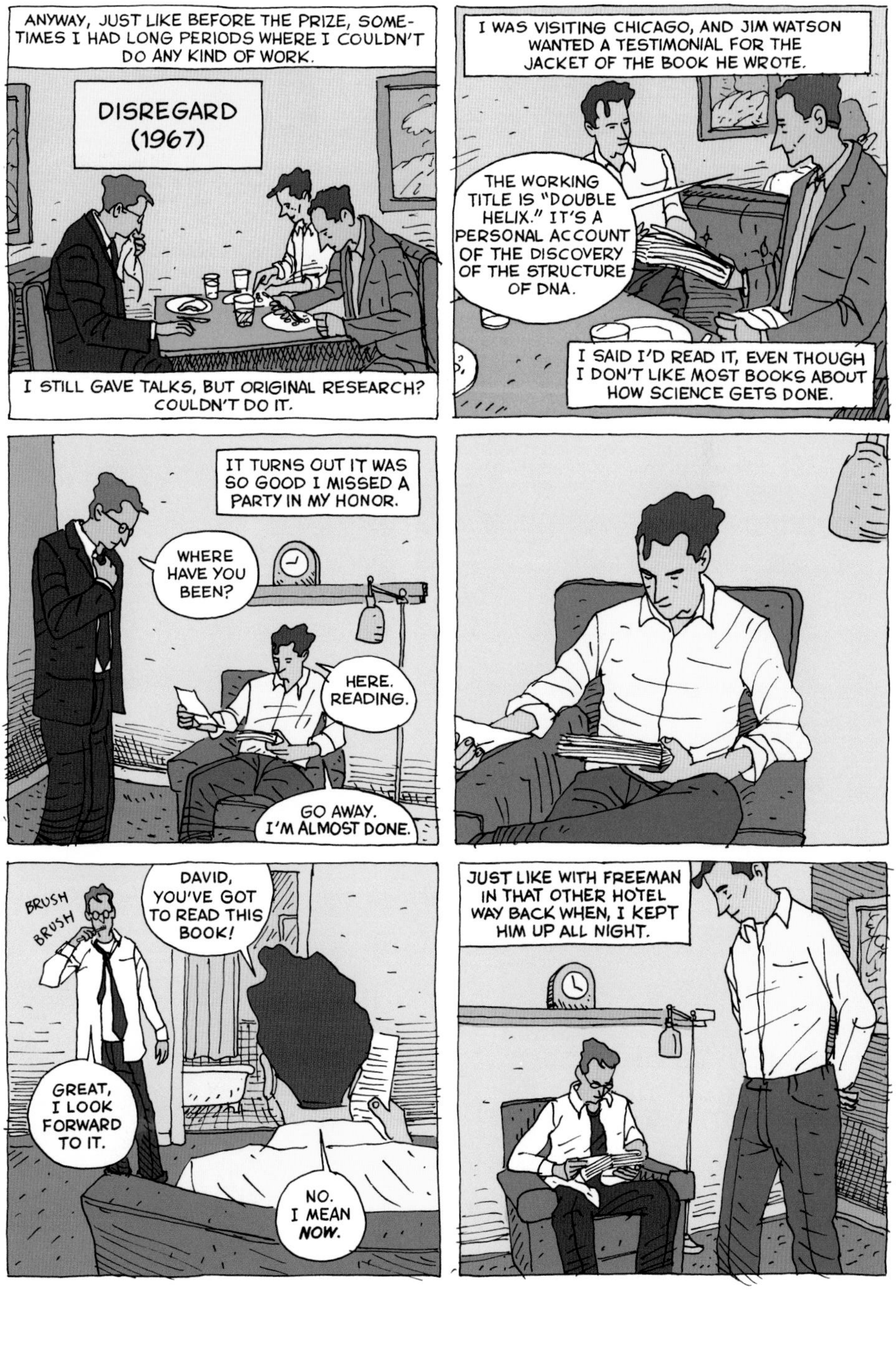
ANYWAY, JUST LIKE BEFORE THE PRIZE, SOMETIMES I HAD LONG PERIODS WHERE I COULDN'T DO ANY KIND OF WORK.
DISREGARD (1967)
I STILL GAVE TALKS, BUT ORIGINAL RESEARCH? COULDN'T DO IT.
I WAS VISITING CHICAGO, AND JIM WATSON WANTED A TESTIMONIAL FOR THE JACKET OF THE BOOK HE WROTE.
THE WORKING TITLE IS "DOUBLE HELIX." IT'S A PERSONAL ACCOUNT OF THE DISCOVERY OF THE STRUCTURE OF DNA.
I SAID I'D READ IT, EVEN THOUGH I DON'T LIKE MOST BOOKS ABOUT HOW SCIENCE GETS DONE.
IT TURNS OUT IT WAS SO GOOD I MISSED A PARTY IN MY HONOR.
WHERE HAVE YOU BEEN?
HERE. READING.
GO AWAY. I'M ALMOST DONE.
BRUSH BRUSH
DAVID, YOU'VE GOT TO READ THIS BOOK!
GREAT, I LOOK FORWARD TO IT.
NO. I MEAN NOW.
JUST LIKE WITH FREEMAN IN THAT OTHER HOTEL WAY BACK WHEN, I KEPT HIM UP ALL NIGHT.

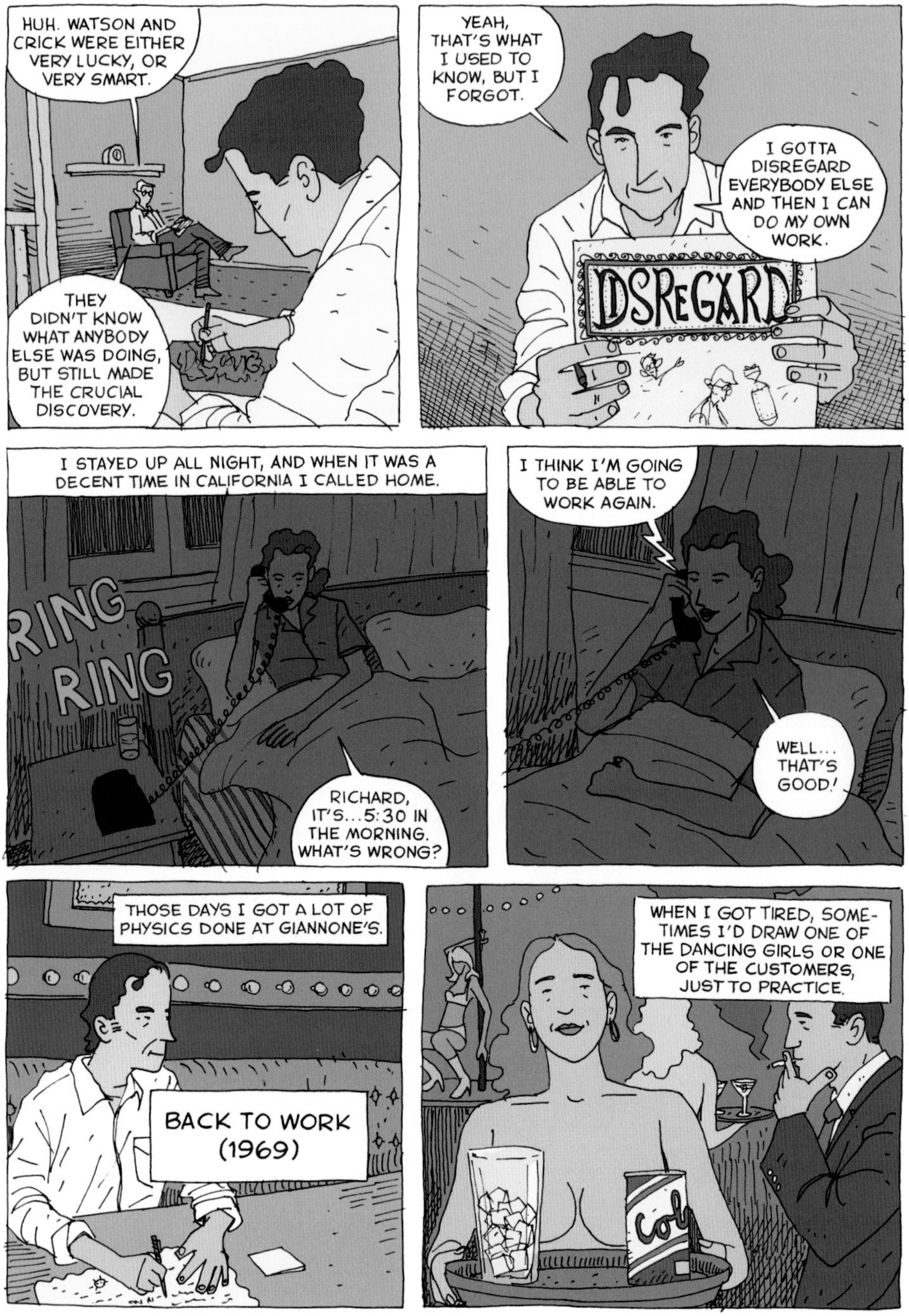
HUH. WATSON AND CRICK WERE EITHER VERY LUCKY, OR VERY SMART.
THEY DIDN'T KNOW WHAT ANYBODY ELSE WAS DOING, BUT STILL MADE THE CRUCIAL DISCOVERY.
YEAH, THAT'S WHAT I USED TO KNOW, BUT I FORGOT.
I GOTTA DISREGARD EVERYBODY ELSE AND THEN I CAN DO MY OWN WORK.
DSREGARD
I STAYED UP ALL NIGHT, AND WHEN IT WAS A DECENT TIME IN CALIFORNIA I CALLED HOME.
RING
RING
RICHARD, IT'S...5:30 IN THE MORNING. WHAT'S WRONG?
I THINK I'M GOING TO BE ABLE TO WORK AGAIN.
WELL... THAT'S GOOD!
THOSE DAYS I GOT A LOT OF PHYSICS DONE AT GIANNONE'S.
BACK TO WORK (1969)
WHEN I GOT TIRED, SOMETIMES I'D DRAW ONE OF THE DANCING GIRLS OR ONE OF THE CUSTOMERS, JUST TO PRACTICE.

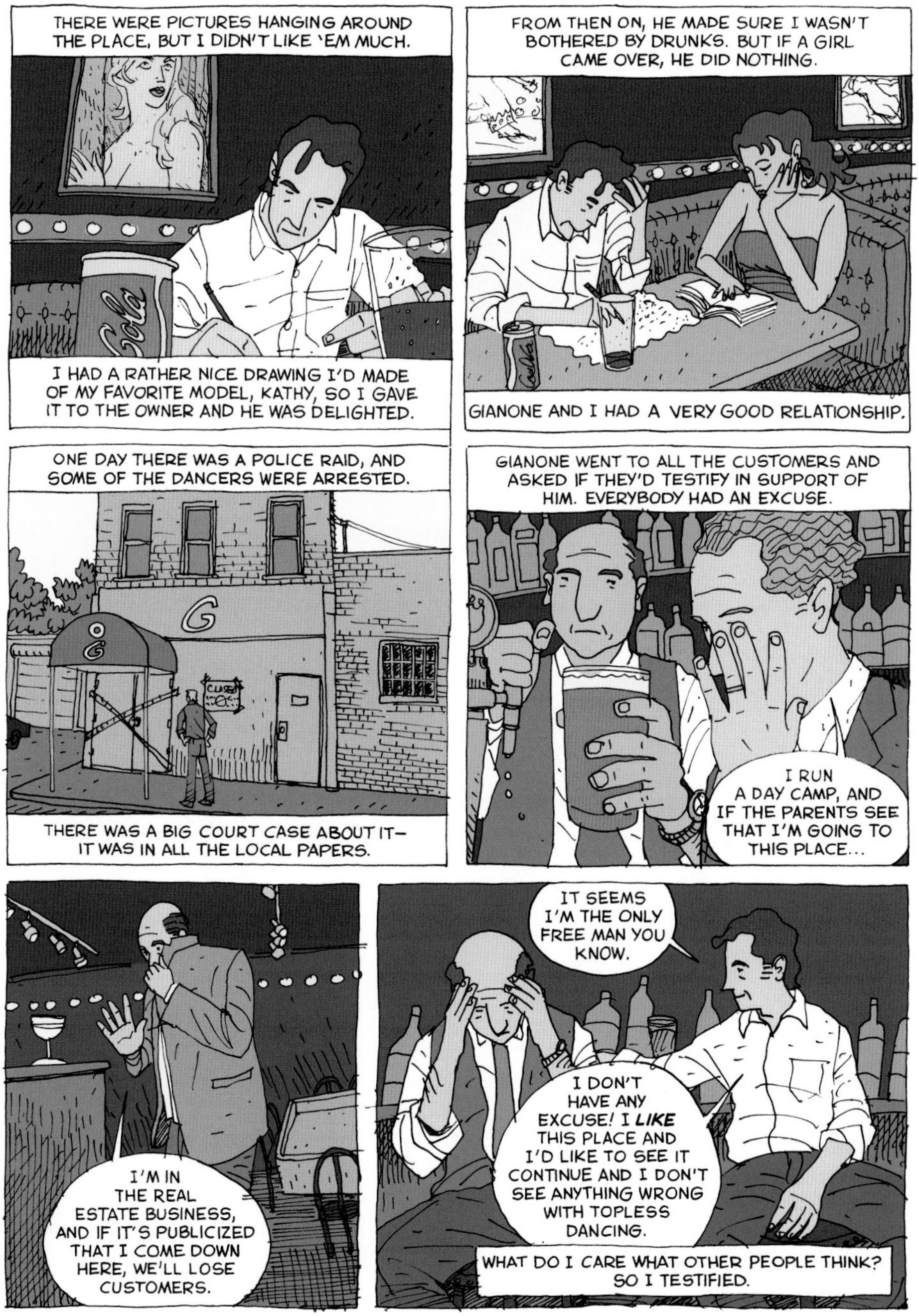
THERE WERE PICTURES HANGING AROUND THE PLACE, BUT I DIDN'T LIKE 'EM MUCH.
Cola
I HAD A RATHER NICE DRAWING I'D MADE OF MY FAVORITE MODEL, KATHY, SO I GAVE IT TO THE OWNER AND HE WAS DELIGHTED.
FROM THEN ON, HE MADE SURE I WASN'T BOTHERED BY DRUNKS. BUT IF A GIRL CAME OVER, HE DID NOTHING.
GIANONE AND I HAD A VERY GOOD RELATIONSHIP.
ONE DAY THERE WAS A POLICE RAID, AND SOME OF THE DANCERS WERE ARRESTED.
G
CLOSED
THERE WAS A BIG COURT CASE ABOUT IT— IT WAS IN ALL THE LOCAL PAPERS.
GIANONE WENT TO ALL THE CUSTOMERS AND ASKED IF THEY'D TESTIFY IN SUPPORT OF HIM. EVERYBODY HAD AN EXCUSE.
I RUN A DAY CAMP, AND IF THE PARENTS SEE THAT I'M GOING TO THIS PLACE...
I'M IN THE REAL ESTATE BUSINESS, AND IF IT'S PUBLICIZED THAT I COME DOWN HERE, WE'LL LOSE CUSTOMERS.
IT SEEMS I'M THE ONLY FREE MAN YOU KNOW.
I DON'T HAVE ANY EXCUSE! I ***LIKE*** THIS PLACE AND I'D LIKE TO SEE IT CONTINUE AND I DON'T SEE ANYTHING WRONG WITH TOPLESS DANCING.
WHAT DO I CARE WHAT OTHER PEOPLE THINK? SO I TESTIFIED.

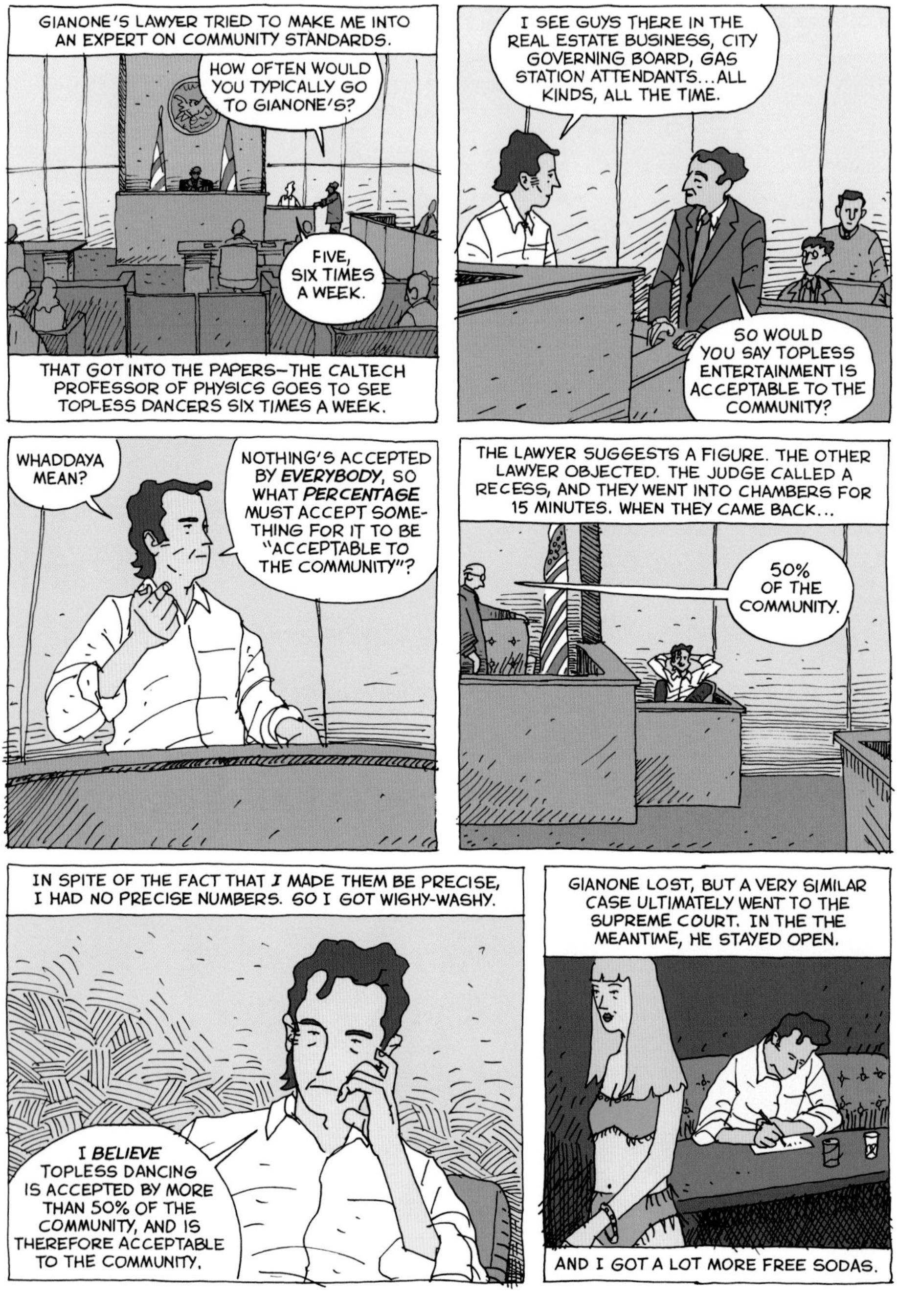
GIANONE'S LAWYER TRIED TO MAKE ME INTO AN EXPERT ON COMMUNITY STANDARDS.
HOW OFTEN WOULD YOU TYPICALLY GO TO GIANONE'S?
FIVE, SIX TIMES A WEEK.
THAT GOT INTO THE PAPERS—THE CALTECH PROFESSOR OF PHYSICS GOES TO SEE TOPLESS DANCERS SIX TIMES A WEEK.
I SEE GUYS THERE IN THE REAL ESTATE BUSINESS, CITY GOVERNING BOARD, GAS STATION ATTENDANTS...ALL KINDS, ALL THE TIME.
SO WOULD YOU SAY TOPLESS ENTERTAINMENT IS ACCEPTABLE TO THE COMMUNITY?
WHADDAYA MEAN?
NOTHING'S ACCEPTED BY *EVERYBODY*, SO WHAT *PERCENTAGE* MUST ACCEPT SOMETHING FOR IT TO BE "ACCEPTABLE TO THE COMMUNITY"?
THE LAWYER SUGGESTS A FIGURE. THE OTHER LAWYER OBJECTED. THE JUDGE CALLED A RECESS, AND THEY WENT INTO CHAMBERS FOR 15 MINUTES. WHEN THEY CAME BACK...
50% OF THE COMMUNITY.
IN SPITE OF THE FACT THAT *I* MADE THEM BE PRECISE, I HAD NO PRECISE NUMBERS. SO I GOT WISHY-WASHY.
I *BELIEVE* TOPLESS DANCING IS ACCEPTED BY MORE THAN 50% OF THE COMMUNITY, AND IS THEREFORE ACCEPTABLE TO THE COMMUNITY.
GIANONE LOST, BUT A VERY SIMILAR CASE ULTIMATELY WENT TO THE SUPREME COURT. IN THE THE MEANTIME, HE STAYED OPEN.
AND I GOT A LOT MORE FREE SODAS.

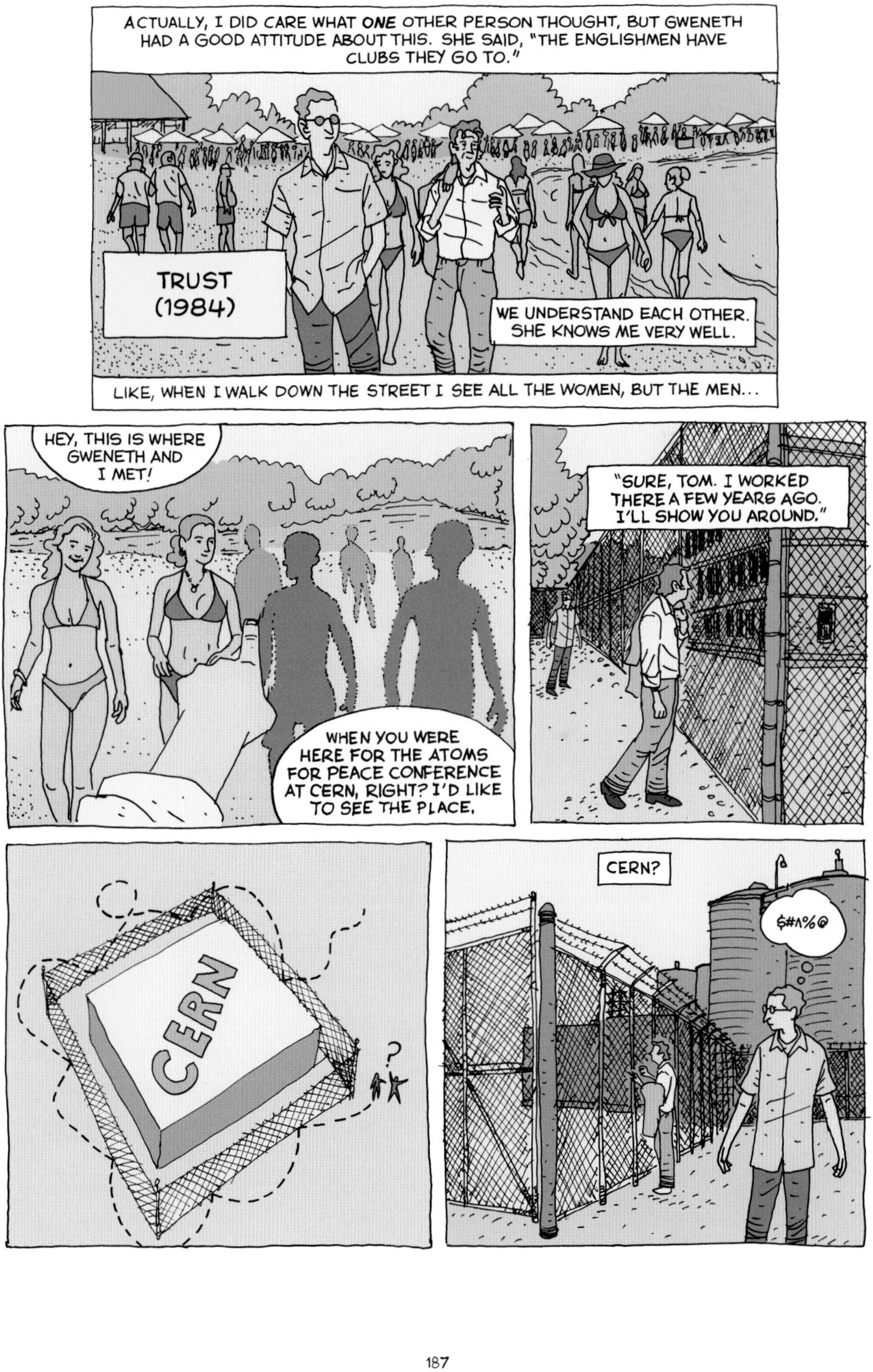
ACTUALLY, I DID CARE WHAT **ONE** OTHER PERSON THOUGHT, BUT GWENETH HAD A GOOD ATTITUDE ABOUT THIS. SHE SAID, "THE ENGLISHMEN HAVE CLUBS THEY GO TO."
TRUST (1984)
WE UNDERSTAND EACH OTHER. SHE KNOWS ME VERY WELL.
LIKE, WHEN I WALK DOWN THE STREET I SEE ALL THE WOMEN, BUT THE MEN...
HEY, THIS IS WHERE GWENETH AND I MET!
WHEN YOU WERE HERE FOR THE ATOMS FOR PEACE CONFERENCE AT CERN, RIGHT? I'D LIKE TO SEE THE PLACE.
"SURE, TOM. I WORKED THERE A FEW YEARS AGO. I'LL SHOW YOU AROUND."
CERN
CERN?
$#^%@

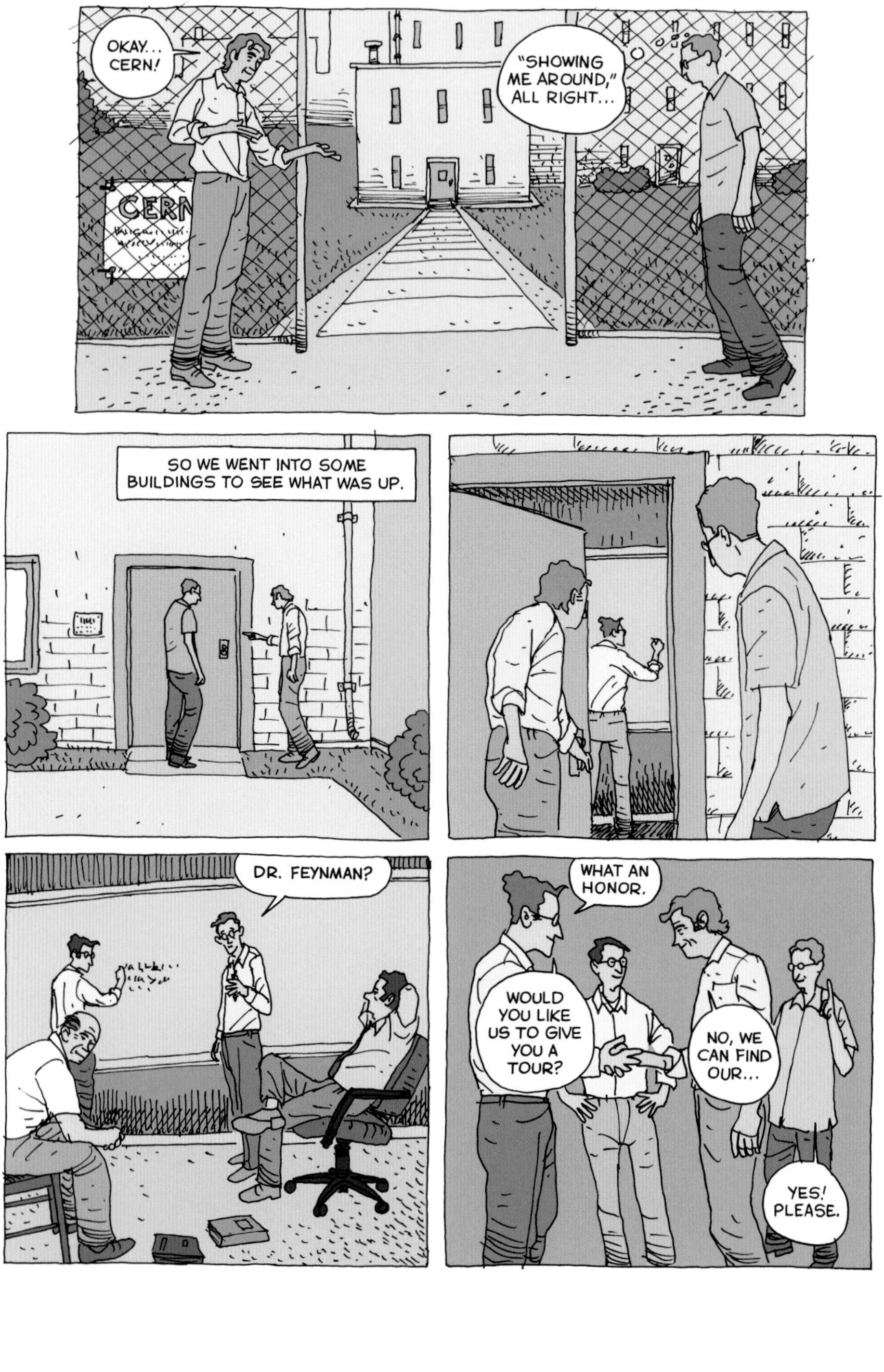
OKAY... CERN!
"SHOWING ME AROUND," ALL RIGHT...
CERN
SO WE WENT INTO SOME BUILDINGS TO SEE WHAT WAS UP.
DR. FEYNMAN?
WHAT AN HONOR.
WOULD YOU LIKE US TO GIVE YOU A TOUR?
NO, WE CAN FIND OUR...
YES! PLEASE.

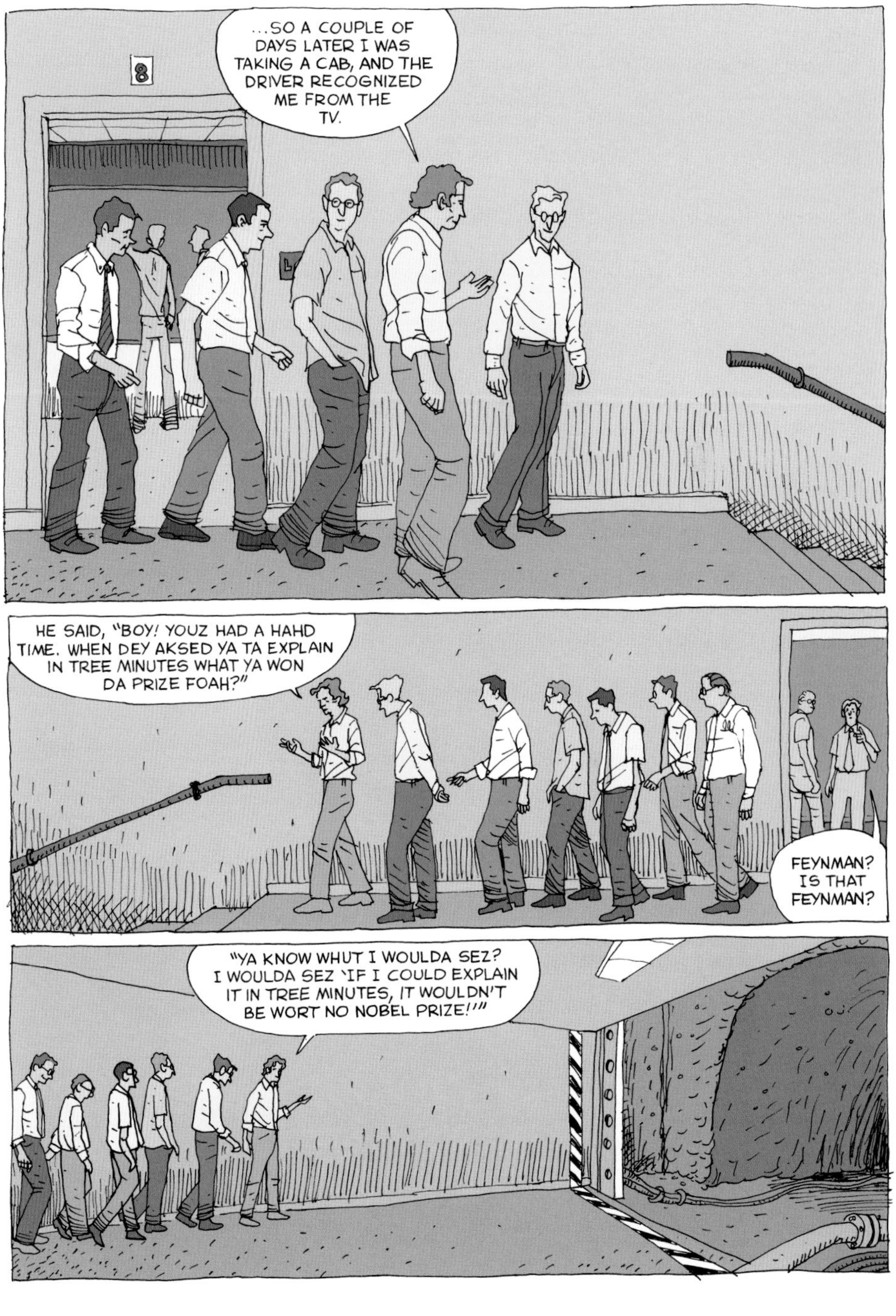
...SO A COUPLE OF DAYS LATER I WAS TAKING A CAB, AND THE DRIVER RECOGNIZED ME FROM THE TV.
HE SAID, "BOY! YOUZ HAD A HAHD TIME. WHEN DEY AKSED YA TA EXPLAIN IN TREE MINUTES WHAT YA WON DA PRIZE FOAH?"
FEYNMAN? IS THAT FEYNMAN?
"YA KNOW WHUT I WOULDA SEZ? I WOULDA SEZ 'IF I COULD EXPLAIN IT IN TREE MINUTES, IT WOULDN'T BE WORT NO NOBEL PRIZE!'"

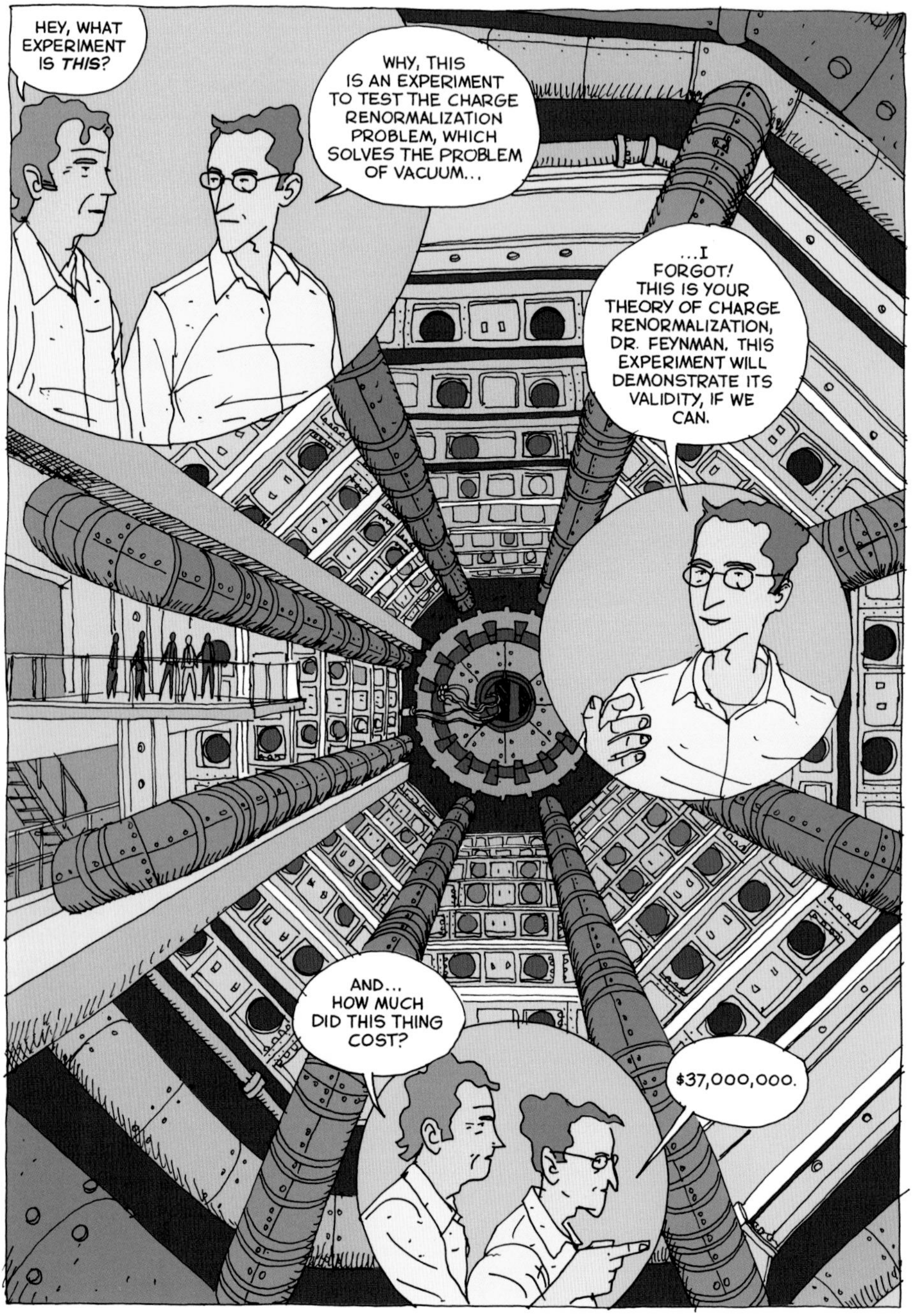
HEY, WHAT EXPERIMENT IS *THIS*?
WHY, THIS IS AN EXPERIMENT TO TEST THE CHARGE RENORMALIZATION PROBLEM, WHICH SOLVES THE PROBLEM OF VACUUM...
...I FORGOT! THIS IS YOUR THEORY OF CHARGE RENORMALIZATION, DR. FEYNMAN. THIS EXPERIMENT WILL DEMONSTRATE ITS VALIDITY, IF WE CAN.
AND... HOW MUCH DID THIS THING COST?
$37,000,000.

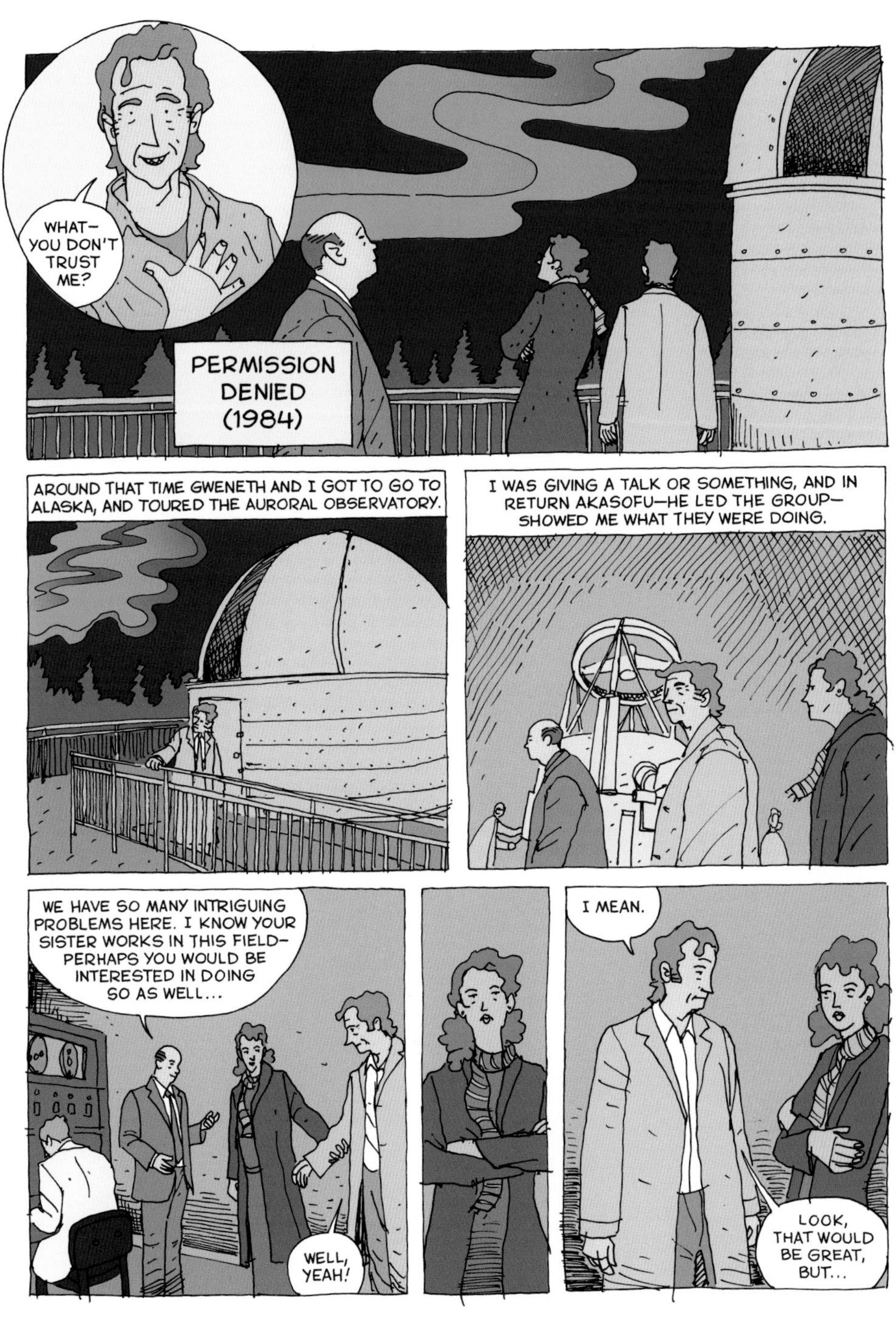
WHAT—YOU DON'T TRUST ME?
PERMISSION DENIED (1984)
AROUND THAT TIME GWENETH AND I GOT TO GO TO ALASKA, AND TOURED THE AURORAL OBSERVATORY.
I WAS GIVING A TALK OR SOMETHING, AND IN RETURN AKASOFU—HE LED THE GROUP—SHOWED ME WHAT THEY WERE DOING.
WE HAVE SO MANY INTRIGUING PROBLEMS HERE. I KNOW YOUR SISTER WORKS IN THIS FIELD—PERHAPS YOU WOULD BE INTERESTED IN DOING SO AS WELL...
WELL, YEAH!
I MEAN.
LOOK, THAT WOULD BE GREAT, BUT...

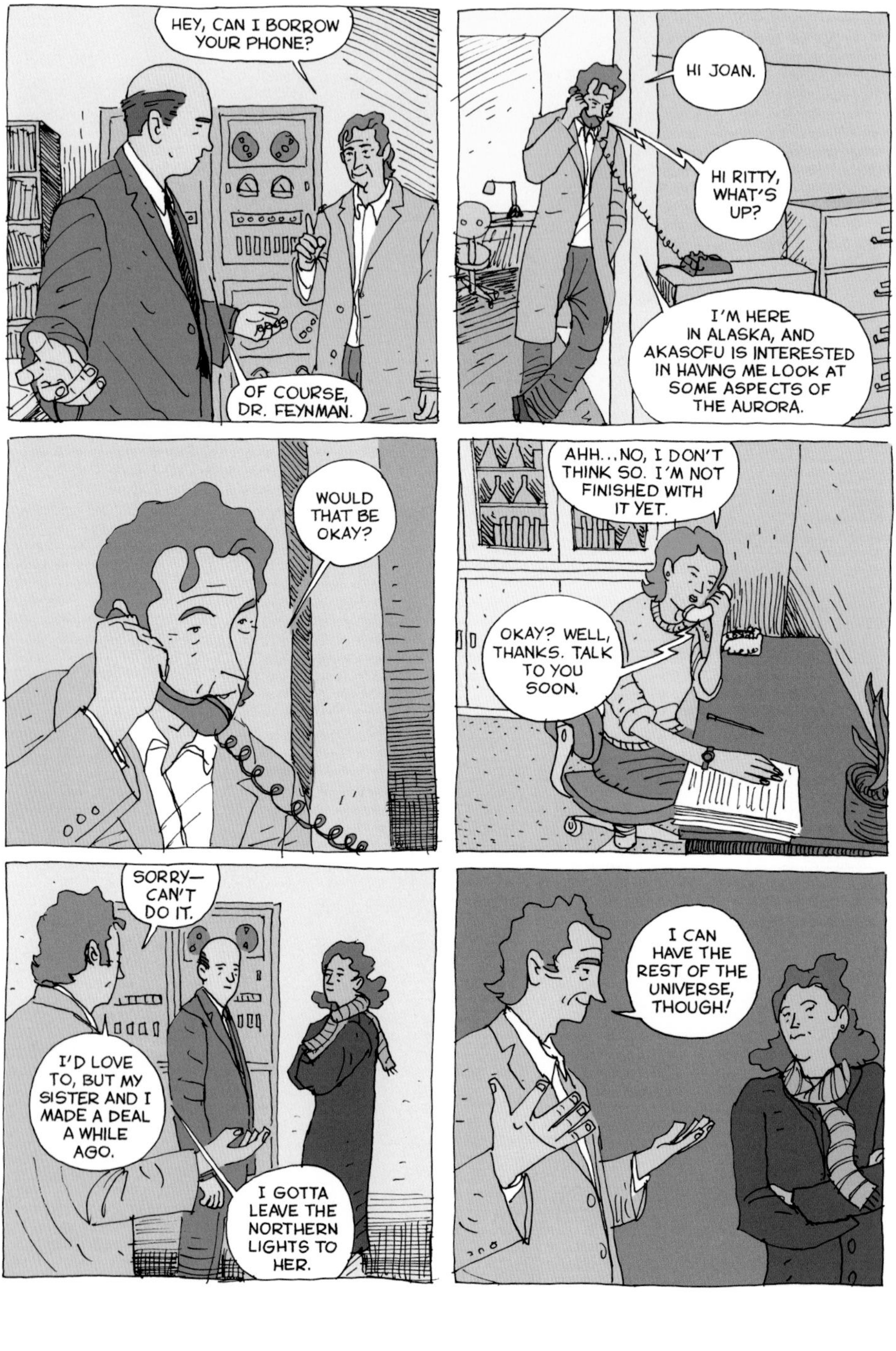
HEY, CAN I BORROW YOUR PHONE?
OF COURSE, DR. FEYNMAN.
HI JOAN.
HI RITTY, WHAT'S UP?
I'M HERE IN ALASKA, AND AKASOFU IS INTERESTED IN HAVING ME LOOK AT SOME ASPECTS OF THE AURORA.
WOULD THAT BE OKAY?
AHH...NO, I DON'T THINK SO. I'M NOT FINISHED WITH IT YET.
OKAY? WELL, THANKS. TALK TO YOU SOON.
SORRY— CAN'T DO IT.
I'D LOVE TO, BUT MY SISTER AND I MADE A DEAL A WHILE AGO.
I GOTTA LEAVE THE NORTHERN LIGHTS TO HER.
I CAN HAVE THE REST OF THE UNIVERSE, THOUGH!

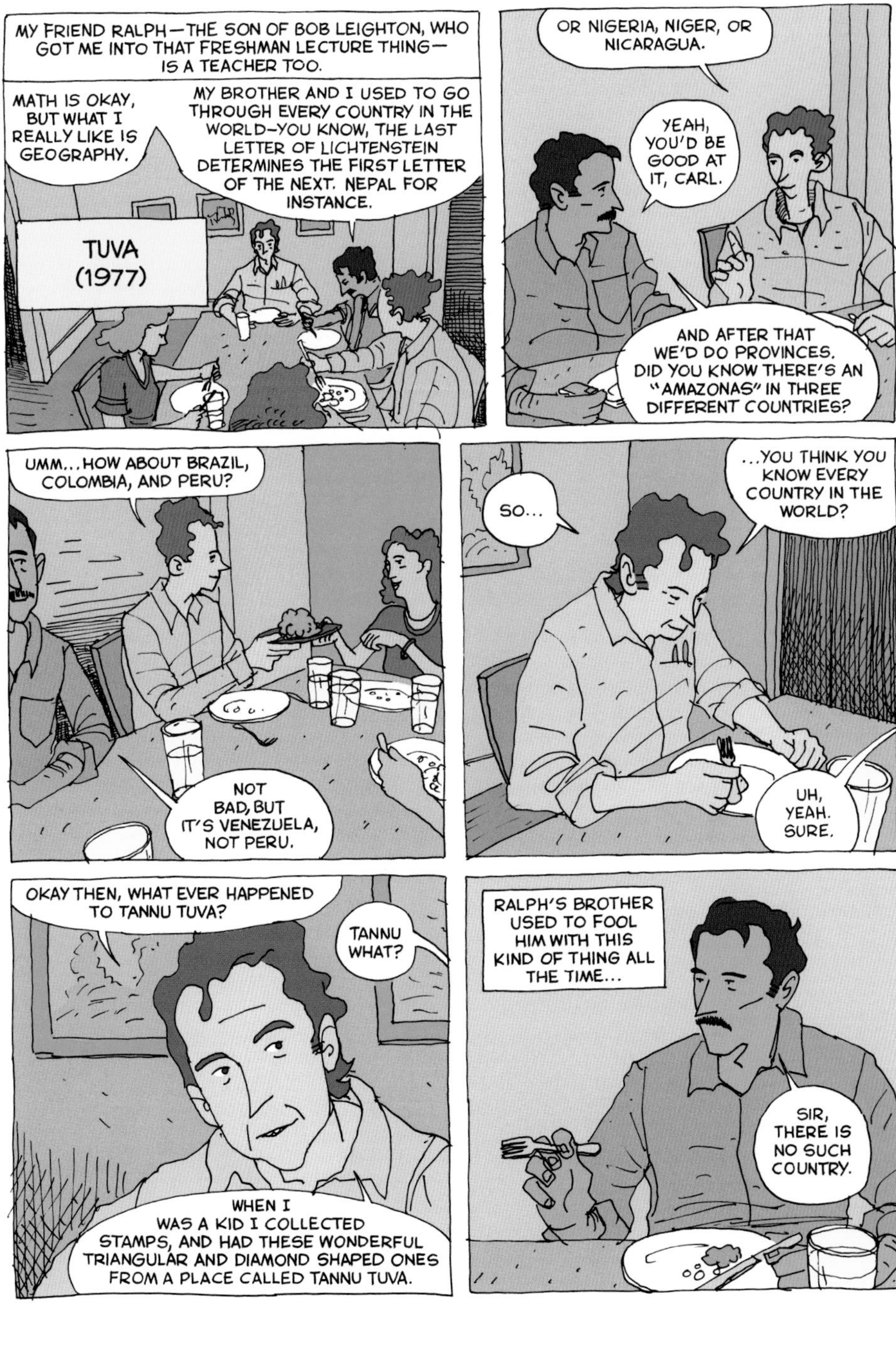
MY FRIEND RALPH—THE SON OF BOB LEIGHTON, WHO GOT ME INTO THAT FRESHMAN LECTURE THING—IS A TEACHER TOO.
MATH IS OKAY, BUT WHAT I REALLY LIKE IS GEOGRAPHY.
MY BROTHER AND I USED TO GO THROUGH EVERY COUNTRY IN THE WORLD–YOU KNOW, THE LAST LETTER OF LICHTENSTEIN DETERMINES THE FIRST LETTER OF THE NEXT. NEPAL FOR INSTANCE.
TUVA (1977)
OR NIGERIA, NIGER, OR NICARAGUA.
YEAH, YOU'D BE GOOD AT IT, CARL.
AND AFTER THAT WE'D DO PROVINCES. DID YOU KNOW THERE'S AN "AMAZONAS" IN THREE DIFFERENT COUNTRIES?
UMM...HOW ABOUT BRAZIL, COLOMBIA, AND PERU?
NOT BAD, BUT IT'S VENEZUELA, NOT PERU.
SO...
...YOU THINK YOU KNOW EVERY COUNTRY IN THE WORLD?
UH, YEAH. SURE.
OKAY THEN, WHAT EVER HAPPENED TO TANNU TUVA?
TANNU WHAT?
WHEN I WAS A KID I COLLECTED STAMPS, AND HAD THESE WONDERFUL TRIANGULAR AND DIAMOND SHAPED ONES FROM A PLACE CALLED TANNU TUVA.
RALPH'S BROTHER USED TO FOOL HIM WITH THIS KIND OF THING ALL THE TIME...
SIR, THERE IS NO SUCH COUNTRY.

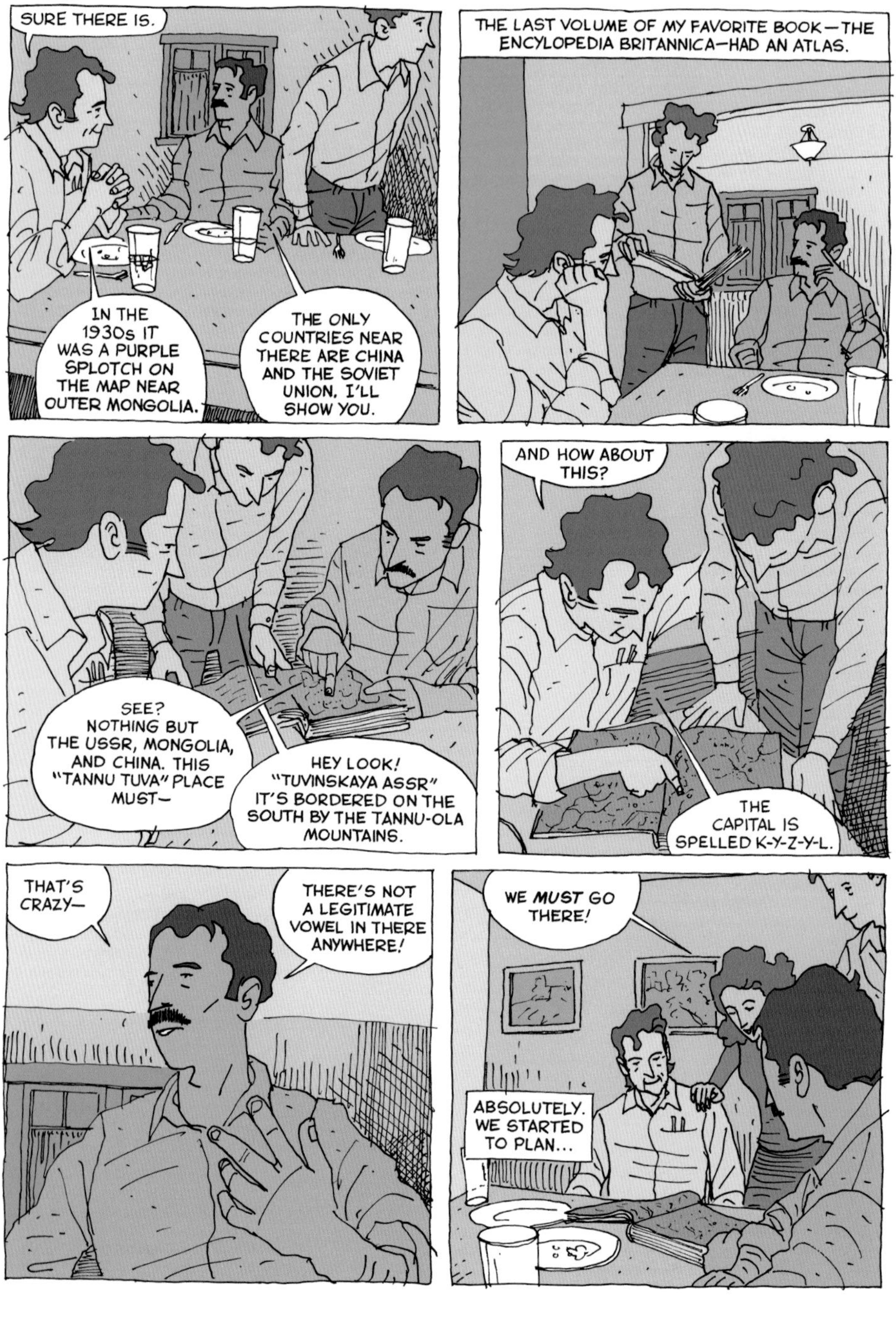
SURE THERE IS.
IN THE 1930s IT WAS A PURPLE SPLOTCH ON THE MAP NEAR OUTER MONGOLIA.
THE ONLY COUNTRIES NEAR THERE ARE CHINA AND THE SOVIET UNION. I'LL SHOW YOU.
THE LAST VOLUME OF MY FAVORITE BOOK—THE ENCYCLOPEDIA BRITANNICA—HAD AN ATLAS.
SEE? NOTHING BUT THE USSR, MONGOLIA, AND CHINA. THIS "TANNU TUVA" PLACE MUST—
HEY LOOK! "TUVINSKAYA ASSR" IT'S BORDERED ON THE SOUTH BY THE TANNU-OLA MOUNTAINS.
AND HOW ABOUT THIS?
THE CAPITAL IS SPELLED K-Y-Z-Y-L.
THAT'S CRAZY—
THERE'S NOT A LEGITIMATE VOWEL IN THERE ANYWHERE!
WE *MUST* GO THERE!
ABSOLUTELY. WE STARTED TO PLAN...

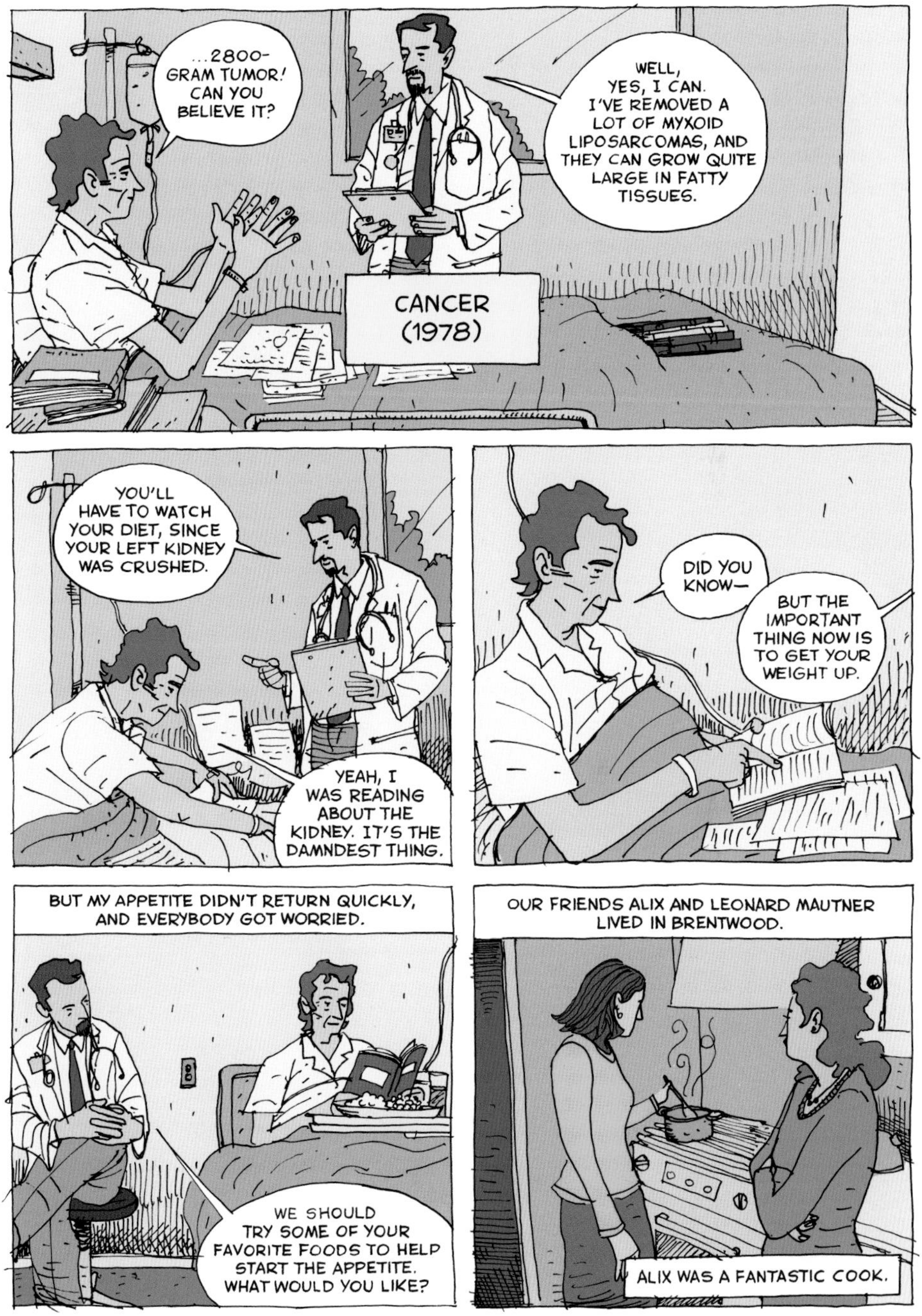
...2800-GRAM TUMOR! CAN YOU BELIEVE IT?
WELL, YES, I CAN. I'VE REMOVED A LOT OF MYXOID LIPOSARCOMAS, AND THEY CAN GROW QUITE LARGE IN FATTY TISSUES.
CANCER (1978)
YOU'LL HAVE TO WATCH YOUR DIET, SINCE YOUR LEFT KIDNEY WAS CRUSHED.
YEAH, I WAS READING ABOUT THE KIDNEY. IT'S THE DAMNDEST THING.
DID YOU KNOW—
BUT THE IMPORTANT THING NOW IS TO GET YOUR WEIGHT UP.
BUT MY APPETITE DIDN'T RETURN QUICKLY, AND EVERYBODY GOT WORRIED.
WE SHOULD TRY SOME OF YOUR FAVORITE FOODS TO HELP START THE APPETITE. WHAT WOULD YOU LIKE?
OUR FRIENDS ALIX AND LEONARD MAUTNER LIVED IN BRENTWOOD.
ALIX WAS A FANTASTIC COOK.

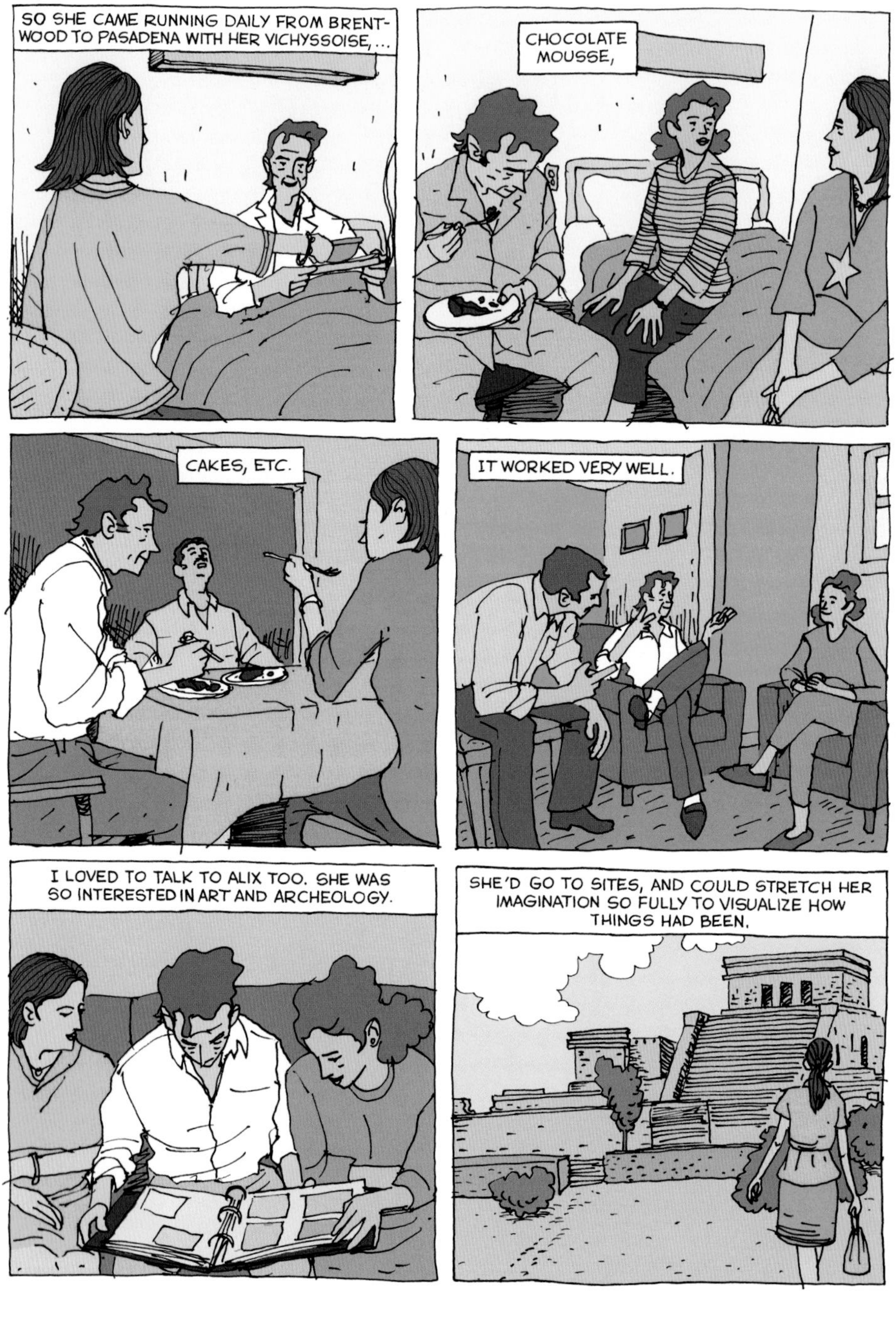
SO SHE CAME RUNNING DAILY FROM BRENT-WOOD TO PASADENA WITH HER VICHYSSOISE, ...
CHOCOLATE MOUSSE,
CAKES, ETC.
IT WORKED VERY WELL.
I LOVED TO TALK TO ALIX TOO. SHE WAS SO INTERESTED IN ART AND ARCHEOLOGY.
SHE'D GO TO SITES, AND COULD STRETCH HER IMAGINATION SO FULLY TO VISUALIZE HOW THINGS HAD BEEN.

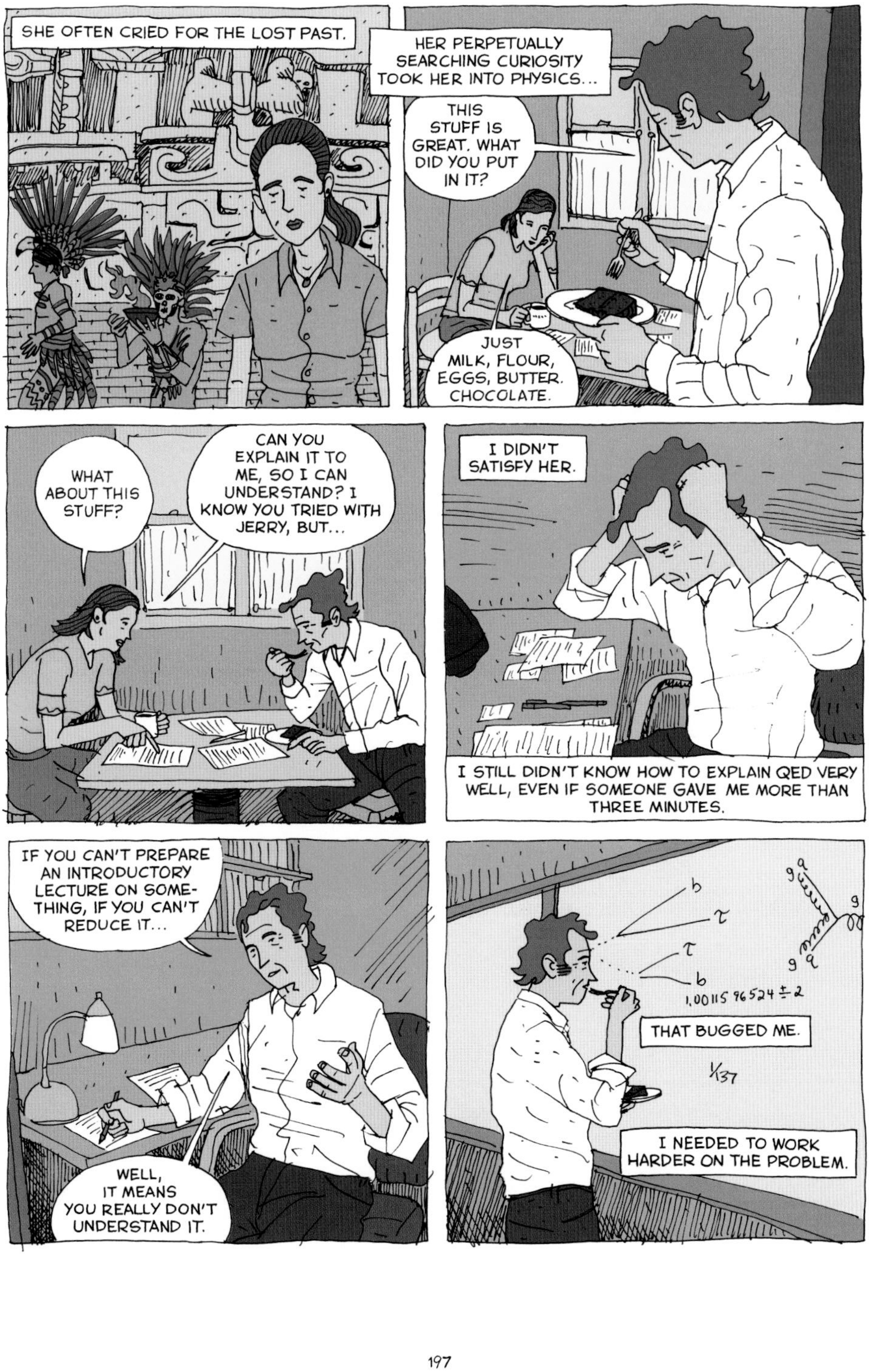
SHE OFTEN CRIED FOR THE LOST PAST.
HER PERPETUALLY SEARCHING CURIOSITY TOOK HER INTO PHYSICS...
THIS STUFF IS GREAT. WHAT DID YOU PUT IN IT?
JUST MILK, FLOUR, EGGS, BUTTER. CHOCOLATE.
WHAT ABOUT THIS STUFF?
CAN YOU EXPLAIN IT TO ME, SO I CAN UNDERSTAND? I KNOW YOU TRIED WITH JERRY, BUT...
I DIDN'T SATISFY HER.
I STILL DIDN'T KNOW HOW TO EXPLAIN QED VERY WELL, EVEN IF SOMEONE GAVE ME MORE THAN THREE MINUTES.
IF YOU CAN'T PREPARE AN INTRODUCTORY LECTURE ON SOME-THING, IF YOU CAN'T REDUCE IT...
WELL, IT MEANS YOU REALLY DON'T UNDERSTAND IT.
1.00115 96524 ± 2
THAT BUGGED ME.
1/137
I NEEDED TO WORK HARDER ON THE PROBLEM.

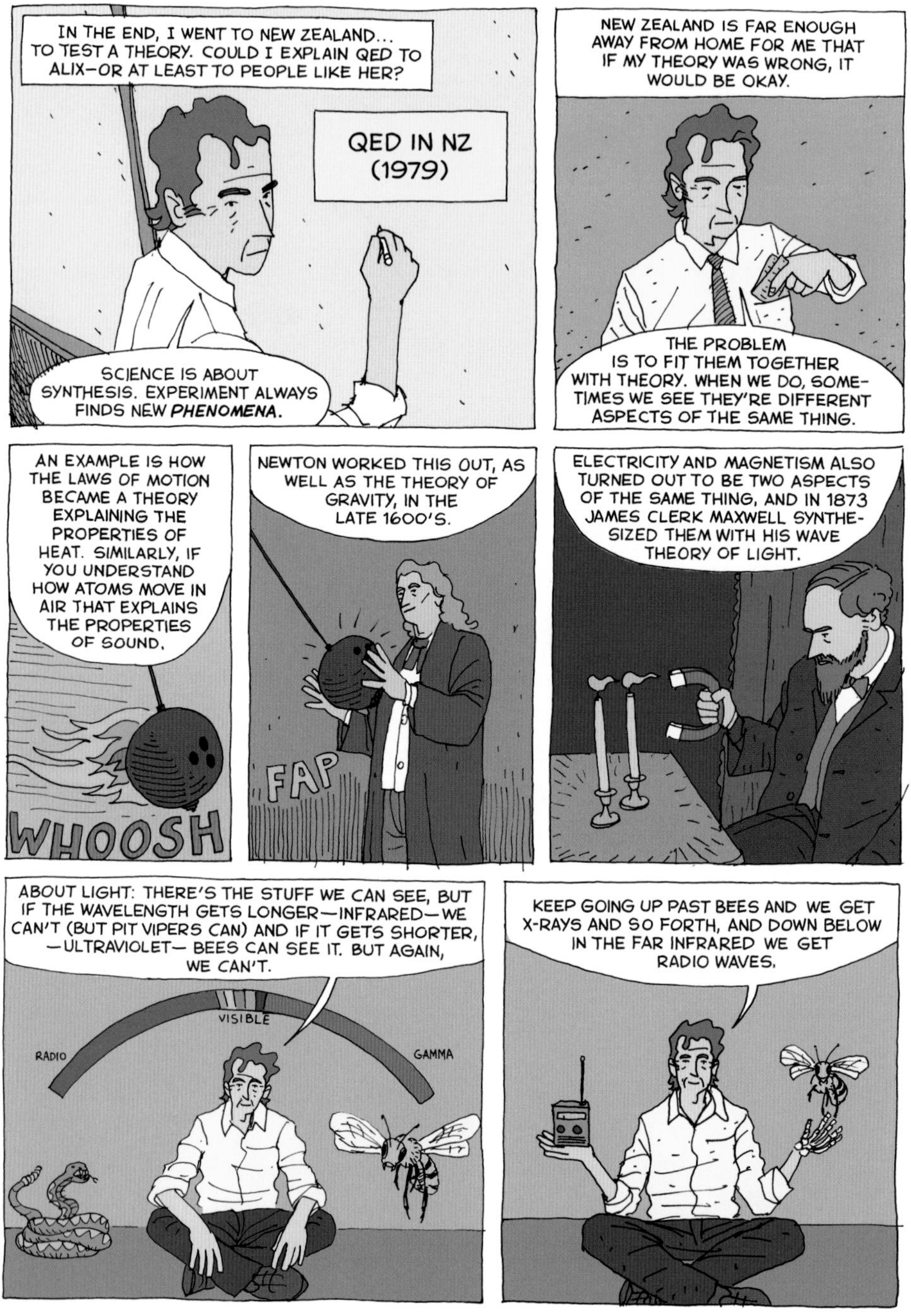
IN THE END, I WENT TO NEW ZEALAND... TO TEST A THEORY. COULD I EXPLAIN QED TO ALIX—OR AT LEAST TO PEOPLE LIKE HER?
QED IN NZ (1979)
SCIENCE IS ABOUT SYNTHESIS. EXPERIMENT ALWAYS FINDS NEW *PHENOMENA*.
NEW ZEALAND IS FAR ENOUGH AWAY FROM HOME FOR ME THAT IF MY THEORY WAS WRONG, IT WOULD BE OKAY.
THE PROBLEM IS TO FIT THEM TOGETHER WITH THEORY. WHEN WE DO, SOMETIMES WE SEE THEY'RE DIFFERENT ASPECTS OF THE SAME THING.
AN EXAMPLE IS HOW THE LAWS OF MOTION BECAME A THEORY EXPLAINING THE PROPERTIES OF HEAT. SIMILARLY, IF YOU UNDERSTAND HOW ATOMS MOVE IN AIR THAT EXPLAINS THE PROPERTIES OF SOUND.
WHOOSH
NEWTON WORKED THIS OUT, AS WELL AS THE THEORY OF GRAVITY, IN THE LATE 1600'S.
FAP
ELECTRICITY AND MAGNETISM ALSO TURNED OUT TO BE TWO ASPECTS OF THE SAME THING, AND IN 1873 JAMES CLERK MAXWELL SYNTHESIZED THEM WITH HIS WAVE THEORY OF LIGHT.
ABOUT LIGHT: THERE'S THE STUFF WE CAN SEE, BUT IF THE WAVELENGTH GETS LONGER—INFRARED—WE CAN'T (BUT PIT VIPERS CAN) AND IF IT GETS SHORTER, —ULTRAVIOLET— BEES CAN SEE IT. BUT AGAIN, WE CAN'T.
VISIBLE
RADIO
GAMMA
KEEP GOING UP PAST BEES AND WE GET X-RAYS AND SO FORTH, AND DOWN BELOW IN THE FAR INFRARED WE GET RADIO WAVES.

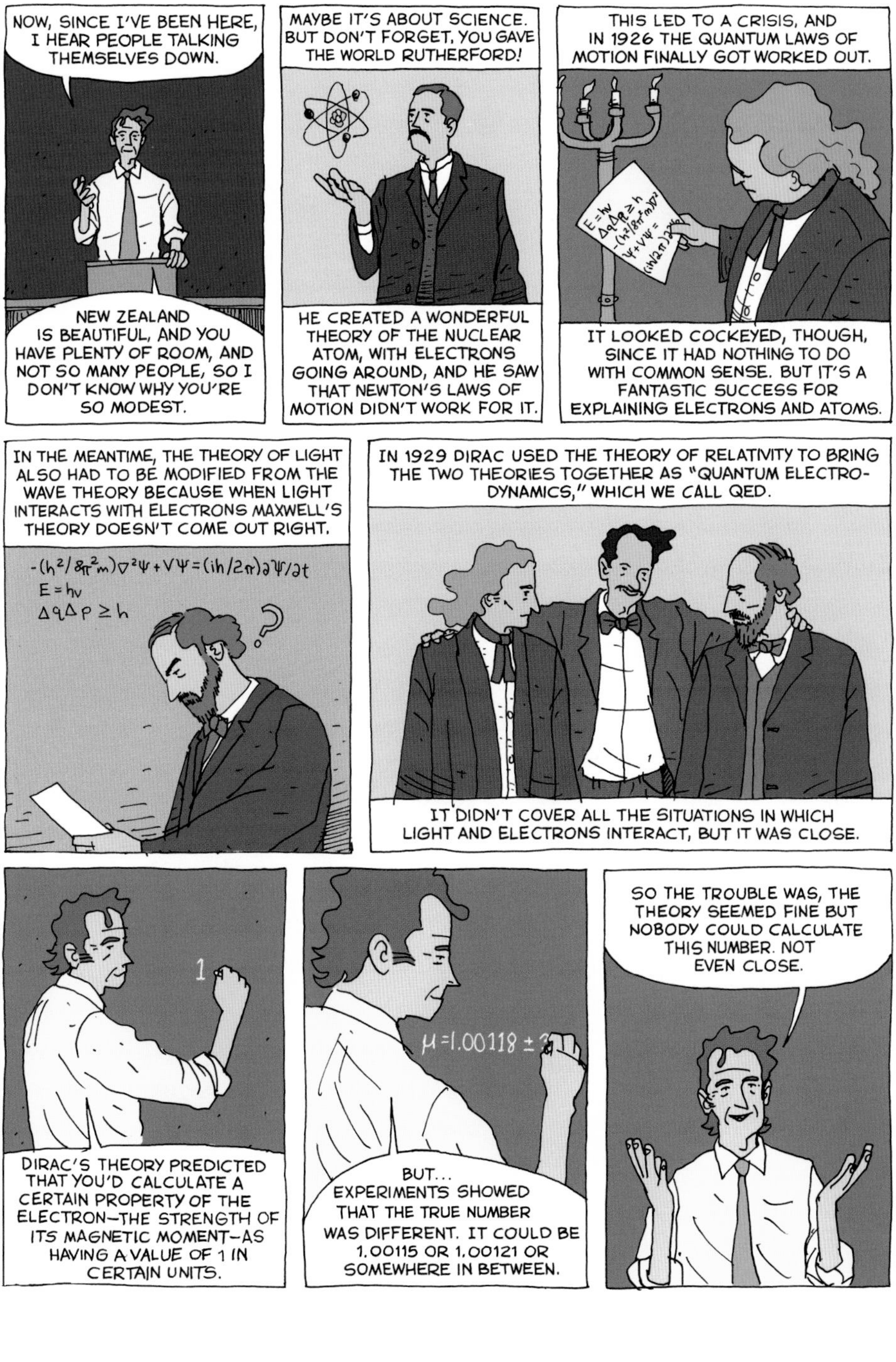

NOW, SINCE I'VE BEEN HERE, I HEAR PEOPLE TALKING THEMSELVES DOWN.
NEW ZEALAND IS BEAUTIFUL, AND YOU HAVE PLENTY OF ROOM, AND NOT SO MANY PEOPLE, SO I DON'T KNOW WHY YOU'RE SO MODEST.
MAYBE IT'S ABOUT SCIENCE. BUT DON'T FORGET, YOU GAVE THE WORLD RUTHERFORD!
HE CREATED A WONDERFUL THEORY OF THE NUCLEAR ATOM, WITH ELECTRONS GOING AROUND, AND HE SAW THAT NEWTON'S LAWS OF MOTION DIDN'T WORK FOR IT.
THIS LED TO A CRISIS, AND IN 1926 THE QUANTUM LAWS OF MOTION FINALLY GOT WORKED OUT.
IT LOOKED COCKEYED, THOUGH, SINCE IT HAD NOTHING TO DO WITH COMMON SENSE. BUT IT'S A FANTASTIC SUCCESS FOR EXPLAINING ELECTRONS AND ATOMS.
IN THE MEANTIME, THE THEORY OF LIGHT ALSO HAD TO BE MODIFIED FROM THE WAVE THEORY BECAUSE WHEN LIGHT INTERACTS WITH ELECTRONS MAXWELL'S THEORY DOESN'T COME OUT RIGHT.
$-(h^2/8\pi^2m)\nabla^2\psi + V\psi = (ih/2\pi)\partial\psi/\partial t$
$E = h\nu$
$\Delta q \Delta p \geq h$
IN 1929 DIRAC USED THE THEORY OF RELATIVITY TO BRING THE TWO THEORIES TOGETHER AS "QUANTUM ELECTRO-DYNAMICS," WHICH WE CALL QED.
IT DIDN'T COVER ALL THE SITUATIONS IN WHICH LIGHT AND ELECTRONS INTERACT, BUT IT WAS CLOSE.
1
DIRAC'S THEORY PREDICTED THAT YOU'D CALCULATE A CERTAIN PROPERTY OF THE ELECTRON—THE STRENGTH OF ITS MAGNETIC MOMENT—AS HAVING A VALUE OF 1 IN CERTAIN UNITS.
μ=1.00118 ±
BUT... EXPERIMENTS SHOWED THAT THE TRUE NUMBER WAS DIFFERENT. IT COULD BE 1.00115 OR 1.00121 OR SOMEWHERE IN BETWEEN.
SO THE TROUBLE WAS, THE THEORY SEEMED FINE BUT NOBODY COULD CALCULATE THIS NUMBER. NOT EVEN CLOSE.

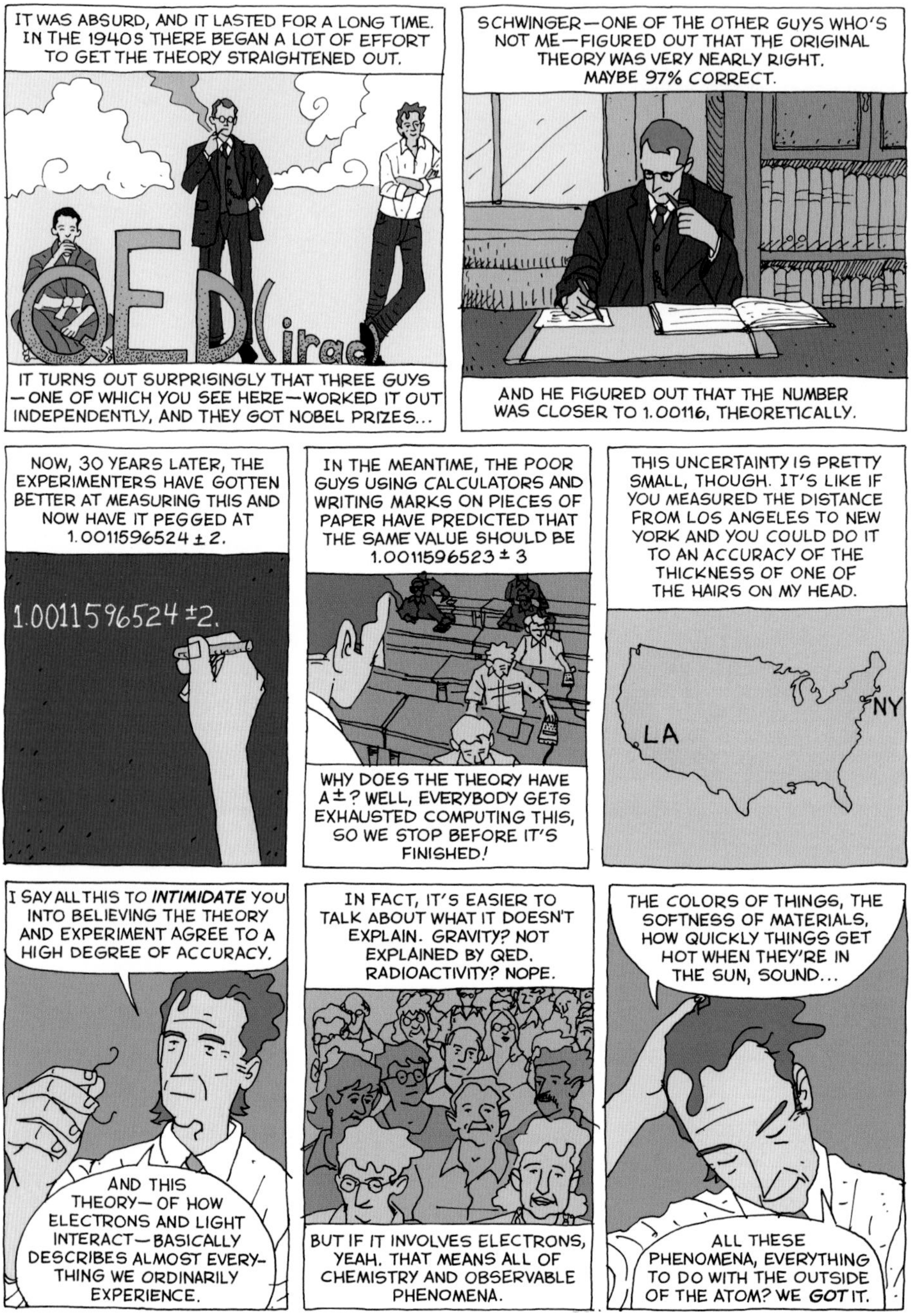

IT WAS ABSURD, AND IT LASTED FOR A LONG TIME. IN THE 1940S THERE BEGAN A LOT OF EFFORT TO GET THE THEORY STRAIGHTENED OUT.
QED
IT TURNS OUT SURPRISINGLY THAT THREE GUYS—ONE OF WHICH YOU SEE HERE—WORKED IT OUT INDEPENDENTLY, AND THEY GOT NOBEL PRIZES...
SCHWINGER—ONE OF THE OTHER GUYS WHO'S NOT ME—FIGURED OUT THAT THE ORIGINAL THEORY WAS VERY NEARLY RIGHT. MAYBE 97% CORRECT.
AND HE FIGURED OUT THAT THE NUMBER WAS CLOSER TO 1.00116, THEORETICALLY.
NOW, 30 YEARS LATER, THE EXPERIMENTERS HAVE GOTTEN BETTER AT MEASURING THIS AND NOW HAVE IT PEGGED AT 1.0011596524 ± 2.
1.0011596524 ±2.
IN THE MEANTIME, THE POOR GUYS USING CALCULATORS AND WRITING MARKS ON PIECES OF PAPER HAVE PREDICTED THAT THE SAME VALUE SHOULD BE 1.0011596523 ± 3
WHY DOES THE THEORY HAVE A ±? WELL, EVERYBODY GETS EXHAUSTED COMPUTING THIS, SO WE STOP BEFORE IT'S FINISHED!
THIS UNCERTAINTY IS PRETTY SMALL, THOUGH. IT'S LIKE IF YOU MEASURED THE DISTANCE FROM LOS ANGELES TO NEW YORK AND YOU COULD DO IT TO AN ACCURACY OF THE THICKNESS OF ONE OF THE HAIRS ON MY HEAD.
LA
NY
I SAY ALL THIS TO INTIMIDATE YOU INTO BELIEVING THE THEORY AND EXPERIMENT AGREE TO A HIGH DEGREE OF ACCURACY.
AND THIS THEORY—OF HOW ELECTRONS AND LIGHT INTERACT—BASICALLY DESCRIBES ALMOST EVERYTHING WE ORDINARILY EXPERIENCE.
IN FACT, IT'S EASIER TO TALK ABOUT WHAT IT DOESN'T EXPLAIN. GRAVITY? NOT EXPLAINED BY QED. RADIOACTIVITY? NOPE.
BUT IF IT INVOLVES ELECTRONS, YEAH. THAT MEANS ALL OF CHEMISTRY AND OBSERVABLE PHENOMENA.
THE COLORS OF THINGS, THE SOFTNESS OF MATERIALS, HOW QUICKLY THINGS GET HOT WHEN THEY'RE IN THE SUN, SOUND...
ALL THESE PHENOMENA, EVERYTHING TO DO WITH THE OUTSIDE OF THE ATOM? WE GOT IT.

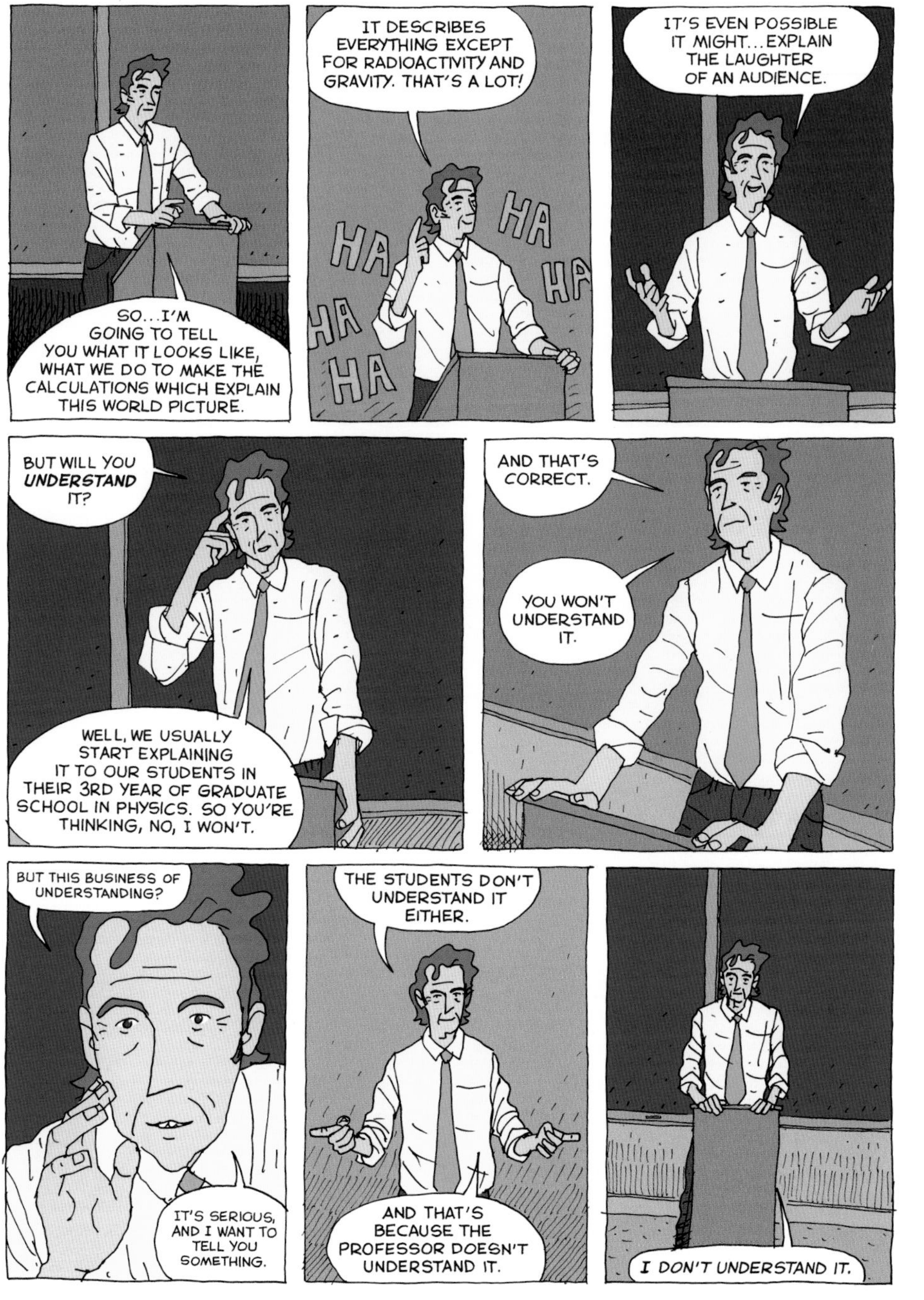
SO...I'M GOING TO TELL YOU WHAT IT LOOKS LIKE, WHAT WE DO TO MAKE THE CALCULATIONS WHICH EXPLAIN THIS WORLD PICTURE.
IT DESCRIBES EVERYTHING EXCEPT FOR RADIOACTIVITY AND GRAVITY. THAT'S A LOT!
HA HA HA HA HA
IT'S EVEN POSSIBLE IT MIGHT...EXPLAIN THE LAUGHTER OF AN AUDIENCE.
BUT WILL YOU UNDERSTAND IT?
WELL, WE USUALLY START EXPLAINING IT TO OUR STUDENTS IN THEIR 3RD YEAR OF GRADUATE SCHOOL IN PHYSICS. SO YOU'RE THINKING, NO, I WON'T.
AND THAT'S CORRECT.
YOU WON'T UNDERSTAND IT.
BUT THIS BUSINESS OF UNDERSTANDING?
IT'S SERIOUS, AND I WANT TO TELL YOU SOMETHING.
THE STUDENTS DON'T UNDERSTAND IT EITHER.
AND THAT'S BECAUSE THE PROFESSOR DOESN'T UNDERSTAND IT.
I DON'T UNDERSTAND IT.

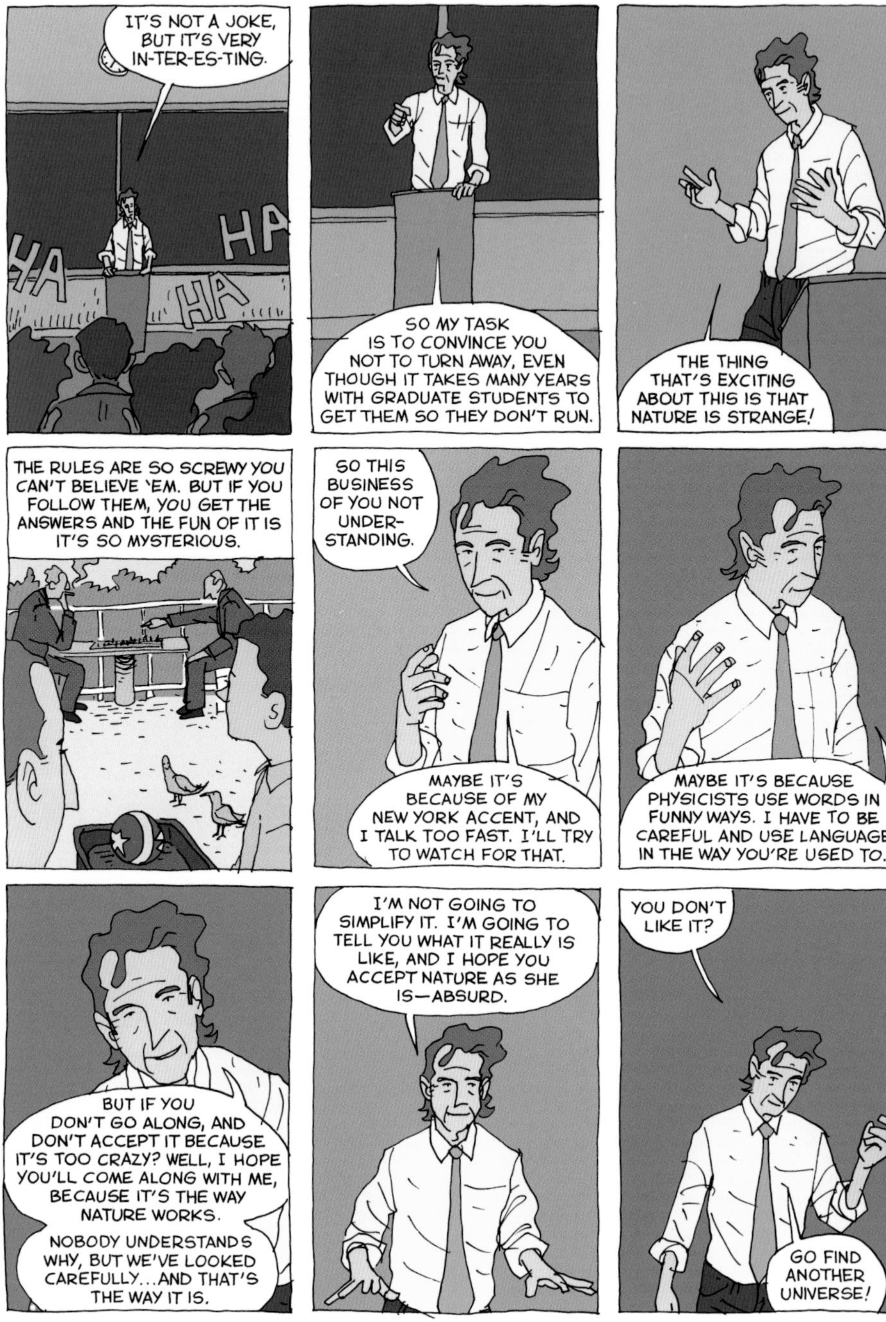

IT'S NOT A JOKE, BUT IT'S VERY IN-TER-ES-TING.
HA
HA
HA
SO MY TASK IS TO CONVINCE YOU NOT TO TURN AWAY, EVEN THOUGH IT TAKES MANY YEARS WITH GRADUATE STUDENTS TO GET THEM SO THEY DON'T RUN.
THE THING THAT'S EXCITING ABOUT THIS IS THAT NATURE IS STRANGE!
THE RULES ARE SO SCREWY YOU CAN'T BELIEVE 'EM. BUT IF YOU FOLLOW THEM, YOU GET THE ANSWERS AND THE FUN OF IT IS IT'S SO MYSTERIOUS.
SO THIS BUSINESS OF YOU NOT UNDER-STANDING.
MAYBE IT'S BECAUSE OF MY NEW YORK ACCENT, AND I TALK TOO FAST. I'LL TRY TO WATCH FOR THAT.
MAYBE IT'S BECAUSE PHYSICISTS USE WORDS IN FUNNY WAYS. I HAVE TO BE CAREFUL AND USE LANGUAGE IN THE WAY YOU'RE USED TO.
BUT IF YOU DON'T GO ALONG, AND DON'T ACCEPT IT BECAUSE IT'S TOO CRAZY? WELL, I HOPE YOU'LL COME ALONG WITH ME, BECAUSE IT'S THE WAY NATURE WORKS.
NOBODY UNDERSTANDS WHY, BUT WE'VE LOOKED CAREFULLY...AND THAT'S THE WAY IT IS.
I'M NOT GOING TO SIMPLIFY IT. I'M GOING TO TELL YOU WHAT IT REALLY IS LIKE, AND I HOPE YOU ACCEPT NATURE AS SHE IS—ABSURD.
YOU DON'T LIKE IT?
GO FIND ANOTHER UNIVERSE!

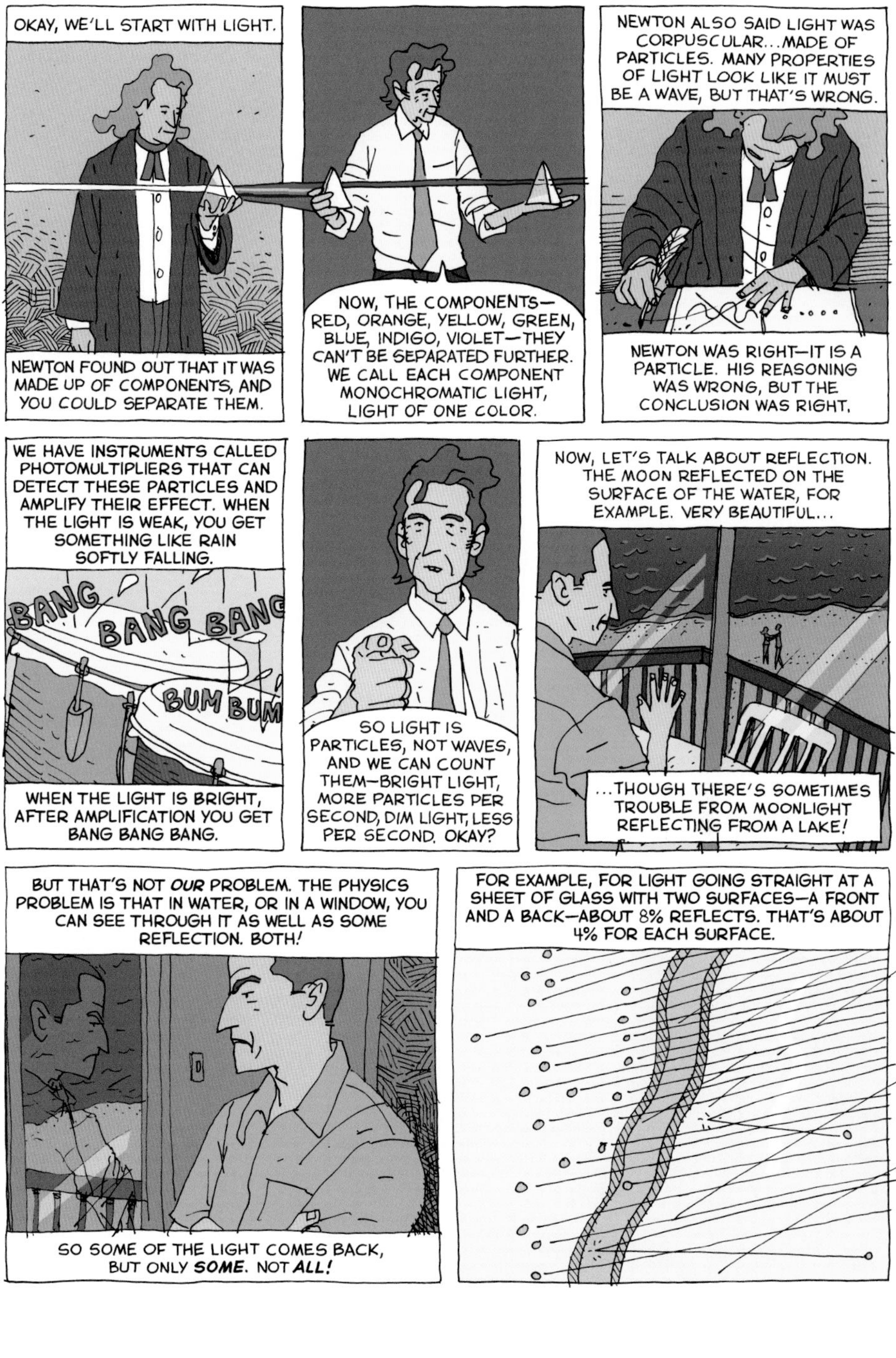
OKAY, WE'LL START WITH LIGHT.
NEWTON FOUND OUT THAT IT WAS MADE UP OF COMPONENTS, AND YOU COULD SEPARATE THEM.
NOW, THE COMPONENTS— RED, ORANGE, YELLOW, GREEN, BLUE, INDIGO, VIOLET—THEY CAN'T BE SEPARATED FURTHER. WE CALL EACH COMPONENT MONOCHROMATIC LIGHT, LIGHT OF ONE COLOR.
NEWTON ALSO SAID LIGHT WAS CORPUSCULAR... MADE OF PARTICLES. MANY PROPERTIES OF LIGHT LOOK LIKE IT MUST BE A WAVE, BUT THAT'S WRONG.
NEWTON WAS RIGHT—IT IS A PARTICLE. HIS REASONING WAS WRONG, BUT THE CONCLUSION WAS RIGHT.
WE HAVE INSTRUMENTS CALLED PHOTOMULTIPLIERS THAT CAN DETECT THESE PARTICLES AND AMPLIFY THEIR EFFECT. WHEN THE LIGHT IS WEAK, YOU GET SOMETHING LIKE RAIN SOFTLY FALLING.
BANG
BANG BANG
BUM BUM
WHEN THE LIGHT IS BRIGHT, AFTER AMPLIFICATION YOU GET BANG BANG BANG.
SO LIGHT IS PARTICLES, NOT WAVES, AND WE CAN COUNT THEM—BRIGHT LIGHT, MORE PARTICLES PER SECOND, DIM LIGHT, LESS PER SECOND. OKAY?
NOW, LET'S TALK ABOUT REFLECTION. THE MOON REFLECTED ON THE SURFACE OF THE WATER, FOR EXAMPLE. VERY BEAUTIFUL...
...THOUGH THERE'S SOMETIMES TROUBLE FROM MOONLIGHT REFLECTING FROM A LAKE!
BUT THAT'S NOT *OUR* PROBLEM. THE PHYSICS PROBLEM IS THAT IN WATER, OR IN A WINDOW, YOU CAN SEE THROUGH IT AS WELL AS SOME REFLECTION. BOTH!
SO SOME OF THE LIGHT COMES BACK, BUT ONLY **SOME**. NOT ***ALL!***
FOR EXAMPLE, FOR LIGHT GOING STRAIGHT AT A SHEET OF GLASS WITH TWO SURFACES—A FRONT AND A BACK—ABOUT 8% REFLECTS. THAT'S ABOUT 4% FOR EACH SURFACE.

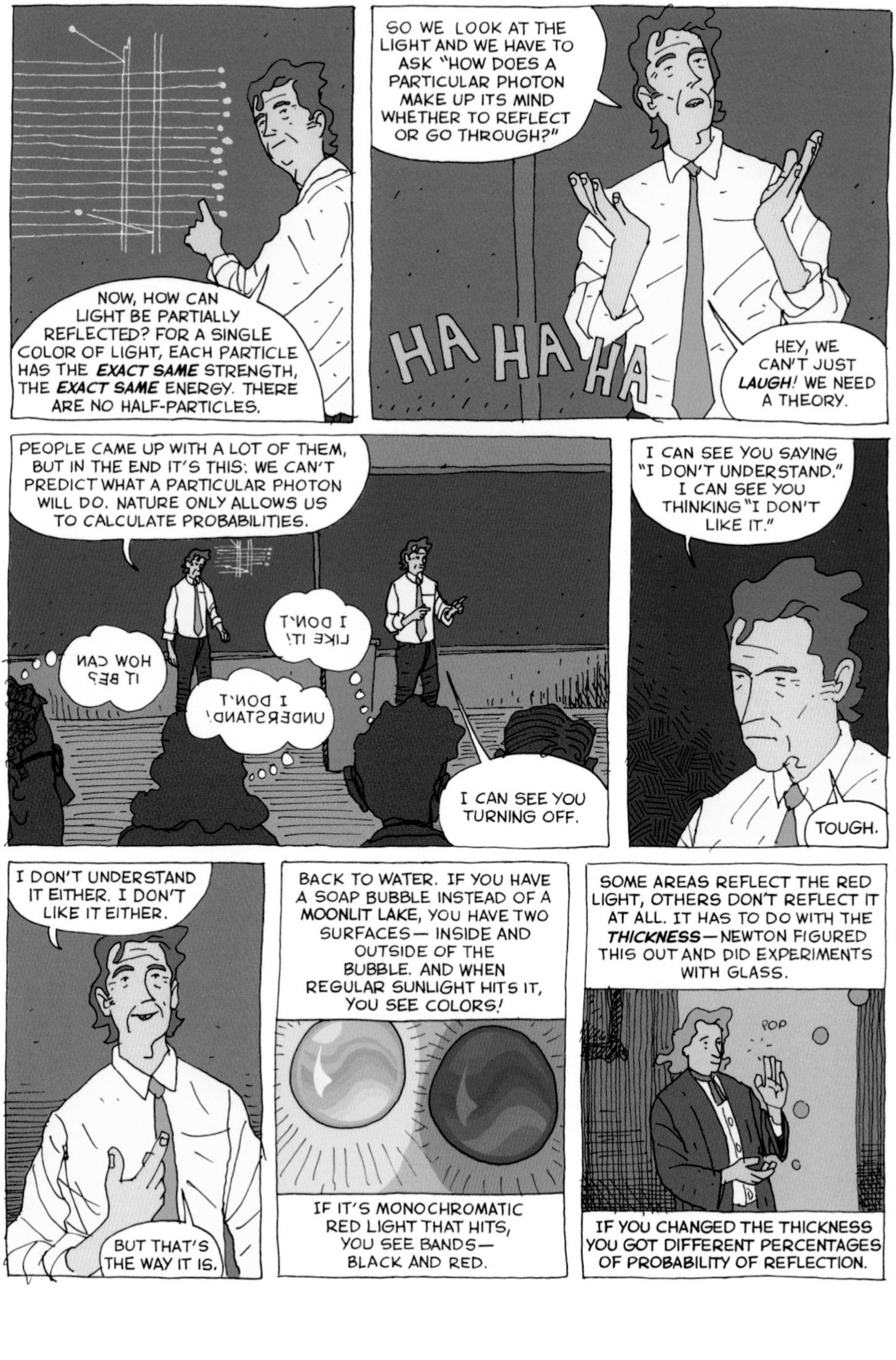
NOW, HOW CAN LIGHT BE PARTIALLY REFLECTED? FOR A SINGLE COLOR OF LIGHT, EACH PARTICLE HAS THE EXACT SAME STRENGTH, THE EXACT SAME ENERGY. THERE ARE NO HALF-PARTICLES.
SO WE LOOK AT THE LIGHT AND WE HAVE TO ASK "HOW DOES A PARTICULAR PHOTON MAKE UP ITS MIND WHETHER TO REFLECT OR GO THROUGH?"
HA HA HA
HEY, WE CAN'T JUST LAUGH! WE NEED A THEORY.
PEOPLE CAME UP WITH A LOT OF THEM, BUT IN THE END IT'S THIS: WE CAN'T PREDICT WHAT A PARTICULAR PHOTON WILL DO. NATURE ONLY ALLOWS US TO CALCULATE PROBABILITIES.
HOW CAN IT BE?
I DON'T LIKE IT!
I DON'T UNDERSTAND!
I CAN SEE YOU TURNING OFF.
I CAN SEE YOU SAYING "I DON'T UNDERSTAND." I CAN SEE YOU THINKING "I DON'T LIKE IT."
TOUGH.
I DON'T UNDERSTAND IT EITHER. I DON'T LIKE IT EITHER.
BUT THAT'S THE WAY IT IS.
BACK TO WATER. IF YOU HAVE A SOAP BUBBLE INSTEAD OF A MOONLIT LAKE, YOU HAVE TWO SURFACES— INSIDE AND OUTSIDE OF THE BUBBLE. AND WHEN REGULAR SUNLIGHT HITS IT, YOU SEE COLORS!
IF IT'S MONOCHROMATIC RED LIGHT THAT HITS, YOU SEE BANDS— BLACK AND RED.
SOME AREAS REFLECT THE RED LIGHT, OTHERS DON'T REFLECT IT AT ALL. IT HAS TO DO WITH THE THICKNESS— NEWTON FIGURED THIS OUT AND DID EXPERIMENTS WITH GLASS.
POP
IF YOU CHANGED THE THICKNESS YOU GOT DIFFERENT PERCENTAGES OF PROBABILITY OF REFLECTION.

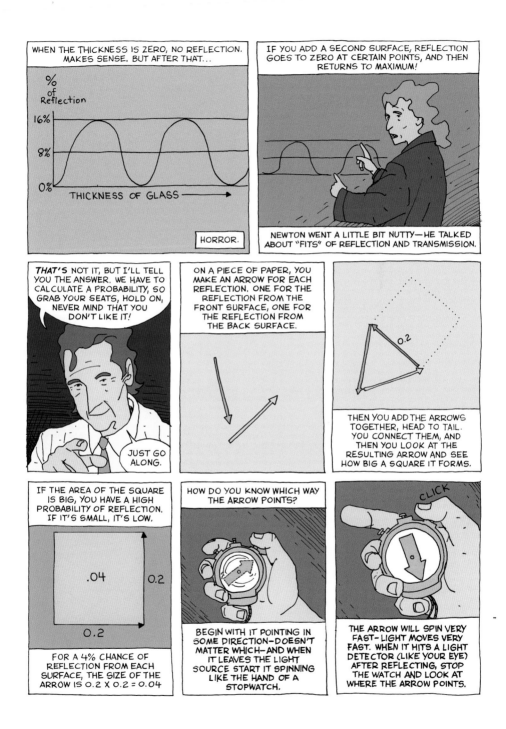

WHEN THE THICKNESS IS ZERO, NO REFLECTION. MAKES SENSE. BUT AFTER THAT...

% of Reflection

16%

8%

0%

THICKNESS OF GLASS →

HORROR.

IF YOU ADD A SECOND SURFACE, REFLECTION GOES TO ZERO AT CERTAIN POINTS, AND THEN RETURNS TO MAXIMUM!

NEWTON WENT A LITTLE BIT NUTTY—HE TALKED ABOUT "FITS" OF REFLECTION AND TRANSMISSION.

THAT'S NOT IT, BUT I'LL TELL YOU THE ANSWER. WE HAVE TO CALCULATE A PROBABILITY, SO GRAB YOUR SEATS, HOLD ON, NEVER MIND THAT YOU DON'T LIKE IT!

JUST GO ALONG.

ON A PIECE OF PAPER, YOU MAKE AN ARROW FOR EACH REFLECTION. ONE FOR THE REFLECTION FROM THE FRONT SURFACE, ONE FOR THE REFLECTION FROM THE BACK SURFACE.

0.2

THEN YOU ADD THE ARROWS TOGETHER, HEAD TO TAIL. YOU CONNECT THEM, AND THEN YOU LOOK AT THE RESULTING ARROW AND SEE HOW BIG A SQUARE IT FORMS.

IF THE AREA OF THE SQUARE IS BIG, YOU HAVE A HIGH PROBABILITY OF REFLECTION. IF IT'S SMALL, IT'S LOW.

.04   0.2

0.2

FOR A 4% CHANCE OF REFLECTION FROM EACH SURFACE, THE SIZE OF THE ARROW IS 0.2 X 0.2 = 0.04

HOW DO YOU KNOW WHICH WAY THE ARROW POINTS?

BEGIN WITH IT POINTING IN SOME DIRECTION—DOESN'T MATTER WHICH—AND WHEN IT LEAVES THE LIGHT SOURCE START IT SPINNING LIKE THE HAND OF A STOPWATCH.

CLICK

THE ARROW WILL SPIN VERY FAST—LIGHT MOVES VERY FAST. WHEN IT HITS A LIGHT DETECTOR (LIKE YOUR EYE) AFTER REFLECTING, STOP THE WATCH AND LOOK AT WHERE THE ARROW POINTS.

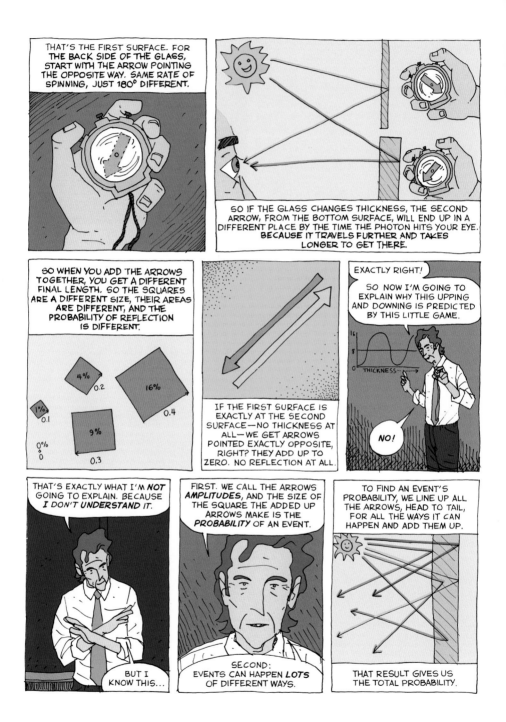

THAT'S THE FIRST SURFACE. FOR **THE BACK SIDE OF THE GLASS,** START WITH THE ARROW POINTING THE OPPOSITE WAY. SAME RATE OF SPINNING, JUST 180° DIFFERENT.

SO IF THE GLASS CHANGES THICKNESS, THE SECOND ARROW, FROM THE BOTTOM SURFACE, WILL END UP IN A DIFFERENT PLACE BY THE TIME THE PHOTON HITS YOUR EYE. BECAUSE IT TRAVELS FURTHER AND TAKES LONGER TO GET THERE.

SO WHEN YOU ADD THE ARROWS TOGETHER, YOU GET A DIFFERENT FINAL LENGTH. SO THE SQUARES ARE A DIFFERENT SIZE, THEIR AREAS ARE DIFFERENT, AND THE PROBABILITY OF REFLECTION IS DIFFERENT.

4% 0.2
1% 0.1
16% 0.4
0% 0
9% 0.3

IF THE FIRST SURFACE IS EXACTLY AT THE SECOND SURFACE—NO THICKNESS AT ALL—WE GET ARROWS POINTED EXACTLY OPPOSITE, RIGHT? THEY ADD UP TO ZERO. NO REFLECTION AT ALL.

EXACTLY RIGHT!

SO NOW I'M GOING TO EXPLAIN WHY THIS UPPING AND DOWNING IS PREDICTED BY THIS LITTLE GAME.

16
8
THICKNESS

NO!

THAT'S EXACTLY WHAT I'M **NOT** GOING TO EXPLAIN. BECAUSE **I DON'T UNDERSTAND IT.**

BUT I KNOW THIS...

FIRST. WE CALL THE ARROWS **AMPLITUDES,** AND THE SIZE OF THE SQUARE THE ADDED UP ARROWS MAKE IS THE **PROBABILITY** OF AN EVENT.

SECOND: EVENTS CAN HAPPEN **LOTS** OF DIFFERENT WAYS.

TO FIND AN EVENT'S PROBABILITY, WE LINE UP ALL THE ARROWS, HEAD TO TAIL, FOR ALL THE WAYS IT CAN HAPPEN AND ADD THEM UP.

THAT RESULT GIVES US THE TOTAL PROBABILITY.

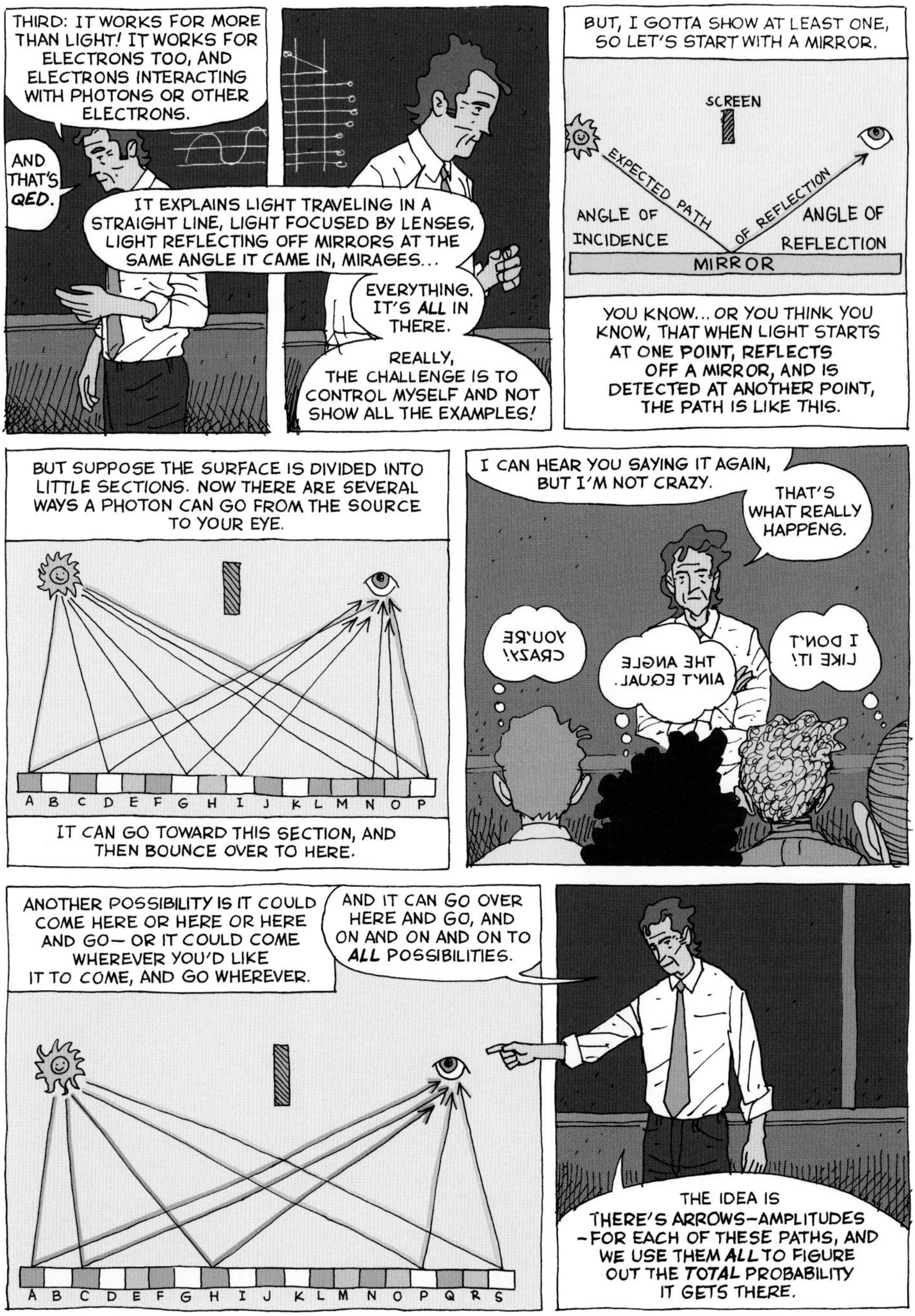
THIRD: IT WORKS FOR MORE THAN LIGHT! IT WORKS FOR ELECTRONS TOO, AND ELECTRONS INTERACTING WITH PHOTONS OR OTHER ELECTRONS.
AND THAT'S **QED**.
IT EXPLAINS LIGHT TRAVELING IN A STRAIGHT LINE, LIGHT FOCUSED BY LENSES, LIGHT REFLECTING OFF MIRRORS AT THE SAME ANGLE IT CAME IN, MIRAGES...
EVERYTHING. IT'S **ALL** IN THERE.
REALLY, THE CHALLENGE IS TO CONTROL MYSELF AND NOT SHOW ALL THE EXAMPLES!
BUT, I GOTTA SHOW AT LEAST ONE, SO LET'S START WITH A MIRROR.
SCREEN
EXPECTED PATH OF REFLECTION
ANGLE OF INCIDENCE
ANGLE OF REFLECTION
MIRROR
YOU KNOW... OR YOU THINK YOU KNOW, THAT WHEN LIGHT STARTS AT ONE POINT, REFLECTS OFF A MIRROR, AND IS DETECTED AT ANOTHER POINT, THE PATH IS LIKE THIS.
BUT SUPPOSE THE SURFACE IS DIVIDED INTO LITTLE SECTIONS. NOW THERE ARE SEVERAL WAYS A PHOTON CAN GO FROM THE SOURCE TO YOUR EYE.
A B C D E F G H I J K L M N O P
IT CAN GO TOWARD THIS SECTION, AND THEN BOUNCE OVER TO HERE.
I CAN HEAR YOU SAYING IT AGAIN, BUT I'M NOT CRAZY.
THAT'S WHAT REALLY HAPPENS.
YOU'RE CRAZY!
THE ANGLE AIN'T EQUAL.
I DON'T LIKE IT!
ANOTHER POSSIBILITY IS IT COULD COME HERE OR HERE OR HERE AND GO— OR IT COULD COME WHEREVER YOU'D LIKE IT TO COME, AND GO WHEREVER.
AND IT CAN GO OVER HERE AND GO, AND ON AND ON AND ON TO **ALL** POSSIBILITIES.
A B C D E F G H I J K L M N O P Q R S
THE IDEA IS THERE'S ARROWS—AMPLITUDES —FOR EACH OF THESE PATHS, AND WE USE THEM **ALL** TO FIGURE OUT THE **TOTAL** PROBABILITY IT GETS THERE.

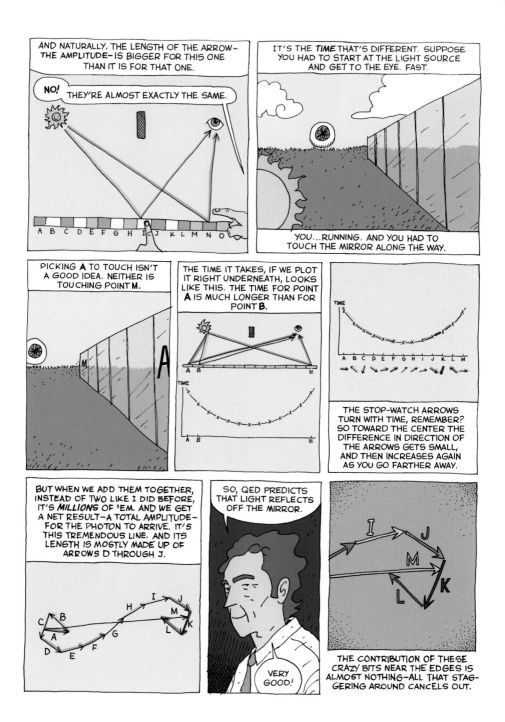

AND NATURALLY, THE LENGTH OF THE ARROW—THE AMPLITUDE—IS BIGGER FOR THIS ONE THAN IT IS FOR THAT ONE.

NO! THEY'RE ALMOST EXACTLY THE SAME.

A B C D E F G H I J K L M N O

IT'S THE *TIME* THAT'S DIFFERENT. SUPPOSE YOU HAD TO START AT THE LIGHT SOURCE AND GET TO THE EYE. FAST.

YOU...RUNNING. AND YOU HAD TO TOUCH THE MIRROR ALONG THE WAY.

PICKING **A** TO TOUCH ISN'T A GOOD IDEA. NEITHER IS TOUCHING POINT **M**.

M A

THE TIME IT TAKES, IF WE PLOT IT RIGHT UNDERNEATH, LOOKS LIKE THIS. THE TIME FOR POINT **A** IS MUCH LONGER THAN FOR POINT **B**.

TIME
A B M

TIME
A B C D E F G H I J K L M

THE STOP-WATCH ARROWS TURN WITH TIME, REMEMBER? SO TOWARD THE CENTER THE DIFFERENCE IN DIRECTION OF THE ARROWS GETS SMALL, AND THEN INCREASES AGAIN AS YOU GO FARTHER AWAY.

BUT WHEN WE ADD THEM TOGETHER, INSTEAD OF TWO LIKE I DID BEFORE, IT'S *MILLIONS* OF 'EM. AND WE GET A NET RESULT—A TOTAL AMPLITUDE—FOR THE PHOTON TO ARRIVE. IT'S THIS TREMENDOUS LINE. AND ITS LENGTH IS MOSTLY MADE UP OF ARROWS D THROUGH J.

C B
A
D E F G H I J M L K

SO, QED PREDICTS THAT LIGHT REFLECTS OFF THE MIRROR.

VERY GOOD!

I J
M
L K

THE CONTRIBUTION OF THESE CRAZY BITS NEAR THE EDGES IS ALMOST NOTHING—ALL THAT STAGGERING AROUND CANCELS OUT.

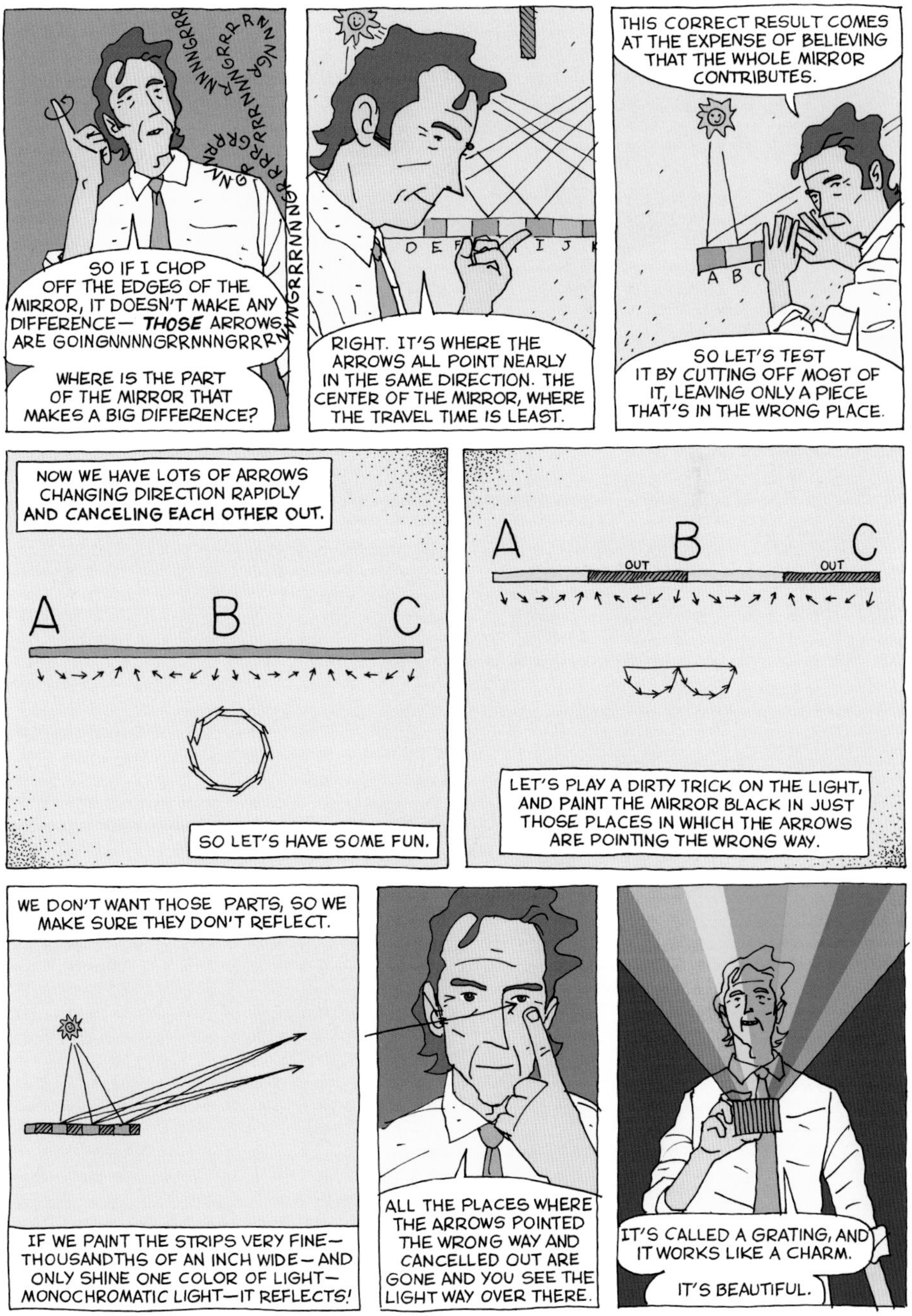
NNNNGRRRNNNGRRRNNNGRRRNNNGRRRNNNGRRRNNNGRR
SO IF I CHOP OFF THE EDGES OF THE MIRROR, IT DOESN'T MAKE ANY DIFFERENCE— **THOSE** ARROWS ARE GOINGNNNNGRRNNNGRRRNNNGRRRNNNGRRR
WHERE IS THE PART OF THE MIRROR THAT MAKES A BIG DIFFERENCE?
D E F I J K
RIGHT. IT'S WHERE THE ARROWS ALL POINT NEARLY IN THE SAME DIRECTION. THE CENTER OF THE MIRROR, WHERE THE TRAVEL TIME IS LEAST.
THIS CORRECT RESULT COMES AT THE EXPENSE OF BELIEVING THAT THE WHOLE MIRROR CONTRIBUTES.
A B C
SO LET'S TEST IT BY CUTTING OFF MOST OF IT, LEAVING ONLY A PIECE THAT'S IN THE WRONG PLACE.
NOW WE HAVE LOTS OF ARROWS CHANGING DIRECTION RAPIDLY AND CANCELING EACH OTHER OUT.
A B C
SO LET'S HAVE SOME FUN.
A B C
OUT
OUT
LET'S PLAY A DIRTY TRICK ON THE LIGHT, AND PAINT THE MIRROR BLACK IN JUST THOSE PLACES IN WHICH THE ARROWS ARE POINTING THE WRONG WAY.
WE DON'T WANT THOSE PARTS, SO WE MAKE SURE THEY DON'T REFLECT.
IF WE PAINT THE STRIPS VERY FINE—THOUSANDTHS OF AN INCH WIDE—AND ONLY SHINE ONE COLOR OF LIGHT—MONOCHROMATIC LIGHT—IT REFLECTS!
ALL THE PLACES WHERE THE ARROWS POINTED THE WRONG WAY AND CANCELLED OUT ARE GONE AND YOU SEE THE LIGHT WAY OVER THERE.
IT'S CALLED A GRATING, AND IT WORKS LIKE A CHARM.
IT'S BEAUTIFUL.

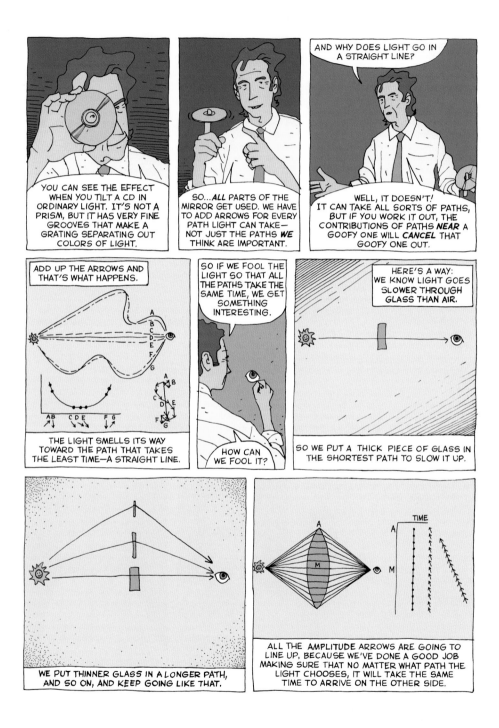

YOU CAN SEE THE EFFECT WHEN YOU TILT A CD IN ORDINARY LIGHT. IT'S NOT A PRISM, BUT IT HAS VERY FINE GROOVES THAT MAKE A GRATING SEPARATING OUT COLORS OF LIGHT.

SO...*ALL* PARTS OF THE MIRROR GET USED. WE HAVE TO ADD ARROWS FOR EVERY PATH LIGHT CAN TAKE—NOT JUST THE PATHS *WE* THINK ARE IMPORTANT.

AND WHY DOES LIGHT GO IN A STRAIGHT LINE?

WELL, IT DOESN'T! IT CAN TAKE ALL SORTS OF PATHS, BUT IF YOU WORK IT OUT, THE CONTRIBUTIONS OF PATHS *NEAR* A GOOFY ONE WILL *CANCEL* THAT GOOFY ONE OUT.

ADD UP THE ARROWS AND THAT'S WHAT HAPPENS.

THE LIGHT SMELLS ITS WAY TOWARD THE PATH THAT TAKES THE LEAST TIME—A STRAIGHT LINE.

SO IF WE FOOL THE LIGHT SO THAT ALL THE PATHS TAKE THE SAME TIME, WE GET SOMETHING INTERESTING.

HOW CAN WE FOOL IT?

HERE'S A WAY: WE KNOW LIGHT GOES SLOWER THROUGH GLASS THAN AIR.

SO WE PUT A THICK PIECE OF GLASS IN THE SHORTEST PATH TO SLOW IT UP.

WE PUT THINNER GLASS IN A *LONGER* PATH, AND SO ON, AND KEEP GOING LIKE THAT.

ALL THE AMPLITUDE ARROWS ARE GOING TO LINE UP, BECAUSE WE'VE DONE A GOOD JOB MAKING SURE THAT NO MATTER WHAT PATH THE LIGHT CHOOSES, IT WILL TAKE THE SAME TIME TO ARRIVE ON THE OTHER SIDE.

210

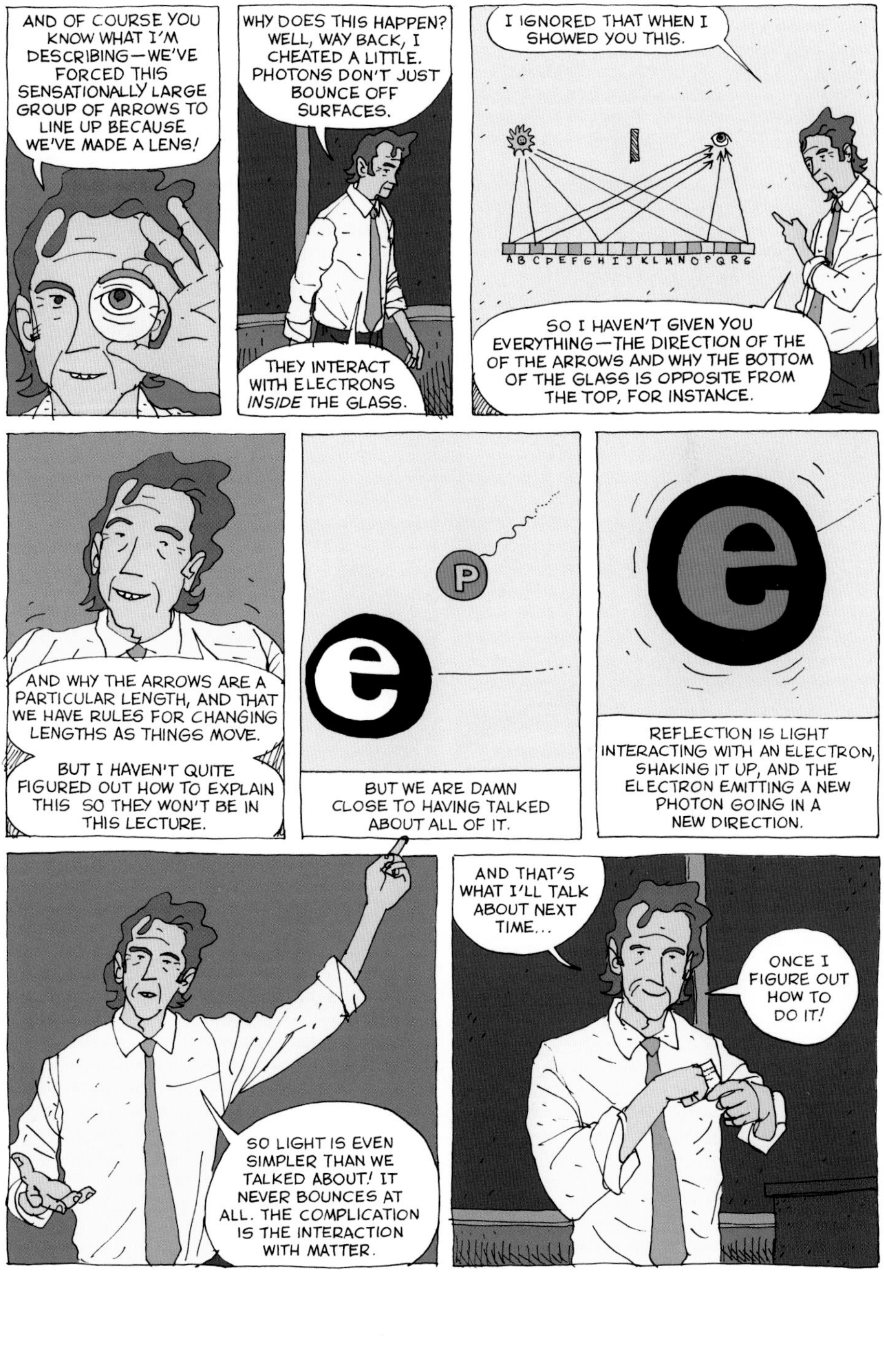
AND OF COURSE YOU KNOW WHAT I'M DESCRIBING—WE'VE FORCED THIS SENSATIONALLY LARGE GROUP OF ARROWS TO LINE UP BECAUSE WE'VE MADE A LENS!
WHY DOES THIS HAPPEN? WELL, WAY BACK, I CHEATED A LITTLE. PHOTONS DON'T JUST BOUNCE OFF SURFACES.
THEY INTERACT WITH ELECTRONS *INSIDE* THE GLASS.
I IGNORED THAT WHEN I SHOWED YOU THIS.
A B C D E F G H I J K L M N O P Q R S
SO I HAVEN'T GIVEN YOU EVERYTHING—THE DIRECTION OF THE OF THE ARROWS AND WHY THE BOTTOM OF THE GLASS IS OPPOSITE FROM THE TOP, FOR INSTANCE.
AND WHY THE ARROWS ARE A PARTICULAR LENGTH, AND THAT WE HAVE RULES FOR CHANGING LENGTHS AS THINGS MOVE.
BUT I HAVEN'T QUITE FIGURED OUT HOW TO EXPLAIN THIS SO THEY WON'T BE IN THIS LECTURE.
P
e
BUT WE ARE DAMN CLOSE TO HAVING TALKED ABOUT ALL OF IT.
e
REFLECTION IS LIGHT INTERACTING WITH AN ELECTRON, SHAKING IT UP, AND THE ELECTRON EMITTING A NEW PHOTON GOING IN A NEW DIRECTION.
SO LIGHT IS EVEN SIMPLER THAN WE TALKED ABOUT! IT NEVER BOUNCES AT ALL. THE COMPLICATION IS THE INTERACTION WITH MATTER.
AND THAT'S WHAT I'LL TALK ABOUT NEXT TIME...
ONCE I FIGURE OUT HOW TO DO IT!

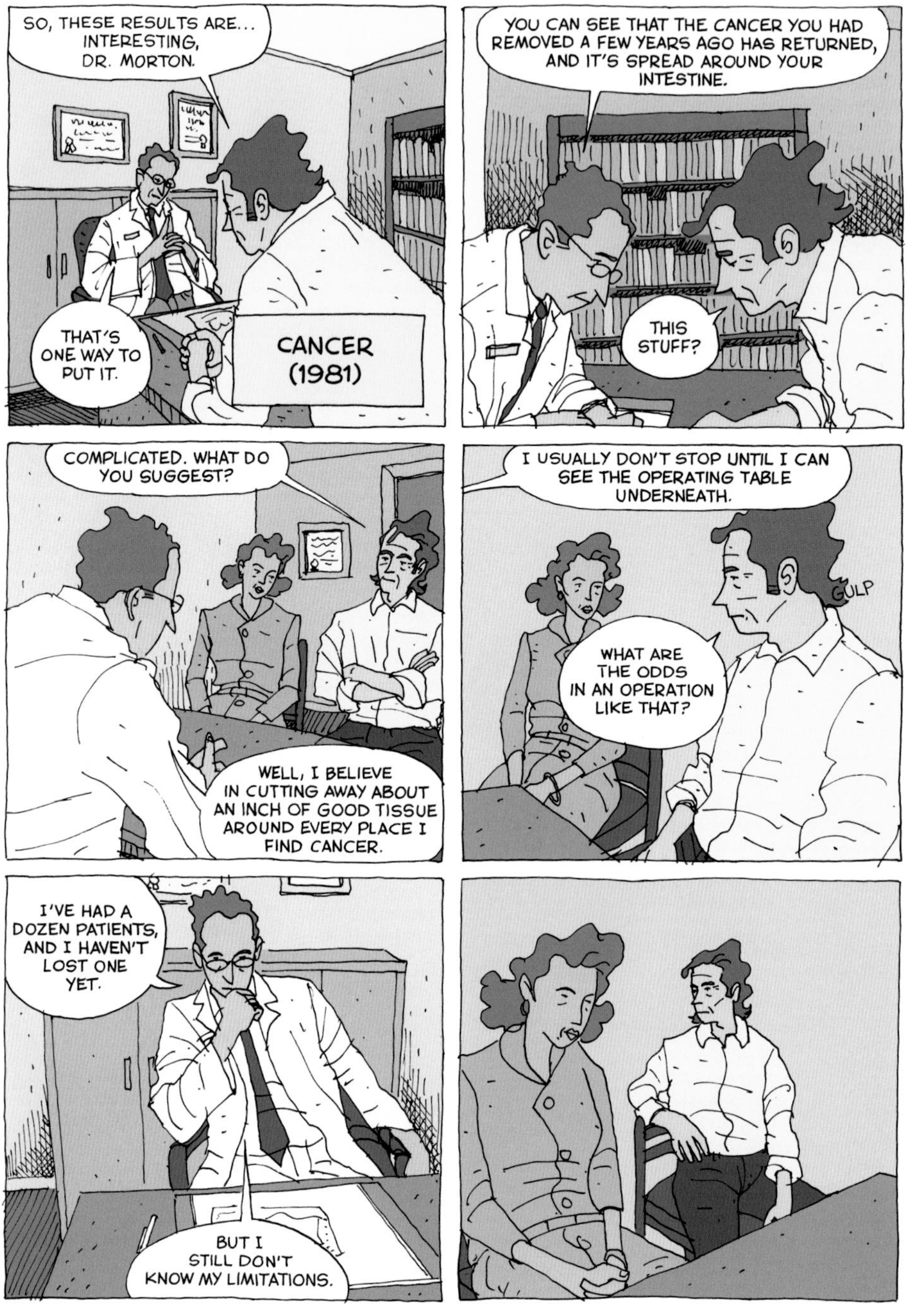
SO, THESE RESULTS ARE... INTERESTING, DR. MORTON.
THAT'S ONE WAY TO PUT IT.
CANCER (1981)
YOU CAN SEE THAT THE CANCER YOU HAD REMOVED A FEW YEARS AGO HAS RETURNED, AND IT'S SPREAD AROUND YOUR INTESTINE.
THIS STUFF?
COMPLICATED. WHAT DO YOU SUGGEST?
WELL, I BELIEVE IN CUTTING AWAY ABOUT AN INCH OF GOOD TISSUE AROUND EVERY PLACE I FIND CANCER.
I USUALLY DON'T STOP UNTIL I CAN SEE THE OPERATING TABLE UNDERNEATH.
GULP
WHAT ARE THE ODDS IN AN OPERATION LIKE THAT?
I'VE HAD A DOZEN PATIENTS, AND I HAVEN'T LOST ONE YET.
BUT I STILL DON'T KNOW MY LIMITATIONS.

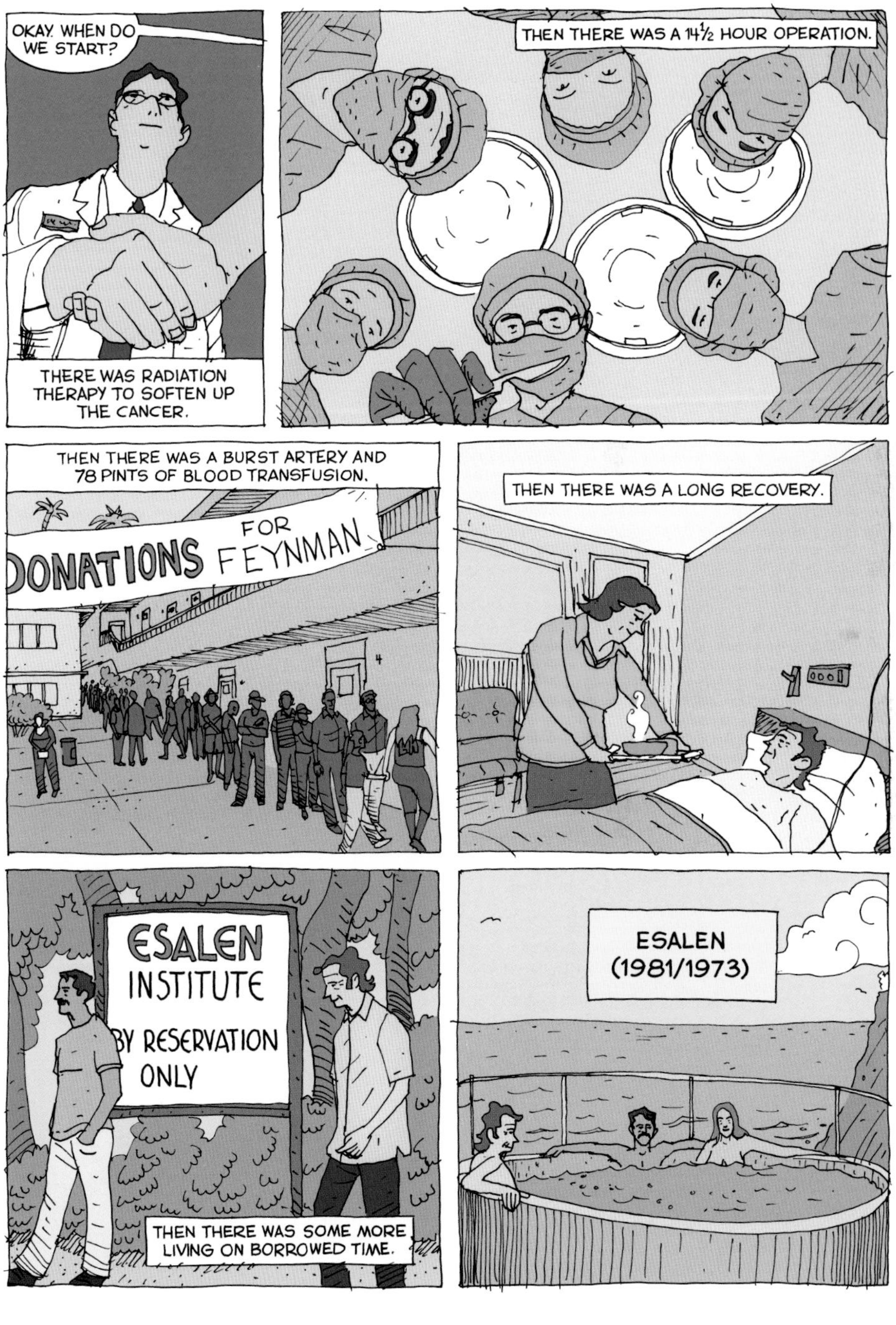
OKAY. WHEN DO WE START?
THERE WAS RADIATION THERAPY TO SOFTEN UP THE CANCER.
THEN THERE WAS A 14½ HOUR OPERATION.
THEN THERE WAS A BURST ARTERY AND 78 PINTS OF BLOOD TRANSFUSION.
DONATIONS FOR FEYNMAN
THEN THERE WAS A LONG RECOVERY.
ESALEN INSTITUTE BY RESERVATION ONLY
THEN THERE WAS SOME MORE LIVING ON BORROWED TIME.
ESALEN (1981/1973)

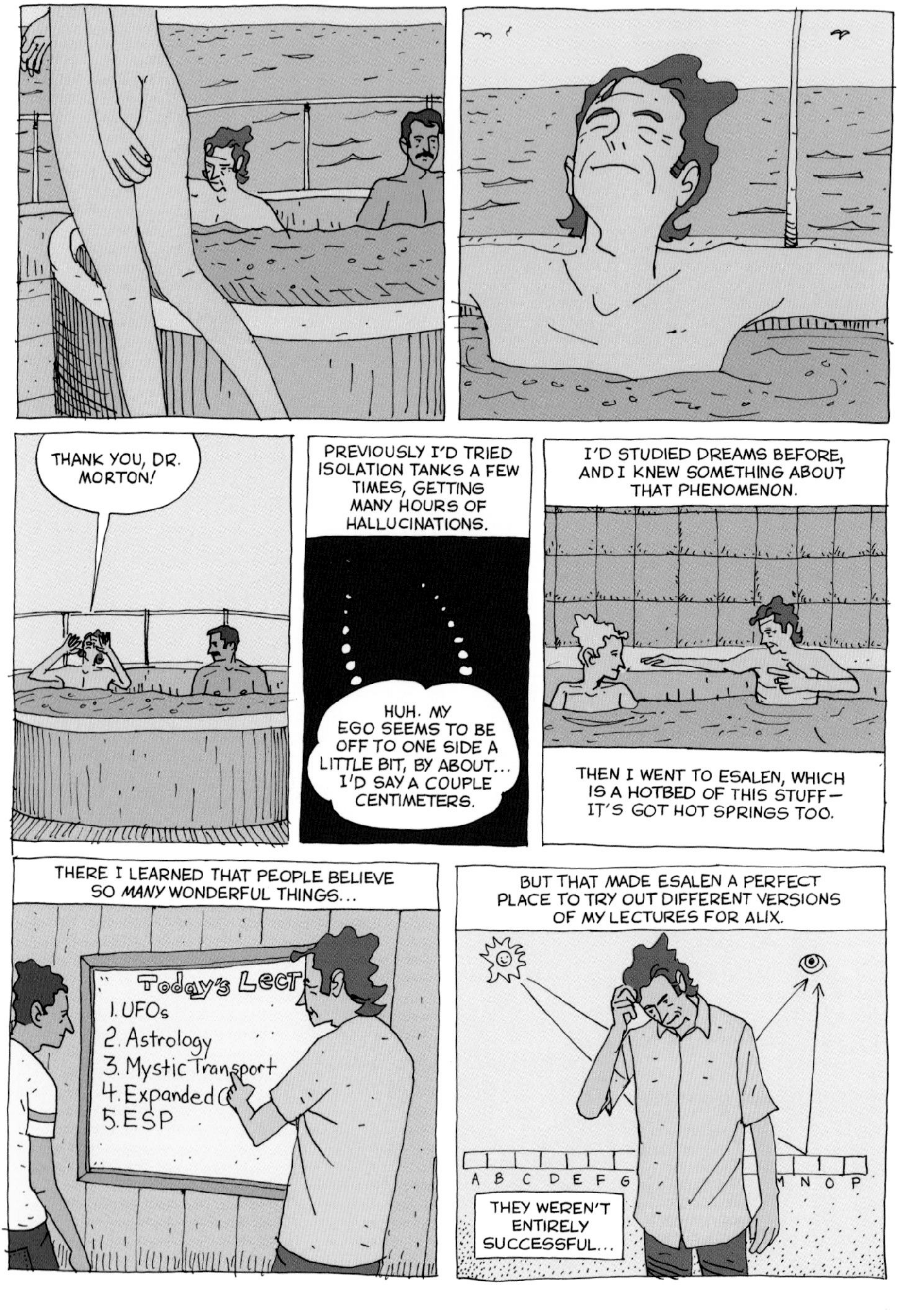
THANK YOU, DR. MORTON!
PREVIOUSLY I'D TRIED ISOLATION TANKS A FEW TIMES, GETTING MANY HOURS OF HALLUCINATIONS.
HUH. MY EGO SEEMS TO BE OFF TO ONE SIDE A LITTLE BIT, BY ABOUT... I'D SAY A COUPLE CENTIMETERS.
I'D STUDIED DREAMS BEFORE, AND I KNEW SOMETHING ABOUT THAT PHENOMENON.
THEN I WENT TO ESALEN, WHICH IS A HOTBED OF THIS STUFF— IT'S GOT HOT SPRINGS TOO.
THERE I LEARNED THAT PEOPLE BELIEVE SO MANY WONDERFUL THINGS...
Today's Lect
1. UFOs
2. Astrology
3. Mystic Transport
4. Expanded C
5. ESP
BUT THAT MADE ESALEN A PERFECT PLACE TO TRY OUT DIFFERENT VERSIONS OF MY LECTURES FOR ALIX.
A B C D E F G
M N O P
THEY WEREN'T ENTIRELY SUCCESSFUL...

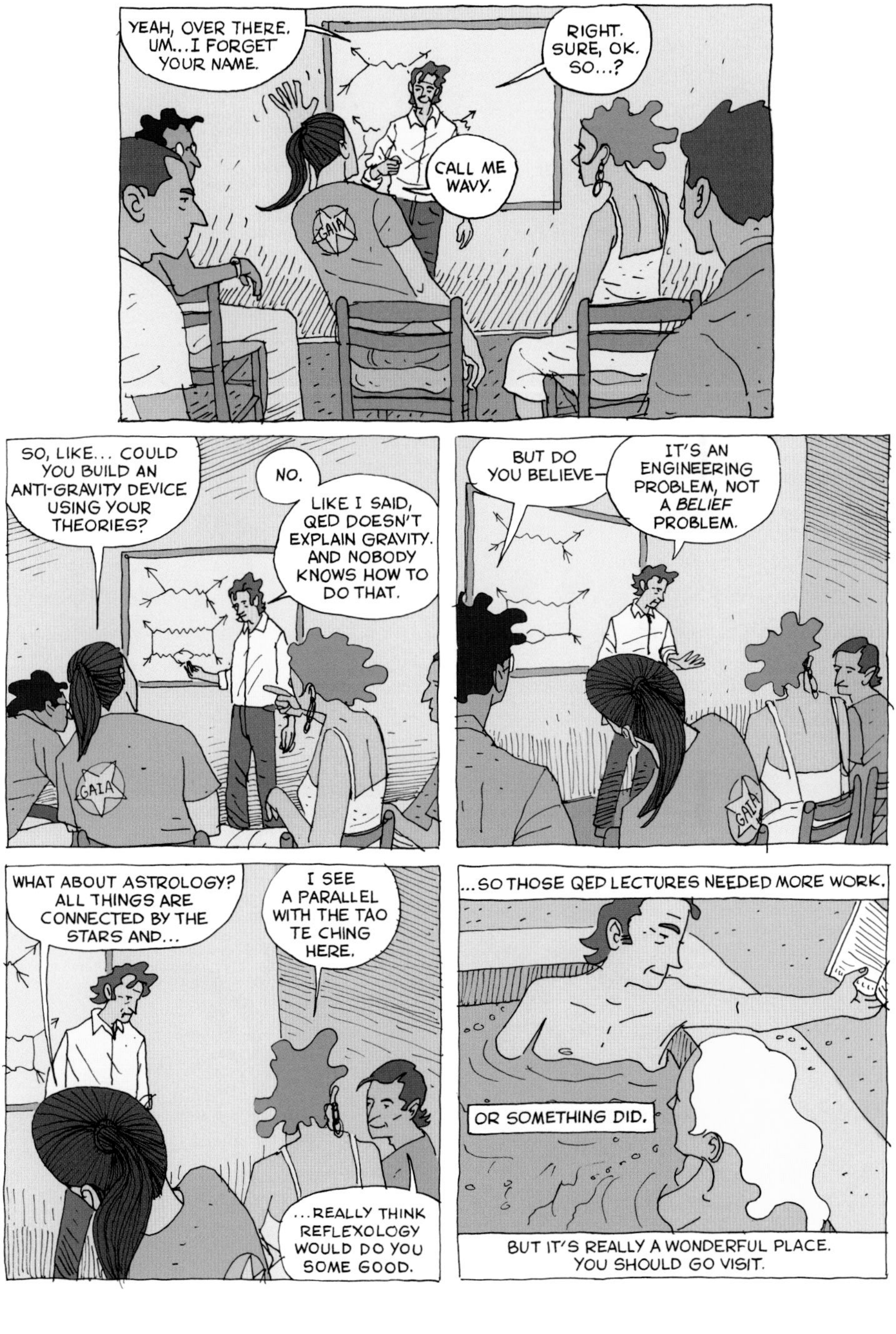
YEAH, OVER THERE. UM...I FORGET YOUR NAME.
RIGHT. SURE, OK. SO...?
CALL ME WAVY.
GAIA
SO, LIKE... COULD YOU BUILD AN ANTI-GRAVITY DEVICE USING YOUR THEORIES?
NO.
LIKE I SAID, QED DOESN'T EXPLAIN GRAVITY. AND NOBODY KNOWS HOW TO DO THAT.
GAIA
BUT DO YOU BELIEVE—
IT'S AN ENGINEERING PROBLEM, NOT A *BELIEF* PROBLEM.
GAIA
WHAT ABOUT ASTROLOGY? ALL THINGS ARE CONNECTED BY THE STARS AND...
I SEE A PARALLEL WITH THE TAO TE CHING HERE.
...REALLY THINK REFLEXOLOGY WOULD DO YOU SOME GOOD.
...SO THOSE QED LECTURES NEEDED MORE WORK.
OR SOMETHING DID.
BUT IT'S REALLY A WONDERFUL PLACE. YOU SHOULD GO VISIT.

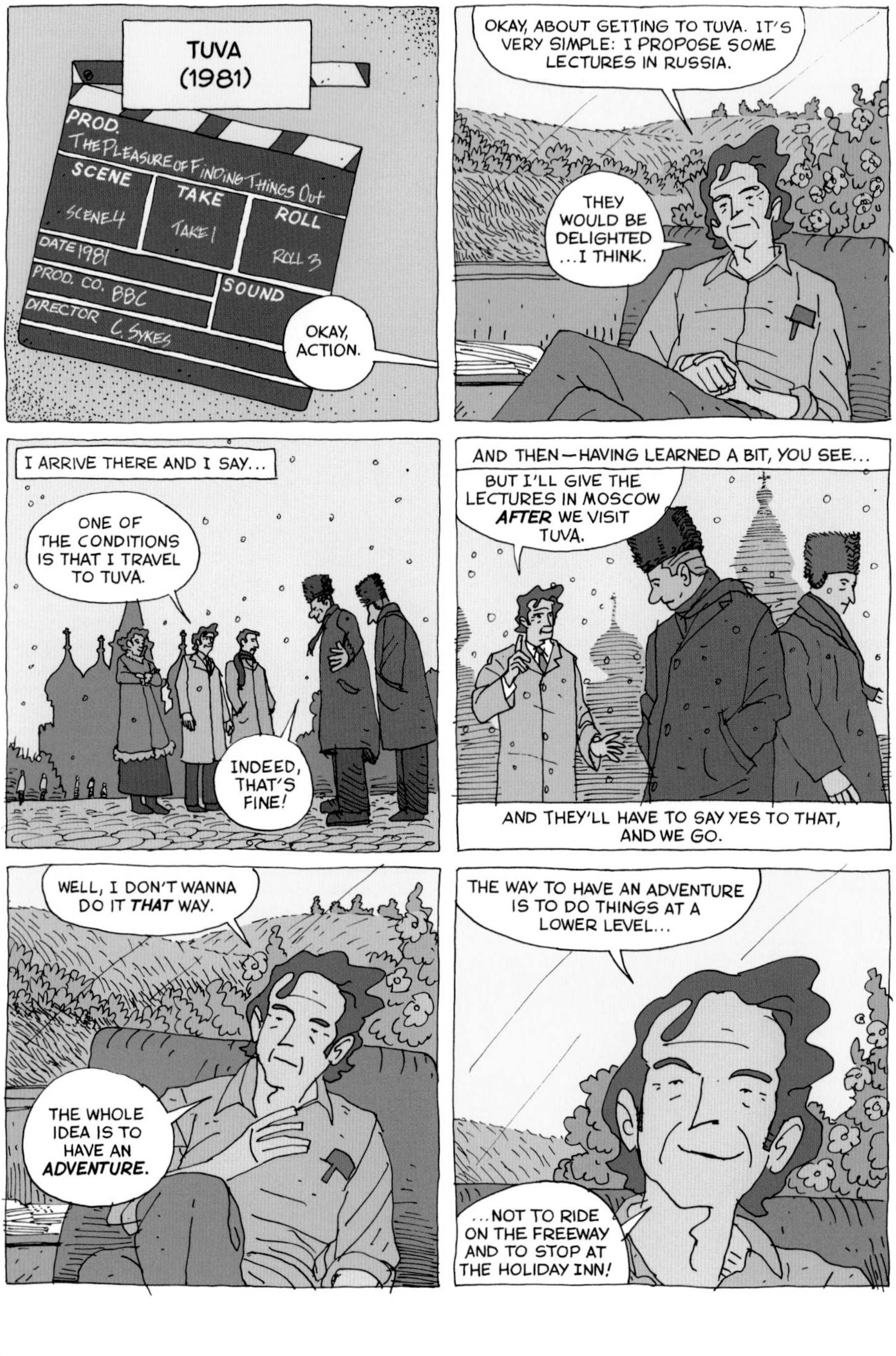
TUVA (1981)
PROD. THE PLEASURE OF FINDING THINGS OUT
SCENE
SCENE 4
TAKE
TAKE 1
ROLL
ROLL 3
DATE 1981
PROD. CO. BBC
SOUND
DIRECTOR C. SYKES
OKAY, ACTION.
OKAY, ABOUT GETTING TO TUVA. IT'S VERY SIMPLE: I PROPOSE SOME LECTURES IN RUSSIA.
THEY WOULD BE DELIGHTED ...I THINK.
I ARRIVE THERE AND I SAY...
ONE OF THE CONDITIONS IS THAT I TRAVEL TO TUVA.
INDEED, THAT'S FINE!
AND THEN—HAVING LEARNED A BIT, YOU SEE...
BUT I'LL GIVE THE LECTURES IN MOSCOW AFTER WE VISIT TUVA.
AND THEY'LL HAVE TO SAY YES TO THAT, AND WE GO.
WELL, I DON'T WANNA DO IT THAT WAY.
THE WHOLE IDEA IS TO HAVE AN ADVENTURE.
THE WAY TO HAVE AN ADVENTURE IS TO DO THINGS AT A LOWER LEVEL...
...NOT TO RIDE ON THE FREEWAY AND TO STOP AT THE HOLIDAY INN!

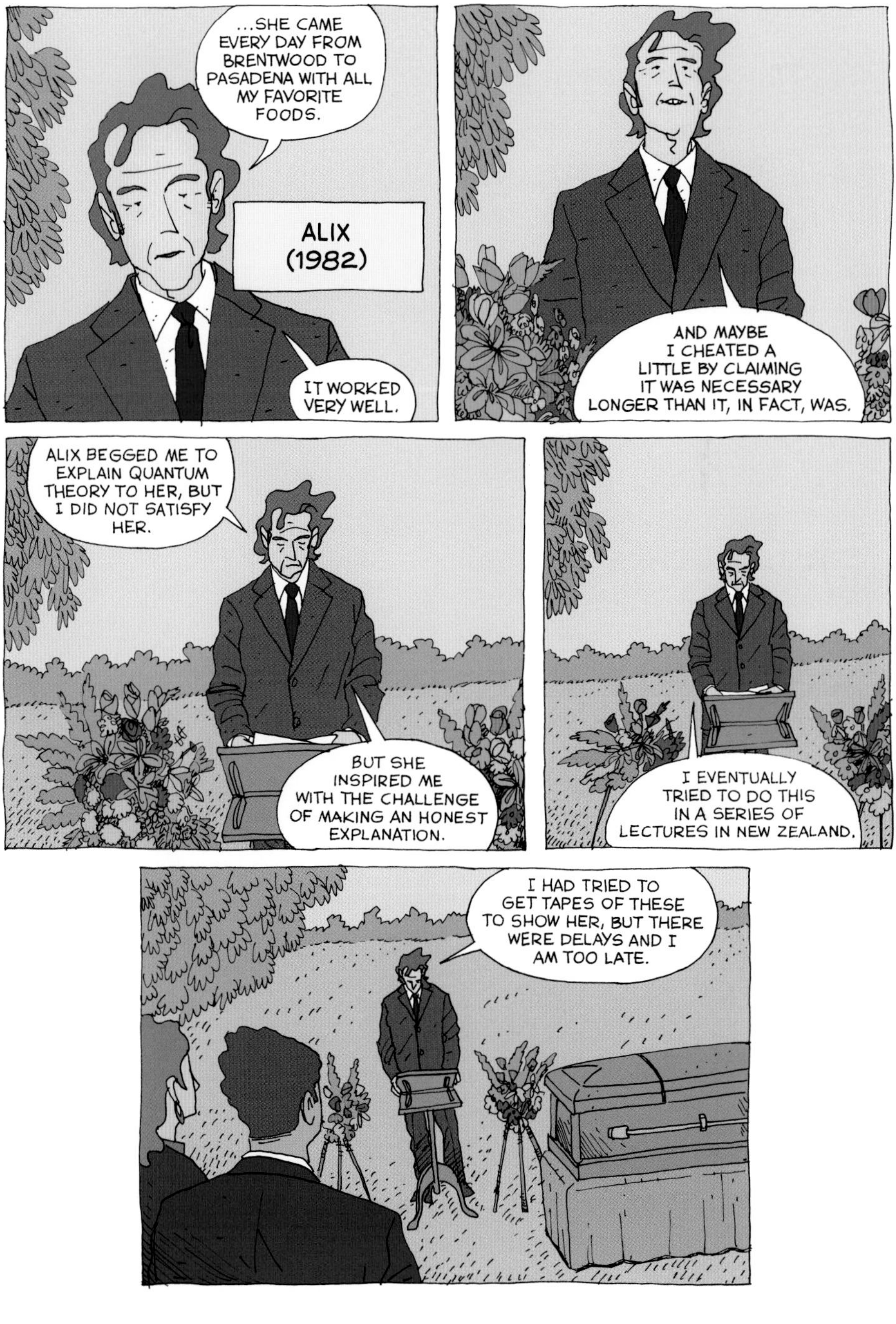
...SHE CAME EVERY DAY FROM BRENTWOOD TO PASADENA WITH ALL MY FAVORITE FOODS.
ALIX (1982)
IT WORKED VERY WELL.
AND MAYBE I CHEATED A LITTLE BY CLAIMING IT WAS NECESSARY LONGER THAN IT, IN FACT, WAS.
ALIX BEGGED ME TO EXPLAIN QUANTUM THEORY TO HER, BUT I DID NOT SATISFY HER.
BUT SHE INSPIRED ME WITH THE CHALLENGE OF MAKING AN HONEST EXPLANATION.
I EVENTUALLY TRIED TO DO THIS IN A SERIES OF LECTURES IN NEW ZEALAND.
I HAD TRIED TO GET TAPES OF THESE TO SHOW HER, BUT THERE WERE DELAYS AND I AM TOO LATE.

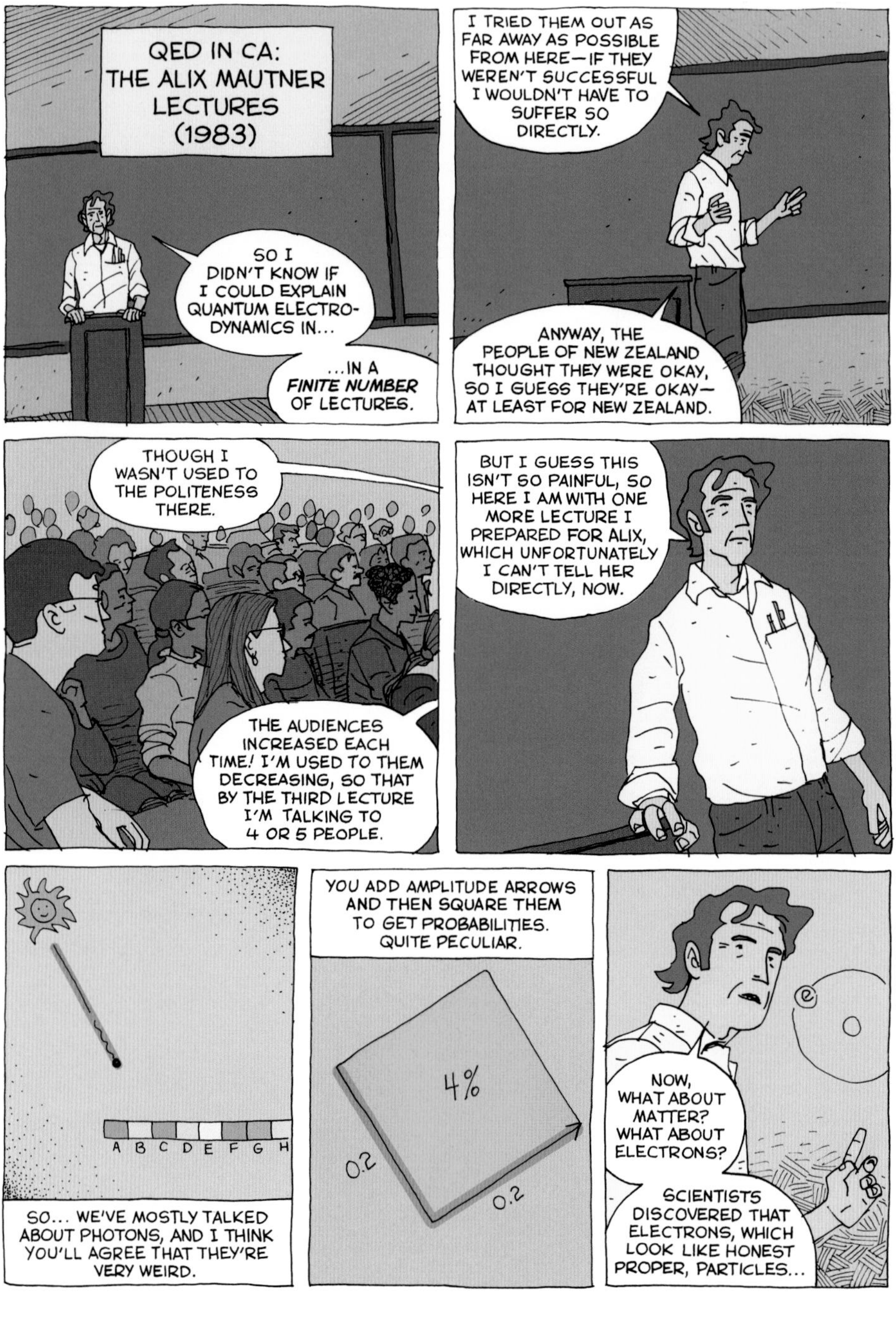

QED IN CA: THE ALIX MAUTNER LECTURES (1983)
SO I DIDN'T KNOW IF I COULD EXPLAIN QUANTUM ELECTRO-DYNAMICS IN...
...IN A *FINITE NUMBER* OF LECTURES.
I TRIED THEM OUT AS FAR AWAY AS POSSIBLE FROM HERE—IF THEY WEREN'T SUCCESSFUL I WOULDN'T HAVE TO SUFFER SO DIRECTLY.
ANYWAY, THE PEOPLE OF NEW ZEALAND THOUGHT THEY WERE OKAY, SO I GUESS THEY'RE OKAY—AT LEAST FOR NEW ZEALAND.
THOUGH I WASN'T USED TO THE POLITENESS THERE.
THE AUDIENCES INCREASED EACH TIME! I'M USED TO THEM DECREASING, SO THAT BY THE THIRD LECTURE I'M TALKING TO 4 OR 5 PEOPLE.
BUT I GUESS THIS ISN'T SO PAINFUL, SO HERE I AM WITH ONE MORE LECTURE I PREPARED FOR ALIX, WHICH UNFORTUNATELY I CAN'T TELL HER DIRECTLY, NOW.
A B C D E F G H
SO... WE'VE MOSTLY TALKED ABOUT PHOTONS, AND I THINK YOU'LL AGREE THAT THEY'RE VERY WEIRD.
YOU ADD AMPLITUDE ARROWS AND THEN SQUARE THEM TO GET PROBABILITIES. QUITE PECULIAR.
4%
0.2
0.2
e
NOW, WHAT ABOUT MATTER? WHAT ABOUT ELECTRONS?
SCIENTISTS DISCOVERED THAT ELECTRONS, WHICH LOOK LIKE HONEST PROPER, PARTICLES...

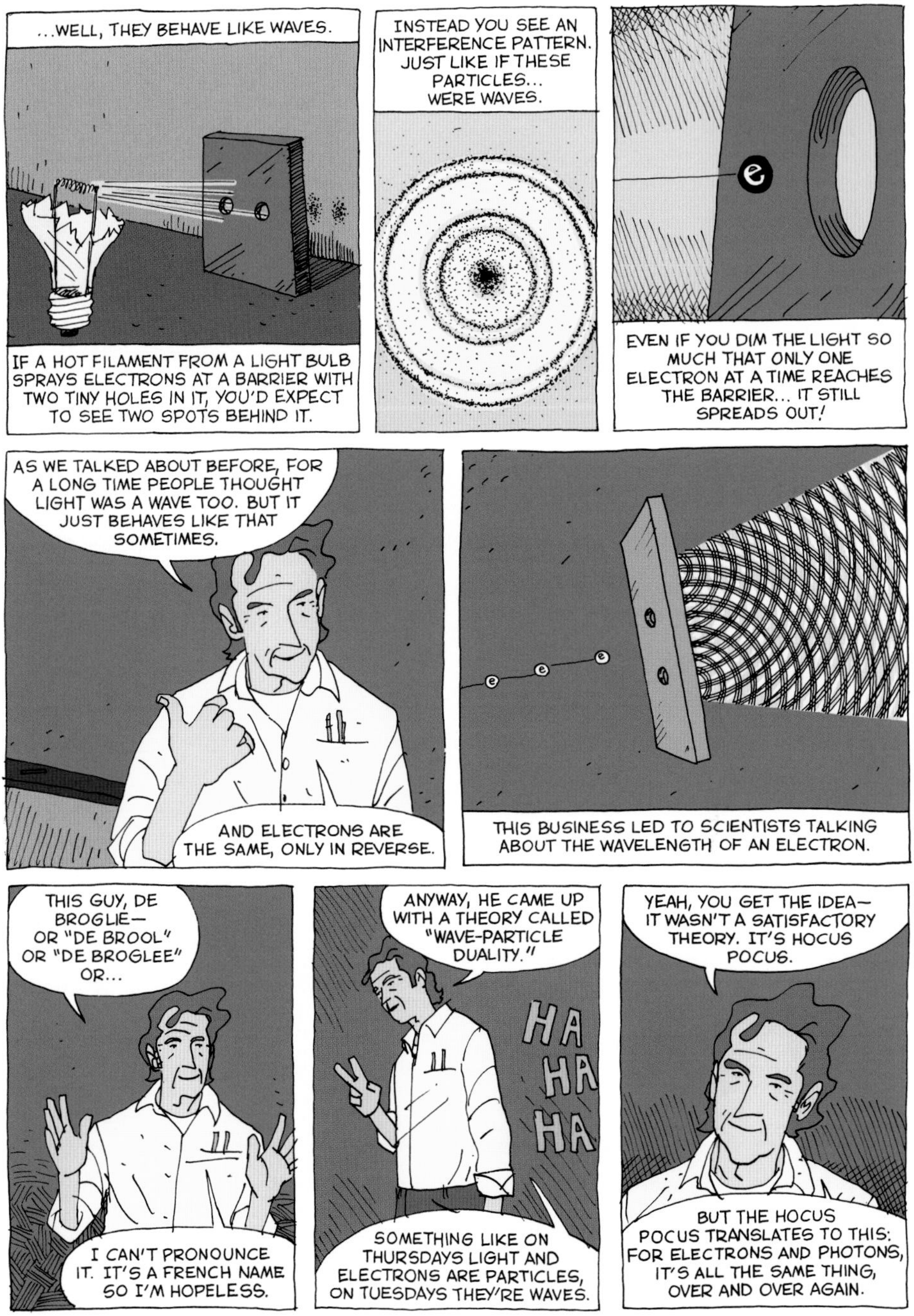

...WELL, THEY BEHAVE LIKE WAVES.
IF A HOT FILAMENT FROM A LIGHT BULB SPRAYS ELECTRONS AT A BARRIER WITH TWO TINY HOLES IN IT, YOU'D EXPECT TO SEE TWO SPOTS BEHIND IT.
INSTEAD YOU SEE AN INTERFERENCE PATTERN. JUST LIKE IF THESE PARTICLES... WERE WAVES.
EVEN IF YOU DIM THE LIGHT SO MUCH THAT ONLY ONE ELECTRON AT A TIME REACHES THE BARRIER... IT STILL SPREADS OUT!
AS WE TALKED ABOUT BEFORE, FOR A LONG TIME PEOPLE THOUGHT LIGHT WAS A WAVE TOO. BUT IT JUST BEHAVES LIKE THAT SOMETIMES.
AND ELECTRONS ARE THE SAME, ONLY IN REVERSE.
THIS BUSINESS LED TO SCIENTISTS TALKING ABOUT THE WAVELENGTH OF AN ELECTRON.
THIS GUY, DE BROGLIE— OR "DE BROOL" OR "DE BROGLEE" OR...
I CAN'T PRONOUNCE IT. IT'S A FRENCH NAME SO I'M HOPELESS.
ANYWAY, HE CAME UP WITH A THEORY CALLED "WAVE-PARTICLE DUALITY."
HA HA HA
SOMETHING LIKE ON THURSDAYS LIGHT AND ELECTRONS ARE PARTICLES, ON TUESDAYS THEY'RE WAVES.
YEAH, YOU GET THE IDEA— IT WASN'T A SATISFACTORY THEORY. IT'S HOCUS POCUS.
BUT THE HOCUS POCUS TRANSLATES TO THIS: FOR ELECTRONS AND PHOTONS, IT'S ALL THE SAME THING, OVER AND OVER AGAIN.

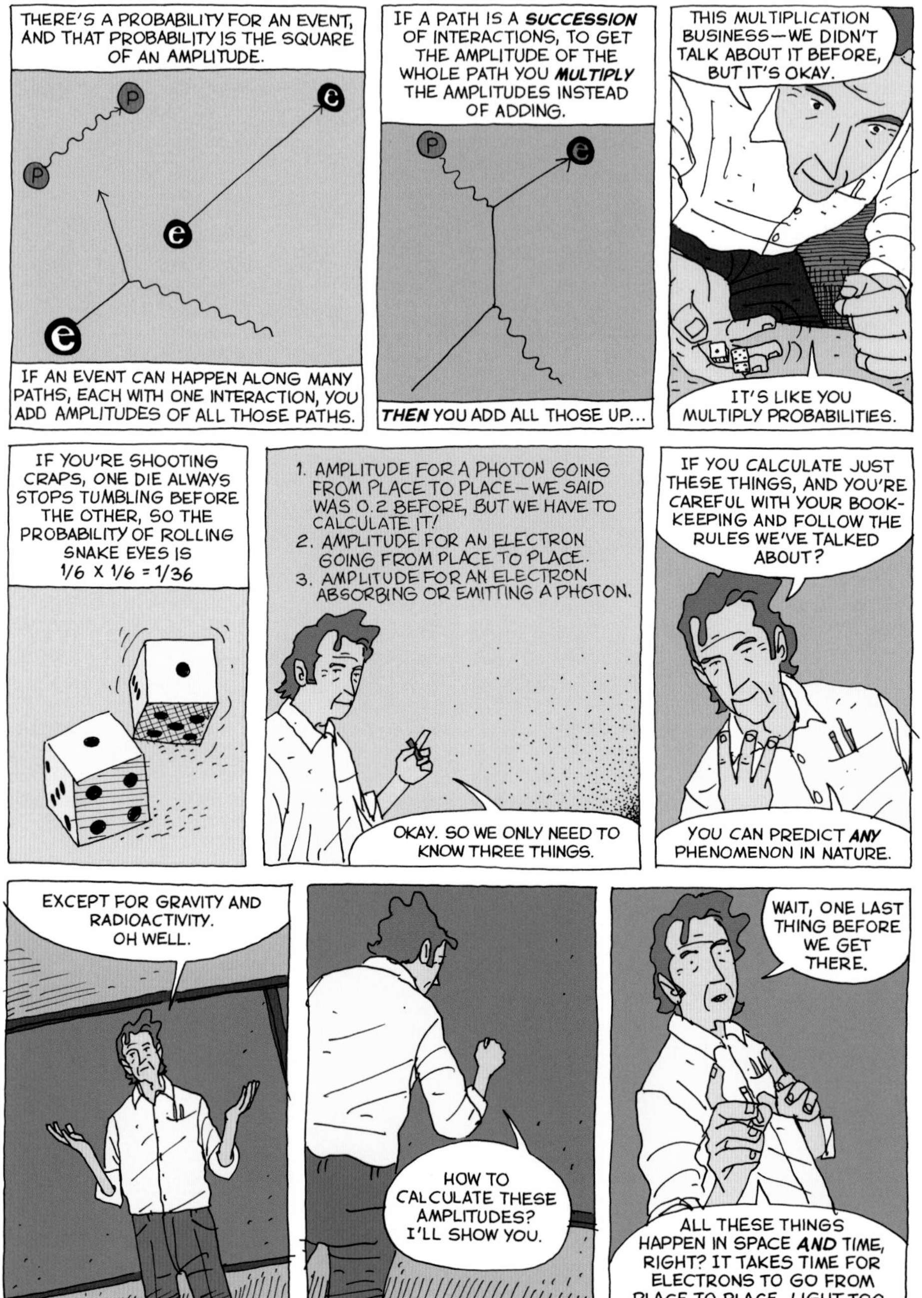
THERE'S A PROBABILITY FOR AN EVENT, AND THAT PROBABILITY IS THE SQUARE OF AN AMPLITUDE.
P
P
e
e
e
IF AN EVENT CAN HAPPEN ALONG MANY PATHS, EACH WITH ONE INTERACTION, YOU ADD AMPLITUDES OF ALL THOSE PATHS.
IF A PATH IS A **SUCCESSION** OF INTERACTIONS, TO GET THE AMPLITUDE OF THE WHOLE PATH YOU **MULTIPLY** THE AMPLITUDES INSTEAD OF ADDING.
P
e
THEN YOU ADD ALL THOSE UP...
THIS MULTIPLICATION BUSINESS—WE DIDN'T TALK ABOUT IT BEFORE, BUT IT'S OKAY.
IT'S LIKE YOU MULTIPLY PROBABILITIES.
IF YOU'RE SHOOTING CRAPS, ONE DIE ALWAYS STOPS TUMBLING BEFORE THE OTHER, SO THE PROBABILITY OF ROLLING SNAKE EYES IS 1/6 X 1/6 = 1/36
1. AMPLITUDE FOR A PHOTON GOING FROM PLACE TO PLACE—WE SAID WAS 0.2 BEFORE, BUT WE HAVE TO CALCULATE IT!
2. AMPLITUDE FOR AN ELECTRON GOING FROM PLACE TO PLACE.
3. AMPLITUDE FOR AN ELECTRON ABSORBING OR EMITTING A PHOTON.
OKAY. SO WE ONLY NEED TO KNOW THREE THINGS.
IF YOU CALCULATE JUST THESE THINGS, AND YOU'RE CAREFUL WITH YOUR BOOK-KEEPING AND FOLLOW THE RULES WE'VE TALKED ABOUT?
YOU CAN PREDICT **ANY** PHENOMENON IN NATURE.
EXCEPT FOR GRAVITY AND RADIOACTIVITY. OH WELL.
HOW TO CALCULATE THESE AMPLITUDES? I'LL SHOW YOU.
WAIT, ONE LAST THING BEFORE WE GET THERE.
ALL THESE THINGS HAPPEN IN SPACE **AND** TIME, RIGHT? IT TAKES TIME FOR ELECTRONS TO GO FROM PLACE TO PLACE. LIGHT TOO.

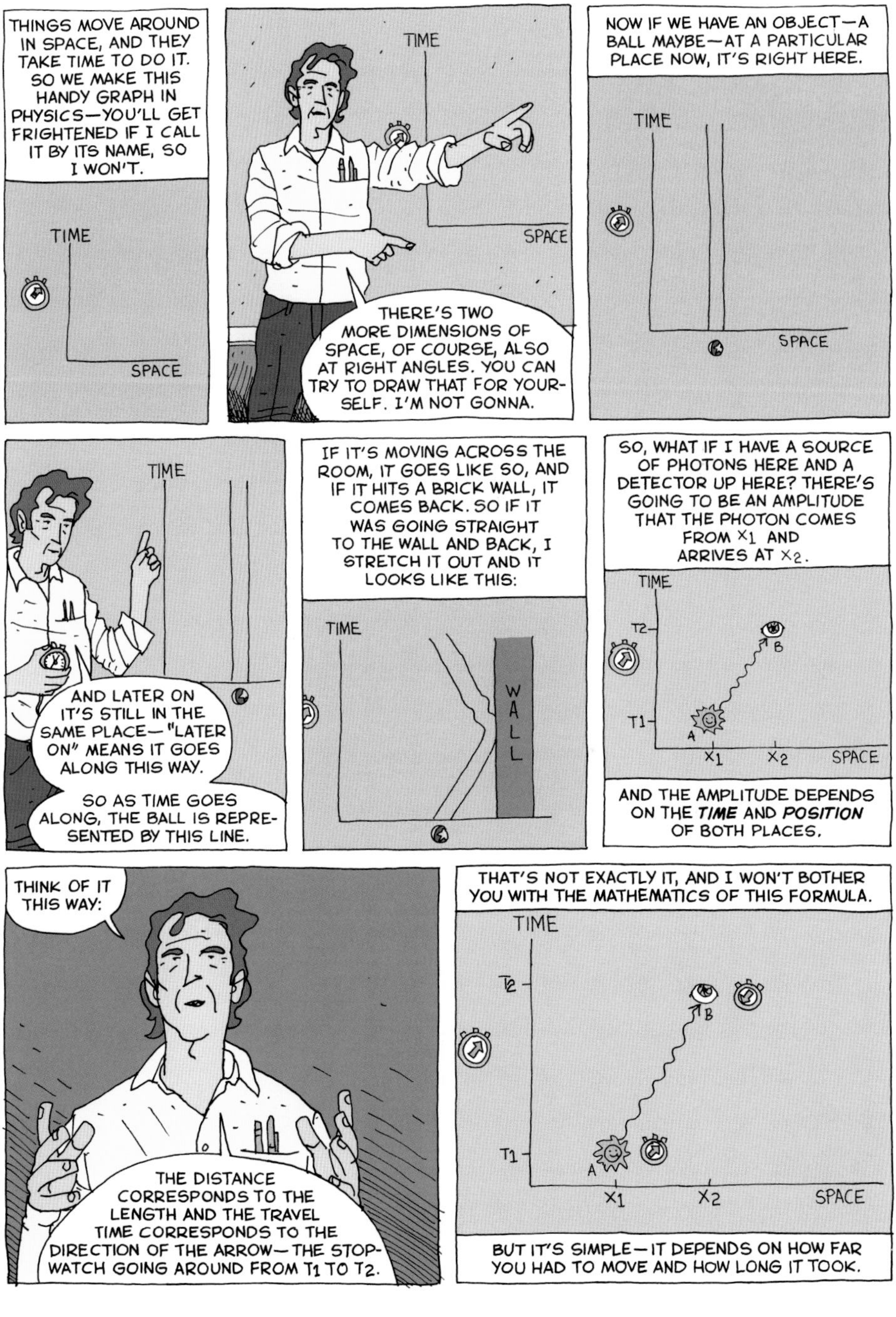
THINGS MOVE AROUND IN SPACE, AND THEY TAKE TIME TO DO IT. SO WE MAKE THIS HANDY GRAPH IN PHYSICS—YOU'LL GET FRIGHTENED IF I CALL IT BY ITS NAME, SO I WON'T.
TIME
SPACE
TIME
SPACE
THERE'S TWO MORE DIMENSIONS OF SPACE, OF COURSE, ALSO AT RIGHT ANGLES. YOU CAN TRY TO DRAW THAT FOR YOURSELF. I'M NOT GONNA.
NOW IF WE HAVE AN OBJECT—A BALL MAYBE—AT A PARTICULAR PLACE NOW, IT'S RIGHT HERE.
TIME
SPACE
TIME
AND LATER ON IT'S STILL IN THE SAME PLACE—"LATER ON" MEANS IT GOES ALONG THIS WAY.
SO AS TIME GOES ALONG, THE BALL IS REPRESENTED BY THIS LINE.
IF IT'S MOVING ACROSS THE ROOM, IT GOES LIKE SO, AND IF IT HITS A BRICK WALL, IT COMES BACK. SO IF IT WAS GOING STRAIGHT TO THE WALL AND BACK, I STRETCH IT OUT AND IT LOOKS LIKE THIS:
TIME
WALL
SO, WHAT IF I HAVE A SOURCE OF PHOTONS HERE AND A DETECTOR UP HERE? THERE'S GOING TO BE AN AMPLITUDE THAT THE PHOTON COMES FROM X1 AND ARRIVES AT X2.
TIME
T2
T1
A
B
X1
X2
SPACE
AND THE AMPLITUDE DEPENDS ON THE TIME AND POSITION OF BOTH PLACES.
THINK OF IT THIS WAY:
THE DISTANCE CORRESPONDS TO THE LENGTH AND THE TRAVEL TIME CORRESPONDS TO THE DIRECTION OF THE ARROW—THE STOPWATCH GOING AROUND FROM T1 TO T2.
THAT'S NOT EXACTLY IT, AND I WON'T BOTHER YOU WITH THE MATHEMATICS OF THIS FORMULA.
TIME
T2
T1
A
B
X1
X2
SPACE
BUT IT'S SIMPLE—IT DEPENDS ON HOW FAR YOU HAD TO MOVE AND HOW LONG IT TOOK.

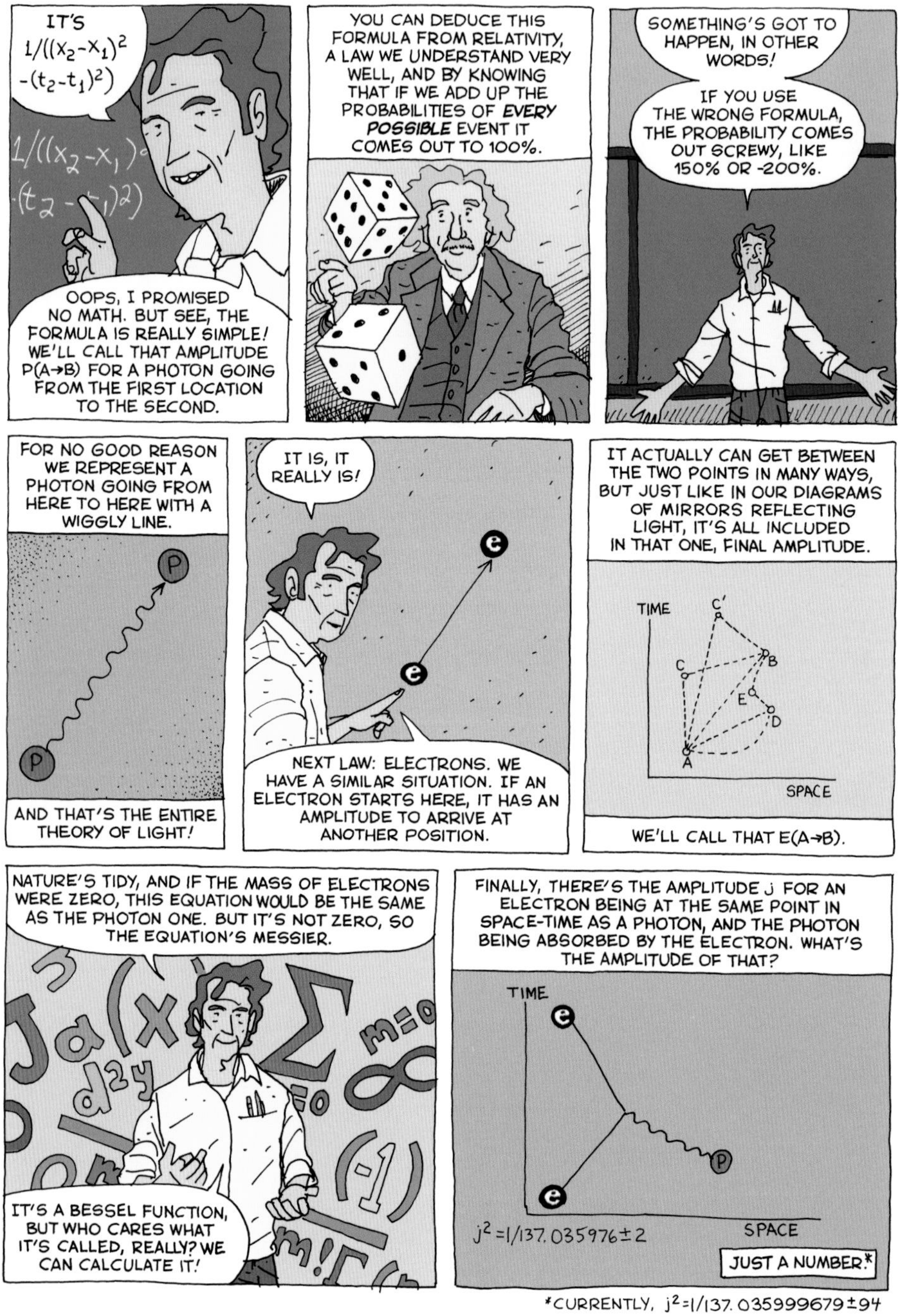
IT'S 1/((X2-X1)2 -(t2-t1)2)
OOPS, I PROMISED NO MATH. BUT SEE, THE FORMULA IS REALLY SIMPLE! WE'LL CALL THAT AMPLITUDE P(A→B) FOR A PHOTON GOING FROM THE FIRST LOCATION TO THE SECOND.
YOU CAN DEDUCE THIS FORMULA FROM RELATIVITY, A LAW WE UNDERSTAND VERY WELL, AND BY KNOWING THAT IF WE ADD UP THE PROBABILITIES OF EVERY POSSIBLE EVENT IT COMES OUT TO 100%.
SOMETHING'S GOT TO HAPPEN, IN OTHER WORDS!
IF YOU USE THE WRONG FORMULA, THE PROBABILITY COMES OUT SCREWY, LIKE 150% OR -200%.
FOR NO GOOD REASON WE REPRESENT A PHOTON GOING FROM HERE TO HERE WITH A WIGGLY LINE.
AND THAT'S THE ENTIRE THEORY OF LIGHT!
IT IS, IT REALLY IS!
NEXT LAW: ELECTRONS. WE HAVE A SIMILAR SITUATION. IF AN ELECTRON STARTS HERE, IT HAS AN AMPLITUDE TO ARRIVE AT ANOTHER POSITION.
IT ACTUALLY CAN GET BETWEEN THE TWO POINTS IN MANY WAYS, BUT JUST LIKE IN OUR DIAGRAMS OF MIRRORS REFLECTING LIGHT, IT'S ALL INCLUDED IN THAT ONE, FINAL AMPLITUDE.
TIME
SPACE
WE'LL CALL THAT E(A→B).
NATURE'S TIDY, AND IF THE MASS OF ELECTRONS WERE ZERO, THIS EQUATION WOULD BE THE SAME AS THE PHOTON ONE. BUT IT'S NOT ZERO, SO THE EQUATION'S MESSIER.
IT'S A BESSEL FUNCTION, BUT WHO CARES WHAT IT'S CALLED, REALLY? WE CAN CALCULATE IT!
FINALLY, THERE'S THE AMPLITUDE j FOR AN ELECTRON BEING AT THE SAME POINT IN SPACE-TIME AS A PHOTON, AND THE PHOTON BEING ABSORBED BY THE ELECTRON. WHAT'S THE AMPLITUDE OF THAT?
TIME
SPACE
j2 =1/137.035976±2
JUST A NUMBER.*

*CURRENTLY, j^2=1/137.035999679±94

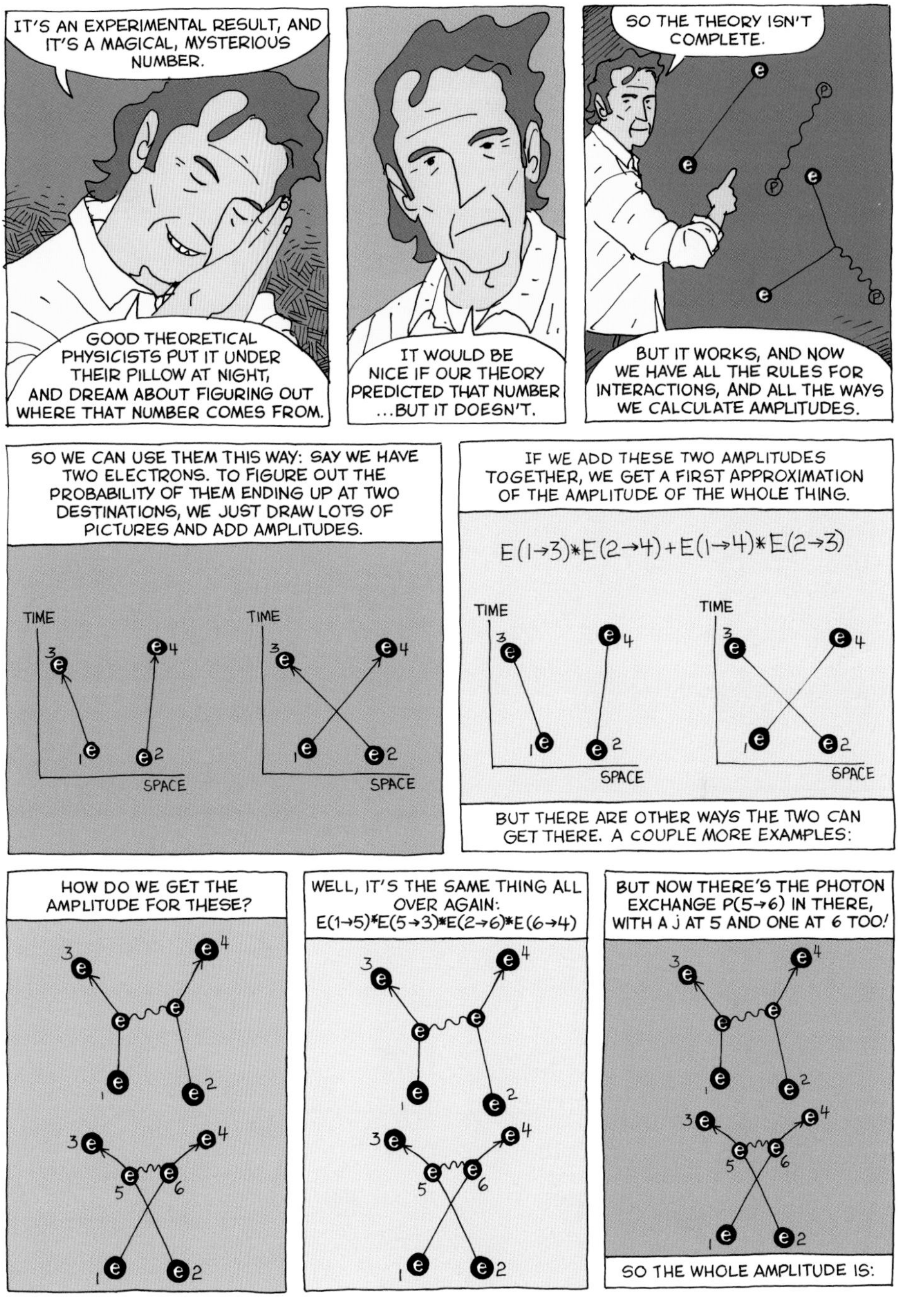

IT'S AN EXPERIMENTAL RESULT, AND IT'S A MAGICAL, MYSTERIOUS NUMBER.
GOOD THEORETICAL PHYSICISTS PUT IT UNDER THEIR PILLOW AT NIGHT, AND DREAM ABOUT FIGURING OUT WHERE THAT NUMBER COMES FROM.
IT WOULD BE NICE IF OUR THEORY PREDICTED THAT NUMBER ...BUT IT DOESN'T.
SO THE THEORY ISN'T COMPLETE.
BUT IT WORKS, AND NOW WE HAVE ALL THE RULES FOR INTERACTIONS, AND ALL THE WAYS WE CALCULATE AMPLITUDES.
SO WE CAN USE THEM THIS WAY: SAY WE HAVE TWO ELECTRONS. TO FIGURE OUT THE PROBABILITY OF THEM ENDING UP AT TWO DESTINATIONS, WE JUST DRAW LOTS OF PICTURES AND ADD AMPLITUDES.
TIME
SPACE
IF WE ADD THESE TWO AMPLITUDES TOGETHER, WE GET A FIRST APPROXIMATION OF THE AMPLITUDE OF THE WHOLE THING.
E(1→3)*E(2→4)+E(1→4)*E(2→3)
BUT THERE ARE OTHER WAYS THE TWO CAN GET THERE. A COUPLE MORE EXAMPLES:
HOW DO WE GET THE AMPLITUDE FOR THESE?
WELL, IT'S THE SAME THING ALL OVER AGAIN: E(1→5)*E(5→3)*E(2→6)*E(6→4)
BUT NOW THERE'S THE PHOTON EXCHANGE P(5→6) IN THERE, WITH A j AT 5 AND ONE AT 6 TOO!
SO THE WHOLE AMPLITUDE IS:

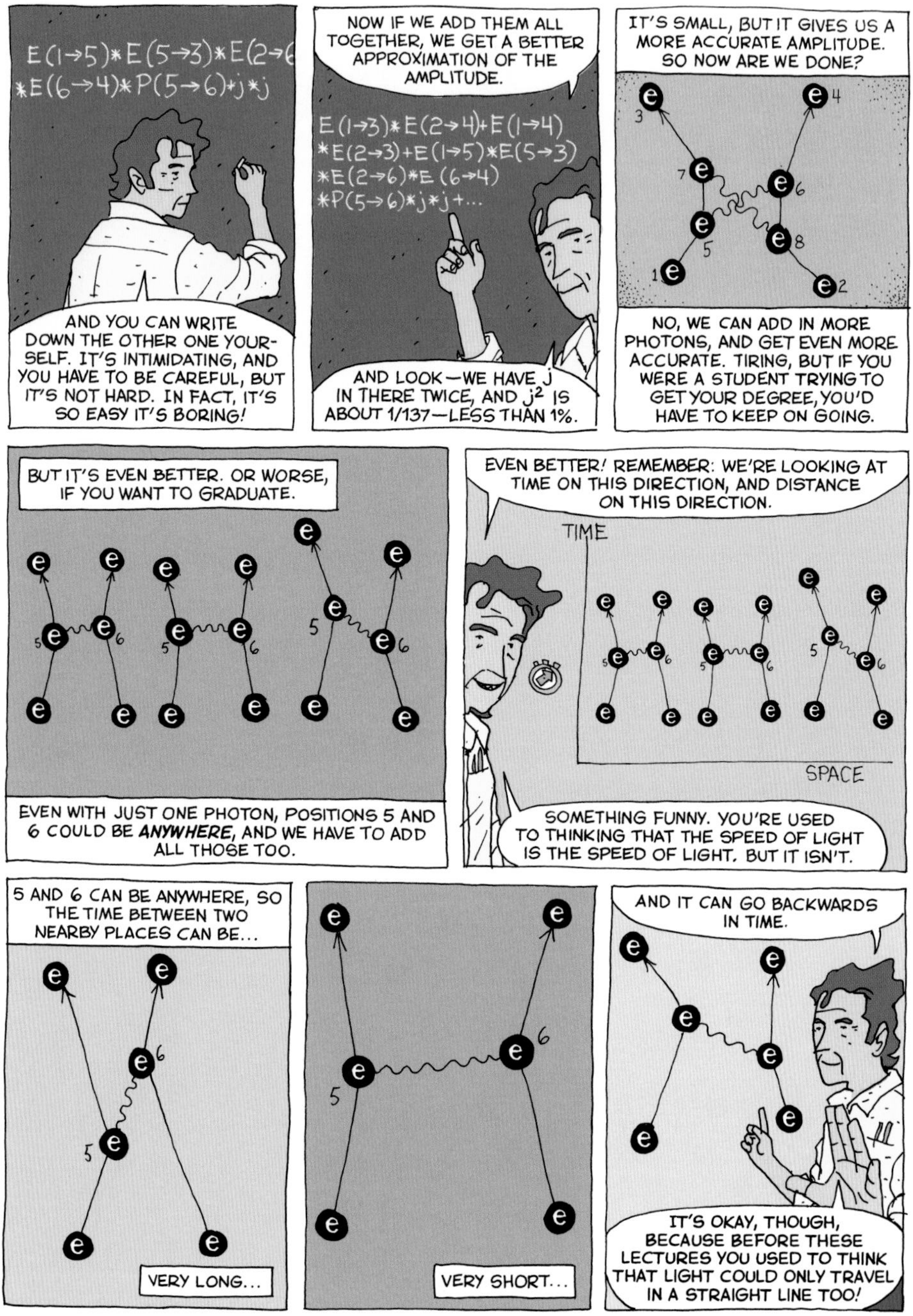
E(1→5)*E(5→3)*E(2→6
*E(6→4)*P(5→6)*j*j
AND YOU CAN WRITE DOWN THE OTHER ONE YOURSELF. IT'S INTIMIDATING, AND YOU HAVE TO BE CAREFUL, BUT IT'S NOT HARD. IN FACT, IT'S SO EASY IT'S BORING!
NOW IF WE ADD THEM ALL TOGETHER, WE GET A BETTER APPROXIMATION OF THE AMPLITUDE.
E(1→3)*E(2→4)+E(1→4)
*E(2→3)+E(1→5)*E(5→3)
*E(2→6)*E(6→4)
*P(5→6)*j*j+...
AND LOOK—WE HAVE j IN THERE TWICE, AND j² IS ABOUT 1/137—LESS THAN 1%.
IT'S SMALL, BUT IT GIVES US A MORE ACCURATE AMPLITUDE. SO NOW ARE WE DONE?
NO, WE CAN ADD IN MORE PHOTONS, AND GET EVEN MORE ACCURATE. TIRING, BUT IF YOU WERE A STUDENT TRYING TO GET YOUR DEGREE, YOU'D HAVE TO KEEP ON GOING.
BUT IT'S EVEN BETTER. OR WORSE, IF YOU WANT TO GRADUATE.
EVEN WITH JUST ONE PHOTON, POSITIONS 5 AND 6 COULD BE ***ANYWHERE,*** AND WE HAVE TO ADD ALL THOSE TOO.
EVEN BETTER! REMEMBER: WE'RE LOOKING AT TIME ON THIS DIRECTION, AND DISTANCE ON THIS DIRECTION.
TIME
SPACE
SOMETHING FUNNY. YOU'RE USED TO THINKING THAT THE SPEED OF LIGHT IS THE SPEED OF LIGHT. BUT IT ISN'T.
5 AND 6 CAN BE ANYWHERE, SO THE TIME BETWEEN TWO NEARBY PLACES CAN BE...
VERY LONG...
VERY SHORT...
AND IT CAN GO BACKWARDS IN TIME.
IT'S OKAY, THOUGH, BECAUSE BEFORE THESE LECTURES YOU USED TO THINK THAT LIGHT COULD ONLY TRAVEL IN A STRAIGHT LINE TOO!

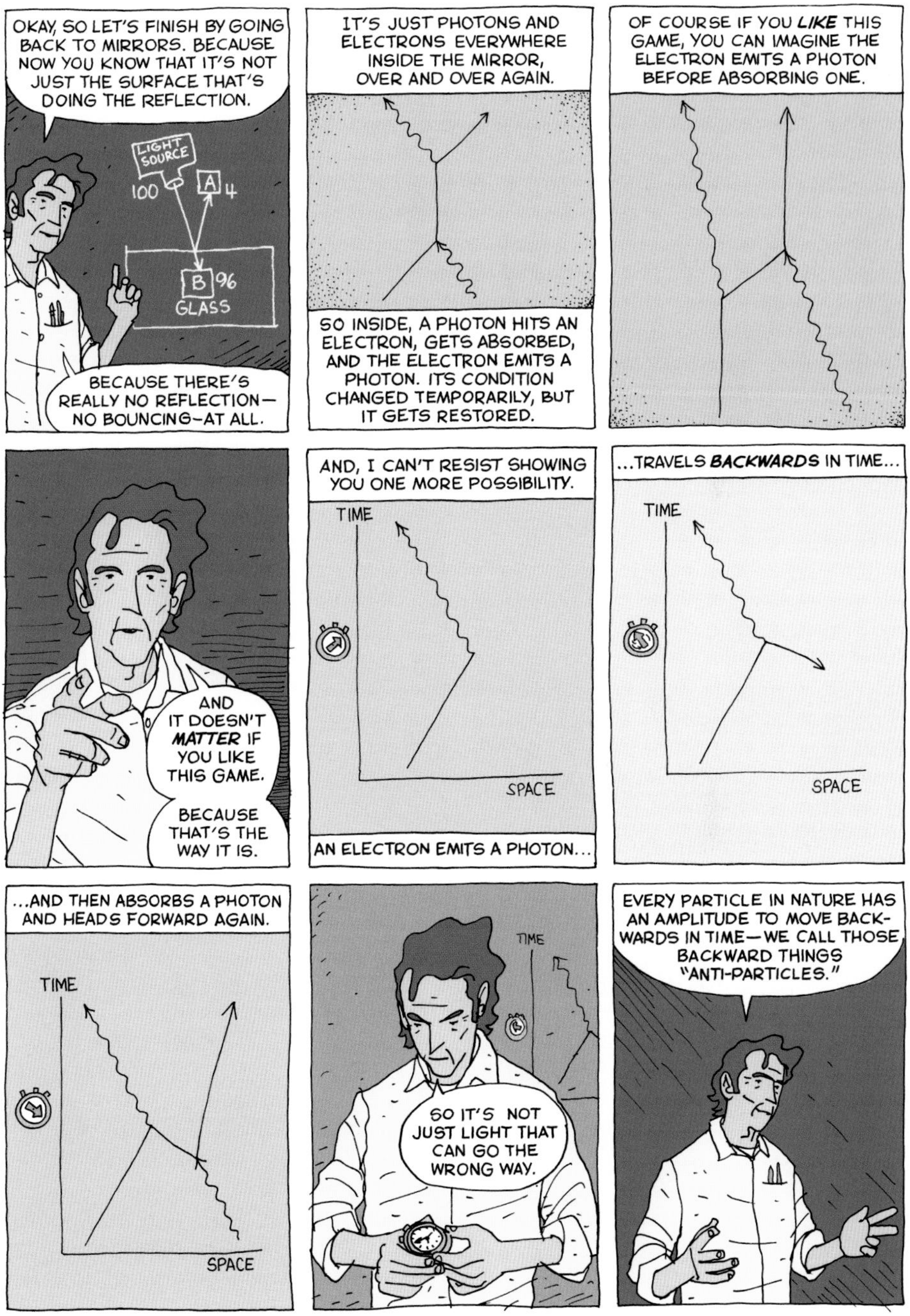

OKAY, SO LET'S FINISH BY GOING BACK TO MIRRORS. BECAUSE NOW YOU KNOW THAT IT'S NOT JUST THE SURFACE THAT'S DOING THE REFLECTION.
LIGHT SOURCE
100
A 4
B 96
GLASS
BECAUSE THERE'S REALLY NO REFLECTION— NO BOUNCING—AT ALL.
IT'S JUST PHOTONS AND ELECTRONS EVERYWHERE INSIDE THE MIRROR, OVER AND OVER AGAIN.
SO INSIDE, A PHOTON HITS AN ELECTRON, GETS ABSORBED, AND THE ELECTRON EMITS A PHOTON. ITS CONDITION CHANGED TEMPORARILY, BUT IT GETS RESTORED.
OF COURSE IF YOU LIKE THIS GAME, YOU CAN IMAGINE THE ELECTRON EMITS A PHOTON BEFORE ABSORBING ONE.
AND IT DOESN'T MATTER IF YOU LIKE THIS GAME.
BECAUSE THAT'S THE WAY IT IS.
AND, I CAN'T RESIST SHOWING YOU ONE MORE POSSIBILITY.
TIME
SPACE
AN ELECTRON EMITS A PHOTON...
...TRAVELS BACKWARDS IN TIME...
TIME
SPACE
...AND THEN ABSORBS A PHOTON AND HEADS FORWARD AGAIN.
TIME
SPACE
TIME
SO IT'S NOT JUST LIGHT THAT CAN GO THE WRONG WAY.
EVERY PARTICLE IN NATURE HAS AN AMPLITUDE TO MOVE BACK-WARDS IN TIME—WE CALL THOSE BACKWARD THINGS "ANTI-PARTICLES."

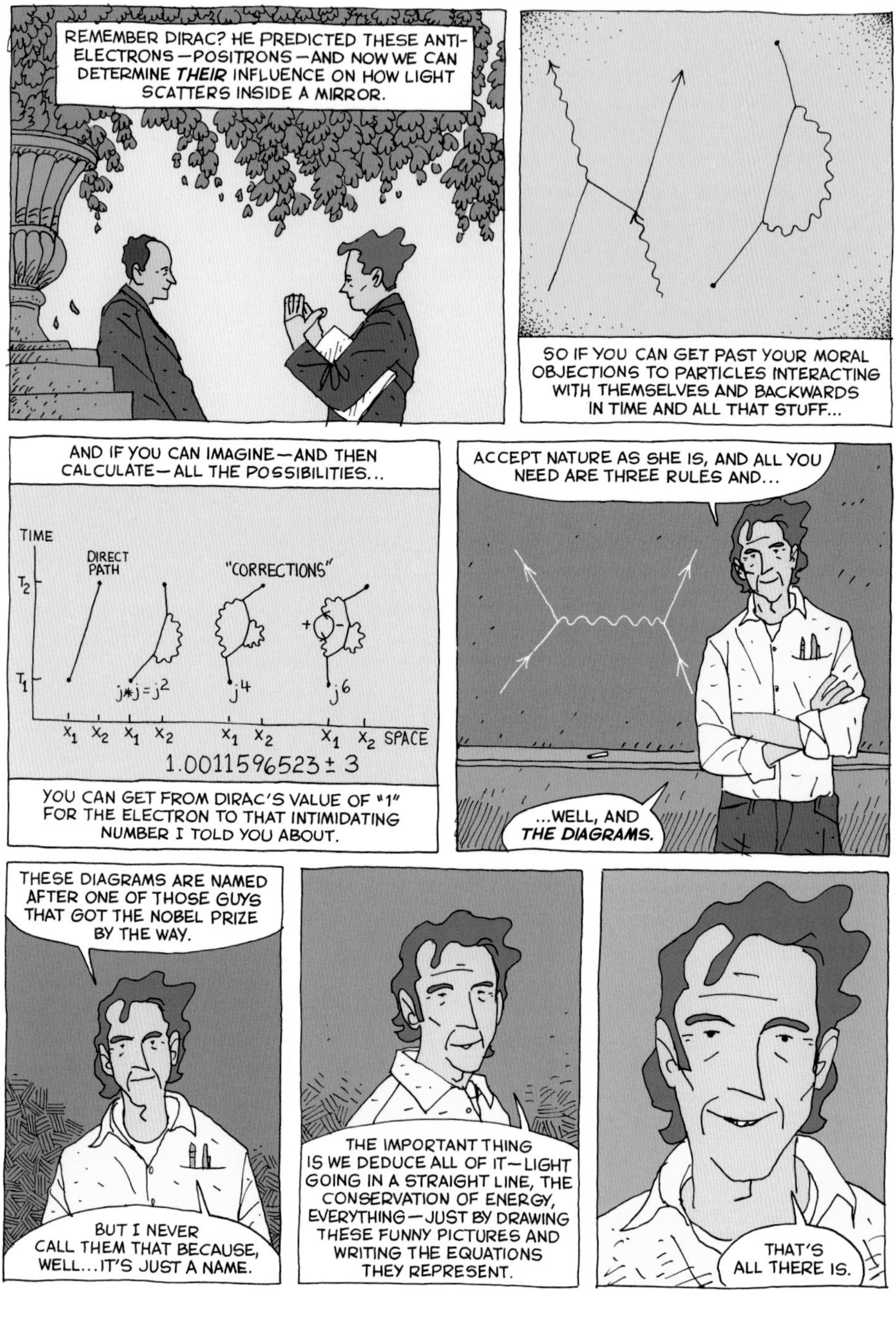

REMEMBER DIRAC? HE PREDICTED THESE ANTI-ELECTRONS—POSITRONS—AND NOW WE CAN DETERMINE THEIR INFLUENCE ON HOW LIGHT SCATTERS INSIDE A MIRROR.
SO IF YOU CAN GET PAST YOUR MORAL OBJECTIONS TO PARTICLES INTERACTING WITH THEMSELVES AND BACKWARDS IN TIME AND ALL THAT STUFF...
AND IF YOU CAN IMAGINE—AND THEN CALCULATE—ALL THE POSSIBILITIES...
TIME
DIRECT PATH
"CORRECTIONS"
j*j = j²
j⁴
j⁶
X1 X2 X1 X2 X1 X2 X1 X2 SPACE
1.0011596523 ± 3
YOU CAN GET FROM DIRAC'S VALUE OF "1" FOR THE ELECTRON TO THAT INTIMIDATING NUMBER I TOLD YOU ABOUT.
ACCEPT NATURE AS SHE IS, AND ALL YOU NEED ARE THREE RULES AND...
...WELL, AND THE DIAGRAMS.
THESE DIAGRAMS ARE NAMED AFTER ONE OF THOSE GUYS THAT GOT THE NOBEL PRIZE BY THE WAY.
BUT I NEVER CALL THEM THAT BECAUSE, WELL... IT'S JUST A NAME.
THE IMPORTANT THING IS WE DEDUCE ALL OF IT—LIGHT GOING IN A STRAIGHT LINE, THE CONSERVATION OF ENERGY, EVERYTHING—JUST BY DRAWING THESE FUNNY PICTURES AND WRITING THE EQUATIONS THEY REPRESENT.
THAT'S ALL THERE IS.

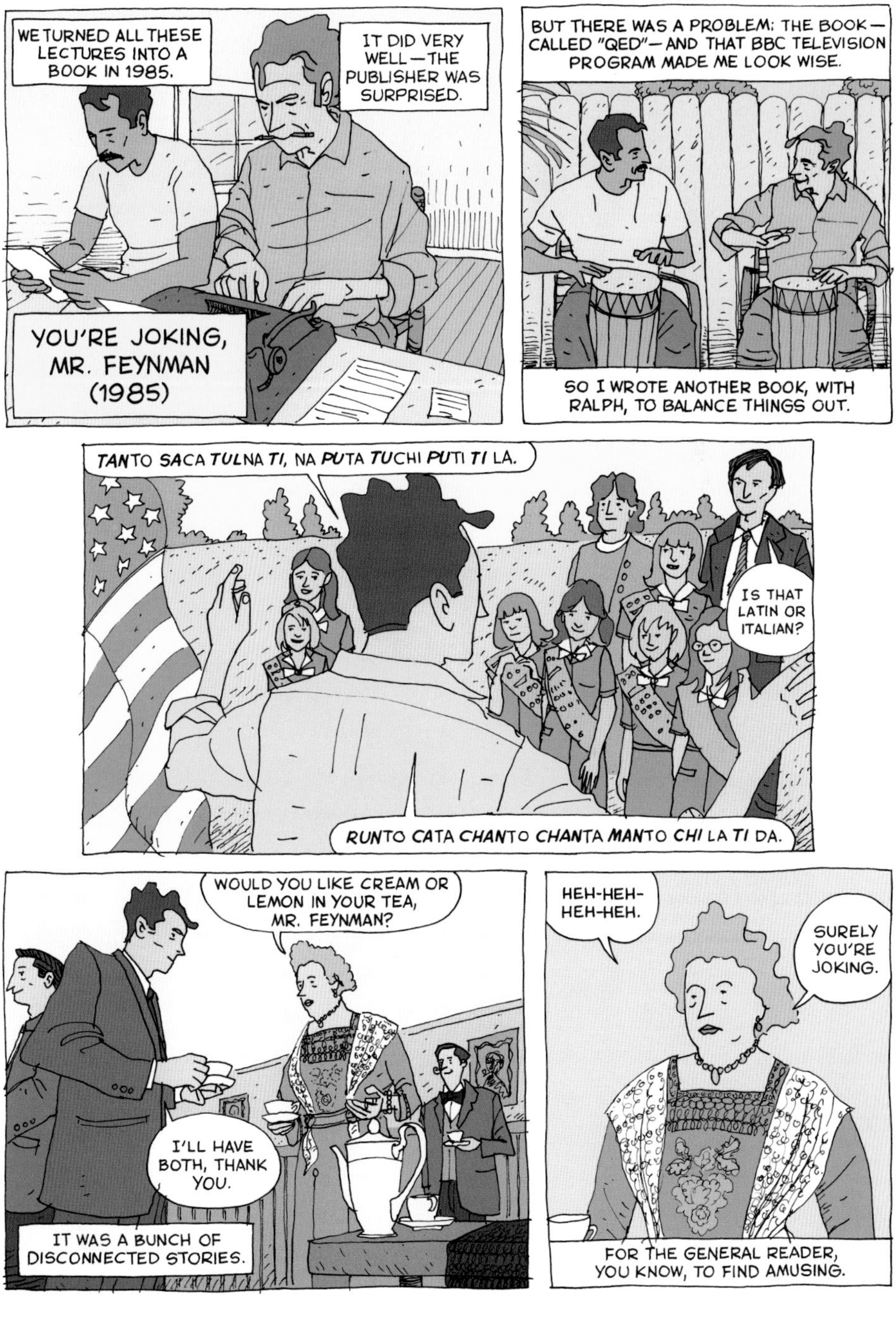

WE TURNED ALL THESE LECTURES INTO A BOOK IN 1985.
IT DID VERY WELL—THE PUBLISHER WAS SURPRISED.
YOU'RE JOKING, MR. FEYNMAN (1985)
BUT THERE WAS A PROBLEM: THE BOOK—CALLED "QED"—AND THAT BBC TELEVISION PROGRAM MADE ME LOOK WISE.
SO I WROTE ANOTHER BOOK, WITH RALPH, TO BALANCE THINGS OUT.
TANTO SACA TULNA TI, NA PUTA TUCHI PUTI TI LA.
IS THAT LATIN OR ITALIAN?
RUNTO CATA CHANTO CHANTA MANTO CHI LA TI DA.
WOULD YOU LIKE CREAM OR LEMON IN YOUR TEA, MR. FEYNMAN?
I'LL HAVE BOTH, THANK YOU.
IT WAS A BUNCH OF DISCONNECTED STORIES.
HEH-HEH-HEH-HEH.
SURELY YOU'RE JOKING.
FOR THE GENERAL READER, YOU KNOW, TO FIND AMUSING.

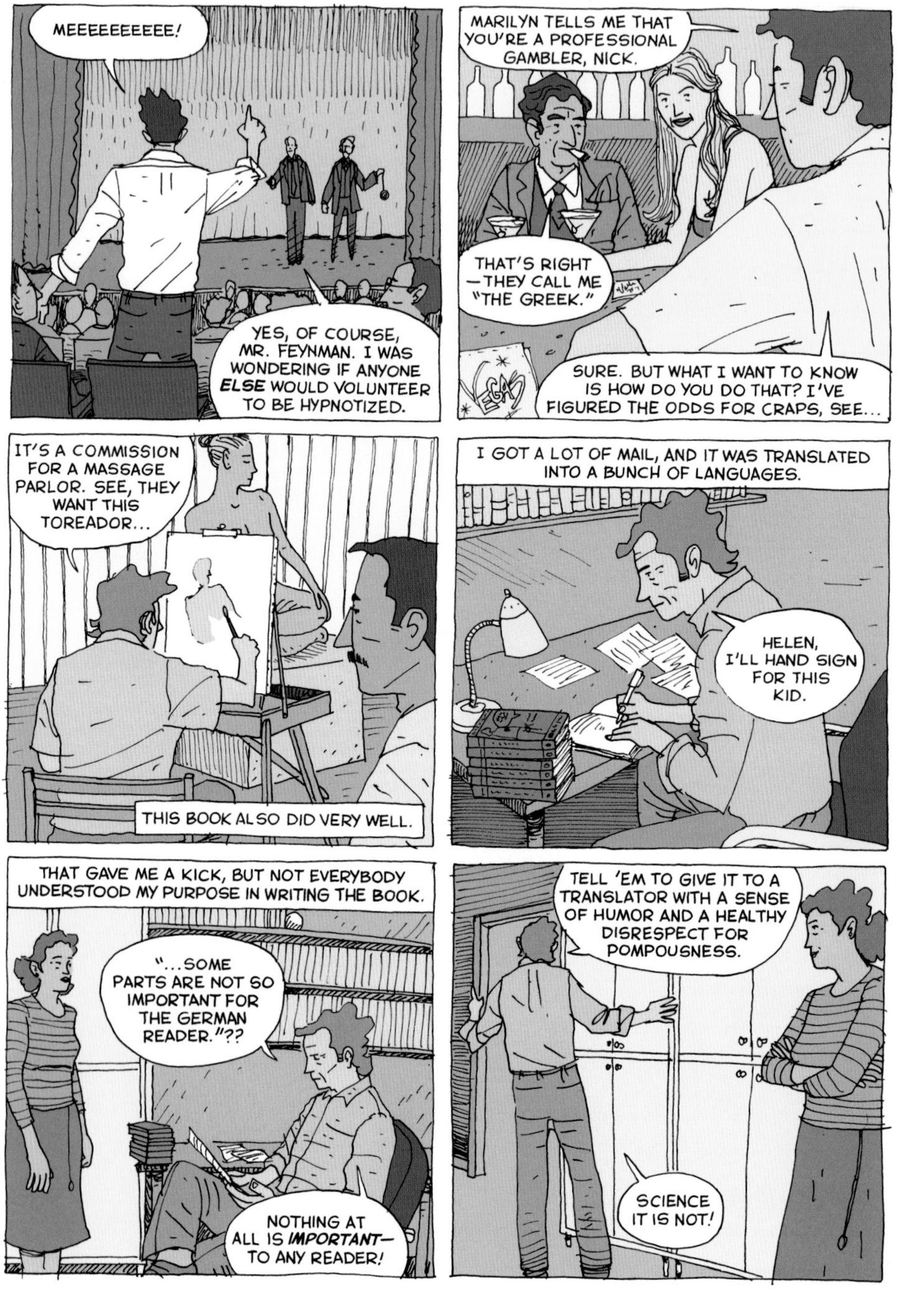
MEEEEEEEEEE!
YES, OF COURSE, MR. FEYNMAN. I WAS WONDERING IF ANYONE ELSE WOULD VOLUNTEER TO BE HYPNOTIZED.
MARILYN TELLS ME THAT YOU'RE A PROFESSIONAL GAMBLER, NICK.
THAT'S RIGHT —THEY CALL ME "THE GREEK."
SURE. BUT WHAT I WANT TO KNOW IS HOW DO YOU DO THAT? I'VE FIGURED THE ODDS FOR CRAPS, SEE...
VEGAS
IT'S A COMMISSION FOR A MASSAGE PARLOR. SEE, THEY WANT THIS TOREADOR...
THIS BOOK ALSO DID VERY WELL.
I GOT A LOT OF MAIL, AND IT WAS TRANSLATED INTO A BUNCH OF LANGUAGES.
HELEN, I'LL HAND SIGN FOR THIS KID.
THAT GAVE ME A KICK, BUT NOT EVERYBODY UNDERSTOOD MY PURPOSE IN WRITING THE BOOK.
"...SOME PARTS ARE NOT SO IMPORTANT FOR THE GERMAN READER."??
NOTHING AT ALL IS IMPORTANT— TO ANY READER!
TELL 'EM TO GIVE IT TO A TRANSLATOR WITH A SENSE OF HUMOR AND A HEALTHY DISRESPECT FOR POMPOUSNESS.
SCIENCE IT IS NOT!

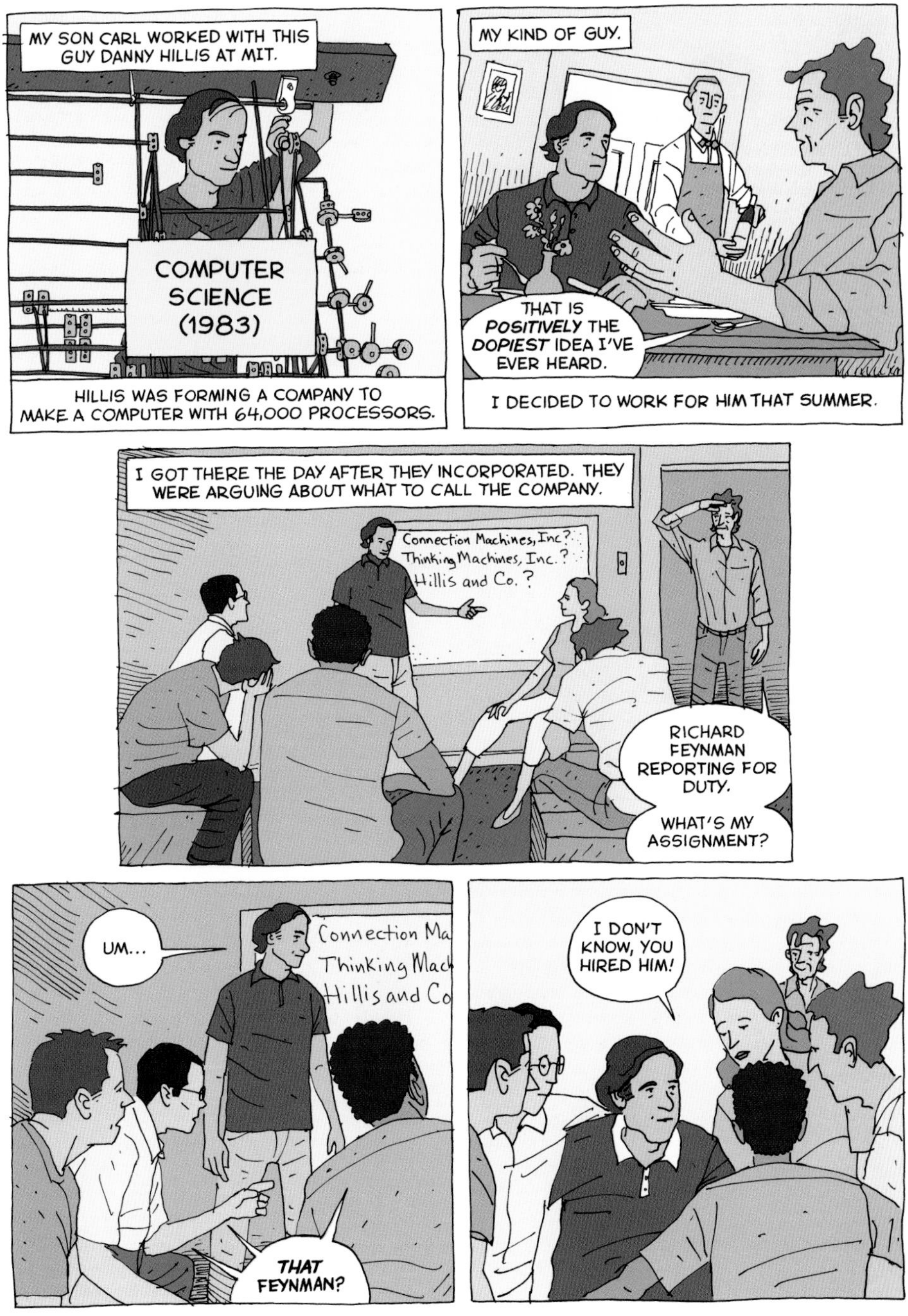

MY SON CARL WORKED WITH THIS GUY DANNY HILLIS AT MIT.
COMPUTER SCIENCE (1983)
HILLIS WAS FORMING A COMPANY TO MAKE A COMPUTER WITH 64,000 PROCESSORS.
MY KIND OF GUY.
THAT IS POSITIVELY THE DOPIEST IDEA I'VE EVER HEARD.
I DECIDED TO WORK FOR HIM THAT SUMMER.
I GOT THERE THE DAY AFTER THEY INCORPORATED. THEY WERE ARGUING ABOUT WHAT TO CALL THE COMPANY.
Connection Machines, Inc.?
Thinking Machines, Inc.?
Hillis and Co.?
RICHARD FEYNMAN REPORTING FOR DUTY.
WHAT'S MY ASSIGNMENT?
UM...
THAT FEYNMAN?
I DON'T KNOW, YOU HIRED HIM!

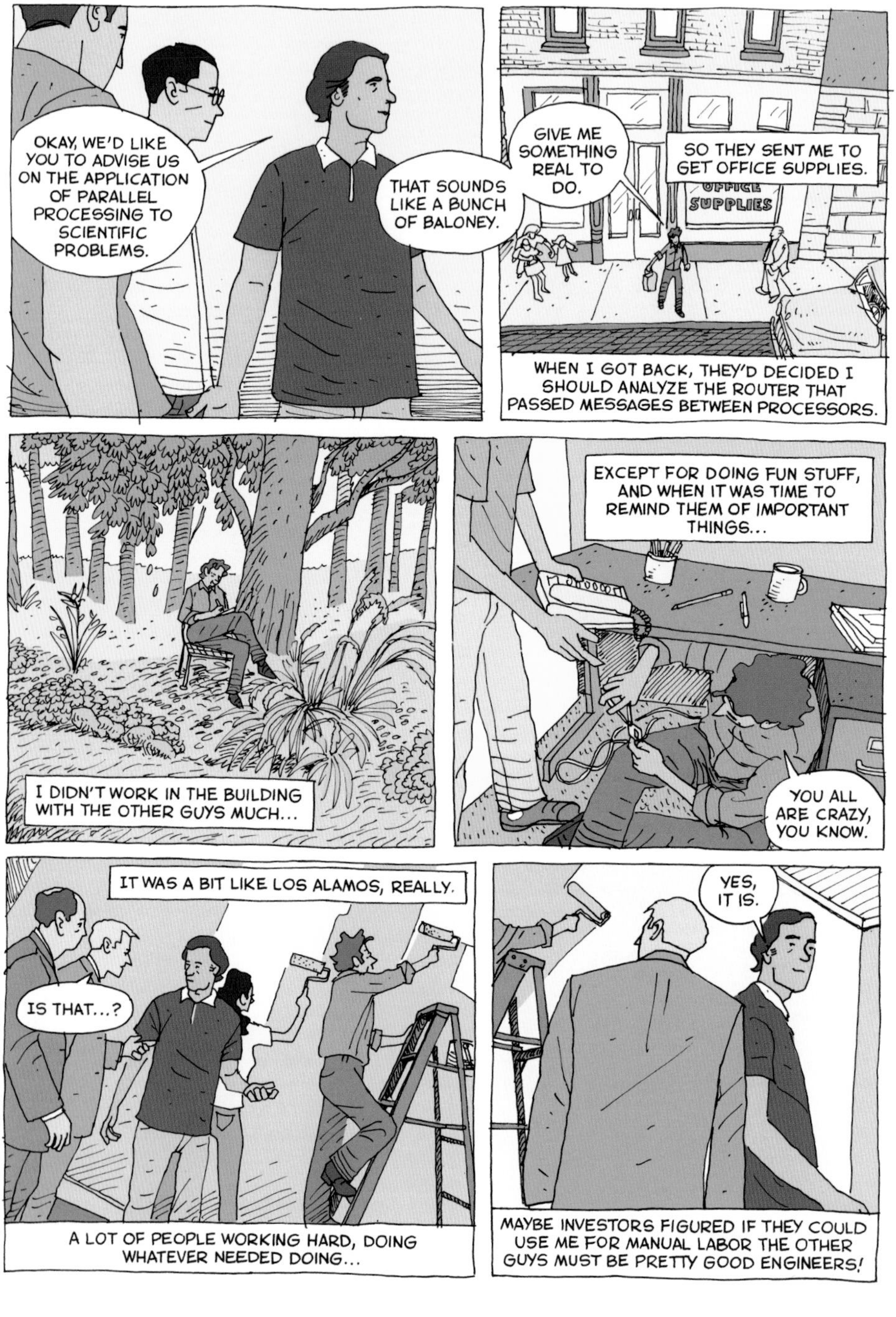
OKAY, WE'D LIKE YOU TO ADVISE US ON THE APPLICATION OF PARALLEL PROCESSING TO SCIENTIFIC PROBLEMS.
THAT SOUNDS LIKE A BUNCH OF BALONEY.
GIVE ME SOMETHING REAL TO DO.
SO THEY SENT ME TO GET OFFICE SUPPLIES.
OFFICE SUPPLIES
WHEN I GOT BACK, THEY'D DECIDED I SHOULD ANALYZE THE ROUTER THAT PASSED MESSAGES BETWEEN PROCESSORS.
I DIDN'T WORK IN THE BUILDING WITH THE OTHER GUYS MUCH...
EXCEPT FOR DOING FUN STUFF, AND WHEN IT WAS TIME TO REMIND THEM OF IMPORTANT THINGS...
YOU ALL ARE CRAZY, YOU KNOW.
IT WAS A BIT LIKE LOS ALAMOS, REALLY.
IS THAT...?
A LOT OF PEOPLE WORKING HARD, DOING WHATEVER NEEDED DOING...
YES, IT IS.
MAYBE INVESTORS FIGURED IF THEY COULD USE ME FOR MANUAL LABOR THE OTHER GUYS MUST BE PRETTY GOOD ENGINEERS!

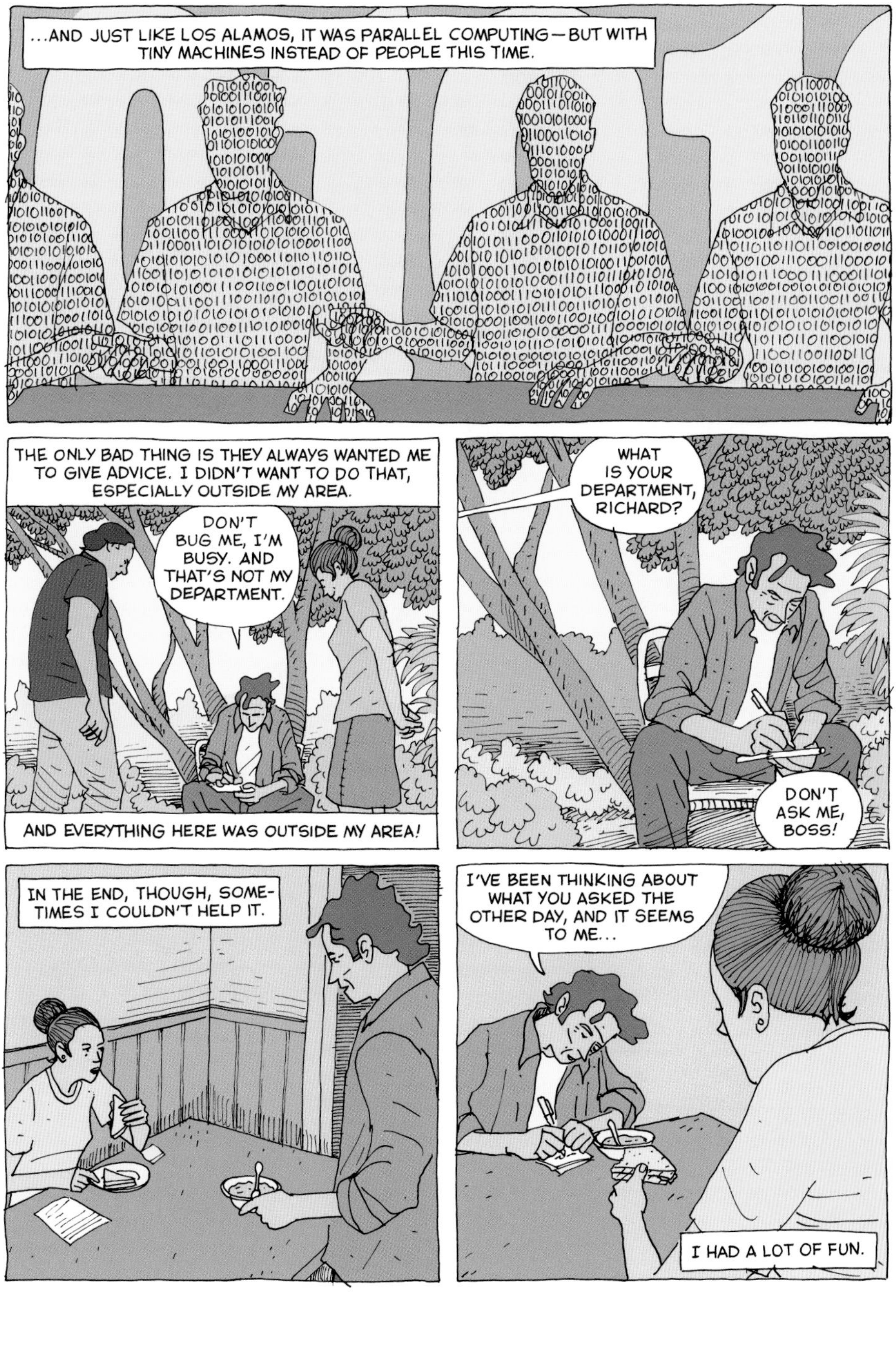
...AND JUST LIKE LOS ALAMOS, IT WAS PARALLEL COMPUTING—BUT WITH TINY MACHINES INSTEAD OF PEOPLE THIS TIME.
THE ONLY BAD THING IS THEY ALWAYS WANTED ME TO GIVE ADVICE. I DIDN'T WANT TO DO THAT, ESPECIALLY OUTSIDE MY AREA.
DON'T BUG ME, I'M BUSY. AND THAT'S NOT MY DEPARTMENT.
AND EVERYTHING HERE WAS OUTSIDE MY AREA!
WHAT IS YOUR DEPARTMENT, RICHARD?
DON'T ASK ME, BOSS!
IN THE END, THOUGH, SOMETIMES I COULDN'T HELP IT.
I'VE BEEN THINKING ABOUT WHAT YOU ASKED THE OTHER DAY, AND IT SEEMS TO ME...
I HAD A LOT OF FUN.

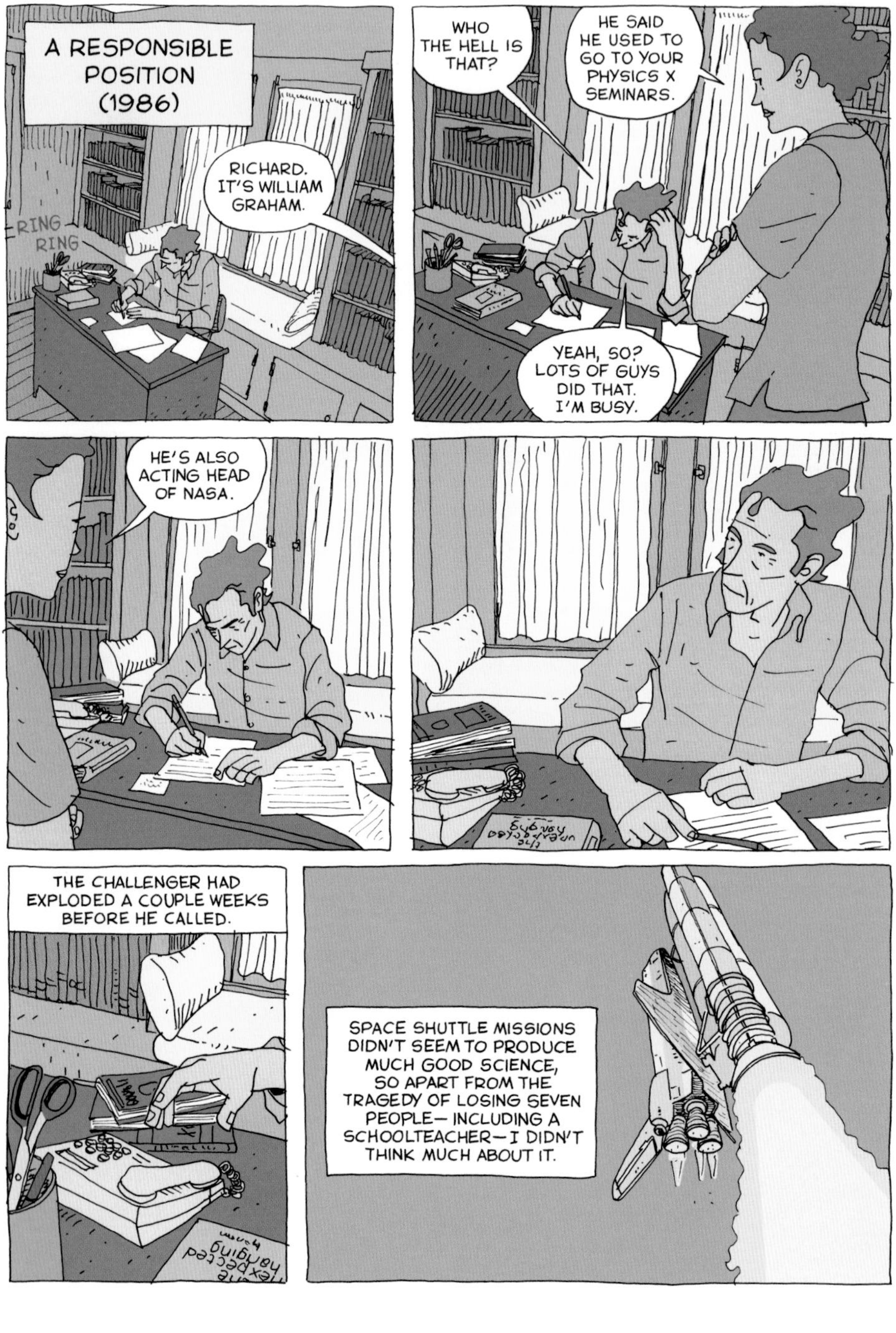
A RESPONSIBLE POSITION (1986)
RING RING
RICHARD. IT'S WILLIAM GRAHAM.
WHO THE HELL IS THAT?
HE SAID HE USED TO GO TO YOUR PHYSICS X SEMINARS.
YEAH, SO? LOTS OF GUYS DID THAT. I'M BUSY.
HE'S ALSO ACTING HEAD OF NASA.
THE CHALLENGER HAD EXPLODED A COUPLE WEEKS BEFORE HE CALLED.
SPACE SHUTTLE MISSIONS DIDN'T SEEM TO PRODUCE MUCH GOOD SCIENCE, SO APART FROM THE TRAGEDY OF LOSING SEVEN PEOPLE— INCLUDING A SCHOOLTEACHER— I DIDN'T THINK MUCH ABOUT IT.

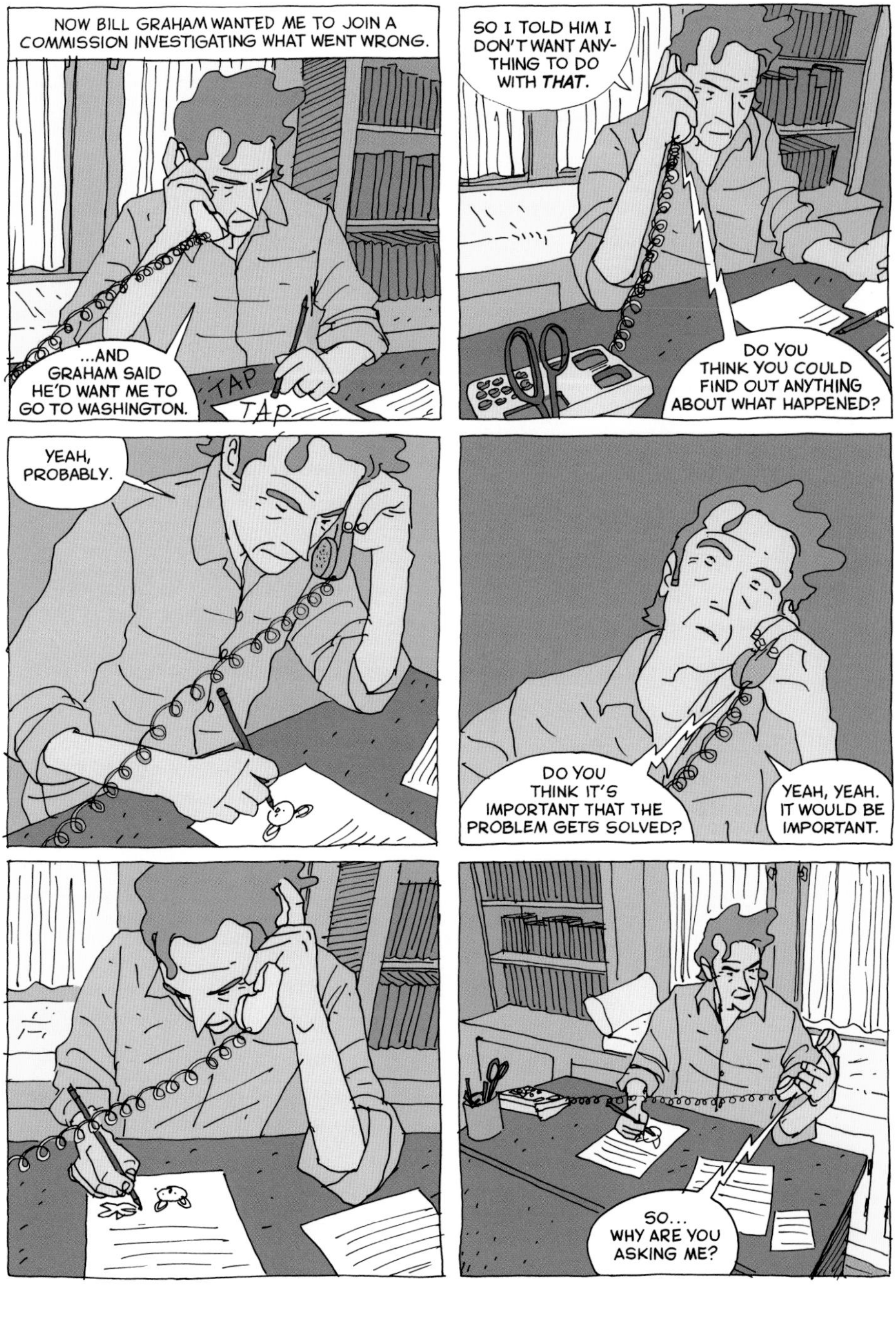
NOW BILL GRAHAM WANTED ME TO JOIN A COMMISSION INVESTIGATING WHAT WENT WRONG.
...AND GRAHAM SAID HE'D WANT ME TO GO TO WASHINGTON.
TAP
TAP
SO I TOLD HIM I DON'T WANT ANY-THING TO DO WITH *THAT*.
DO YOU THINK YOU COULD FIND OUT ANYTHING ABOUT WHAT HAPPENED?
YEAH, PROBABLY.
DO YOU THINK IT'S IMPORTANT THAT THE PROBLEM GETS SOLVED?
YEAH, YEAH. IT WOULD BE IMPORTANT.
SO... WHY ARE YOU ASKING ME?

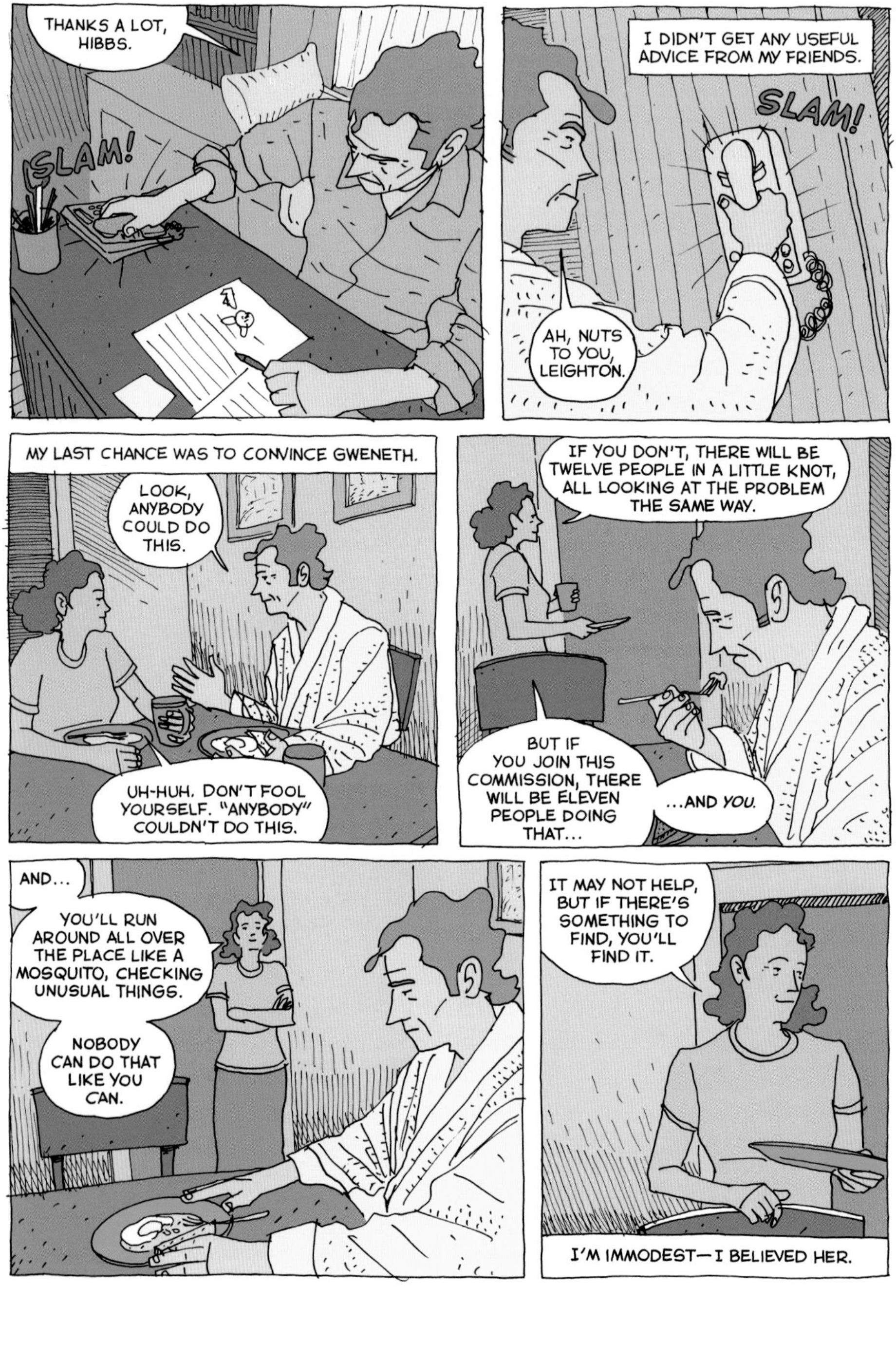
THANKS A LOT, HIBBS.
SLAM!
I DIDN'T GET ANY USEFUL ADVICE FROM MY FRIENDS.
SLAM!
AH, NUTS TO YOU, LEIGHTON.
MY LAST CHANCE WAS TO CONVINCE GWENETH.
LOOK, ANYBODY COULD DO THIS.
UH-HUH. DON'T FOOL YOURSELF. "ANYBODY" COULDN'T DO THIS.
IF YOU DON'T, THERE WILL BE TWELVE PEOPLE IN A LITTLE KNOT, ALL LOOKING AT THE PROBLEM THE SAME WAY.
BUT IF YOU JOIN THIS COMMISSION, THERE WILL BE ELEVEN PEOPLE DOING THAT...
...AND *YOU*.
AND...
YOU'LL RUN AROUND ALL OVER THE PLACE LIKE A MOSQUITO, CHECKING UNUSUAL THINGS.
NOBODY CAN DO THAT LIKE YOU CAN.
IT MAY NOT HELP, BUT IF THERE'S SOMETHING TO FIND, YOU'LL FIND IT.
I'M IMMODEST—I BELIEVED HER.

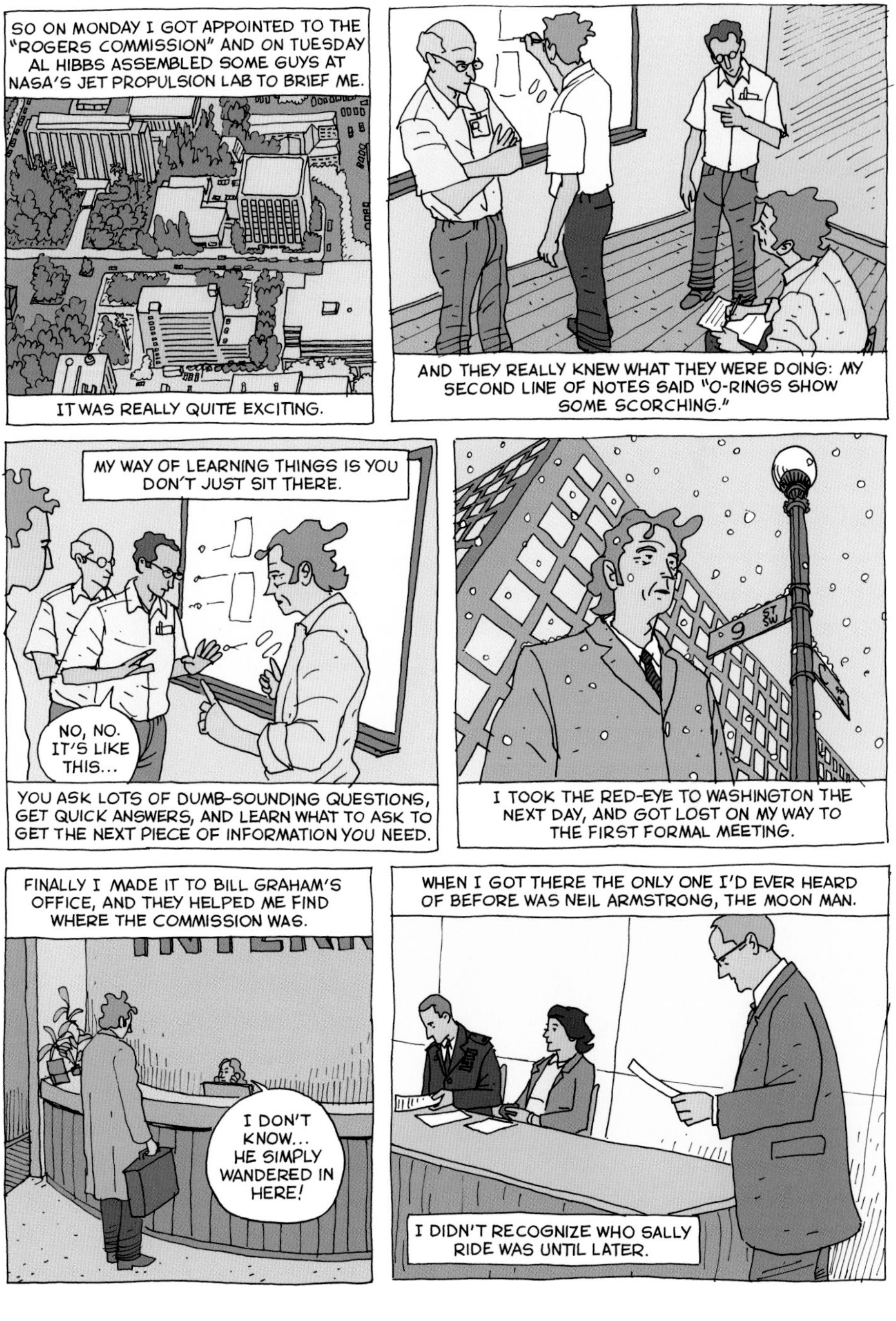
SO ON MONDAY I GOT APPOINTED TO THE "ROGERS COMMISSION" AND ON TUESDAY AL HIBBS ASSEMBLED SOME GUYS AT NASA'S JET PROPULSION LAB TO BRIEF ME.
IT WAS REALLY QUITE EXCITING.
AND THEY REALLY KNEW WHAT THEY WERE DOING: MY SECOND LINE OF NOTES SAID "O-RINGS SHOW SOME SCORCHING."
MY WAY OF LEARNING THINGS IS YOU DON'T JUST SIT THERE.
NO, NO. IT'S LIKE THIS...
YOU ASK LOTS OF DUMB-SOUNDING QUESTIONS, GET QUICK ANSWERS, AND LEARN WHAT TO ASK TO GET THE NEXT PIECE OF INFORMATION YOU NEED.
9 ST SW
I TOOK THE RED-EYE TO WASHINGTON THE NEXT DAY, AND GOT LOST ON MY WAY TO THE FIRST FORMAL MEETING.
FINALLY I MADE IT TO BILL GRAHAM'S OFFICE, AND THEY HELPED ME FIND WHERE THE COMMISSION WAS.
I DON'T KNOW... HE SIMPLY WANDERED IN HERE!
WHEN I GOT THERE THE ONLY ONE I'D EVER HEARD OF BEFORE WAS NEIL ARMSTRONG, THE MOON MAN.
I DIDN'T RECOGNIZE WHO SALLY RIDE WAS UNTIL LATER.

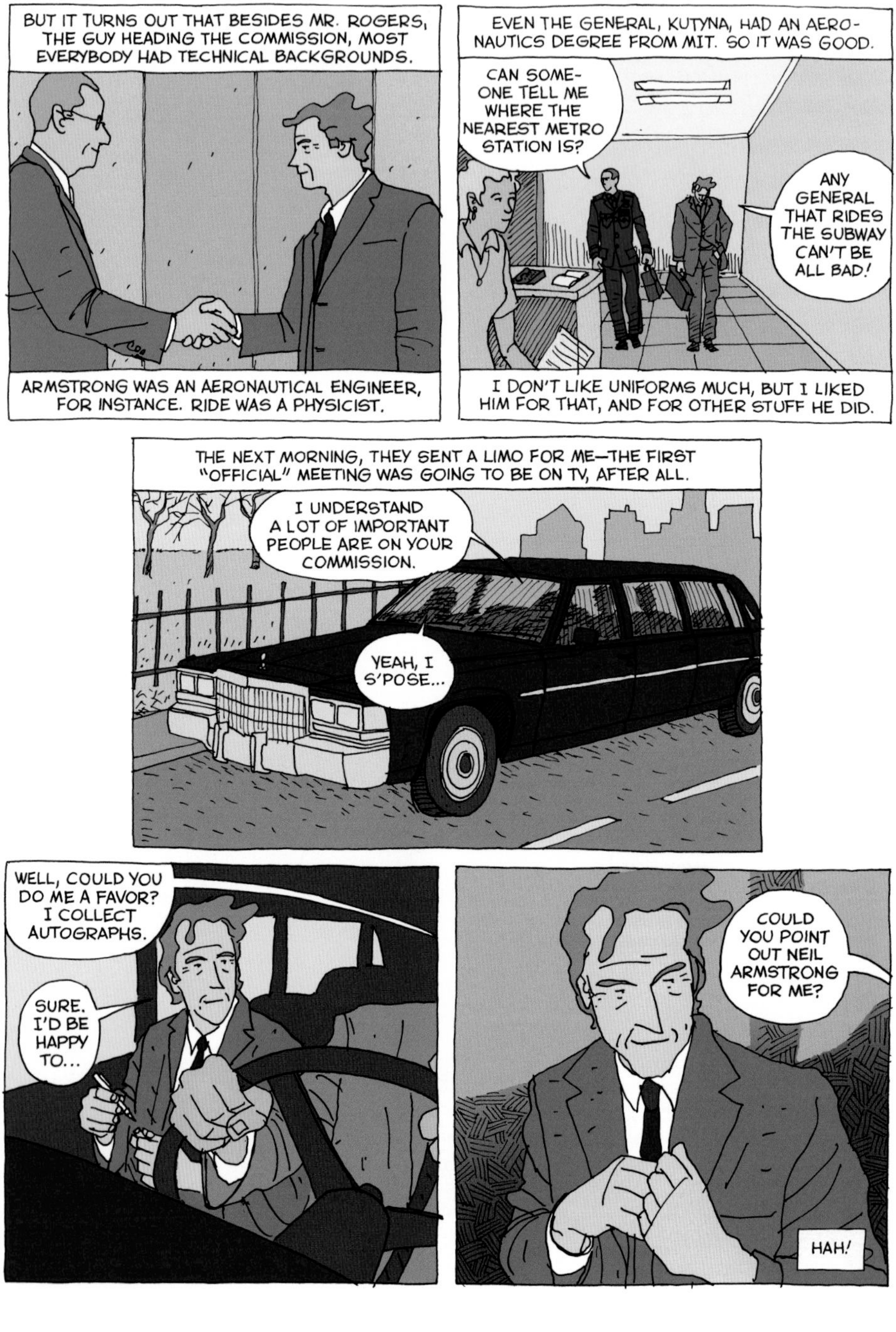
BUT IT TURNS OUT THAT BESIDES MR. ROGERS, THE GUY HEADING THE COMMISSION, MOST EVERYBODY HAD TECHNICAL BACKGROUNDS.
ARMSTRONG WAS AN AERONAUTICAL ENGINEER, FOR INSTANCE. RIDE WAS A PHYSICIST.
EVEN THE GENERAL, KUTYNA, HAD AN AERONAUTICS DEGREE FROM MIT. SO IT WAS GOOD.
CAN SOMEONE TELL ME WHERE THE NEAREST METRO STATION IS?
ANY GENERAL THAT RIDES THE SUBWAY CAN'T BE ALL BAD!
I DON'T LIKE UNIFORMS MUCH, BUT I LIKED HIM FOR THAT, AND FOR OTHER STUFF HE DID.
THE NEXT MORNING, THEY SENT A LIMO FOR ME—THE FIRST "OFFICIAL" MEETING WAS GOING TO BE ON TV, AFTER ALL.
I UNDERSTAND A LOT OF IMPORTANT PEOPLE ARE ON YOUR COMMISSION.
YEAH, I S'POSE...
WELL, COULD YOU DO ME A FAVOR? I COLLECT AUTOGRAPHS.
SURE. I'D BE HAPPY TO...
COULD YOU POINT OUT NEIL ARMSTRONG FOR ME?
HAH!

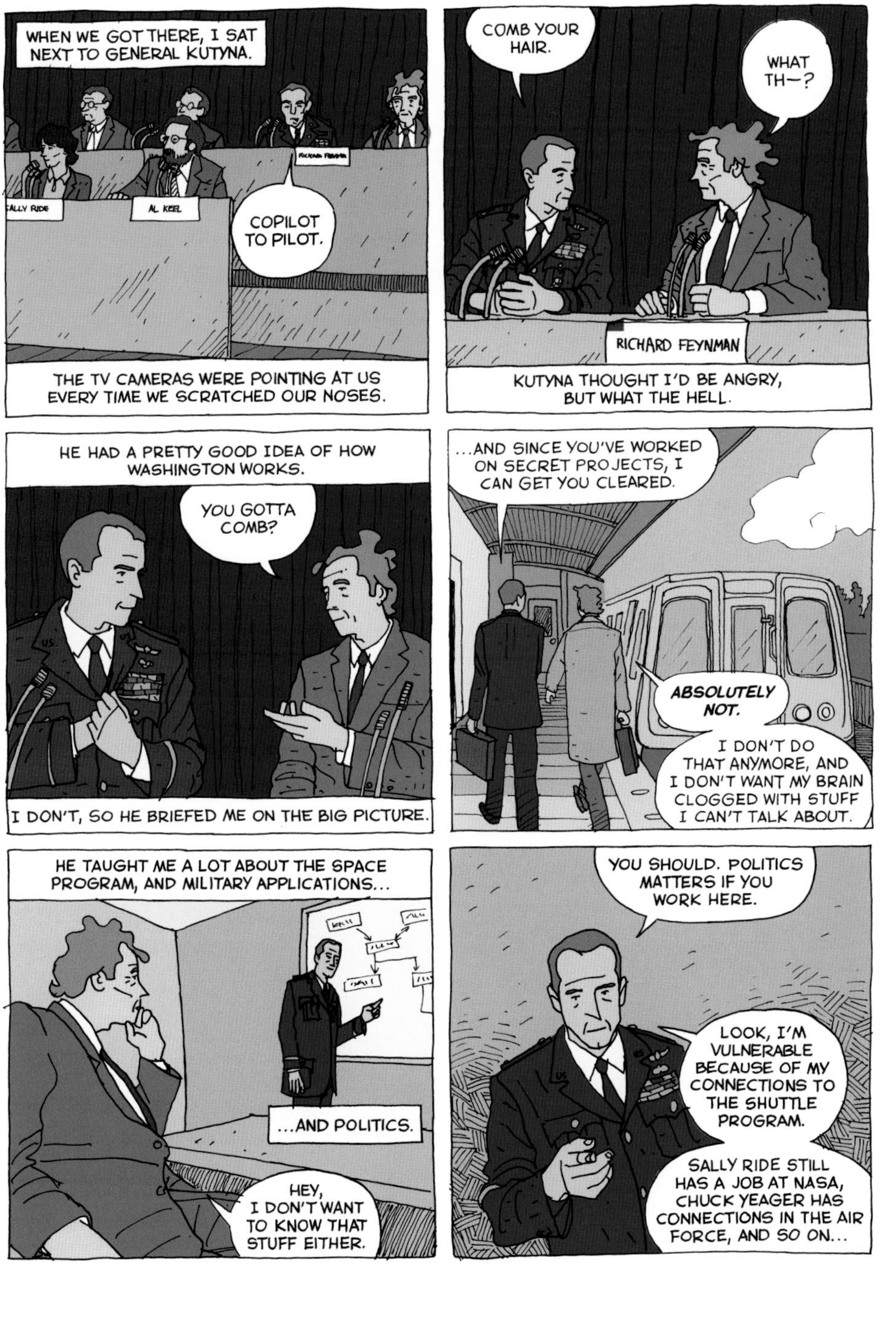
WHEN WE GOT THERE, I SAT NEXT TO GENERAL KUTYNA.
SALLY RIDE
AL KEEL
RICHARD FEYNMAN
COPILOT TO PILOT.
THE TV CAMERAS WERE POINTING AT US EVERY TIME WE SCRATCHED OUR NOSES.
COMB YOUR HAIR.
WHAT TH—?
RICHARD FEYNMAN
KUTYNA THOUGHT I'D BE ANGRY, BUT WHAT THE HELL.
HE HAD A PRETTY GOOD IDEA OF HOW WASHINGTON WORKS.
YOU GOTTA COMB?
I DON'T, SO HE BRIEFED ME ON THE BIG PICTURE.
...AND SINCE YOU'VE WORKED ON SECRET PROJECTS, I CAN GET YOU CLEARED.
ABSOLUTELY NOT.
I DON'T DO THAT ANYMORE, AND I DON'T WANT MY BRAIN CLOGGED WITH STUFF I CAN'T TALK ABOUT.
HE TAUGHT ME A LOT ABOUT THE SPACE PROGRAM, AND MILITARY APPLICATIONS...
...AND POLITICS.
HEY, I DON'T WANT TO KNOW THAT STUFF EITHER.
YOU SHOULD. POLITICS MATTERS IF YOU WORK HERE.
LOOK, I'M VULNERABLE BECAUSE OF MY CONNECTIONS TO THE SHUTTLE PROGRAM.
SALLY RIDE STILL HAS A JOB AT NASA, CHUCK YEAGER HAS CONNECTIONS IN THE AIR FORCE, AND SO ON...

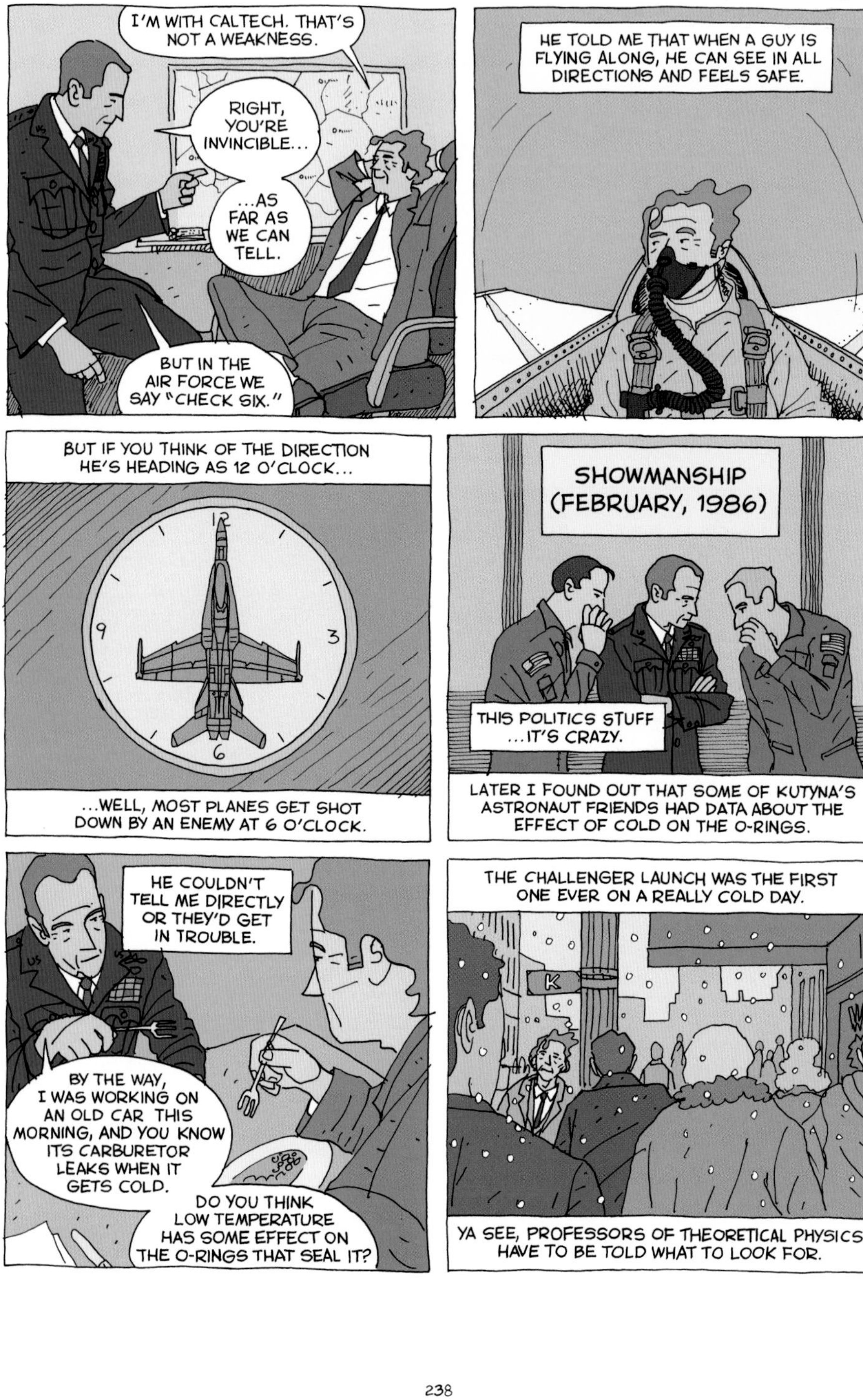
I'M WITH CALTECH. THAT'S NOT A WEAKNESS.
RIGHT, YOU'RE INVINCIBLE...
...AS FAR AS WE CAN TELL.
BUT IN THE AIR FORCE WE SAY "CHECK SIX."
HE TOLD ME THAT WHEN A GUY IS FLYING ALONG, HE CAN SEE IN ALL DIRECTIONS AND FEELS SAFE.
BUT IF YOU THINK OF THE DIRECTION HE'S HEADING AS 12 O'CLOCK...
12
3
6
9
...WELL, MOST PLANES GET SHOT DOWN BY AN ENEMY AT 6 O'CLOCK.
SHOWMANSHIP (FEBRUARY, 1986)
THIS POLITICS STUFF ...IT'S CRAZY.
LATER I FOUND OUT THAT SOME OF KUTYNA'S ASTRONAUT FRIENDS HAD DATA ABOUT THE EFFECT OF COLD ON THE O-RINGS.
HE COULDN'T TELL ME DIRECTLY OR THEY'D GET IN TROUBLE.
BY THE WAY, I WAS WORKING ON AN OLD CAR THIS MORNING, AND YOU KNOW ITS CARBURETOR LEAKS WHEN IT GETS COLD.
DO YOU THINK LOW TEMPERATURE HAS SOME EFFECT ON THE O-RINGS THAT SEAL IT?
THE CHALLENGER LAUNCH WAS THE FIRST ONE EVER ON A REALLY COLD DAY.
K
YA SEE, PROFESSORS OF THEORETICAL PHYSICS HAVE TO BE TOLD WHAT TO LOOK FOR.

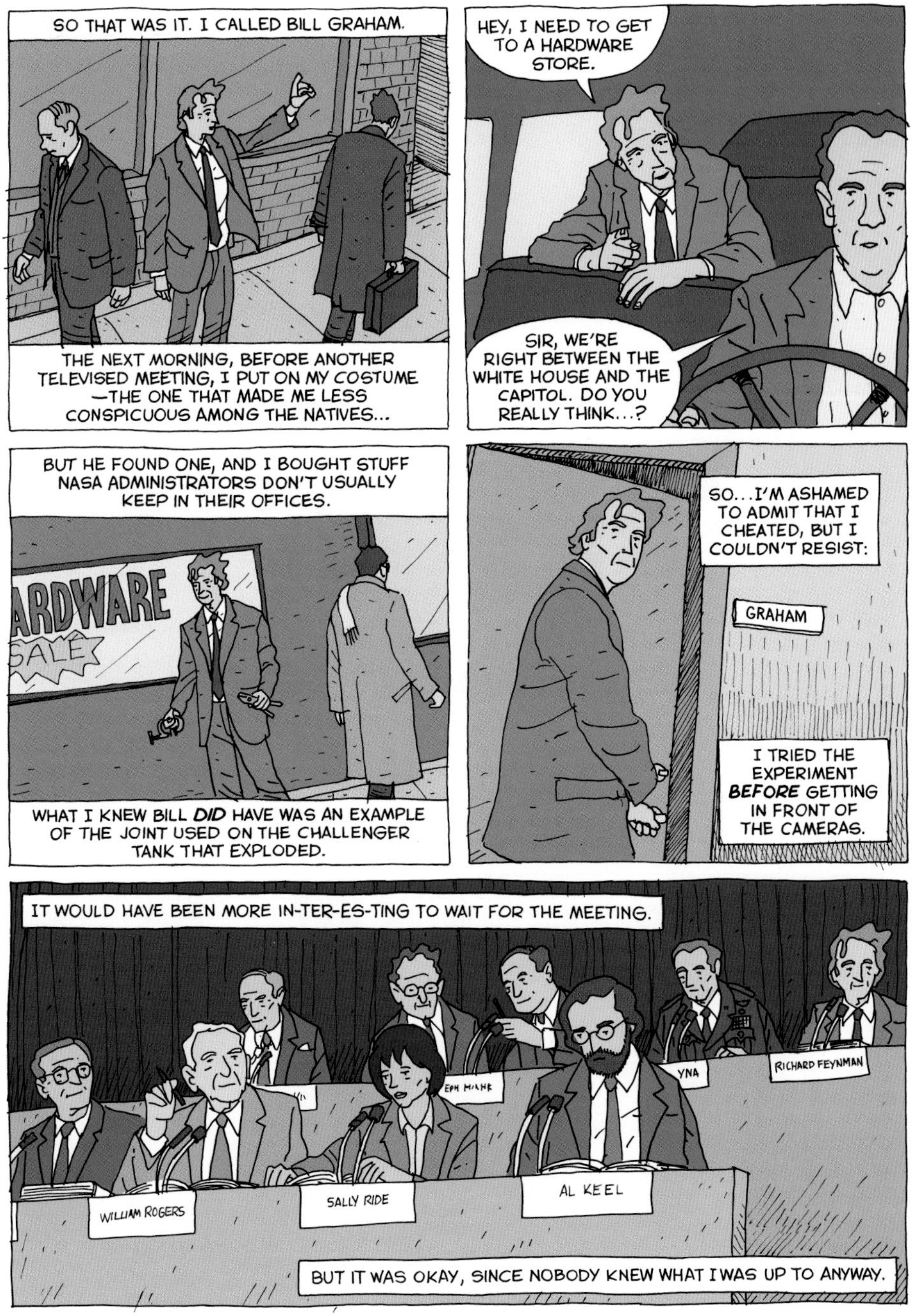

SO THAT WAS IT. I CALLED BILL GRAHAM.
THE NEXT MORNING, BEFORE ANOTHER TELEVISED MEETING, I PUT ON MY *COSTUME* —THE ONE THAT MADE ME LESS CONSPICUOUS AMONG THE NATIVES...
HEY, I NEED TO GET TO A HARDWARE STORE.
SIR, WE'RE RIGHT BETWEEN THE WHITE HOUSE AND THE CAPITOL. DO YOU REALLY THINK...?
BUT HE FOUND ONE, AND I BOUGHT STUFF NASA ADMINISTRATORS DON'T USUALLY KEEP IN THEIR OFFICES.
ARDWARE
SALE
WHAT I KNEW BILL ***DID*** HAVE WAS AN EXAMPLE OF THE JOINT USED ON THE CHALLENGER TANK THAT EXPLODED.
SO... I'M ASHAMED TO ADMIT THAT I CHEATED, BUT I COULDN'T RESIST:
GRAHAM
I TRIED THE EXPERIMENT ***BEFORE*** GETTING IN FRONT OF THE CAMERAS.
IT WOULD HAVE BEEN MORE IN-TER-ES-TING TO WAIT FOR THE MEETING.
RICHARD FEYNMAN
WILLIAM ROGERS
SALLY RIDE
AL KEEL
BUT IT WAS OKAY, SINCE NOBODY KNEW WHAT I WAS UP TO ANYWAY.

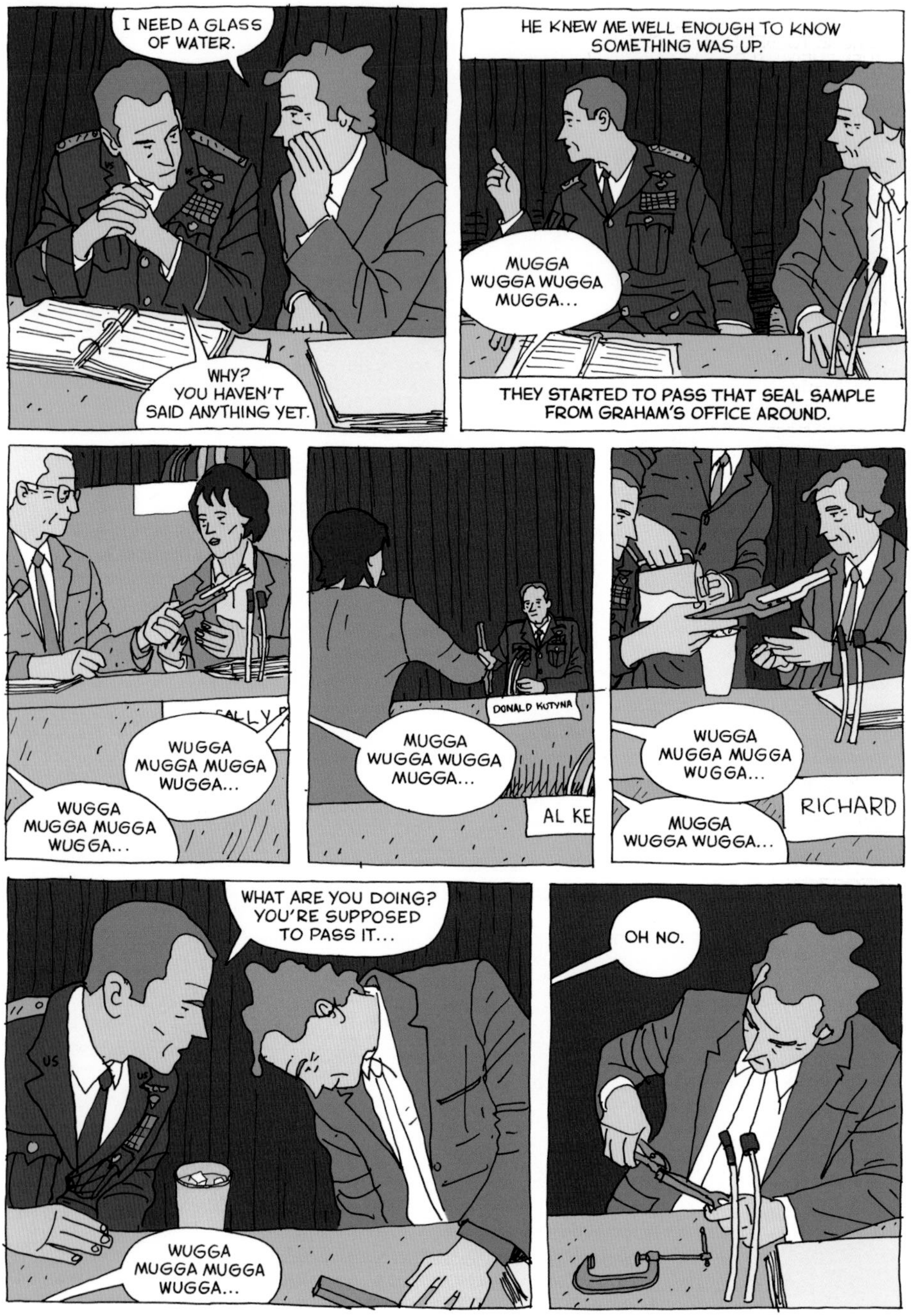
I NEED A GLASS OF WATER.
WHY? YOU HAVEN'T SAID ANYTHING YET.
HE KNEW ME WELL ENOUGH TO KNOW SOMETHING WAS UP.
MUGGA WUGGA WUGGA MUGGA...
THEY STARTED TO PASS THAT SEAL SAMPLE FROM GRAHAM'S OFFICE AROUND.
WUGGA MUGGA MUGGA WUGGA...
WUGGA MUGGA MUGGA WUGGA...
DONALD KUTYNA
MUGGA WUGGA WUGGA MUGGA...
AL KE
WUGGA MUGGA MUGGA WUGGA...
RICHARD
MUGGA WUGGA WUGGA...
WHAT ARE YOU DOING? YOU'RE SUPPOSED TO PASS IT...
WUGGA MUGGA MUGGA WUGGA...
OH NO.

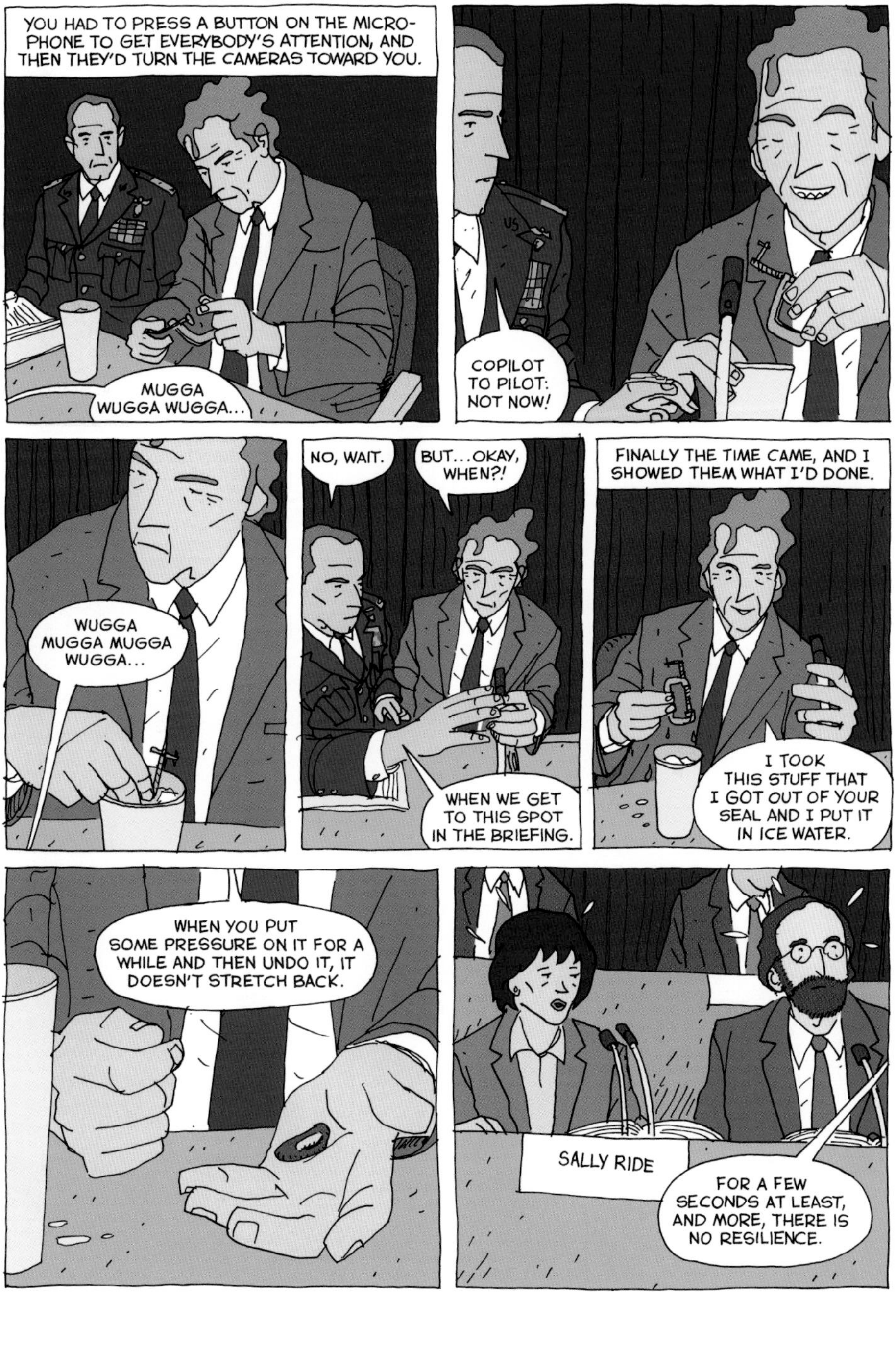
YOU HAD TO PRESS A BUTTON ON THE MICROPHONE TO GET EVERYBODY'S ATTENTION, AND THEN THEY'D TURN THE CAMERAS TOWARD YOU.
MUGGA WUGGA WUGGA...
COPILOT TO PILOT: NOT NOW!
WUGGA MUGGA MUGGA WUGGA...
NO, WAIT.
BUT...OKAY, WHEN?!
WHEN WE GET TO THIS SPOT IN THE BRIEFING.
FINALLY THE TIME CAME, AND I SHOWED THEM WHAT I'D DONE.
I TOOK THIS STUFF THAT I GOT OUT OF YOUR SEAL AND I PUT IT IN ICE WATER.
WHEN YOU PUT SOME PRESSURE ON IT FOR A WHILE AND THEN UNDO IT, IT DOESN'T STRETCH BACK.
SALLY RIDE
FOR A FEW SECONDS AT LEAST, AND MORE, THERE IS NO RESILIENCE.

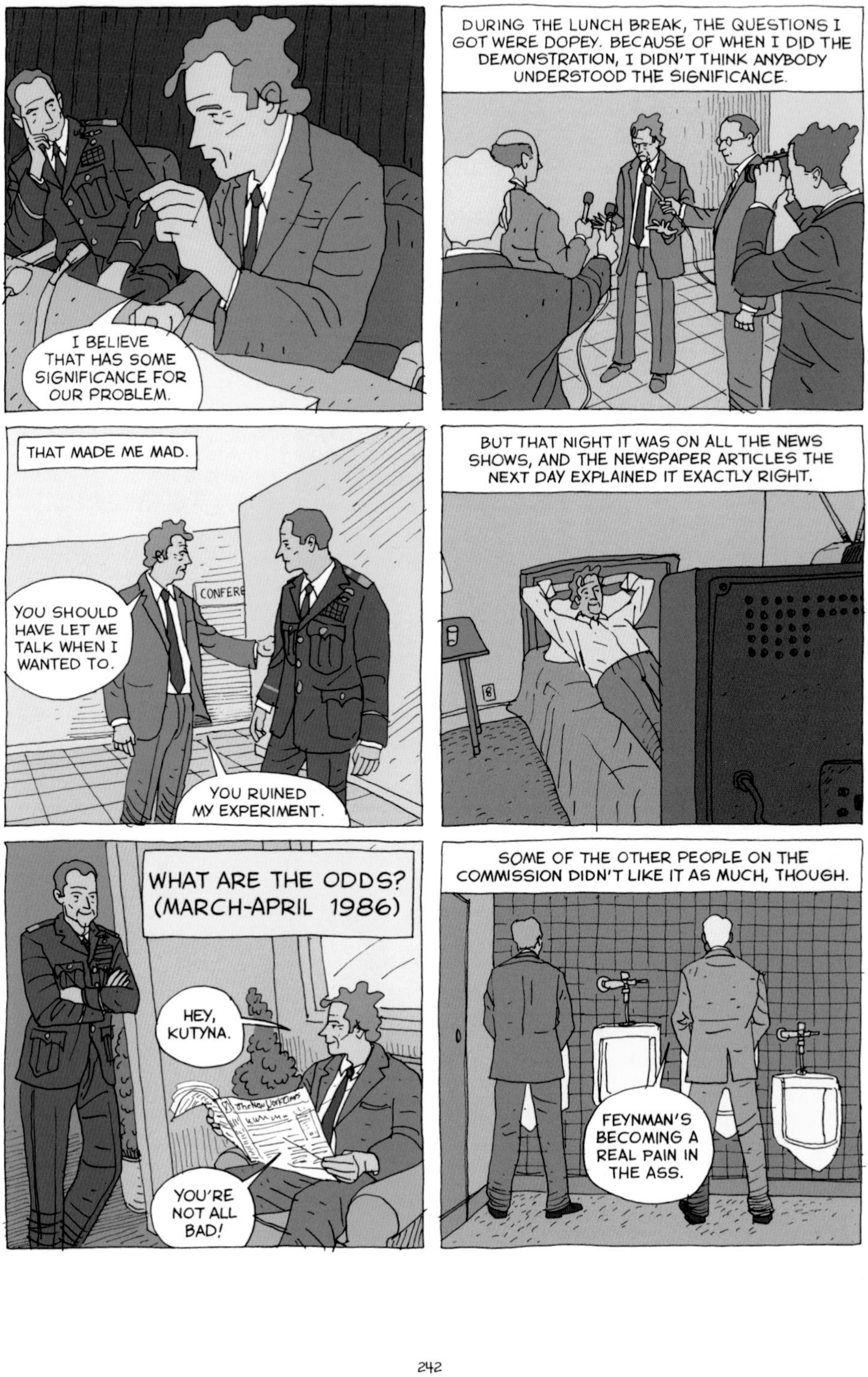
I BELIEVE THAT HAS SOME SIGNIFICANCE FOR OUR PROBLEM.
DURING THE LUNCH BREAK, THE QUESTIONS I GOT WERE DOPEY. BECAUSE OF WHEN I DID THE DEMONSTRATION, I DIDN'T THINK ANYBODY UNDERSTOOD THE SIGNIFICANCE.
THAT MADE ME MAD.
CONFERE
YOU SHOULD HAVE LET ME TALK WHEN I WANTED TO.
YOU RUINED MY EXPERIMENT.
BUT THAT NIGHT IT WAS ON ALL THE NEWS SHOWS, AND THE NEWSPAPER ARTICLES THE NEXT DAY EXPLAINED IT EXACTLY RIGHT.
WHAT ARE THE ODDS? (MARCH-APRIL 1986)
HEY, KUTYNA.
YOU'RE NOT ALL BAD!
SOME OF THE OTHER PEOPLE ON THE COMMISSION DIDN'T LIKE IT AS MUCH, THOUGH.
FEYNMAN'S BECOMING A REAL PAIN IN THE ASS.

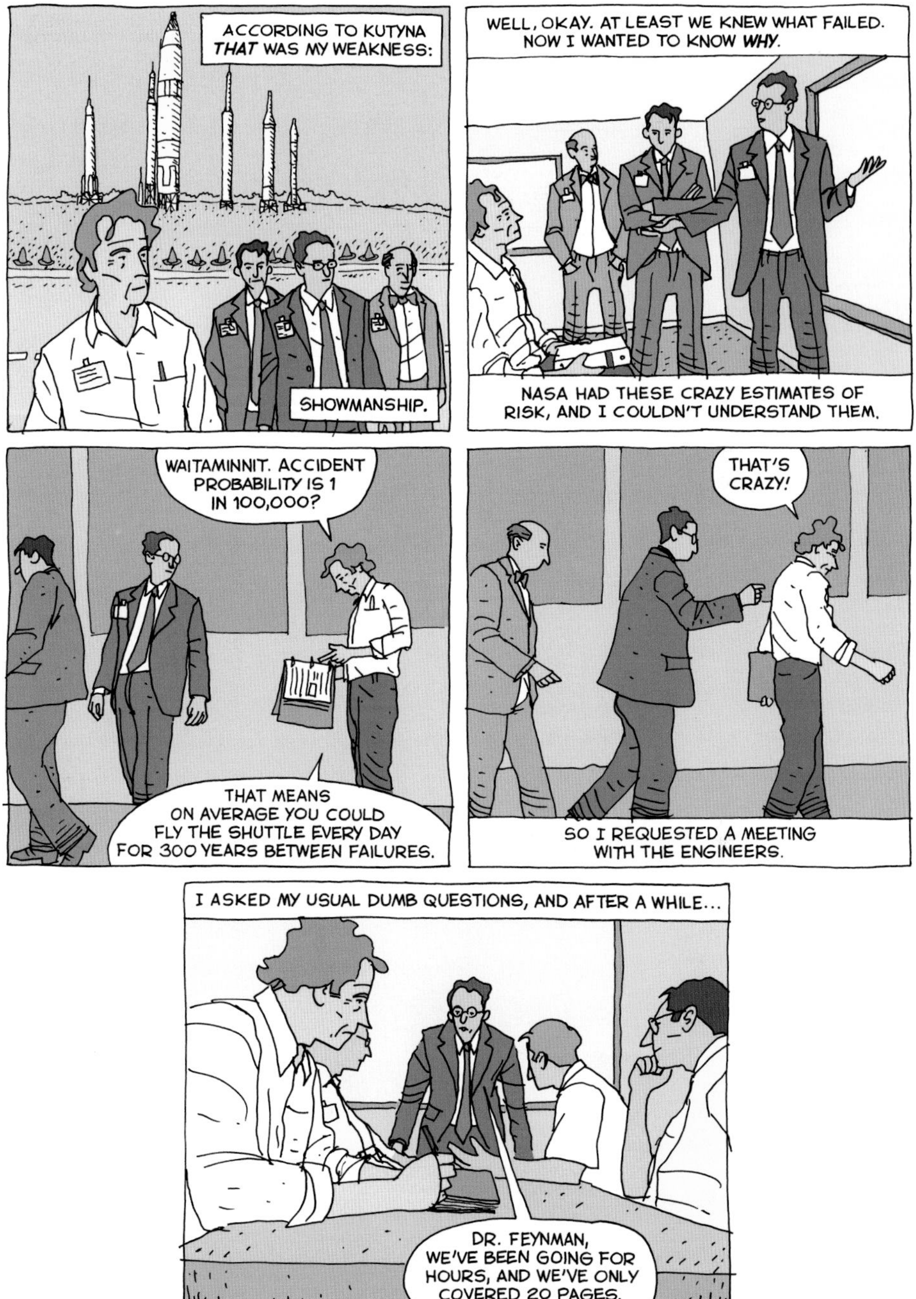
ACCORDING TO KUTYNA *THAT* WAS MY WEAKNESS:
SHOWMANSHIP.
WELL, OKAY. AT LEAST WE KNEW WHAT FAILED. NOW I WANTED TO KNOW ***WHY***.
NASA HAD THESE CRAZY ESTIMATES OF RISK, AND I COULDN'T UNDERSTAND THEM.
WAITAMINNIT. ACCIDENT PROBABILITY IS 1 IN 100,000?
THAT MEANS ON AVERAGE YOU COULD FLY THE SHUTTLE EVERY DAY FOR 300 YEARS BETWEEN FAILURES.
THAT'S CRAZY!
SO I REQUESTED A MEETING WITH THE ENGINEERS.
I ASKED MY USUAL DUMB QUESTIONS, AND AFTER A WHILE...
DR. FEYNMAN, WE'VE BEEN GOING FOR HOURS, AND WE'VE ONLY COVERED 20 PAGES.

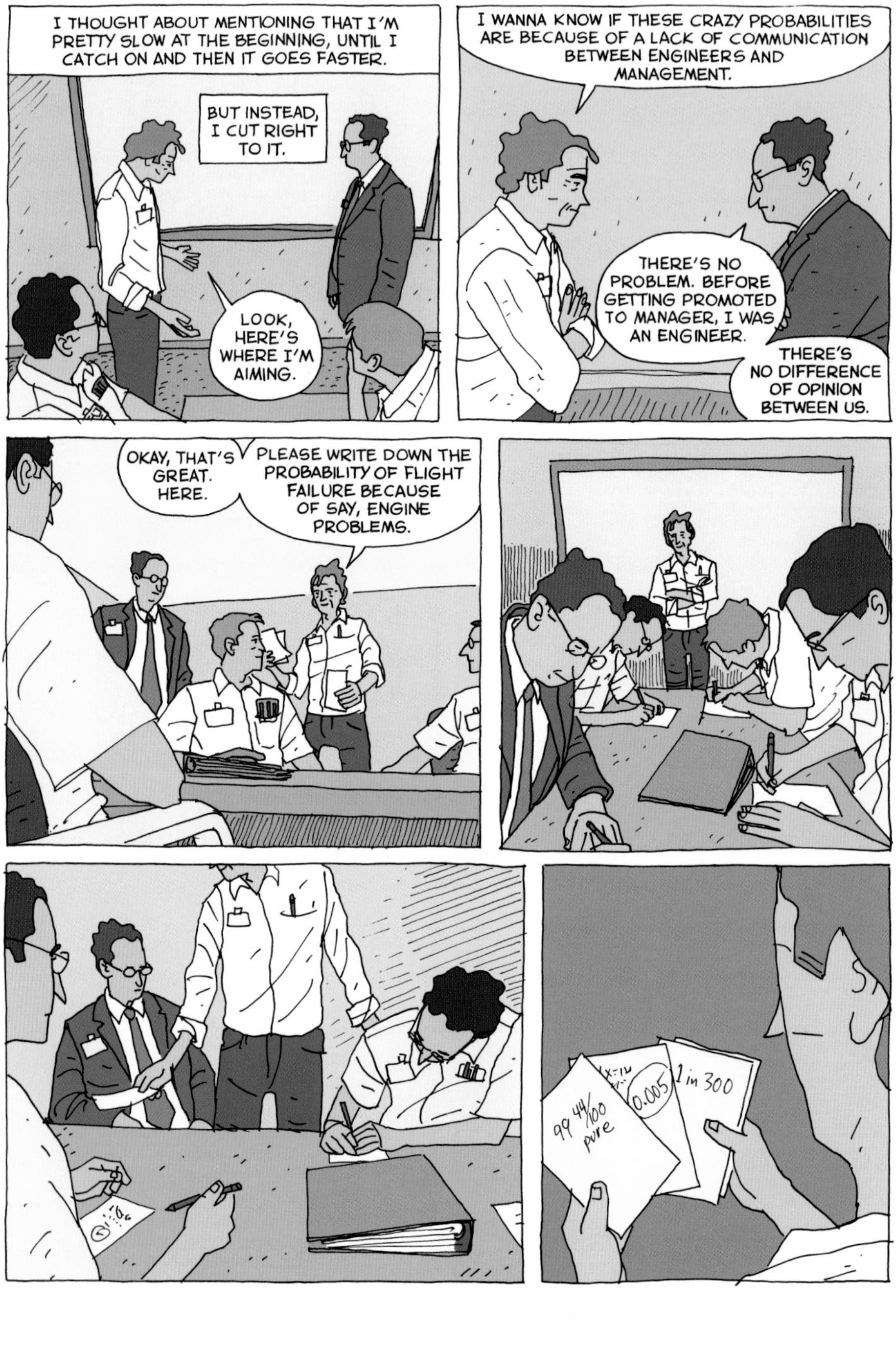
I THOUGHT ABOUT MENTIONING THAT I'M PRETTY SLOW AT THE BEGINNING, UNTIL I CATCH ON AND THEN IT GOES FASTER.
BUT INSTEAD, I CUT RIGHT TO IT.
LOOK, HERE'S WHERE I'M AIMING.
I WANNA KNOW IF THESE CRAZY PROBABILITIES ARE BECAUSE OF A LACK OF COMMUNICATION BETWEEN ENGINEERS AND MANAGEMENT.
THERE'S NO PROBLEM. BEFORE GETTING PROMOTED TO MANAGER, I WAS AN ENGINEER.
THERE'S NO DIFFERENCE OF OPINION BETWEEN US.
OKAY, THAT'S GREAT. HERE.
PLEASE WRITE DOWN THE PROBABILITY OF FLIGHT FAILURE BECAUSE OF SAY, ENGINE PROBLEMS.
99 44/100 pure
0.005
1 in 300

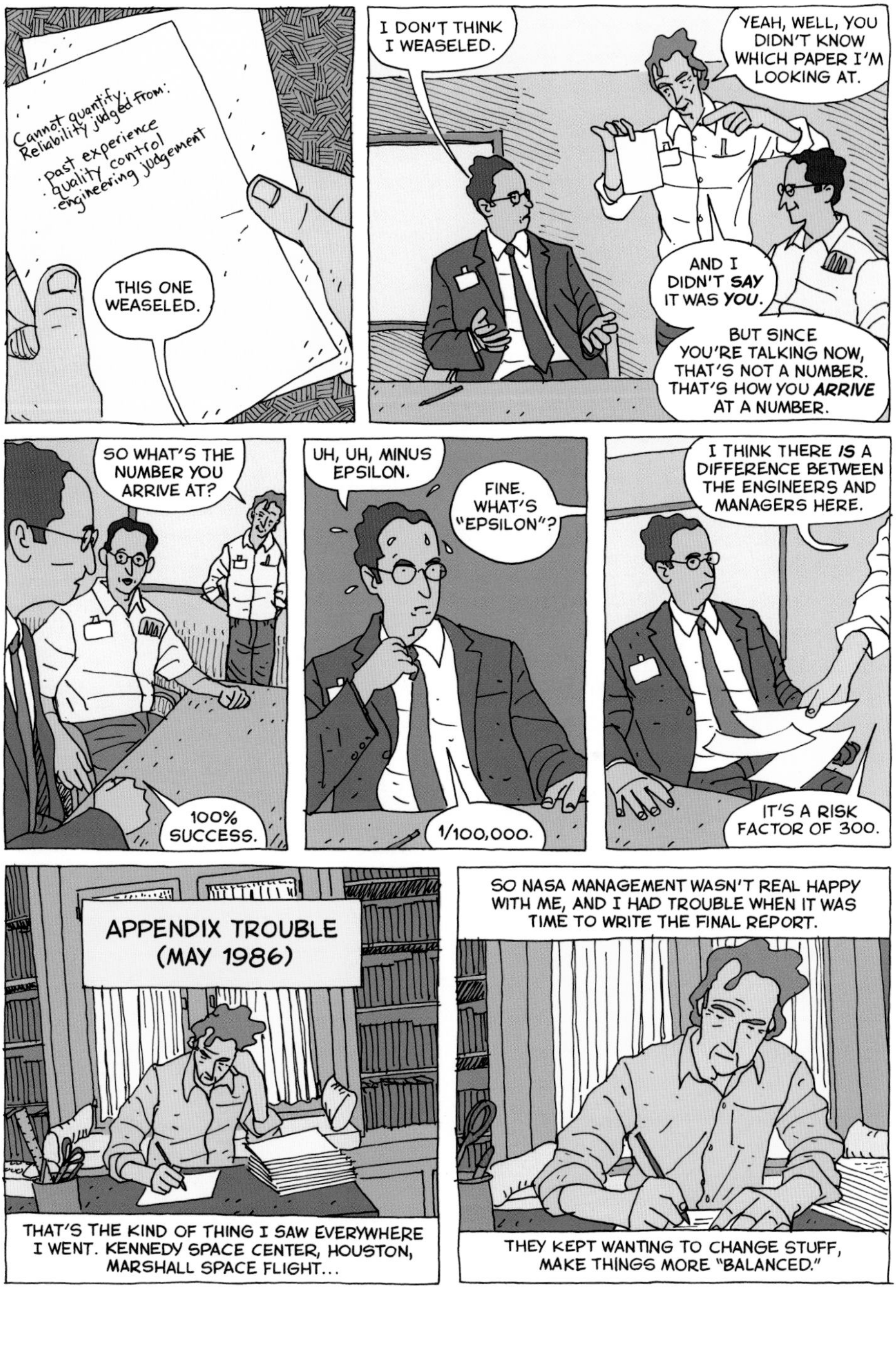
Cannot quantify.
Reliability judged from:
· past experience
· quality control
· engineering judgement
THIS ONE WEASELED.
I DON'T THINK I WEASELED.
YEAH, WELL, YOU DIDN'T KNOW WHICH PAPER I'M LOOKING AT.
AND I DIDN'T SAY IT WAS YOU.
BUT SINCE YOU'RE TALKING NOW, THAT'S NOT A NUMBER. THAT'S HOW YOU ARRIVE AT A NUMBER.
SO WHAT'S THE NUMBER YOU ARRIVE AT?
100% SUCCESS.
UH, UH, MINUS EPSILON.
FINE. WHAT'S "EPSILON"?
1/100,000.
I THINK THERE IS A DIFFERENCE BETWEEN THE ENGINEERS AND MANAGERS HERE.
IT'S A RISK FACTOR OF 300.
APPENDIX TROUBLE (MAY 1986)
THAT'S THE KIND OF THING I SAW EVERYWHERE I WENT. KENNEDY SPACE CENTER, HOUSTON, MARSHALL SPACE FLIGHT...
SO NASA MANAGEMENT WASN'T REAL HAPPY WITH ME, AND I HAD TROUBLE WHEN IT WAS TIME TO WRITE THE FINAL REPORT.
THEY KEPT WANTING TO CHANGE STUFF, MAKE THINGS MORE "BALANCED."

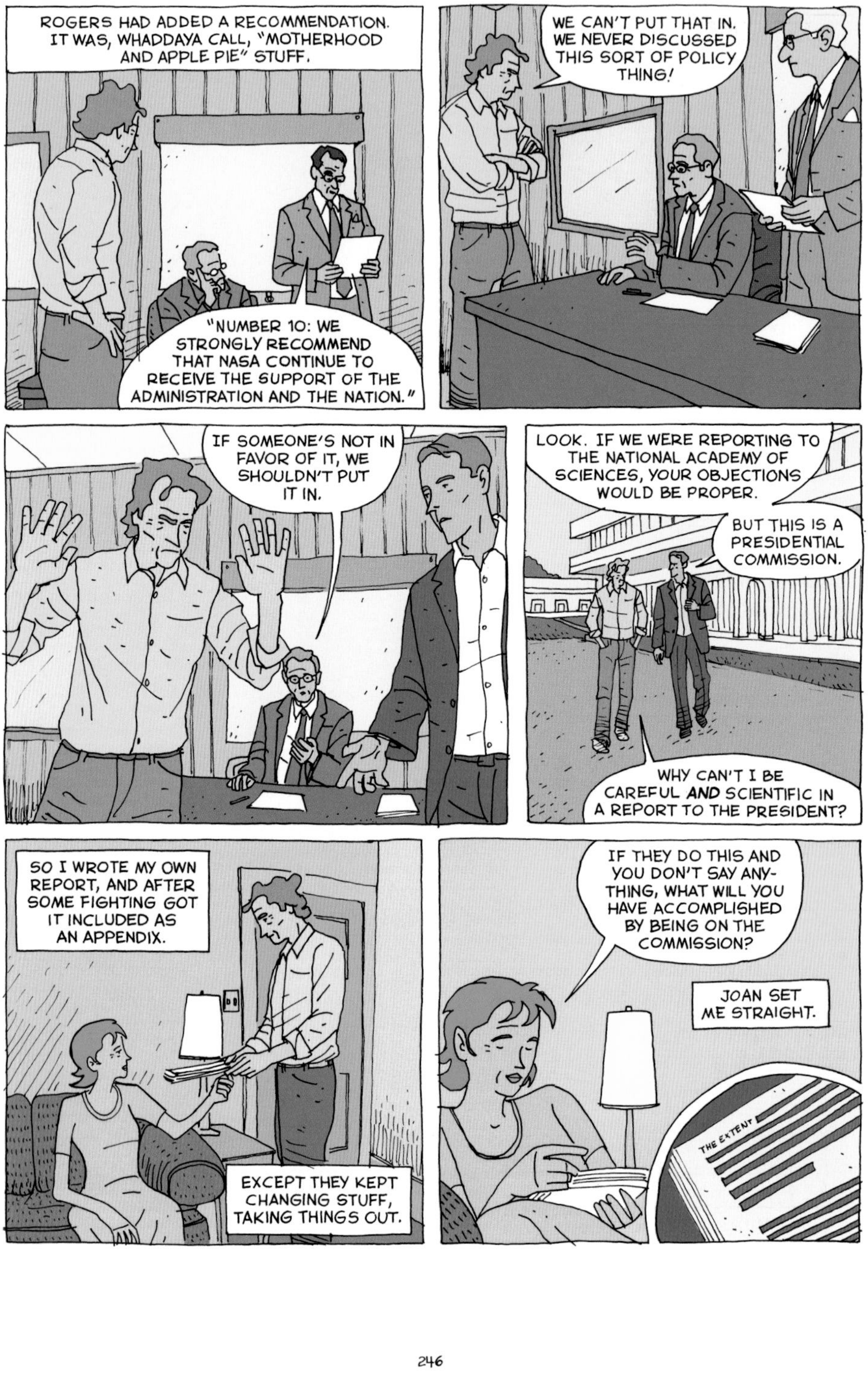
ROGERS HAD ADDED A RECOMMENDATION. IT WAS, WHADDAYA CALL, "MOTHERHOOD AND APPLE PIE" STUFF.
"NUMBER 10: WE STRONGLY RECOMMEND THAT NASA CONTINUE TO RECEIVE THE SUPPORT OF THE ADMINISTRATION AND THE NATION."
WE CAN'T PUT THAT IN. WE NEVER DISCUSSED THIS SORT OF POLICY THING!
IF SOMEONE'S NOT IN FAVOR OF IT, WE SHOULDN'T PUT IT IN.
LOOK. IF WE WERE REPORTING TO THE NATIONAL ACADEMY OF SCIENCES, YOUR OBJECTIONS WOULD BE PROPER.
BUT THIS IS A PRESIDENTIAL COMMISSION.
WHY CAN'T I BE CAREFUL **AND** SCIENTIFIC IN A REPORT TO THE PRESIDENT?
SO I WROTE MY OWN REPORT, AND AFTER SOME FIGHTING GOT IT INCLUDED AS AN APPENDIX.
EXCEPT THEY KEPT CHANGING STUFF, TAKING THINGS OUT.
IF THEY DO THIS AND YOU DON'T SAY ANY-THING, WHAT WILL YOU HAVE ACCOMPLISHED BY BEING ON THE COMMISSION?
JOAN SET ME STRAIGHT.
THE EXTENT

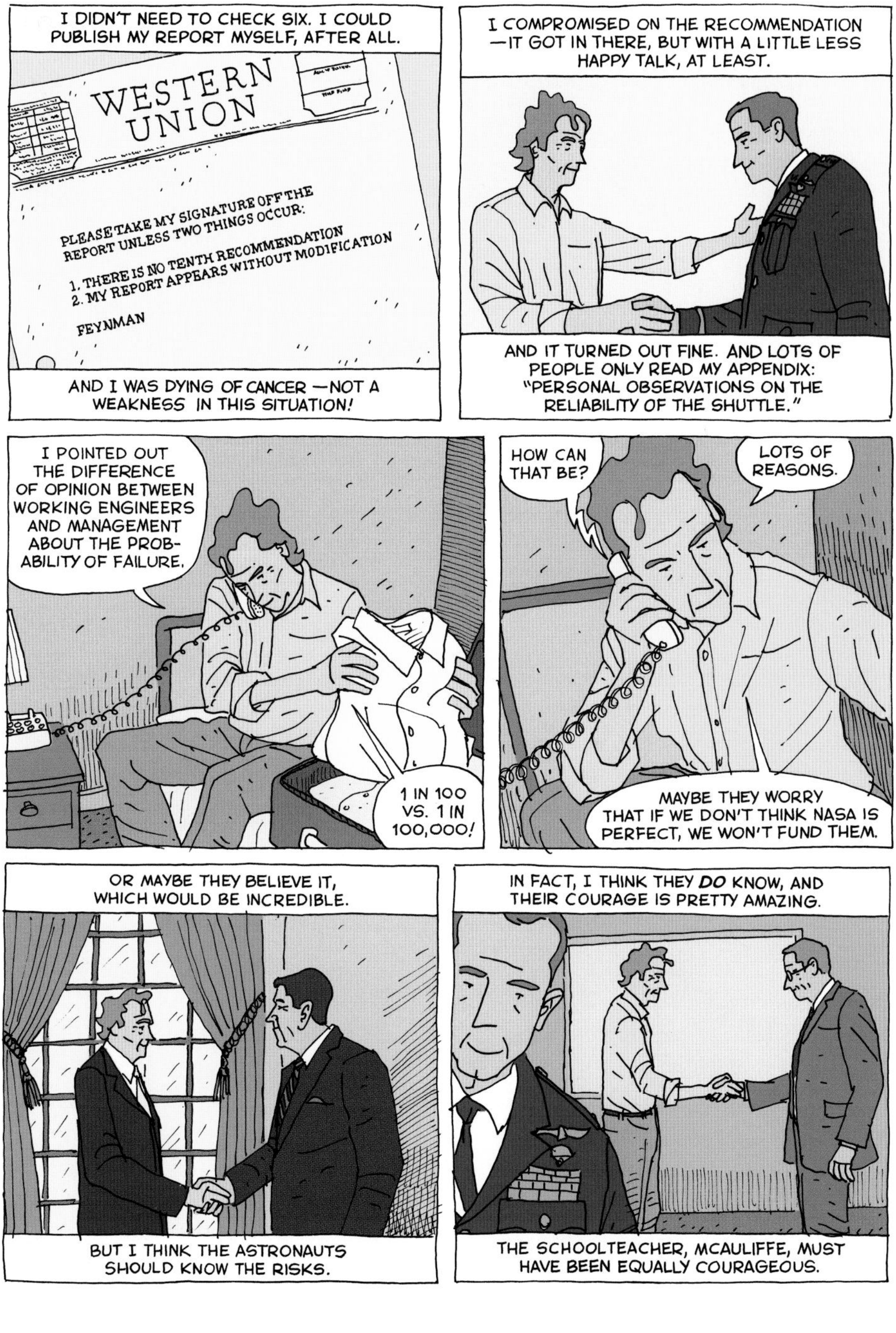
I DIDN'T NEED TO CHECK SIX. I COULD PUBLISH MY REPORT MYSELF, AFTER ALL.
WESTERN UNION
PLEASE TAKE MY SIGNATURE OFF THE REPORT UNLESS TWO THINGS OCCUR:
1. THERE IS NO TENTH RECOMMENDATION
2. MY REPORT APPEARS WITHOUT MODIFICATION
FEYNMAN
AND I WAS DYING OF CANCER —NOT A WEAKNESS IN THIS SITUATION!
I COMPROMISED ON THE RECOMMENDATION —IT GOT IN THERE, BUT WITH A LITTLE LESS HAPPY TALK, AT LEAST.
AND IT TURNED OUT FINE. AND LOTS OF PEOPLE ONLY READ MY APPENDIX: "PERSONAL OBSERVATIONS ON THE RELIABILITY OF THE SHUTTLE."
I POINTED OUT THE DIFFERENCE OF OPINION BETWEEN WORKING ENGINEERS AND MANAGEMENT ABOUT THE PROBABILITY OF FAILURE.
1 IN 100 VS. 1 IN 100,000!
HOW CAN THAT BE?
LOTS OF REASONS.
MAYBE THEY WORRY THAT IF WE DON'T THINK NASA IS PERFECT, WE WON'T FUND THEM.
OR MAYBE THEY BELIEVE IT, WHICH WOULD BE INCREDIBLE.
BUT I THINK THE ASTRONAUTS SHOULD KNOW THE RISKS.
IN FACT, I THINK THEY DO KNOW, AND THEIR COURAGE IS PRETTY AMAZING.
THE SCHOOLTEACHER, MCAULIFFE, MUST HAVE BEEN EQUALLY COURAGEOUS.

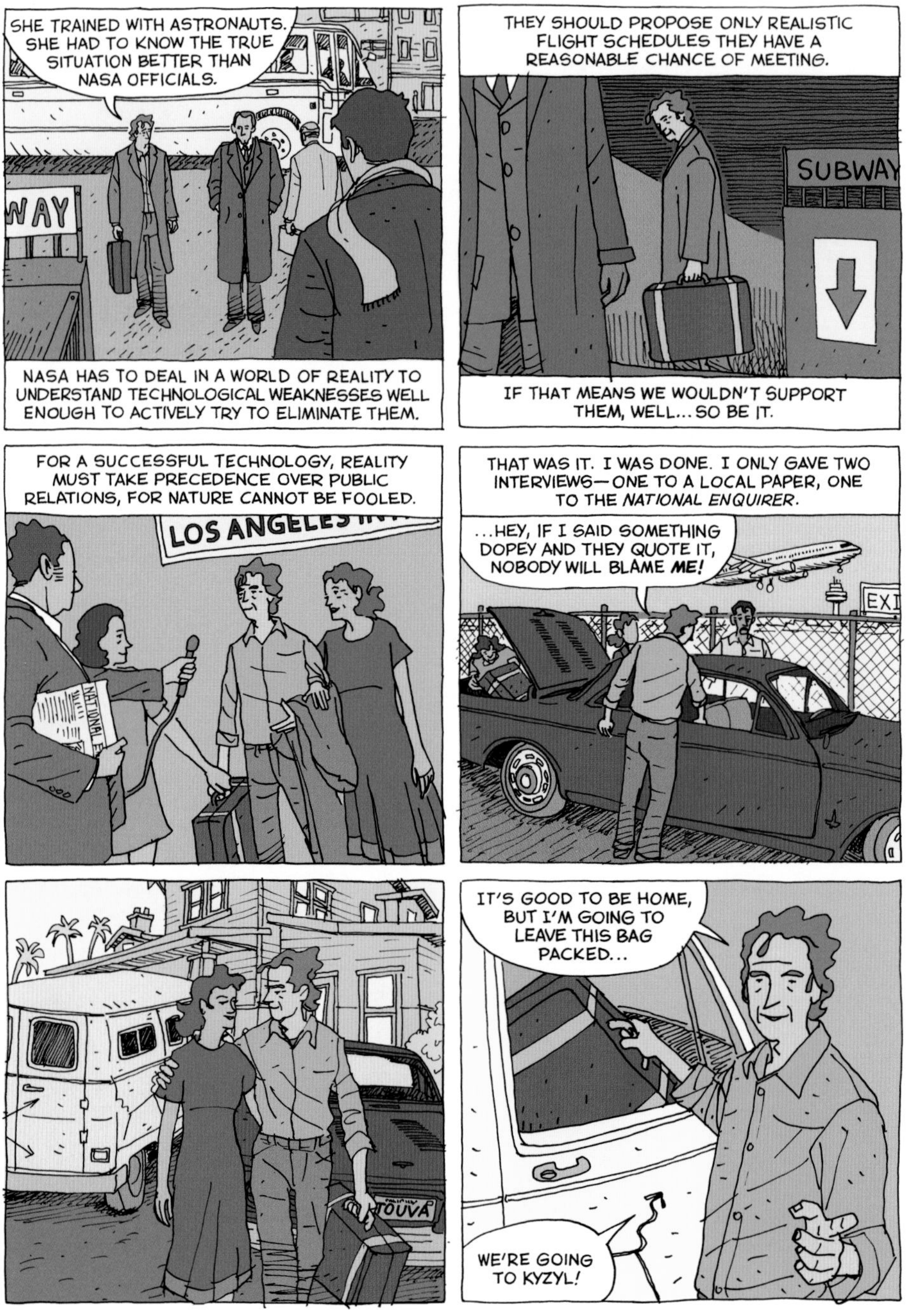
SHE TRAINED WITH ASTRONAUTS. SHE HAD TO KNOW THE TRUE SITUATION BETTER THAN NASA OFFICIALS.
WAY
NASA HAS TO DEAL IN A WORLD OF REALITY TO UNDERSTAND TECHNOLOGICAL WEAKNESSES WELL ENOUGH TO ACTIVELY TRY TO ELIMINATE THEM.
THEY SHOULD PROPOSE ONLY REALISTIC FLIGHT SCHEDULES THEY HAVE A REASONABLE CHANCE OF MEETING.
SUBWAY
IF THAT MEANS WE WOULDN'T SUPPORT THEM, WELL... SO BE IT.
FOR A SUCCESSFUL TECHNOLOGY, REALITY MUST TAKE PRECEDENCE OVER PUBLIC RELATIONS, FOR NATURE CANNOT BE FOOLED.
LOS ANGELES
NATIONAL E
THAT WAS IT. I WAS DONE. I ONLY GAVE TWO INTERVIEWS—ONE TO A LOCAL PAPER, ONE TO THE *NATIONAL ENQUIRER*.
...HEY, IF I SAID SOMETHING DOPEY AND THEY QUOTE IT, NOBODY WILL BLAME **ME!**
EXI
TOUVA
IT'S GOOD TO BE HOME, BUT I'M GOING TO LEAVE THIS BAG PACKED...
WE'RE GOING TO KYZYL!

RICHARD FEYNMAN DIDN'T MAKE IT TO TUVA.
THE OFFICIAL, FORMAL INVITATION FROM MOSCOW PERMITTING THE VISIT
WAS DATED FEBRUARY 19, 1988—FOUR DAYS AFTER HE DIED.

HIS LAST WORDS, UPON FORCING HIMSELF AWAKE FROM A COMA:

"I'D HATE TO DIE TWICE. IT'S SO BORING."

(ALMOST COMPLETE*) BIBLIOGRAPHY & EARLY SKETCHES

OUR STACK OF FEYNMAN MATERIAL IS OVER A METER HIGH AND RISING. YOURS MAY NOT GROW THAT TALL, BUT IF YOU LIKED THIS STORY, YOU'LL ENJOY THE FOLLOWING...AND BE SURPRISED AT HOW MUCH WE HAD TO LEAVE OUT. SO PLEASE READ MORE, STARTING WITH THE BOOKS BY FEYNMAN HIMSELF:

THE CHARACTER OF PHYSICAL LAW (CAMBRIDGE, MA: MIT PRESS, 1965).
>THE PERFECT BOOK TO READ TO GET A FEEL FOR QUANTUM PHYSICS. OUR OPENING SCENE CAME FROM AN ANECDOTE RECOUNTED IN JAMES GLEICK'S INTRODUCTION TO THE 1994 MODERN LIBRARY EDITION, SO SEEK THAT ONE OUT AS WELL.

CLASSIC FEYNMAN: ALL THE ADVENTURES OF A CURIOUS CHARACTER (NY: W.W. NORTON, 2006).
>I STILL LIKE THE ORIGINAL VOLUMES—*SURELY YOU'RE JOKING, MR. FEYNMAN!* (1985) AND *WHAT DO <u>YOU</u> CARE WHAT OTHER PEOPLE THINK?* (1988)—SINCE READING FEYNMAN STORIES OUT OF ORDER IS A GREAT WAY TO EXPERIENCE A LIFE THE WAY IT IS: MESSY, AND WITHOUT A DISCERNABLE PLOT! BUT THIS MAY JUST BE BIAS RESULTING FROM THOSE BEING MY FIRST EXPOSURE TO THEM, SO IF YOU ONLY GET ONE BOOK, THE SPLENDID *CLASSIC FEYNMAN* IS YOUR BEST BET. IT NOT ONLY HAS ALL THE STORIES FROM ITS PREDECESSORS, BUT ALSO

INCLUDES A CD OF FEYNMAN TALKING ABOUT "LOS ALAMOS FROM BELOW." CARL AND MICHELLE FEYNMAN CONTRIBUTE TO THIS OMNIBUS EDITION AS WELL, AS DO FREEMAN DYSON AND ALAN ALDA. IT'S GREAT.

THE FEYNMAN LECTURES ON PHYSICS, VOLUMES I-III, BY RICHARD FEYNMAN, ROBERT LEIGHTON, AND MATTHEW SANDS (READING, MA: ADDISON-WESLEY, 1964).

ALONG WITH *QED* (PAGE 255), WITH THIS YOU HAVE ALL OF PHYSICS, RIGHT THERE WAITING FOR YOU. OKAY, NOT REALLY ALL, BUT CLOSE ENOUGH! THE *LECTURES* ARE TOUGHER SLEDDING, AND I'VE ONLY DIPPED INTO THEM HERE AND THERE MYSELF. BUT THEY'RE WELL WORTH THE EFFORT. (I RECOMMEND VOLUME II'S CHAPTER 19, "THE PRINCIPLE OF LEAST ACTION", AS YOUR FIRST SAMPLE.)

FEYNMAN'S TIPS ON PHYSICS, BY RICHARD FEYNMAN, MICHAEL A. GOTTLIEB, AND RALPH LEIGHTON (SAN FRANCISCO: PEARSON ADDISON-WESLEY, 2006).

NOT ONLY DOES FEYNMAN PROVIDE TIPS ON SOLVING PHYSICS PROBLEMS, HE ALSO PROVIDES TIPS ON HOW TO DEAL WITH BEING BELOW AVERAGE. THE LATTER ARE SURPRISINGLY GOOD, GIVEN HE HAD NO PRACTICAL EXPERIENCE WITH THAT POSITION. MATT SANDS' INTRODUCTION, DESCRIBING THE ORIGINS OF THE FEYNMAN LECTURES, IS ALSO ENTERTAINING.

THE MEANING OF IT ALL: THOUGHTS OF A CITIZEN SCIENTIST
(READING, MA: PERSEUS BOOKS, 1998).

> THESE LECTURES ARE SOME OF THE RESULTS OF THE
> MIDDLE-AGED DISEASE FEYNMAN TALKED ABOUT ON
> PAGE 156. THEY'RE REALLY NOT BAD! THESE CLEANED
> UP VERSIONS MAKE AN INTERESTING COUNTERPOINT TO
> THE RAW TRANSCRIPTIONS AVAILABLE IN THE CALTECH
> ARCHIVES—WHICH I ENJOYED READING AND CONSULTING
> EVEN MORE, AND SOME OF WHICH MADE THEIR WAY INTO
> THIS BOOK.

NOBEL ACCEPTANCE AND "THE DEVELOPMENT OF THE SPACE-TIME
VIEW OF QUANTUM ELECTRODYNAMICS," FROM *LES PRIX NOBEL EN
1965* (STOCKHOLM: IMPRIMERIE ROYALE P.A. NORSTEDT & SÖNER,
1966).

> EVEN CONSTRAINED BY A TUXEDO AND ROYALTY, FEYNMAN
> REMAINED QUOTABLE.

THE PLEASURE OF FINDING THINGS OUT, EDITED BY JEFFREY
ROBBINS (CAMBRIDGE, MA: HELIX BOOKS, 1999).

> THESE ESSAYS ARE ALL TERRIFIC, AS IS FREEMAN DYSON'S
> FOREWORD. READ THEM ALL, BUT IF YOU'RE SHORT ON TIME
> THE BEST ONES ARE "THE PLEASURE OF FINDING THINGS
> OUT," "THERE'S PLENTY OF ROOM AT THE BOTTOM," "THE
> SMARTEST MAN IN THE WORLD," "CARGO CULT SCIENCE,"
> AND...AW, WHO AM I KIDDING? YOU SHOULD READ THEM ALL!

PERFECTLY REASONABLE DEVIATIONS FROM THE BEATEN TRACK (NY:
BASIC BOOKS, 2005).

> BEFORE E-MAIL, PEOPLE WROTE LETTERS. AND IN 1968
> RICHARD FEYNMAN STARTED DONATING HIS PERSONAL
> PAPERS TO THE CALTECH ARCHIVES. THANK GOODNESS.
> MICHELLE FEYNMAN EDITED THE COLLECTION AND ADDED
> COMMENTARY AS WELL AS MANY LOVELY PHOTOGRAPHS.

QED: THE STRANGE THEORY OF LIGHT AND MATTER (PRINCETON,
NJ: PRINCETON UNIVERSITY PRESS, 1985).

> FEYNMAN WROTE THIS SPECIFICALLY FOR PEOPLE—LIKE ALIX
> MAUTNER—WITH NO BACKGROUND IN MATH OR PHYSICS.
> YOU CAN SEE FEYNMAN DELIVER EARLY VERSIONS OF THE
> LECTURES IN NEW ZEALAND THANKS TO THE VEGA SCIENCE
> TRUST: *HTTP://WWW.VEGA.ORG.UK/VIDEO/SUBSERIES/8/*.

PAULI & EINSTEIN

THE REASON FOR ANTIPARTICLES, FROM "ELEMENTARY PARTICLES AND THE LAWS OF PHYSICS: THE 1986 DIRAC MEMORIAL LECTURES" (ALEXANDRIA, VA: SCIENTIFIC CONSULTING SERVICES, 1997) AND *ELEMENTARY PARTICLES AND THE LAWS OF PHYSICS: THE 1986 DIRAC MEMORIAL LECTURES*, BY RICHARD FEYNMAN AND STEVEN WEINBERG (CAMBRIDGE: CAMBRIDGE UNIVERSITY PRESS, 1987).

    IN THIS VIDEO/BOOK PAIRING FEYNMAN'S VISUAL, AND PHYSICAL, LECTURE STYLE IS IN FULL FORCE DURING HIS DEMONSTRATION OF THE IMPORTANCE OF SPIN AND ROTATION IN PHYSICAL THEORIES.

SIX EASY PIECES: ESSENTIALS OF PHYSICS EXPLAINED BY ITS MOST BRILLIANT TEACHER (READING, MA: HELIX BOOKS, 1994) AND *SIX NOT-SO-EASY PIECES: EINSTEIN'S RELATIVITY, SYMMETRY, AND SPACE-TIME* (READING, MA: HELIX BOOKS, 1997).

    THOUGH THE SECOND BOOK MAY CHALLENGE YOU WITH ITS COMPLEX IDEAS, BOTH ARE STILL A WHOLE LOT OF FUN! I RECOMMEND THE AUDIO VERSIONS TO GET THE FULL FLAVOR OF FEYNMAN'S STYLE.

I ALSO RECOMMEND THE FOLLOWING SECONDARY SOURCES, MANY BY PEOPLE WHO KNEW FEYNMAN WELL:

THE BEAT OF A DIFFERENT DRUM: THE LIFE AND SCIENCE OF RICHARD FEYNMAN, BY JAGDISH MEHRA (OXFORD: OXFORD UNIVERSITY PRESS, 1994).

PLENTY OF PHYSICS, AND MATHEMATICAL PHYSICS AT THAT, BUT THIS IS STILL MY FAVORITE BIOGRAPHY OF FEYNMAN BOTH BECAUSE OF ITS COMPLETENESS AND THE OBVIOUS CARE MEHRA TOOK IN INTERVIEWING FEYNMAN AND WEAVING TOGETHER THE WHOLE OF HIS LIFE.

THE BEST MIND SINCE EINSTEIN, PRODUCED BY CHRISTOPHER SYKES AND MELANIE WALLACE (BOSTON: WGBH/PBS/NOVA SERIES, ORIGINALLY BROADCAST DECEMBER 21, 1993).

A TERRIFIC OVERVIEW OF FEYNMAN'S LIFE AND CAREER IN SCIENCE. WATCH THIS IF YOU CAN'T GET HOLD OF *NO ORDINARY GENIUS* (SEE PAGE 259), SINCE THIS IS A CONDENSED VERSION OF THAT PROGRAM.

CLIMBING THE MOUNTAIN: THE SCIENTIFIC BIOGRAPHY OF JULIAN SCHWINGER, BY JAGDISH MEHRA AND KIMBALL A. MILTON (OXFORD: OXFORD UNIVERSITY PRESS, 2000).

AN EXCELLENT COMPANION TO MEHRA'S *THE BEAT OF A DIFFERENT DRUM*.

DISTURBING THE UNIVERSE, BY FREEMAN DYSON (NY: HARPER AND ROW, 1979).

DYSON'S COMPARISON OF FEYNMAN TO JOF THE JUGGLER WILL STRIKE A CHORD WITH YOU. READ THE WHOLE BOOK, THOUGH. IT'S TERRIFIC.

FEYNMAN'S RAINBOW: A SEARCH FOR BEAUTY IN PHYSICS AND IN LIFE, BY LEONARD MLODINOW (NY: WARNER BOOKS, 2003).

SLIM AND SLIGHT, BUT I ENJOYED READING ABOUT MLODINOW'S BRIEF, PERSONAL, AND ALMOST INCIDENTAL ENCOUNTERS WITH FEYNMAN. I MARKED A NUMBER OF QUOTES AS I READ IT, BUT DIDN'T END UP USING THEM FOR THIS STORY. IT'S THAT KIND OF BOOK.

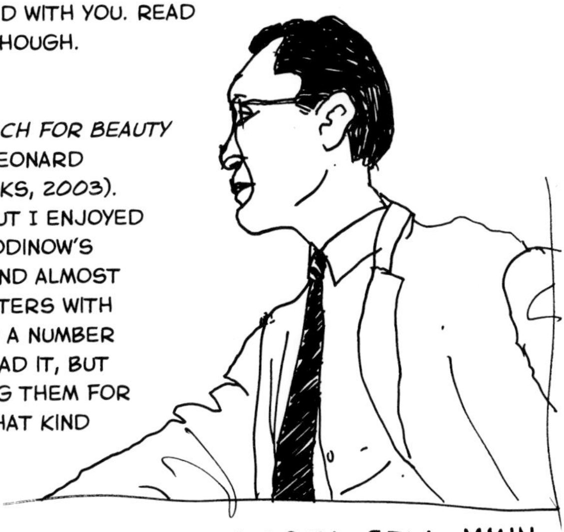

MURRAY GELL-MANN

FROM EROS TO GAIA, BY FREEMAN DYSON (NY: PENGUIN BOOKS, 1992).
GREAT THROUGHOUT, BUT DYSON'S FIRST-PERSON
RECOLLECTIONS OF FEYNMAN'S DISCOVERY OF QED
AND THE ROAD TRIP THAT RESULTED IN DYSON HIMSELF
BECOMING THE TRANSLATOR OF FEYNMAN'S IDIOSYNCRATIC
THEORY TO THE REST OF THE PHYSICS COMMUNITY MAKE IT
INDISPENSABLE.

FUN TO IMAGINE, PRODUCED BY CHRISTOPHER SYKES (BBC2,
ORIGINALLY BROADCAST IN 1983).
SNIPPETS OF PHYSICS AS VIEWED THROUGH THE FEYNMAN
LENS. THEY'RE QUICK AND A LOT OF FUN.

GENIUS: THE LIFE AND SCIENCE OF RICHARD FEYNMAN, BY JAMES
GLEICK (NY: PANTHEON, 1992).
THIS RIVALS MEHRA'S BOOK, BUT WITH LESS DEPTH IN THE
PHYSICS. IT'S STILL WORTH READING, BUT I DIDN'T FIND
MYSELF CONSULTING IT DURING THE PRODUCTION OF THIS
BOOK.

JULIAN SCHWINGER: THE PHYSICIST, THE TEACHER, AND THE MAN,
EDITED BY Y. JACK NG (NJ: WORLD SCIENTIFIC, 1996).
SCHWINGER, THOUGH FAR LESS FAMOUS THAN HIS
COUNTERPART, IS REMEMBERED JUST AS FONDLY BY THOSE
WHO KNEW HIM.

THE LAST JOURNEY OF A GENIUS/THE QUEST FOR TANNU TUVA,
PRODUCED BY CHRISTOPHER SYKES (BBC2 HORIZON SPECIAL/WGBH/
PBS/NOVA, ORIGINALLY BROADCAST JANUARY 24, 1989).
THE STORY OF FEYNMAN AND RALPH LEIGHTON'S EFFORTS
TO GET TO TUVA, WITH MANY ADVENTURES—INCLUDING THE
CHALLENGER INVESTIGATION—ALONG THE WAY. MOVING, SAD,
AND INSPIRING.

"MOST OF THE GOOD STUFF": MEMORIES OF RICHARD FEYNMAN,
EDITED BY LAURIE M. BROWN AND JOHN S. RIGDEN (NY: AMERICAN
INSTITUTE OF PHYSICS, 1993).
MISTITLED: THIS IS *GREAT* STUFF, INCLUDING DANNY HILLIS'
STORIES OF KNOWING, WORKING WITH, AND BEFRIENDING
FEYNMAN. HILLIS CONTINUES TO DO INTERESTING WORK, WITH
THE CLOCK OF THE LONG NOW (*WWW.LONGNOW.ORG*) BEING
MY FAVORITE OF HIS MOST RECENT PROJECTS.

ROBERT OPPENHEIMER

NANO: THE EMERGING SCIENCE OF NANOTECHNOLOGY, BY ED REGIS (BOSTON: LITTLE BROWN, 1995).

> REGIS IS AN ENGAGING WRITER, AND HIS ACCOUNT OF THE LEAD-UP AND IMMEDIATE AFTERMATH OF FEYNMAN'S "PLENTY OF ROOM AT THE BOTTOM" SPEECH GIVES YOU THE FLAVOR OF THE TIMES.

NO ORDINARY GENIUS, PRODUCED BY CHRISTOPHER SYKES (BBC, ORIGINALLY BROADCAST JANUARY 24, 1993).

> THE BEST FEYNMAN OVERVIEW ON VIDEO THAT I CAN THINK OF, BAR NONE. IT AIRED IN THE U.S. AS *THE BEST MIND SINCE EINSTEIN,* BUT THE NOVA/PBS VERSION IS ONLY HALF AS LONG, MEANING IT'S ONLY HALF AS GOOD.

NO ORDINARY GENIUS: THE ILLUSTRATED RICHARD FEYNMAN, BY CHRISTOPHER SYKES (NY: W.W. NORTON, 1994).

> THIS BOOK HAS MEMORIES FROM FEYNMAN'S COLLEAGUES, FAMILY, AND FRIENDS ALONG WITH GREAT ILLUSTRATIONS AND PHOTOGRAPHS. BASED ON THE VIDEO OF THE SAME NAME, IT'S ALSO WONDERFUL.

ARLINE

THE PLEASURE OF FINDING THINGS OUT, PRODUCED BY
CHRISTOPHER SYKES (BBC2 HORIZON/NOVA, ORIGINALLY
BROADCAST 1981).

> MORE FEYNMAN, LIVE, AND ALMOST AS GOOD AS *NO
> ORDINARY GENIUS.*

PHYSICS TODAY, FEBRUARY 1989.

> A SPECIAL ISSUE DEVOTED TO FEYNMAN, WITH ARTICLES BY
> FREEMAN DYSON, MURRAY GELL-MANN, DAVID GOODSTEIN,
> JULIAN SCHWINGER, JOHN WHEELER, AND OTHERS. IT
> CONTAINS MANY PERSONAL AND PROFESSIONAL MEMORIES.

*QED AND THE MEN WHO MADE IT: DYSON, FEYNMAN, SCHWINGER,
AND TOMONAGA*, BY SILVAN S. SCHWEBER (PRINCETON, NJ:
PRINCETON UNIVERSITY PRESS, 1994).

> THIS DETAILED ACCOUNT OF HOW THE PROBLEMS IN
> QED GOT SWEPT UNDER THE RUG SUCCESSFULLY, AS
> FEYNMAN PUT IT, IS WRITTEN FOR A TECHNICAL AUDIENCE.
> IT HAS MANY GREAT ANECDOTES AS WELL, THOUGH, AND
> EXAMPLES OF HOW FEYNMAN'S SPEAKING AND WRITING
> STYLE LED TO A GREATER UNDERSTANDING OF THE
> FUNDAMENTAL PHYSICS BEHIND THE PROBLEM.

RICHARD FEYNMAN: A LIFE IN SCIENCE, BY JOHN GRIBBIN AND MARY GRIBBIN (NY: DUTTON, 1997).
> BESIDES MEHRA'S BOOK, MY FAVORITE BIOGRAPHY OF FEYNMAN, SINCE IT GIVES GENERAL READERS JUST ENOUGH OF A TASTE OF HIS PHYSICS TO MAKE THEM WANT TO LEARN MORE.

SIN-ITIRO TOMONAGA: LIFE OF A JAPANESE PHYSICIST, EDITED BY MAKINOSUKE MATSUI AND HIROSHI EZAWA, TRANSLATED FROM THE JAPANESE BY CHERYL FUJIMOTO AND TAKAKO SANO (TOKYO: MYU, 1995).
> THE ONLY ENGLISH-LANGUAGE BOOK ABOUT TOMONAGA THAT I FOUND READILY ACCESSIBLE. FORTUNATELY, IT'S VERY GOOD!

TAKE THE WORLD FROM ANOTHER POINT OF VIEW (YORKSHIRE TELEVISION, INTERVIEW SEGMENTS BY SIMON WELFARE 1973).
> FEYNMAN, INTERVIEWED IN MILIBACK, HIGH IN THE YORKSHIRE PENNINES, WHILE ON HOLIDAY NEAR GWENETH'S HOME. MANY OF THE STORIES HERE SHOW UP IN OTHER PLACES, BUT THE TAP-TAP-TAP ANECDOTE—AS TOLD TO THE FAMED ASTRONOMER SIR FRED HOYLE—MADE TRACKING THIS ONE DOWN WORTHWHILE. YOU CAN READ A TRANSCRIPT AT *HTTP://CALTECHES.LIBRARY.CALTECH.EDU/35/2/POINTOFVIEW.HTM*

TUVA OR BUST, BY RALPH LEIGHTON (NY: W.W. NORTON, 1991).
> HOW FEYNMAN ALMOST GOT TO TUVA. THIS IS AN ADVENTURE STORY WITH A SAD ENDING—GWENETH FEYNMAN ALSO DIED BEFORE SHE COULD MAKE IT TO KYZYL—BUT READING IT WILL STILL MAKE YOU HAPPY.

* AND...THERE'S EVEN MORE. WE CALLED THIS BIBLIOGRAPHY "ALMOST COMPLETE" BECAUSE DOZENS OF OTHER BOOKS AND ARTICLES AND EVEN A FEW WEBSITES ALSO HELPED US BUILD A PICTURE OF THE CENTURY FEYNMAN RACED ACROSS ON HIS WAY TO BECOMING A LEGEND. IF YOU'D LIKE TO START DOWN THAT PATH, I RECOMMEND THE EXCELLENT *A TALE OF TWO CONTINENTS* BY ABRAHAM PAIS AND *THE MAKING OF THE ATOMIC BOMB* BY RICHARD RHODES, BOTH OF WHICH CAN PROVIDE YOU WITH BACKGROUND ON FEYNMAN'S ERA IN PHYSICS AND THE WORLD. AND IF YOU WANT AN EVEN MORE TECHNICAL OVERVIEW THAN YOU GOT FROM MEHRA'S BOOK, YOU CAN CONSULT THE MASSIVE *SELECTED PAPERS OF RICHARD FEYNMAN* (RIVER EDGE, NJ: WORLD SCIENTIFIC, 2000).

AND IN ADDITION TO THINGS YOU CAN EASILY GET AT LIBRARIES OR BOOKSTORES OR ON THE WEB, WE ALSO INCORPORATED INFORMATION, STORIES, AND MANY FACTS FROM UNPUBLISHED MATERIALS HELD IN THE CALIFORNIA INSTITUTE OF TECHNOLOGY'S ARCHIVES. WE OWE JUDITH GOODSTEIN AND HER STAFF OUR THANKS FOR THE ASSISTANCE AND HOSPITALITY THEY PROVIDED DURING MY VISIT THERE. THEY EVEN RECOMMENDED A FINE MOTEL, RIGHT ON ROUTE 66. (I DIDN'T BARBEQUE, THOUGH.) OUR FAVORITE FIND WAS THE TRANSCRIPT FOR HIS 1965 "ADDRESS TO FAR ROCKAWAY HIGH SCHOOL AUDIENCE." MUCH OF THE SEQUENCE SHOWING FEYNMAN IN HIGH SCHOOL COMES FROM THIS SPEECH. EVEN IF CALTECH WASN'T SUCH A LOVELY PLACE, AND EVEN IF THEIR STUDENT UNION DIDN'T OFFER TERRIFIC VEGETARIAN PIZZA BY THE SLICE, I'D RECOMMEND A TRIP JUST TO READ IT.

FINALLY, ON PAGE 178 WE PROMISED YOU THE STORY OF WHY TOMONAGA MISSED THE NOBEL CEREMONY. SO, IN HIS OWN WORDS AS QUOTED ON PAGE 271 OF *SIN-ITIRO TOMONAGA: LIFE OF A JAPANESE PHYSICIST* (MATSUI, ED.):

> ALTHOUGH I SENT A LETTER SAYING THAT I WOULD BE "PLEASED TO ATTEND," I LOATHED THE THOUGHT OF GOING, THINKING THAT THE COLD WOULD BE SEVERE, AS THE CEREMONY WAS TO BE HELD IN DECEMBER, AND THAT THE INEVITABLE FORMALITIES WOULD BE TIRESOME.

AFTER I WAS NAMED A NOBEL PRIZE AWARDEE, MANY PEOPLE CAME TO VISIT, BRINGING LIQUOR. I HAD BARRELS OF IT. ONE DAY, MY FATHER'S YOUNGER BROTHER, WHO LOVED WHISKEY, HAPPENED TO STOP BY AND WE BOTH BEGAN DRINKING GLEEFULLY. WE DRANK A LITTLE TOO MUCH, AND THEN, SEIZING THE OPPORTUNITY THAT MY WIFE HAD GONE OUT SHOPPING, I ENTERED THE BATHROOM TO TAKE A BATH. THERE I SLIPPED AND FELL DOWN, BREAKING SIX OF MY RIBS...IT WAS A PIECE OF GOOD LUCK IN THAT UNHAPPY INCIDENT.

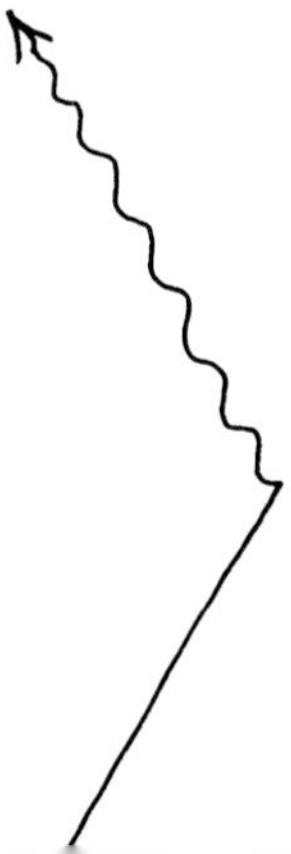

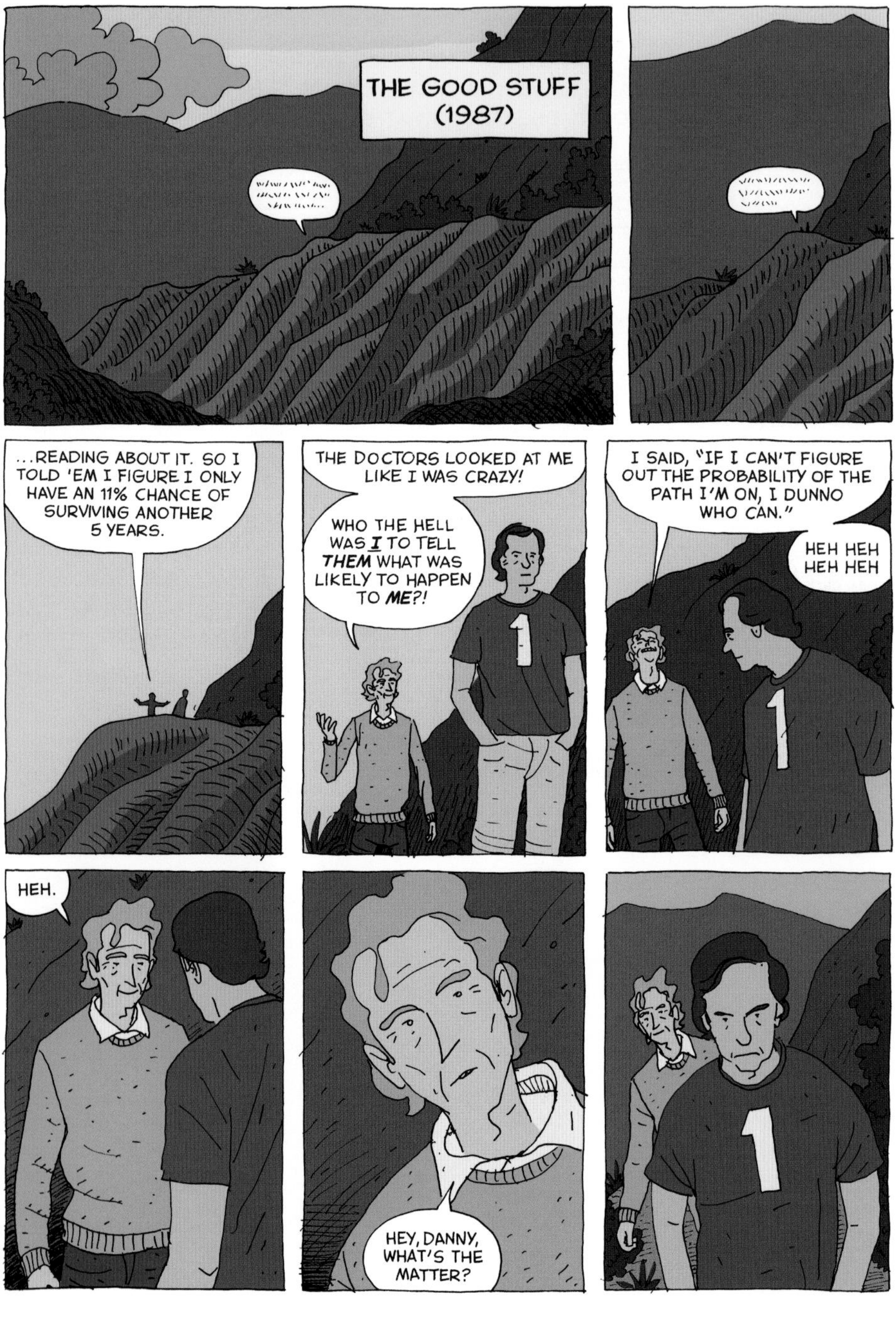
THE GOOD STUFF (1987)
...READING ABOUT IT. SO I TOLD 'EM I FIGURE I ONLY HAVE AN 11% CHANCE OF SURVIVING ANOTHER 5 YEARS.
THE DOCTORS LOOKED AT ME LIKE I WAS CRAZY!
WHO THE HELL WAS *I* TO TELL *THEM* WHAT WAS LIKELY TO HAPPEN TO *ME*?!
I SAID, "IF I CAN'T FIGURE OUT THE PROBABILITY OF THE PATH I'M ON, I DUNNO WHO CAN."
HEH HEH HEH HEH
HEH.
HEY, DANNY, WHAT'S THE MATTER?

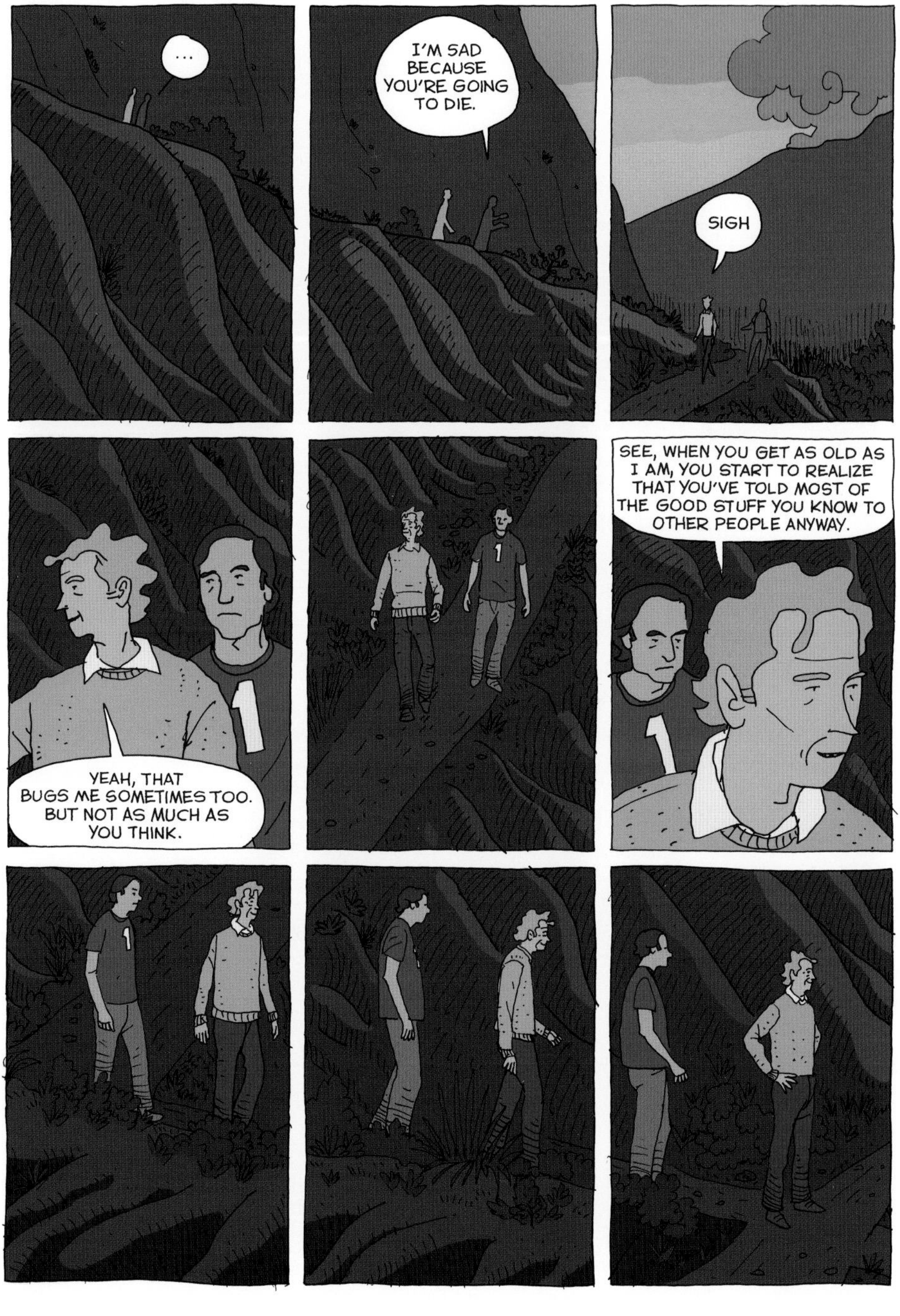
...
I'M SAD BECAUSE YOU'RE GOING TO DIE.
SIGH
YEAH, THAT BUGS ME SOMETIMES TOO. BUT NOT AS MUCH AS YOU THINK.
SEE, WHEN YOU GET AS OLD AS I AM, YOU START TO REALIZE THAT YOU'VE TOLD MOST OF THE GOOD STUFF YOU KNOW TO OTHER PEOPLE ANYWAY.

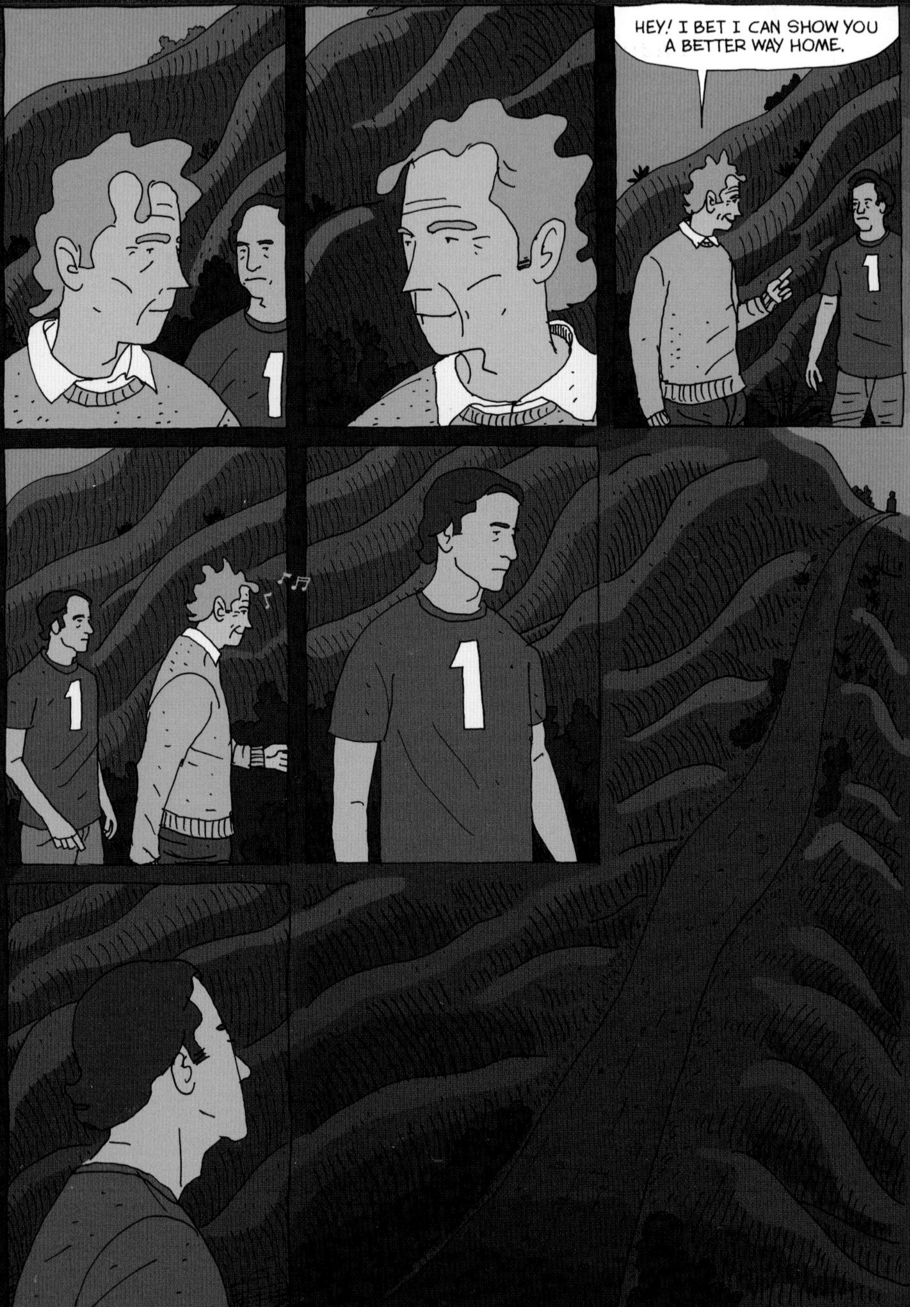
HEY! I BET I CAN SHOW YOU A BETTER WAY HOME.